P9-EEL-792

TECHNOLOGY AND SOCIETY

TECHNOLOGY AND SOCIETY

THE TELEPHONE

AND

TELEPHONE EXCHANGES

THEIR INVENTION AND DEVELOPMENT

J[ohn] E. Kingsbury

ARNO PRESS
A NEW YORK TIMES COMPANY
New York • 1972

Reprint Edition 1972 by Arno Press Inc.

Reprinted from a copy in The Wesleyan
University Library

Technology and Society
ISBN for complete set: 0-405-04680-4
See last pages of this volume for titles.

Manufactured in the United States of America

Library of Congress Cataloging in Publication Data

Kingsbury, John E
The telephone and telephone exchanges.

(Technology and society)
Reprint of the 1915 ed.
1. Telephone--History. I. Title. II. Series.
TK6015.K5 1972 621.385'09 72-5057
ISBN 0-405-04709-6

THE TELEPHONE
AND
TELEPHONE EXCHANGES

THE TELEPHONE

AND

TELEPHONE EXCHANGES

THEIR INVENTION AND DEVELOPMENT

BY

J. E. KINGSBURY, M.I.E.E.

WITH ILLUSTRATIONS.

LONGMANS, GREEN, AND CO.
39 PATERNOSTER ROW, LONDON
FOURTH AVENUE & 30TH STREET, NEW YORK
BOMBAY, CALCUTTA, AND MADRAS
1915

PREFACE

In the following pages the attempt has been made so to relate the inventions and developments in the telephone field that the record may constitute in effect a short history of the telephone industry. The original intention was to write a history, but I was advised on authority which in the book world is entitled to respect not to exceed the limits of a single volume. As that advice has been followed it is considered that the word 'history' as a title should be held in reserve for the use of the author or authors who may produce the comprehensive treatise which the subject deserves.

The original intention has, however, been followed in an abbreviated form. The principal inventions have been selected. The circumstances leading up to them, the developments resulting from them, and the influences bearing on them, have been considered with all the detail that space permits. But telephone exchange service is not merely a matter of plant and invention. Technical, commercial, and political threads compose the fabric, and they are interwoven in the record. The recital of the prior causes has served in most cases to make clear the technical features.

In recounting the progress of Bell's invention reference is frequently made to the evidence given by him in various suits, noted for the sake of brevity as 'Deposition.' The full title is 'The Bell Telephone: The Deposition of Alexander Graham Bell in the suit brought by the United States to annul the Bell patents.' The suit was abandoned, but the deposition (in which is included evidence taken in other suits) was printed in 1908 by the American Bell Telephone Co. 'because of its historical value and scientific interest.' This book of 469 pages, though not published in the ordinary sense, is available for reference in most technical libraries.

Some information regarding early exchange service has been taken from the 'Boston Electrical Handbook,' which was prepared under the auspices of the American Institute of Electrical Engineers as 'a guide for visitors from abroad attending the International Electrical Congress, St. Louis, Mo., September 1904.' This work though printed only for such private circulation may also be found in most technical libraries.

More frequent references are made to the reports of the National Telephone Exchange Association. The word 'National' has been adopted frequently in connection with the telephone—there was, for example, the National Bell Co. in the United States and the National Telephone Co. in Great Britain—so that it may be well to state here that the National Telephone Exchange Association was an association of Telephone Companies in the United States who were licensees under the Bell patents. The reports of the Association constitute a most valuable record of the annual—sometimes semi-annual—discussions of the active participants in the early developments of a new industry.

After the first decade progress was less represented in publications. Committees of experts discussed the problems and originated improvements, but their reports were not published. Descriptions of the apparatus eventually put into use are readily available, but they do not indicate what I was desirous of including—the evolutionary stages of progress in each section. In order to obtain authoritative information on these points I applied to the President of the Western Electric Co. for permission to examine and make extracts from the Reports of the Committees and Conferences of experts of that Company and of the American Telephone Companies who considered from time to time the changes desirable in the art. The obligation which I am under to Mr. H. B. Thayer for kindly placing those voluminous documents at my disposal will be shared by the reader. The more important developments in Switchboards, Cables, and Exchange service generally are here given from the records of the deliberations of those experts, and the information is now published for the first time.

My obligations to my friend Mr. T. D. Lockwood cannot be concisely stated. Some are mentioned in the course of the book, but I am also indebted to him for the loan of documents from which much of the information regarding the early applications in the United

States related in Chapters VIII. and IX. is derived; for practical help in solving doubts or difficulties in some dates and facts; and for much appreciated encouragement in the prosecution of the work.

My indebtedness to the Institution of Electrical Engineers and the American Institute of Electrical Engineers will be obvious by the references to their respective publications.

My thanks are due also to the Editors of various other publications for permission, readily and cordially given, to reproduce illustrations, the sources of which are duly noted in their respective places. In general, these illustrations are limited to early examples of historic interest. Central office interiors, for instance, of later date are well known to telephone readers and easily accessible to others.

I am indebted to my son for assistance in the preparation of the illustrations for the press.

J. E. KINGSBURY.

7 SERJEANTS' INN, TEMPLE,
LONDON, E.C.
October 1915.

CONTENTS

THE TELEPHONE AND TELEPHONE EXCHANGES

CHAPTER I

INTRODUCTORY

THE telephone to which this book relates is the electric speaking telephone that came into the world at the beginning of the last quarter of the nineteenth century.

The name itself is much older than the instrument to which it is now exclusively applied, for before the advent of the electric speaking telephone there were appliances called telephones through which words were spoken though these appliances were not operated electrically, and there were telephones operated electrically which were not available for conversation. Acoustic telegraphs have been called telephones, and, on the other hand, telephones have been designated speaking telegraphs.

The efforts to extend the distance to which speech might be transmitted were, in earlier times, limited to the use of a form of trumpet to increase the power of transmission or reception—the sound waves through the intervening space being subject to the attenuating effect of the atmosphere—or of a tube between speaker and listener, within which tube the sound waves were confined.

A specimen of the ear trumpet was exhibited at the Royal Society in 1668, and of it a contemporary writes :—

I did try the use of the Otacousticon, which was only a great glass bottle broke at the bottom, putting the neck to my ear, and there I did plainly hear the dancing of the oars of the boats in the Thames to Arundel Gallery window, which, without it, I could not in the least do.[1]

[1] *Diary of Samuel Pepys*, April 2, 1668.

The speaking trumpet was developed by Sir Samuel Morland, Master of Mechanics to Charles II. Morland published an elaborate treatise on the subject. The following is a literal transcription of the title page :—

Tuba Stentoro-Phonica an Instrument of Excellent Use as well at sea as at Land; Invented and variously experimented in the year 1670 and Humbly Presented to the King's most Excellent Majesty Charles II in the year 1671 By S. Morland. The Instruments (or speaking Trumpets) of all Sizes and Dimensions, are made and Sold by Mr. Simon Beal, one of His Majesties Trumps in Suffolk Street. London. Printed by W. Godbid and are to be sold by M. Pitt at the White Hart in Little Britain, 1671.

The work describes and illustrates a number of trumpets; contains the copy of a letter from Fr. Digby to Lord Arlington, Principal Secretary of State, which shows that the King took great interest in the trumpets since the letter was written 'as the best way of satisfying His Majesty concerning them'; and includes also 'A short discourse Touching the *Nature of Sounds*, and the manner how (as I conceive) they are magnified, or rather multiplied, by the *Tuba Stentoro-Phonica*.'

The practical results are recorded in the course of a contemporary review or abstract, such as might be found in *Science Abstracts* of the present day, which appeared in the *Philosophical Transactions* :—

The Author of this instrument relates first the several trials made with it, of which the most considerable was, that the largest of those that have been as yet employed turned trumpet-wise being 5 feet 6 inches long, and of 21 inches diameter at the great end, and 2 inches at the less; when by His Majesty's special command it was tried at Deal Castle by the Governor thereof, the voice was plainly heard off at Sea as far as the King's Ships usually ride, which is between two and three miles, at a time when the wind blew from the shore.[1]

It is to Morland's speaking trumpet that Samuel Butler refers in the third part of 'Hud bras,' which was published in 1678 :—

I heard a formidable noise,
Loud as the stent'rophonic voice,
That roared far off.[2]

There was controversy over the claim to the invention of the speaking trumpet as there was in after years over the invention of the speaking telephone. Butler elsewhere describes it as a

[1] *Philosophical Transactions*, No. 79, p. 3056. [2] *Hudibras*, III. i. 251.

'new nicked-named old invention.'[1] But the contention mainly lay between Morland and Kircher. Dr. Derham (1657–1735), an eminent English divine and natural philosopher, gave the credit to the latter.

A century later the evidence was carefully considered by Professor Beckmann of the University of Gottingen in his 'History of Inventions and Discoveries translated by William Johnston and published by J. Bell No. 148 Oxford Street, London' in 1797. Beckmann relates that 'Derham refuses the invention to his countryman and gives it to Kircher.' But Beckmann returns the compliment. He refuses the invention to his own countryman and gives it to Morland.

Kircher [he says] pretends that so early as the year 1649 he had caused such a machine to be set up in the Jesuits College. But supposing this to be true, it can only be said that he then approached very near to the invention of the speaking trumpet, by an instrument however which in reality was calculated to strengthen the hearing and not the voice, and therefore only the half is true of what he advanced in his preface in 1673, that twenty years before he had described in his *Musurgia* the trumpet invented in England.[2]

An important distinction is thus drawn between a speaking trumpet intended to strengthen the sounds transmitted and an ear trumpet for increasing the audibility of sounds received. Beckmann was disposed to give Kircher credit for the invention of a receiver and to deny him the merit of inventing a transmitter. But the judicial summing-up and verdict on the rival claims have their principal interest in the emphasis laid on the test of application to practical utility :—

When I unite all the evidence in favour of Kircher it appears to be certain that he made known and employed the ear trumpet earlier than the portable speaking trumpet ; that he, however, approached very near to the invention of the latter, but did not cause one to be constructed before Sir [Samuel] Morland, to whom the honour belongs of having first brought it to that state as to be of real use.[3]

However efficacious the speaking trumpet may have been for the use of His Majesty's navy in the seventeenth century, it was not equal to the requirements of the service in the nineteenth. Flags and semaphores had a longer reach, though the need of codes made them less eloquent. Before he became a writer of matchless fiction Captain Marryat was 'the inventor (1817) of the completest

[1] *An Heroical Epistle of Hudibras to Sidrophel*, l. 21.
[2] *History of Inventions*, i. 161.
[3] *Ibid.* i. 164.

B 2

code of signals ever introduced,' which was in use in the royal and mercantile service of Great Britain as well as in navies of foreign nations. But, perfect as a code might be, flags and semaphores were only able to transmit its meanings in a clear atmosphere.

In 1845 another captain, one John Taylor, invented an instrument 'for conveying signals during foggy weather by sounds produced by means of compressed air forced through trumpets.' This instrument was called 'The Telephone,' and is briefly described in the 'Year Book of Facts in Science and Art,' 1845, p. 55.

The same publication in 1854 (p. 130) describes a 'New Telephone.' This was the invention of M. Sudre of Paris, and did not vary materially in principle from that of Taylor; but M. Sudre had probably developed a more complete code. There were three notes, and in a practical demonstration, a sentence having been written down, the notes were struck by M. Sudre

> alternately, according to his method, when a third person, without any previous knowledge of the writing, repeated the words merely from hearing the notes.

The French Institute and other scientific bodies were said to have passed 'high encomiums' upon this invention.

In the London Exhibition of 1851 Francis Whishaw of 9 John Street, Adelphi, exhibited a 'Gutta Percha Telephone,'[1] which is not described, but the prior use of the word 'telephone' in connection with instruments of the trumpet type justifies the inference that it was a speaking trumpet. The same manufacturer exhibited the telekouphonon, which is described and illustrated. It is the familiar domestic speaking tube, 'or speaking telegraph, consisting of gutta percha, glass, metal, or other proper tubing.' In an article on gutta percha and its manufactures in 'Stories of Inventors and Discoverers,' 1860 (p. 328), it is asserted that Mr. Whishaw had

> early discovered the valuable property which Gutta Percha possesses for the conveyance of sound, and accordingly made of it the Telakouphanon or speaking trumpet through which, by simply whispering, the voice could be audibly conducted for a distance of three quarters of a mile, and a conversation by this means be kept up.

The author has not correctly transcribed the name, but is not so far out as the Jury whose Report has telekerephona.[2]

[1] *Official Description and Illustrated Catalogue of the Great Exhibition*, 1851, i. 455.
[2] *Reports of the Juries*, p. 598.

Since the telekouphonon, as the catalogue has it, was a speaking tube for domestic use and not a speaking trumpet, it is probable that the name was incorrectly introduced into the article quoted. What was probably intended to be referred to was a gutta percha speaking trumpet, and this was catalogued as a 'Gutta percha telephone.'

Though applied to signalling trumpets the word 'telephone' does not appear to have come into use in England in connection with speaking tubes as it had in Germany. The examiner in the United States Patent Office gives two references of the year 1869 and one of 1871 in proof that 'the ordinary speaking tube was known in Germany as the "Telephone."'[1]

The *Didaskalia*, published at Frankfort, in its issue of September 28, 1854, referred to Bourseul's suggestion[2] for the electrical transmission of speech under the heading of 'Electrische Telephonie,' and Reis applied the name of The Telephone to the apparatus which he devised in 1861.[3] The still earlier use of the word 'telephone' in connection with apparatus devised by Wheatstone is referred to later in this chapter.

While the speaking trumpet is given a section to itself in Beckmann's history, speaking machines are ominously included in a chapter entitled 'Jugglers.' The early machines of this kind were either so constructed as to conceal a confederate or were devices connected by a speaking tube with a distant point where such confederate was more securely hidden. The latter was the plan generally adopted when these deceptions were practised for religious purposes.

> Whether [says Beckmann] the head of Orpheus spoke in the island of Lesbos, or, what is more probable, the answers were conveyed to it by the priests, as was the case with the tripod at Delphi, cannot with certainty be determined. That the impostor Alexander, however, caused his Æsculapius to speak in this manner is expressly related by Lucian.[4]

Those speaking machines which purported to answer various questions submitted to them were comprised within a figure or sometimes only a head placed upon a box, the front of which 'for the better deception' was filled with 'a pair of bellows, a sounding board, cylinder and pipes supposed to represent the organs of speech.' The popular opinion of these machines varied. Some affirmed that the voice issued from the machine, others that the juggler himself answered by speaking as ventriloquists do, and

[1] *The Speaking Telephone Interferences*, p. 4.
[2] *Applications de l'Électricité*, Du Moncel, 1854.
[3] See Chapter XII.
[4] *History of Inventions*, iii. 333.

some believed that the answers were given by a man somewhere concealed. When the illusion was detected the populace imagined they had a right to avenge themselves for being imposed on, which the historian evidently regarded as being very unreasonable. He says :—

For my part I do not see why a juggler, with a speaking machine, is a more culpable impostor than he who pretends to breathe out flames and to swallow boiling oil, or to make puppets speak, as in the Chinese shadows. The spectators pay for the pleasure which they receive from a well-concealed deception, and with greater satisfaction the more difficult it is to discover it.[1]

In the eighteenth century much interest was taken in automata. Chess players, singing birds, and other mechanical contrivances were constructed for the edification or amusement of European courts. Useless in themselves, they probably tended to develop the skill of the mechanicians, whose aspirations rose from chess players and singing birds to speaking machines. Towards the end of the century the scientific interest in speech increased, and it is to the Imperial Academy of St. Petersburg (or Petrograd) that credit has to be given for the first substantial encouragement to inquiry. In the year 1779 the Academy proposed for the annual prize two questions. Firstly, what was the nature and character of the vowels a e i o u, 'so different from each other'; and secondly, could an instrument be constructed like the *vox humana* pipes of the organ, which should accurately express the sounds of the vowels? The prize was awarded to Professor Kratzenstein, who

constructed a series of tubes, which, when applied to an organ bellows, imitated with tolerable accuracy the five vowel sounds required. These tubes were of the most grotesque and complicated forms; for which no reason was offered except that experience had shown these forms to be the best adapted to the production of the sounds in question. Some of the pipes were rendered vocal by the application of vibrating reeds, like those in clarionets; while others were open in the manner of a common organ pipe.[2]

In 1791 De Kempelen of Vienna published a description of his talking machine in which the bellows and pipes were not supplied for the purpose of deception, but, like those of Kratzenstein, were really used. De Kempelen attempted to imitate not only vowel sounds but also consonants.

Mr. Willis of Cambridge (1829) was less ambitious but more

[1] *History of Inventions*, iii. 331.

[2] *Saturday Magazine*, February 11, 1843. Wheatstone (*Scientific Papers*, p. 352) gives a more abbreviated account, but apparently from the same source.

thorough. He limited his efforts to vowel sounds and formed the foundation of the scientific inquiry which Wheatstone and Helmholtz continued. A contemporary of Wheatstone's in the analysis of speech sounds was Dr. Rush of Philadelphia, who wrote a work entitled 'The Philosophy of the Human Voice.'

Mr. Reale exhibited before the American Philosophical Society in 1843 a machine capable of enunciating various letters and words. It is said that the instrument took him sixteen years to construct, and that he destroyed it 'in a frenzy,' presumably of despair; and Professor Faber of Vienna produced a machine, considered to be more successful than any of its predecessors, which was exhibited in London in 1846.

In all the machines of this type the designers sought to produce speech. They succeeded only within a strictly limited range, and very imperfectly even then. The so-called speaking machines which spoke were really speaking tubes with a talker at one end and a listener at the other, the conducting medium being air, the waves which reached the ear being the waves directly produced by the voice, but retaining their power to a greater distance than in the atmosphere because the walls of the tube reduced the loss of energy from spherical action.

The utilisation of a solid medium for transmission of sound to a distance was developed by Wheatstone, who, in the introduction to his paper 'On the transmission of musical sounds through solid linear conductors, and on their subsequent reciprocation,' published in the 'Journal of the Royal Institution,' 1831, vol. 2, wrote:—

The fact of the transmission of sound through solid bodies, as when a stick or a metal rod is placed with one extremity to the ear, and is struck or scratched at the other end, did not escape the observations of the ancient philosophers; but it was for a long time erroneously supposed that an aeriform medium was alone capable of receiving sonorous impressions; and in conformity with this opinion, Lord Bacon, when noticing this experiment, assumes that the sound is propagated by spirits contained within the pores of the body. The first correct observations on this subject appear to have been made by Dr. Hooke in 1667, who made an experiment with a distended wire of sufficient length to observe that the same sound was propagated far swifter through the wire than through the air. Professor Wunsch, of Berlin, made, in 1788, a similar experiment, substituting 1728 feet of connected wooden laths for the wire, and confirmed Dr. Hooke's results.[1]

Beckmann makes more definite mention of the ancient philosophers in connection with this subject. 'It had been remarked,' he

[1] Wheatstone's *Scientific Papers*, p. 47.

says, 'even in Pliny's time, that the least touching of a beam of wood could be heard when one placed one's ear at the other end.'[1]

The year 1667 given by Wheatstone as that of Dr. Hooke's observations has been adopted generally by later writers. In a note to his paper Wheatstone quotes the Preface to Hooke's 'Micrographia.' The edition of this work published in 1667 was the second. The first appeared in 1665, having been 'ordered to be printed by the Council of the Royal Society, November 23, 1664.' The paragraph partly quoted by Wheatstone appears in the first edition (1665) on the eighth page of the Preface, the complete paragraph being as follows :—

And as Glasses have highly promoted our seeing, so 'tis not improbable, but that there may be found many Mechanical Inventions to improve our other Senses, of hearing, smelling, tasting, touching. 'Tis not impossible to hear a whisper a furlong's distance, it having been already done ; and perhaps the nature of the thing would not make it more impossible, though that furlong should be ten times multiply'd. And though some famous Authors have affirm'd it impossible to hear through the thinnest plate of Muscovy-glass ; yet I know a way, by which 'tis easie enough to hear one speak through a wall a yard thick. It has not yet been thoroughly examin'd, how far Otocousticons may be improv'd, nor what other wayes there may be of quickning our hearing, or conveying sound through other bodies then the Air : for that that is not the only medium, I can assure the Reader, that I have, by the help of a distended wire, propagated the sound to a very considerable distance in an instant, or with as seemingly quick a motion as that of light, at least, incomparably swifter then that, which at the same time was propagated through the Air ; and this not only in a straight line, or direct, but in one bended in many angles.

Ten years before writing his Royal Institution paper Charles Wheatstone, in 1821, whilst an assistant to his uncle, a musical instrument maker, devised a method of transmitting music by means of wooden rods. The rod at the transmitting end was attached to a piano and at the receiving end to a sounding board designed to represent an ancient lyre, the conventional representative of music in art and literature. The device attracted considerable attention in scientific circles and amongst the public. Accounts of the performance were given in the *Repository of Arts*, September 1, and the *Literary Gazette*, September 15, 1821, but the principle of operation was not disclosed. In his paper 'New Experiments on Sound,' published in Thomson's 'Annals of Philosophy,' 1823, Wheatstone described it fully.

[1] *History of Inventions*, i. 152.

In my first experiments on this subject [he says] I placed a tuning fork or a chord extended on a bow, on the extremity of a glass or metallic rod five feet in length, communicating with a sounding board; the sound was heard as instantaneously as when the fork was in immediate contact; and it immediately ceased when the rod was removed from the sounding board or the fork from the rod. From this it is evident that the vibrations, inaudible in their transmission, being multiplied by meeting with a sonorous body, become very sensibly heard. Pursuing my investigations on this subject, I have discovered means for transmitting, through rods of much greater lengths and of very considerable thicknesses, the sounds of all musical instruments dependent on the vibrations of solid bodies, and of many descriptions of wind instruments. It is astonishing how all the varieties of tune, quality, and audibility, and all the combinations of harmony, are thus transmitted unimpaired, and again rendered audible by communication with an appropriate receiver. One of the practical applications of this discovery has been exhibited in London about two years, under the appellation of 'The Enchanted Lyre.' So perfect was the illusion in this instance from the intense vibratory state of the reciprocating instrument, and from the interception of the sounds of the distant exciting one, that it was universally imagined to be one of the highest efforts of ingenuity in musical mechanism.[1]

That Wheatstone called the enchanted lyre a telephone is a statement which is to be found in most of the books on the subject, but, so far as I have been able to examine them, without any reference to source or authority. The inference has very generally been drawn that the enchanted lyre was called a telephone from its inception; which if correct would carry the word back to 1821. But there were good reasons why no such descriptive name should have been given to it then.

A certain degree of mystery was necessary in order to arouse public interest. Wheatstone used to go through the form of winding up the lyre with a key, but neither of the magazine writers was misled thereby. The winding up was regarded as 'evidently a mere *ruse*.' Moreover, the *Repository of Arts* thought 'proper to add' the statement of Mr. Wheatstone that the exhibition was 'the application of a general principle for conducting sound, which principle he professed himself to be capable of carrying to a much greater extent.' The mild deception practised was akin to that of the conjurer who excites interest by mystification, but it serves to show that it was not then Wheatstone's desire to explain so much as would have been explained if he had applied the word 'telephone' to the contrivance. The name of 'The Enchanted Lyre' was too simple to stand alone and needed supplementing

[1] Wheatstone's *Scientific Papers*, p. 7.

by a compound from a dead language. This alternative name was 'The Acoucryptophone,'[1] in which the mysterious or secret feature was comprised. But not only in this early stage did Wheatstone refrain from calling the apparatus a telephone. Neither in Thomson's 'Annals of Philosophy' of 1823, where the principle of operation was first disclosed, nor in any subsequent description included in his collected Papers does Wheatstone use the word 'telephone.' It is just such a word as he might have coined, and probably did coin, but whether before the electric telegraph had come into operation is doubtful.

Wheatstone collaborated with Cooke in perfecting and introducing their electric telegraph in 1837. Three years later, in consequence of a misunderstanding which had prevailed respecting their 'relative positions in connection with the invention,' the arbitration of Sir M. Isambard Brunel and Professor J. F. Daniel was sought. In the course of this arbitration, which was agreed upon on November 16, 1840, the word 'telephone' is frequently used, and this is probably its earliest use.

In a printed pamphlet constituting his case Professor Wheatstone says :—

> The subject of telegraphic communication has for a long series of years occupied my thoughts. When I made in 1823 the discovery that sounds of all kinds might be transmitted perfectly and powerfully through solid wires and rods, and might be reproduced in distant places, I thought that I had an efficient and economical means of establishing a telegraphic (or rather a telephonic) communication between two distant places.[2]

In the same 'case' he refers to his proposal for a telegraph with two wires, in which electric sparks and his revolving mirror were employed. He describes this briefly, and adds :—

> I have not continued these experiments, but the principles employed in them have been of great use to me in my most recent investigations. My rhythmical telephone, invented last year [1839], is but a modification of it, in which the strokes of a bell are substituted for the sparks, and the voltaic current for the electric discharge.[3]

While Wheatstone himself uses the word 'telephone' only in connection with an electric bell, and the word 'telephonic'[4] in

[1] *Min. Proc. Institution of Civil Engineers*, xlvii. 284.

[2] *The Electric Telegraph—was it invented by Prof. Wheatstone*, Part II. p. 81.

[3] *Ibid.* p. 84.

[4] *The Imperial Dictionary* published in 1854 contains the word 'telephonic,' but not the word 'telephone.'

a parenthetic way for the purpose of verbal accuracy, his opponents in the arbitration proceedings frequently describe his apparatus as the telephone. In the address of Mr. Wilson laid before the arbitrators on February 27, 1841, as an introduction to the evidence to be adduced on Mr. Cooke's behalf, repeated reference is made to 'the telephone':—

I maintain, gentlemen, that it was then [after Cooke's introduction to him] *and not till then*, that Professor Wheatstone became connected with the Practical Electric Telegraph; and that up to that time he had done nothing in any respect more practical than his telegraph with common electricity and a revolving mirror, or his idea, to which his letter refers, of a telephone between London and Edinburgh.[1]

It is evident that Professor Wheatstone had turned his mind to the Telephone and to the Electric Telegraph for fourteen or fifteen years without any practical result.[2]

There are other references of a similar nature, but further quotation is unnecessary. The word 'telephone' is here used in 1841 not by Wheatstone himself but by the legal representative of the opposing party, and it was used with greater frequency than would seem to have been really necessary. The reason must be sought in the nature of the inquiry submitted to arbitration: To whom was due the credit for the introduction of the Practical Electric Telegraph? The frequent use of the word 'telephone' by Mr. Cooke's solicitor is probably to be explained by his desire to emphasise the unpractical nature of Wheatstone's idea. 'The telephone' was in fact used as a sort of term of reproach. It represented something unpractical leading to nothing, whilst the electric telegraph, on the other hand, was the 'Practical Electric Telegraph,' and Cooke's work thereon had resulted in great public advantage.

This line of argument in view of the purpose of the arbitration was not unfair. Valuable as it was scientifically, the enchanted lyre experiment had never been carried practically beyond the room-to-room stage. But whatever may have been the prior use, as to which some uncertainty may be felt, after this period the term 'telephone' was very generally applied to it.

In 1855 there was a demonstration at the Polytechnic Institution before Queen Victoria, which was referred to by Mr. C. K. Salaman in a letter to the 'Choir' as a 'telephone concert,'[3] and an illustration accompanying a lecture-table modification of this device is

[1] *The Electric Telegraph—was it invented by Prof. Wheatstone*, Part II. p. 121.

[2] *Ibid.* p. 124.

[3] *Post and Telegraphs*, Tegg, 1878, p. 290.

entitled 'The Miniature Telephonic Concert' in Pepper's 'Cyclopædic Science Simplified,' published in 1869.[1] Tyndall utilised the illustration in the course of a series of lectures on Sound at the Society of Arts. The Society's Journal of May 5, 1871,[2] states that 'among many other interesting illustrations was an example of the "telephone concert" of Sir Charles Wheatstone.' And in the 'Handbook to the Special Loan Collection of Scientific Apparatus,' 1876,[3] it is said that the transmission of sound in wood 'was ingeniously demonstrated by Wheatstone. His telephone consisted of long rods of light pine,' etc.

Wheatstone's suggestion had nothing in common with the telephone now in general use. He aimed at the mechanical or molecular transmission of acoustic vibrations, and even in that (except for purely local demonstration) had not gone beyond the stage of suggestion. It is from the theoretical, rather than the practical, standpoint that we should consider the suggestion that Mr. Cooke's solicitor called a telephone.

The conclusion of Wheatstone's paper, of which the introductory passage was quoted on page 7, is as follows :—

When sound is allowed to diffuse itself in all directions as from a centre, its intensity, according to theory, decreases as the square of the distance increases; but if it be confined to one rectilinear direction, no diminution of intensity ought to take place. But this is on the supposition that the conducting body possesses perfect homogeneity, and is uniform in its structure, conditions which never obtain in our actual experiments. Could any conducting substance be rendered perfectly equal in density and elasticity so as to allow the undulations to proceed with a uniform velocity without any reflections or interferences, it would be as easy to transmit sounds through such conductors from Aberdeen to London as it is now to establish a communication from one chamber to another. Whether any substance can be rendered thus homogeneous and uniform remains for future philosophers to determine.[4]

Wheatstone saw clearly the limitations of any molecular method of transmission, and equally clearly the benefits which must result from speech transmission if a practicable method could be found. He continues :—

The transmission to distant places, and the multiplication of musical performances, are objects of far less importance than the conveyance of the articulations of speech. I have found by experiment that all these articulations, as well as the musical inflexions

[1] P. 525.
[2] *Journal Society of Arts*, xix. 510.
[3] *Handbook, etc.*, p. 99.
[4] Wheatstone's *Scientific Papers*, p. 62.

of the voice, may be perfectly, though feebly transmitted to any of the previously described reciprocating instruments by connecting the conductor, either immediately with some part of the neck or head contiguous to the larynx, or with the sounding board to which the mouth of the speaker or singer is closely applied. The almost hopeless difficulty of communicating sounds produced in air with sufficient intensity to solid bodies might induce us to despair of further success; but could articulations similar to those enounced by the human organs of speech be produced immediately in solid bodies, their transmission might be effected with any required degree of intensity. Some recent investigations lead us to hope that we are not far from effecting these desiderata; and if all the articulations were once thus obtained, the construction of a machine for the arrangement of them into syllables, words, and sentences would demand no knowledge beyond that we already possess.[1]

We are left in doubt as to the nature of the machine to be constructed. The 'previously described reciprocating instruments' were of the sounding board or lyre type, comparatively simple in construction, but something much more complicated is suggested by a machine whose function it should be to arrange the transmitted vibrations 'into syllables, words, and sentences.'

It seems to have been assumed that for the reproduction of speech, complex contrivances capable of artificially producing speech would be required. The suggestion of Bourseul (1854) and the experimental efforts of Reis (1861) did but little to dissipate the idea. If the question were asked who, amongst the predecessors of the real inventor, should have produced the telephone the unhesitating answer would be—Charles Wheatstone. He was the earliest practical experimenter in this direction. His studies in acoustics were original and profound. He was the first to enunciate the theory of vowel tones. He was equally well equipped in his knowledge of the science of electricity and of its practical applications. He was renowned for his ingenious mechanical inventions, and was provided with a staff of workers specially trained to produce, in the form of highly finished instruments, the creations of his brain. Yet the idea which his paper of 1831 conveys, that some elaborate machines were essential for speech transmission, was not withdrawn. Notwithstanding Wheatstone's sanguine views regarding the possibility of producing such machines, no surprise is felt that inventors were deterred from entering upon an enterprise in which the experiments must be costly and the expectations of success but slight. Even in the circles of science there was still a mystery and a magic in human speech.

[1] Wheatstone's *Scientific Papers*, p. 62.

CHAPTER II

THE SPOKEN WORD

THE magic and the mystery of the spoken word have cast a spell over mankind from the earliest times through all the ages; the greatest homage being paid perhaps by the philosophers of Hermopolis, who, materialising the intangible, deified speech and attributed to it the power of a Creator. To them there was a very real meaning in the phrase 'In the beginning was the Word.' To them it was 'speech, and above all the simple emission of the voice which gave the world the form it now bears.'

The magic and the mystery of speech have appealed not the less powerfully to those who have regarded it simply as a means of transferring thought, of communicating intelligence, or of stirring the hearts of men. Whether it were the exhortations of the priests, the appeals of the orators of ancient times, or the addresses of more recent politicians, the medium of communication was the same and the power was the power of speech.

The stirring times of the nineteenth century revived the power of the platform which had been on the wane [1] and gave a new interest to the cultivation of the powers of speech, until, as the century advanced, professors of elocution were numerous. Some of them published books, and the impression which most of these books produce is that in the main these teachers of oratory devoted little attention to the mystery and mechanism of speech, but dwelt

[1] 'It is somewhat hard to realise in the present day, when the platform is so great a power in the land, how completely non-existent it was as a factor in political life, as the nineteenth century dawned upon the country. Events of world-wide importance were occurring abroad; mighty movements were beginning at home; but the Government had closed the avenues to discussion, and the people were compelled to silence. Matter enough had they for thought; problems enough to perplex them; grievances and suffering enough to make them cry out; but the articulate voice came not, was not permitted to come; and though the Press, trammelled and terrorised, acted to some extent as a vehicle for the expression of the thoughts of the people, the public voice as spoken from the platform was dumb.'—*The Platform*, Jephson, i. 297 (Macmillan. 1892).

much on the effects of gesture and of emphasis. Alexander Bell of 25 Norton Street, Portland Place, London, was a professor of elocution who published in 1835 'The Practical Elocutionist,' which he dedicated to Lord Brougham. He subsequently published other works, all of an elocutionary character rather than analytical of speech.

His son Alexander Melville Bell followed him in the same profession, practising in Edinburgh. Alexander Melville Bell made a more complete study of the mechanism of speech than any professor of elocution before him. From Edinburgh he removed to London, and was lecturer on elocution in University College. He published works on 'The Principles of Speech, and Cure of Stammering,' 'Letters and Sounds—An Introduction to English Reading on an entirely new plan,' 'Observations on Stammering and the Principles of Elocution,' and other works of a similar character; but internal evidence shows that in the view of the author his *magnum opus* was 'Visible Speech, the Science of Universal Alphabetics.' In the preface to this work he says—

The scientific interest attaching to the invention of *visible speech* has alone induced me to consent to the publication of the system under copyright. My desire was that this invention—the applications of which are as universal as speech itself—should at its inauguration have been made free from restrictions; but my endeavours to effect an arrangement for this purpose have been frustrated. I wish to put on record here a statement of the facts concerning my offer of the invention to the British Government, and the reception of the offer.

The author offered to relinquish all copyright in the explanatory work as well as all exclusive property in the system and its applications if the expense of casting the new types and publishing the theory of the system were defrayed from public resources. 'This request was made in vain. The subject did not lie within the province of any of the existing State Departments, and the memorial was, on this ground, politely bowed out from one after the other of the executive offices.'

This work was published in 1867. It was preceded in 1865 by a book with a similar title, 'Visible Speech, a New Fact demonstrated.' That the propagation of the system as a measure of public usefulness was still a cherished dream of the author in 1875 is evidenced by a letter to him from his now famous son, Alexander Graham Bell, who wrote on March 18 of that year recording the progress achieved with his harmonic telegraph. There will be occasion to refer to this letter again, and for the moment only the thoughtfulness of the son for the cherished work of the father need be noted. He says: 'Whenever I am free to dispose of my interest in the invention I shall do so, and then you may expect to see Visible Speech go ahead.' Both father and son were to be disappointed. The invention then under consideration was not destined to be directly productive, but the encouraging development recorded was an important aid to the achievement of a still greater invention. Yet though that greater invention has been completed, and though its use has been beyond the dreams of the most sanguine, Visible Speech has not 'gone ahead' in the practical way which was contemplated. Its value in connection with the science of language may be gauged from the remarks made by Henry Sweet, M.A., in the preface of his work, 'A History of English Sounds,'[1] wherein the author says that his investigations were due to the combined influence of Bell's 'Visible Speech,' Ellis's 'Early English Pronunciation,' and the German School of comparative and historical philology; his use of the revised visible speech notation for exact purposes required, he said, no justification. 'Although far from perfect, it is the only system which is universal in its application and at the same time capable of being worked practically.' It was to him a source of some pride that, just as Henry Nicol and himself were the first to take up Bell's visible speech and apply it to linguistic investigation and the practical study of language, so also were they the first to welcome the revolutionary investigations in Ellis's 'Early English Pronunciation,' and he says in conclusion, 'My debt to Mr. Bell speaks for itself.'

It is not, however, our province now to follow in any detail the indebtedness of the philologist to Alexander Melville Bell, but to record the important part which 'Visible Speech' and its demonstrations served in the education and development of the author's son in the direction of acoustics and speech analysis. On these grounds it is necessary to give a brief account of 'Visible Speech.'

In the first place it must be said that the title has led some writers to assume that the subject has reference to the rendering visible the aerial vibrations arising from speech. The work has no relation to the mechanical effects of speech, nor is it a treatise on the instruction of deaf mutes, as another writer suggests.[2] The utility of the system for the deaf and dumb and also for the blind is subsidiary to the main purpose which, as its sub-title indicates, is the introduction of a universal alphabet adapted to all languages and all dialects. It is in fact a phonetic system as Isaac Pitman's phonography was a phonetic system, but it permitted the putting in graphic form the most varied utterances and the most delicate

[1] Oxford Clarendon Press, 1888 edition.
[2] Munro, *Heroes of the Telegraph*, p. 185.

nuances of speech. Instead of adopting arbitrary characters as Pitman did, for an excellent reason, in his phonography, or of adapting existing types as in phonetic spelling, Melville Bell based his alphabet or signs upon a system which analysed the method of speech production, as for instance a circle indicating 'The Throat Open (aspirate),' and an oval indicating 'The Throat Contracted (whisper).' Other characters were based upon the position of the lips, the tongue, and so on, the principal object being to produce a written character which should be so definitely phonetic as to represent any variety of speech sound and thus be available for international usage.

The author enumerates ten special uses of his invention, relating mainly to the teaching or recording of language; but there are two which require to be quoted because of their bearing on the subsequent work of Alexander Graham Bell :—

> 3. The teaching of the Deaf and Dumb to speak. In this department very striking results may be confidently anticipated. The Deaf and Dumb possess all the organs of speech and only require to be directed *visibly* in their use. The *feeling* of organic action will probably be developed by practice to a keenness corresponding to that which the sense of touch acquires among the Blind.
>
> 7. The Telegraphic communication of messages in any language through all countries, without translation. Visible Speech does not interfere with the use of ordinary alphabets in literature, etc., but for international purposes it may very advantageously supplant all local alphabets. Roman letters have been fully tried and found sadly wanting in Telegraphy.[1]

The effect of 'Visible Speech' on the telegraphic art has been important but indirect rather than direct.

Having designed his system, Mr. Melville Bell was eager to demonstrate its efficacy, and for this purpose taught its principles to his two sons, Edward Charles and Alexander Graham. The former died in his nineteenth year (May 17, 1867) and to his memory the work is dedicated. The latter happily still lives, and in the meantime has achieved much. Amongst the persons before whom the demonstrations were made was Mr. Alexander J. Ellis, F.R.S., who wrote an account of them for the *Reader* of September 3, 1864. He recounts how the two sons who were to read the writing were sent out of the room whilst he dictated slowly and distinctly to Mr. Bell the sounds which he wished to be written.

> The result was perfectly satisfactory—that is, Mr. Bell wrote down my queer and purposely exaggerated pronunciations and

[1] *Visible Speech*, pp. 20, 21.

mispronunciations and delicate distinctions in such a manner that his sons, not having heard them, so uttered them as to surprise me by the extremely correct echo of my own voice.

Mr. Ellis wrote a second letter to the same publication (August 5, 1865) indicating his very great interest in, and his expectations of valuable results from the system. The demonstrations have had no lasting results on their particular subject. But the acquaintance of Alexander Graham Bell with Alexander Ellis was of especial importance, for Ellis was not only a high authority on the phonetics of speech, but was also the translator of Helmholtz's book, 'On the Sensations of Tone,' the work that first disclosed in their entirety the acoustic principles which underlie the production, the aerial transmission, and the audibility of speech. One of the direct results of that acquaintanceship may be seen from the statement made by Alexander Graham Bell that it was not until after his interview with Mr. Ellis in London that he really took up the study of the subject of electricity.[1]

Another acquaintance of Bell's early days was Sir Charles Wheatstone.

[1] *Deposition*, p. 207.

CHAPTER III

THE GROWTH OF AN IDEA

ALEXANDER GRAHAM BELL relates that Sir Charles Wheatstone lent to his (Bell's) father the work on 'Le Mécanisme de la Parole,' by Baron de Kempelen, giving a full description and plates of his celebrated automaton speaking machine, 'and Sir Charles Wheatstone himself showed me before I came to this country [the United States] a reproduction of Baron de Kempelen's speaking machine which he had made, and I saw it operated by his own hands and heard it speak.'[1]

Thus was Bell early brought into communication with men who were most highly endowed with information on the subject which interested him. The son and grandson of professors of elocution he was himself destined for the same profession. His father had branched out beyond the ordinary boundaries of the teacher's art, had extended his interests into adjoining fields and saw the advantage which must come from scientific knowledge. Alexander Graham Bell was educated at the Royal High School of the Scottish capital, where he passed through the whole curriculum of the school. He attended lectures on classical subjects at the Edinburgh University, and a course on anatomy at University College, London, where he matriculated as an undergraduate of the London University in the year 1867. From his boyhood he had been specially educated by his father, at home, on subjects relating to sound and the mechanism of speech, as his father intended him to follow his own profession and become a teacher of articulation.

He also received at a very early age training in music from Signor Auguste Bertini, after whose death his musical education was carried on, on the Bertini method, by his mother.[2]

From his earliest childhood his attention was directed to the study of acoustics, and especially to the subject of speech, and he was urged by his father to study everything relating to these subjects, as they would have an important bearing upon what was to be his

[1] *Deposition*, p. 207. [2] *Ibid.* pp. 6–7.

professional work. His father also encouraged him to experiment, and offered a prize for the successful construction of a speaking machine. Alexander Graham Bell made a machine of this kind, as a boy, and was able to make it articulate a few words.[1] He read all the books that his father had in his library upon these subjects, including his father's and grandfather's works upon speech, also 'The Real Character,' by Bishop Wilkins, published about 1680; *Prosodia Rationalis*, by Joshua Steele, and the work of de Kempelen already referred to.[2] He recognised that he had exceptional advantages for knowing of works on speech, and that he was probably acquainted with a considerable portion of the literature upon that subject. His studies in acoustics had gone so far in 1864 or 1865, when teacher of elocution and music in Weston House Academy, Elgin, Morayshire, that he made experiments upon his own mouth to determine the resonance pitches of the mouth cavities during the production of the vowel sounds. He describes these experiments as follows:—

Considering the instrument of speech as a tube, extending from the vocal cords to the lips, I saw that the constriction of the mouth passage anywhere divided that tube into two bottle-shaped cavities, placed neck to neck. I came to the conclusion that these cavities, like ordinary bottles, should be capable of producing, by resonance, musical tones, and that the pitch of the tone produced by each cavity should be dependent upon its size, shape, etc. I placed the side of a lead-pencil against my cheek and tapped it forcibly, while I assumed with my mouth the positions for the vowel sounds. The agitation of the air in the front cavity of the mouth produced a hollow sound, like that occasioned by tapping against the side of an empty bottle; and this sound had the element of pitch. I found that the pitch of the front cavity of the mouth was different for every vowel. Upon then placing the pencil against my throat and tapping it forcibly, as before, I caused an agitation of the air in the back cavity of the mouth; and in this way was able to study the resonance pitches of the back cavity during the assumption of the vowel positions. I thus discovered that most vowel positions were capable of yielding a double resonance, the front cavity of the mouth yielding a different tone from that produced by the back cavity. Having familiarised myself with the proper tones of the mouth cavities, I experimented upon myself to ascertain whether I could perceive these tones while I actually uttered vowel sounds. Upon singing all the vowels upon the same pitch of voice, I was delighted to find that I could perceive the characteristic tones of the mouth cavities as feeble musical effects, mingling faintly with the voice; and I made what I then deemed to be an original discovery—that vowel quality was produced by

[1] *Deposition*, p. 7. [2] *Ibid.* p. 207.

the resonance tones of the mouth cavities mingling faintly with the tone of the voice. I made an elaborate series of experiments to determine the resonance pitches of the mouth cavities in uttering different vowel sounds, and I communicated the results to the late Mr. Alexander J. Ellis, of London, England. These experiments were made in Elgin, Scotland, somewhere about the year 1864 or 1865. Mr. Ellis informed me that the experiments, which I had thought to be original with myself, had already been made by Helmholtz, and that Helmholtz had demonstrated the compound nature of the vowel sounds by producing them artificially by a synthetical process. For example, he would cause the simultaneous vibration of three tuning forks of different pitches—one of these would represent the pitch of the voice—and this fork he caused to vibrate in front of a resonator tuned to its own pitch so as to cause it to produce a loud musical tone. The other two forks corresponded in pitch to the front and back cavities of the mouth in uttering some vowel sound. These forks were caused to resound very faintly. The simultaneous vibration of the three forks produced one loud, fundamental sound, and the two higher partial tones. The effect upon the ear was as though someone sang a vowel sound. In an interview with Mr. Ellis, he attempted to describe to me the apparatus used by Helmholtz. Helmholtz kept his forks in vibration by means of electro-magnets and a voltaic battery; but I found that I had not sufficient electrical knowledge to understand the arrangement used by Helmholtz. I therefore determined to study electricity; for I felt it was my duty as a student of speech to study Helmholtz's researches and to repeat his experiments. I do not remember when I first commenced the study of electricity, but I know that during my stay in the city of Bath, in England, I was practically experimenting with ordinary telegraph apparatus, and trying in vain to cause the continuous vibration of a tuning fork by means of electro-magnets. I think the date of my stay in Bath was 1867, but am not quite sure. Helmholtz's researches were not published in the English language before my arrival in America; and I could not, unfortunately, read his work in the original German; but very shortly before my leaving Great Britain, in 1870, I procured a copy of the French edition of his work.[1]

Helmholtz's work was first published in German in 1863, in French in 1868. The English translation did not appear until 1875. The progress made by the last-mentioned date indicates that Bell must have been indebted to his knowledge of French for acquiring in detail that insight into vowel tones which was an essential foundation for his subsequent work. Helmholtz says:—

The vowels of speech are in reality tones produced by membranous tongues (the vocal chords) with a resonance chamber (the

[1] *Deposition*, p. 7.

mouth) capable of altering in length, width, and pitch of resonance, and hence capable also of reinforcing at different times different partials of the compound tone to which it is applied.[1]

In a footnote he adds :—

The theory of vowel tones was first enunciated by Wheatstone in a criticism, unfortunately little known, on Willis's experiments. The latter are described in the *Transactions of the Cambridge Philosophical Society*, vol. iii. p. 231, and Poggendorff's *Annalen der Physik*, vol. xxiv. p. 397. Wheatstone's report on them is contained in the *London and Westminster Review* for October 1837.

This paper of Wheatstone's is also published in his collected Papers, from which the following is quoted :—

Mr. Willis finally concludes, from his experiments, that the vowel quality, added to any sound, is merely the co-existence of its peculiar note with that sound ; this accompanying note being excited by the successive reflections of the original wave of the reed at the extremities of the added tube.

This view of the matter naturally associates the phenomena of vowel sounds with those of multiple resonance, a subject first investigated by Professor Wheatstone.

The phenomena of simple or unisonant resonance are so well known that we need only call attention to one or two of the most striking facts. If a vibrating body be brought near a column or volume of air, which would be capable of producing the same sound were it immediately caused to sound as an organ-pipe or otherwise, then the sound of the vibrating body is greatly reinforced, as when an harmonica glass is brought before an unisonant cavity, or when a tuning fork is placed at the embouchure of a flute, the apertures of which are stopped, so that, if blown into, the flute would sound the same note ; in the latter case the experiment is more remarkable as the sound of the tuning fork is scarcely itself audible. The same effect takes place when the cavity of the mouth is adjusted so as to be in unison with the tuning fork.

We now come to the new facts of resonance. A column of air will not only enter into vibration when it is capable of producing the *same* sound as the vibrating body which causes the resonance, but also when the number of the vibrations which it is capable of making is any simple multiple of that of the original sounding body, or, in other words, if the sound to which the tube is fitted is any harmonic of the original sound.[2]

The resonance effects are more fully dealt with by Wheatstone in his paper 'On the Resonances, or Reciprocated Vibrations

[1] *Sensations of Tone*, Helmholtz, 2nd edition, p. 103.
[2] Wheatstone's *Scientific Papers*, p. 358.

of Columns of Air,' in the *Quarterly Journal of Science*, 1828, volume 3.[1]

It will be seen from the foregoing statement of Bell that his serious study of electricity was undertaken with a view to understanding and repeating Helmholtz's experiments in the artificial production of vowel tones. The knowledge of electricity was another essential to the accomplishment of his subsequent work, but the application of his acoustical knowledge to the production of a telephone was not yet. The suggestion of Bourseul, the attempts of Reis, were so far unknown to him. It is probable that at this period Bell was amongst those who supposed that the transmission of speech required the production of a speaking machine such as he had in his youth attempted to develop and in a small measure succeeded in producing. A rough and weary road had yet to be travelled before he reached the goal of an idea.

In August 1870 Mr. Melville Bell left England with his family and went to reside at Tutelo Heights, near Brantford, in Ontario, Canada.[2] In April 1871 Alexander Graham Bell was invited by the Board of Education of the City of Boston to make experiments in the city school for deaf mutes, to ascertain whether these children could be taught to speak by means of his father's system of visible speech. He remained two months conducting the inquiries at the above-mentioned and other institutions. During the following year, whilst residing in Canada, he paid occasional visits to the United States, and on October 1, 1872, began his permanent residence there. He opened, at 35 West Newton Street, Boston, a school of vocal physiology and received as pupils deaf mutes, teachers of the deaf and dumb, and hearing persons with defective speech.[3]

One of his pupils was a very young deaf child, named George Sanders; another was a young lady named Miss Hubbard. The father of George Sanders and the father of Miss Hubbard jointly provided the funds for Bell's experiments in multiple telegraphy, and Miss Hubbard became the wife of the inventor of the telephone.

Helmholtz's apparatus for the artificial production of vowel sounds seems to have had a sort of fascination for Bell. The experiment determined conclusively the compound nature of vowel sounds, the value of which Bell's previous studies enabled him to appreciate; but he was impressed also with the means employed for transmitting the sounds. His knowledge of electricity was less than his knowledge of acoustics, and to him there was something new in employing one tuning fork to interrupt an electric current and utilising that electric current to vibrate another tuning fork.

[1] Wheatstone's *Scientific Papers*, p. 36. [2] *Deposition*, p. 9.
[3] *Ibid.* p. 11.

We are probably indebted as much to the interest of the versatile Samuel Pepys in music as in natural philosophy for his record of a conversation with Dr. Hooke, whose observation on the transmission of sounds through solid bodies was quoted in Chapter I. The relationship between the pitch of a tone and the frequency of the vibrations was explained by Hooke in a conversation with Pepys which took place on August 8, 1666, and is thus recorded:—

Discoursed with Mr. Hooke about the nature of sounds and he did make me understand the nature of musical sounds made by strings mighty prettily, and told me that having come to a certain number of vibrations proper to make any tone he is able to tell how many strokes a fly makes with her wings (those flies that hum in their flying), by the note that it answers to in music, during their flying. That I suppose is a little too much refined; but his discourse, in general, of sound, was mighty fine.[1]

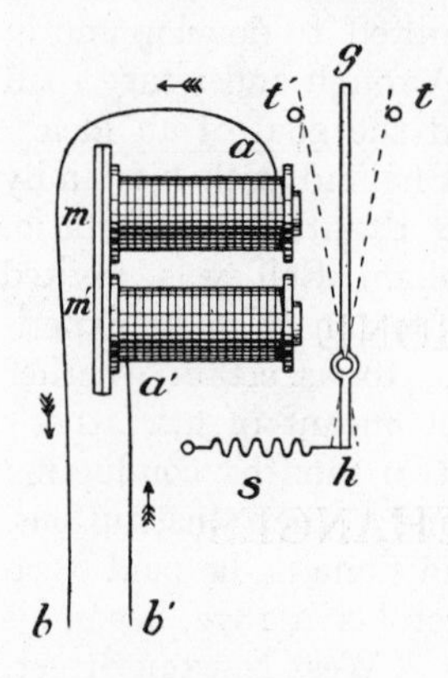

FIG. 1.—Froment's Vibrator.

One of the early examples of the electrical production of sound was that used for the purpose of determining the frequency with which an electro-magnet could be energised and de-energised. As 'Mr. Hooke' explained to Mr. Pepys that the wing flaps of a fly could be counted by the sound given forth, so M. Froment of Paris devised in 1846[2] an instrument in which an electro-magnet attracted a tongue or armature at every pulsation of current sent through the line, and a spring restored the tongue to its former position at every interruption of the current. The primary object of the device was not to create a sound, but to demonstrate the responsive nature of the electro-magnet and to determine by the sound resulting the number of times per second the magnet had been energised. The illustration of the apparatus in fig. 1 is not identical with that shown by Pouillet, but indicates the principle more clearly.[3]

It is unnecessary to quote the description because at this time the operation is generally known, but the following extract may be given for its reference to the effect of residual magnetism:—

The stop, t', is so placed as to prevent the absolute contact of

[1] *Diary of Samuel Pepys*, August 8, 1666.
[2] *Éléments de Physique Expérimentale*, Pouillet, i. § 248.
[3] *The Electric Telegraph*, Lardner, 1855, p. 198.

the arm of the lever with the electro-magnet, but to allow it to approach the latter very closely. Absolute contact is to be avoided, because it is found that in that case the arm adheres to the magnet with a certain force after the current ceases to flow, but so long as absolute contact is prevented, it is immediately brought back by the spring, *s*, when the current is suspended.[1]

By means of this apparatus it was shown

that by the marvellously subtle action of the electric current, the motion of a pendulum is produced, by which a single second of time is divided into from twelve to fourteen thousand equal parts.[2]

Froment's experiment required that the armature or tongue should respond to any rate of vibration. Consequently it was pivoted as a pendulum and had no note of its own. From Froment to Helmholtz was a wide gap acoustically. Helmholtz elaborated the theory of resonance or sympathetic vibration, and his use of electricity was not to transmit a varied series of notes but to send such pulsations over the line as would operate a particular tuning fork, using electricity only as a medium of communication and relying on the principle of sympathetic vibration or resonance.

A description of this principle can best be given in Helmholtz's own simple words. They were part of a popular lecture delivered during the winter of 1857 at Bonn, 'the native town of Beethoven, the mightiest among the heroes of harmony,' where no subject seemed to Helmholtz to be 'better adapted for a popular audience than music itself.'

You will all have observed [he says] the phenomena of the sympathetic production of tones in musical instruments, especially stringed instruments. The string of a pianoforte when the damper is raised begins to vibrate as soon as its proper tone is produced in its neighbourhood with sufficient force by some other means. When this foreign tone ceases the tone of the string will be heard to continue for some little time longer. If we put little paper riders on the string they will be jerked off when its tone is thus produced in the neighbourhood. This sympathetic action of the string depends on the impact of the vibrating particles of air against the string and its sounding board.

Each *separate* wave-crest (or condensation) of air which passes by the string is, of course, too weak to produce a sensible movement in it. But when a long series of wave-crests (or condensations) strike the string in such a manner that each succeeding one increases the slight tremor which resulted from the action of its predecessors, the effect finally becomes sensible. It is a process of exactly the same nature as the swinging of a heavy bell. A powerful man can

[1] *The Electric Telegraph*, Lardner, 1855, p. 199. [2] *Ibid.* p. 201.

scarcely move it sensibly by a single impulse. A boy, by pulling the rope at regular intervals corresponding to the time of its oscillations, can gradually bring it into violent motion.

This particular reinforcement of vibration depends entirely on the rhythmical application of the impulse. When the bell has once been made to vibrate as a pendulum in a very small arc, and the boy always pulls the rope as it falls, and at a time that his pull augments the existing velocity of the bell, this velocity, increasing slightly at each pull, will gradually become considerable. But if the boy apply his power at irregular intervals, sometimes increasing and sometimes diminishing the motion of the bell, he will produce no sensible effect.

In the same way that a mere boy is thus enabled to swing a heavy bell, the tremors of light and mobile air suffice to set in motion the heavy and solid mass of steel contained in a tuning fork, provided that the tone which is excited in the air is exactly in unison with that of the fork, because in this case also every impact of a wave of air against the fork increases the motions excited by the like previous blows.[1]

There was an element of novelty to Bell in combining acoustics and electricity. He was impressed with the employment of an agent to transfer that peculiar power of resonance to distance which seems infinity when compared with the possibilities of the air medium. In a word, he was intensely sympathetic to these sympathetic vibrations. His interest in music prompted him to consider the possibilities from a musical point of view.

While reflecting upon the possibilities of the production of sound by electrical means, it struck me that the principle of vibrating a tuning fork by the intermittent attraction of an electro-magnet might be applied to the electrical production of music.

I imagined to myself a series of tuning forks of different pitches arranged to vibrate automatically in the manner shown by Helmholtz, each fork interrupting at every vibration a voltaic current; and the thought occurred: 'Why should not the depression of a key like that of a piano direct the interrupted current from any one of these forks, through a telegraph wire, to a series of electro-magnets operating the strings of a piano or other musical instrument, in which case a person might play the tuning fork in one place and the music be audible from the electro-magnetic piano in a distant city?

The more I reflected upon this arrangement the more feasible did it seem to me; indeed, I saw no reason why the depression of a number of keys at the tuning fork end of the circuit should not be followed by the audible production of a full chord from the piano with which it was in unison.[2]

[1] *Popular Scientific Lectures*, Helmholtz, p. 80.
[2] *Journal of the Society of Telegraph Engineers*, 1877, vi. 387.

As Wheatstone, a musical instrument maker, first applied his acoustic discoveries to the transmission of music, so Bell, an amateur of music, had first the idea of utilising electricity for similar ends. In both cases also the development proceeded from the pleasurable art of music to the useful art of the communication of intelligence. His interest in electricity led Bell to study the various systems of telegraphy in Great Britain and in America. He was much struck with the simplicity of the Morse alphabet, and with the fact that it could be read by sound. He remarked that—

Instead of having the dots and dashes recorded upon paper, the operators were in the habit of observing the duration of the click of the instruments, and in this way were enabled to distinguish by ear the various signals. It struck me that in a similar manner the duration of a musical note might be made to represent the dot or dash of the telegraph code, so that a person might operate one of the keys of the tuning-fork piano referred to above, and the duration of the sound proceeding from the corresponding string of the distant piano be observed by an operator stationed there. It seemed to me that in this way a number of distinct telegraph messages might be sent simultaneously from the tuning-fork piano to the other end of the circuit, by operators each manipulating a different key of the instrument. These messages would be read by operators stationed at the distant piano, each receiving operator listening for signals of a certain definite pitch, and ignoring all others. In this way could be accomplished the simultaneous transmission of a number of telegraphic messages along a single wire, the number being limited only by the delicacy of the listener's ear. The idea of increasing the carrying power of a telegraph wire in this way took complete possession of my mind, and it was this practical end that I had in view when I commenced my researches in Electric Telephony.[1]

While the general idea of such an invention had originated earlier, Bell's systematic experiments in this direction did not begin until his residence in Boston in October 1872.[2] The form of the apparatus constructed at that time consisted of tuning forks arranged substantially after the manner of Helmholtz.[3] The transmitting tuning fork was placed in a local circuit.

Upon causing the wire to vibrate, the wire attached to the prong was alternately lifted out of the mercury and depressed into it again. The circuit of which the fork formed a part was thus made and broken at every vibration of the fork. The poles of the electro-magnet attracted the prongs of the tuning fork at each

[1] *Journal of the Society of Telegraph Engineers*, 1877, vi. 387.
[2] *Deposition*, p. 12.
[3] *Sensations of Tone*, fig. 33, 1875 edition, p. 178; 1885 edition, p. 122.

make of the circuit, and released them when the circuit was broken. The intermittent attraction of the electro-magnet thus caused the transmitting fork to remain in continuous vibration, emitting continuously its musical tone. By the depression of a telegraph key, the current rendered intermittent by the vibration of the transmitting fork was directed to a line wire which passed to a receiving instrument consisting of an electro-magnet between the poles of which appeared the prongs of a tuning fork. Every time the prong of the transmitting fork made contact with the mercury below it, the prongs of the receiving fork were attracted by the poles of the electro-magnet, between which they were placed; and every time the prong of the transmitting fork broke contact with the mercury below, the prongs of the receiving fork were no longer attracted by the electro-magnet, but were allowed to move freely in the manner of a tuning fork left to itself. Thus, at every vibration of the transmitting fork, the prongs of the receiving fork were attracted by the receiving electro-magnet and released. When the receiving fork had normally the same pitch as the transmitting fork, the intermittent attraction of the electro-magnet would cause it to be thrown into vigorous vibration, thus producing a musical sound of similar pitch to that occasioned by the vibration of the transmitting fork.[1]

I proposed to use a number of transmitting forks of different pitches, and a number of receiving forks, each tuned to the pitch of one of the transmitting forks; and I aimed so to arrange the pitches of these instruments that no transmitting fork should be able to cause vigorous vibration in the prongs of a receiving fork of different pitch from its own. Each transmitting fork was to be provided with a key, by the depression of which electrical impulses from the fork could be sent on to a line wire extending to some distant place where the receivers were to be located. The depression of any key would thus cause on the line wire intermittent impulses of electricity, which would succeed one another with the frequency of a musical sound; with the frequency, in fact, of the vibration of the transmitting fork connected with that key. This intermittent current, passing through all of the coils of all of the electro-magnets of all of the receiving instruments, would occasion the forcible vibration of that fork alone which was the unison of the transmitting fork employed. That is, the receiving fork having the pitch of the transmitter employed would continuously emit its fundamental tone all the time the key at the other end was depressed, but the moment the key at the transmitting end was raised, the receiving fork would stop. Thus, if the key connected to any of the transmitters should be operated in the manner of an ordinary Morse key, long and short musical signals, corresponding to dots and dashes, would be emitted by that receiving fork which had the same pitch as the transmitter employed, but the other receivers would not be thrown into vigorous vibration. Thus,

[1] *Deposition*, p. 13.

the manipulation of any key would cause the vigorous vibration of one, and only one, of the receiving instruments, so that a Morse message transmitted by the manipulation of one of the keys would be received by one, and one only, of the receivers. If two or more of the keys should be depressed simultaneously, then the two or more receiving forks corresponding in pitch to the transmitting forks employed would be set into vigorous vibration; each fork responding to the signal sent from one, and one only, of the transmitters. Thus, upon my method, two or more telegraphic messages could be sent simultaneously along a single line wire, and received separately upon distinct receiving instruments.[1]

After a year of experimenting the tuning forks were superseded (November 1873) by vibrating armatures consisting of single flat plates—really musical reeds. It was apparatus of this type which was chosen by Bell as an illustration of his experiments in the course of his lecture before the Society of Telegraph Engineers in London on October 31, 1877. In the transmitting instrument a steel reed was employed which was kept in continuous vibration by the action of an electro-magnet and local battery. In the course of its vibration the reed struck alternately against two fixed points and so completed alternately a local and a main circuit. When the key was depressed an intermittent current from the main battery was directed to the line wire, and passed through the electro-magnet of a receiving instrument at the distant end of the circuit, and thence to the ground. The steel reed was placed in front of the receiving magnet, and when its normal rate of vibration was the same as the reed of the transmitting instrument it was thrown into powerful vibration, emitting a musical tone of a similar pitch to that produced by the reed of the transmitting instrument, but if it were normally of a different pitch it remained silent.

Subsequent illustrations in the same lecture showed the arrangement of such instruments upon a telegraphic circuit, and after describing them Bell proceeds:—

Without going into details, I shall merely say that the great defects of this plan of multiple telegraphy were found to consist, firstly, in the fact that the receiving operators were required to possess a good musical ear in order to discriminate the signals; and secondly, that the signals could only pass in one direction along the line (so that two wires would be necessary in order to complete communication in both directions).[2]

Here were two difficulties to be overcome. The first was accomplished by a device which was termed a 'vibratory circuit breaker'

[1] *Deposition*, p. 14.
[2] *Journal of the Society of Telegraph Engineers*, vi. 394.

permitting the signals produced by the vibration of the musical reed to be reproduced upon an ordinary telegraphic instrument; the second by passing the intermittent current from the transmitting instruments through the primary wires of an induction coil, and placing the receiving instruments in circuit with the secondary wire. Bell explained that—

In this way free earth communication is secured at either end of the circuit, and the musical signals produced by the manipulation of any key are received at all the stations upon the line. The great objection to this plan is the extreme complication of the parts, and the necessity of employing local and main batteries at every station. It was also found by practical experiment that it was difficult, if not impossible, upon either of the plans here shown to transmit simultaneously the number of musical tones that theory showed to be feasible. Mature consideration revealed the fact that this difficulty lay in the nature of the electrical current employed, and was finally obviated by the invention of the undulatory current.[1]

[1] *Journal of the Society of Telegraph Engineers*, vi. 396.

CHAPTER IV

THE UNDULATORY CURRENT

THE success which eventually attended Bell where others had failed may be traced to his inherited interest and special training, together with the thoroughness with which he investigated the various problems which arose and the completeness of the mastery which he sought over the principles of operation. The conception of the undulatory current was a necessity of telephonic transmission, but, as will have been seen from the extract at the end of the preceding chapter, an undulatory current was first conceived as a means of overcoming one of the difficulties which confronted him in his harmonic telegraph. Having substituted a reed for the tuning fork on the receiving instrument, the reed was next magnetised so that its vibrations should be produced by alternate attractions and repulsions. It was to overcome a difficulty experienced in the transmitter which interrupted a battery current that Bell conceived the idea of turning the receiving reed into the transmitter. He says :—

It then occurred to me that the permanently magnetised reed of a receiving instrument should, if caused to vibrate by mechanical means, itself occasion electrical impulses in the coils of its electro-magnet of the kind required. For I was aware of the fact that when a permanent magnet is moved towards the pole of an electro-magnet, a current of electricity appears in the coil of the electro-magnet ; and that when the permanent magnet is moved from the electro-magnet, a current of opposite kind is induced in the coils.

I had no doubt, therefore, that a permanent magnet, like the reed of one of my receiving instruments, vibrating with the frequency of a musical sound in front of the poles of an electro-magnet, should induce in the coils of the latter alternately positive and negative impulses corresponding in frequency to the vibrations of the reed, and that these reversed impulses would come at equal distances apart.[1]

[1] *Deposition*, p. 21.

The method of operation contemplated is illustrated in fig. 2.

M and M′ are permanent magnets, E and E′ electro-magnets, A and A′ vibrating reeds of similar pitch, B and B′ also vibrating reeds of similar pitch but differing from A and A′. The reeds are attached to bases in the manner usual with reed organs. The instruments are arranged to operate either as transmitters or receivers. 'In this arrangement,' Bell says, 'voltaic batteries, current interrupters, and induction coils became unnecessary, and I was fascinated with the simplicity of the arrangement of circuit.'[1] But the generation of the electric current by such means was novel, and it was yet too soon for Bell to be bold enough to rely upon its use.

It seemed extremely doubtful whether a magneto-electric current generated by the vibration of a magnetised reed in front of an

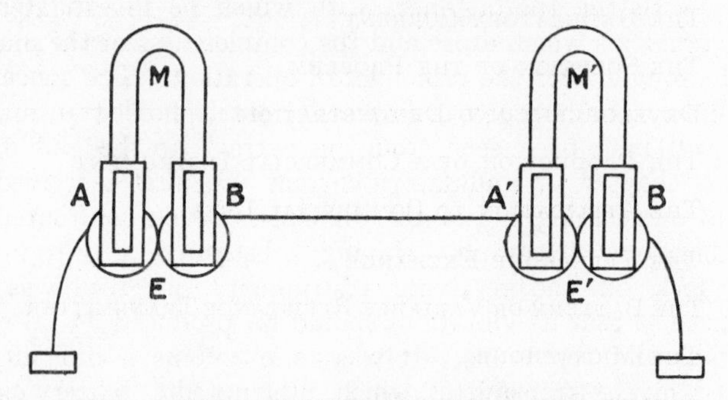

FIG. 2.—Harmonic Telegraph. (Summer of 1874.)

electro-magnet would be sufficiently powerful to produce at the receiving end of the circuit a vibration sufficiently intense to be utilised practically, on real lines, for the purposes of multiple telegraphy. The arrangement seemed to fulfil perfectly the condition I desired, namely, a succession of alternately positive and negative impulses which should be equally distant apart. But I was so impressed with the idea that the magneto currents generated in the way proposed would be too feeble to be utilised on actual working lines, that I sought for some more powerful means of producing reversed impulses of this kind.[1]

Here was an important stage in the development of the invention—a means of creating a current of a particular character in a very simple way. But it was new and in the absence of experience it was unreasonable to expect effective results in practice. Bell was impressed with the idea that the currents so generated would

[1] *Deposition*, p. 22.

be too feeble to be utilised under actual working conditions, and the line of experiment was diverted to what then seemed a more effective method.

But before that further stage was reached he had considered the use of a series of these vibrating reeds together with their accompanying electro-magnets.

I saw that the simultaneous vibration of any number of the reeds would produce upon the line wire a single electrical effect, the resultant of the effects induced in the different coils ; and that the curve that should graphically express this resultant would also represent in a graphical manner the resultant aerial effect produced while the reeds considered were emitting their musical tones. This led to the conception of an improvement in the form of the apparatus considered. The thought occurred—instead of an electro-magnet coil for each reed employed, why not have one electro-magnet for the whole, and let the resultant effect be induced in the coil of the

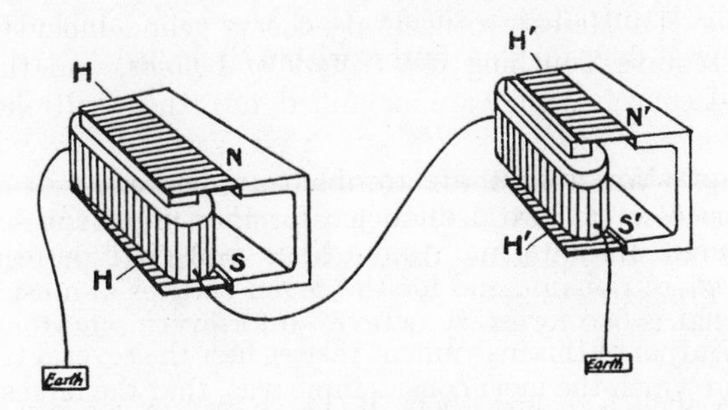

FIG. 3.—Harmonic telegraph or harp telephone. (Summer of 1874.)

electro-magnet itself, instead of inducing separate impulses in distinct and separate electro-magnets, and then combining the whole into a resultant upon the circuit ? In this way originated in my mind the conception of the 'harp' apparatus shown in fig. 16 of my preliminary statement in the 'Harmonic Telegraph Interferences.'[1]

Fig. 16 of the preliminary statement is fig. 19 of the Society of Telegraph Engineers lecture, reproduced here in fig. 3.

My idea of the action of the apparatus, shown in fig. [3], was this : Utter a sound in the neighbourhood of the harp H, and certain of the rods would be thrown into vibrations with different amplitudes. At the other end of the circuit the corresponding rods of the harp H′ would vibrate with their proper relations of force, and the

[1] *Deposition*, p. 34.

timbre of the sound would be reproduced. The expense of constructing such an apparatus as that shown in fig. [3] deterred me from making the attempt, and I sought to simplify the apparatus before venturing to have it made.[1]

Though called a harp the instrument thus suggested might be more accurately compared with the comb of a musical box whose teeth should be set in vibration sympathetically instead of by pins on a barrel. A compound sound being uttered in the neighbourhood of these teeth they would analyse that sound, and those teeth whose vibrations corresponded to the fundamental tone and the respective harmonics of the compound would respond with 'their proper relations of force,' thus originating a current of varying strength which would operate an identical apparatus at the receiving end, with the result that the originating compound sound would be reproduced. It is no wonder that Bell was deterred from constructing such an apparatus and sought a simpler method. Helmholtz enunciated a theory that the ear's ability to analyse sounds was due to the fibres within it called 'Corti's arches,' of which,

according to Waldeyer, there are about 4500 outer arch fibres in the human cochlea. If we deduct 300 for the simple tones which lie beyond musical limits, and cannot have their pitch perfectly apprehended, there remain 4200 for the seven octaves of musical instruments, that is 600 for every octave, 50 for every semi-tone.[2]

Upon such a theory, then, Bell's 'harp' or comb would have needed between four thousand and five thousand 'rods' or teeth to transmit speech, though a smaller number might have been practicable if a pre-determined pitch could have been adopted and retained by a speaker.

In introducing his description of this plan at his lecture before the Society of Telegraph Engineers (I.E.E.) in London, Bell said: 'I therefore devised the apparatus shown in fig. [3], which was my first form of articulating telephone'; whilst a few lines farther on he used the words already quoted: 'The expense of constructing such an apparatus as that shown in fig. [3] deterred me from making the attempt, and I sought to simplify the apparatus before venturing to have it made.'

In this lecture Bell but lightly sketched his preliminary work, and the use of the word 'devised' is probably responsible for this apparatus having been described as the first telephone. For example :—

[1] *Journal of the Society of Telegraph Engineers*, vi. 403.
[2] *Sensations of Tone*, p. 147.

The first form of this instrument *constructed* by Professor Graham Bell, in 1876, is shown in fig. [3]. A harp of steel rods was attached to the poles of a permanent magnet.[1]

Bell's *first instrument* was too far from any form which has come into use to merit more than the briefest description here. Transmitter and receiver were similar and of harplike character, depending for their action upon the principle of resonance.[2]

The harp arrangement was never constructed. Bell so stated in his lecture and more specifically in his legal evidence. He was asked, 'Did you ever actually construct a harp apparatus, such as you have described in connection with fig. 16 of your preliminary statement in the 'Harmonic Telegraph Interferences,' and if not, what mechanical difficulties were obvious ?' To this question Bell replied: 'I did not, because it seemed to me impracticable, as a matter of fact, to construct an apparatus with the multitudinous reeds that theory required.'[3]

The real status of the harp apparatus is not that of the first telephone but the first form which occurred to Bell's mind. He says in his evidence: 'In this way I realised, in the summer of 1874, the conception of a speaking telephone, and the apparatus shown in fig. 16 [the harp form, fig. 3] is the first form of speaking telephone that occurred to my mind.'[4]

It should be added that this conception is not represented by the number of reeds shown in the illustration, as he also says:—

I further saw that if the reeds were multitudinous in number, with very slight differences of pitch between adjoining reeds, the reproduction of quality would be exact, and not simply approximate, so that whatever sound should be made or uttered in the neighbourhood of harp H would be echoed in *facsimile* from harp H′, and if we spoke words to harp H, the reeds of harp H′ would utter words and reproduce the articulation of the speaker.[5]

The harp arrangement assumed the analysis of a complex sound into its components by the appropriate individual reeds and the creation by those reeds of an electric current which should be the equivalent in wave form of the originating sound. The apparatus proposed had been simplified to the extent that only one electromagnet was to be employed instead of several, but the reeds which were to transform sound into current were still numerous. The quotations already given suffice to show that Bell fully realised that the current sent to line was a resultant just as the air wave was a resultant. The analysis by numerous reeds was an

[1] *The Telephone*, Preece & Maier, p. 21. [2] *The Telephone*, Hopkins, p. 11.
[3] *Deposition*, p. 37. [4] *Ibid.* p. 37. [5] *Ibid.* p. 37.

unnecessary operation if a single reed could be so vibrated by the air as to produce a current which should be graphically the equivalent of the air wave. Reflecting on these lines it seemed to Bell that

All that appeared to be necessary was, that one of the reeds, instead of being arranged as a free tuned reed having a rate of vibration of its own, should be forced to move as the air moved, during the production of a sound.[1]

Experiments were then in progress with a phonautograph in which the membrane of a human ear was used. Struck with the disproportion in size and weight between the membrane and the bones that were moved by it, Bell asked himself, 'Why should not a larger and stouter membrane be able to move a piece of steel in the manner I desired? At once the conception of a membrane speaking telephone became complete in my mind; for I saw that a similar instrument to that used as a transmitter could also be employed as a receiver.'[2]

The form which this conception assumed was that indicated in fig. 4, but at this period (the summer of 1874) the membrane form, like the harp form, was only a conception and not a construction. After describing the harp idea in his preliminary statement in the Harmonic Telegraph Interferences,' Bell says:—

Fearing that ridicule would be attached to the idea of transmitting vocal sounds telephonically, especially by those who were unacquainted with Helmholtz's experiments, I said little or nothing of this plan. Indeed, reflection convinced me that, however feasible the scheme looked upon paper, it was impracticable, as the induced currents would be far too feeble to overcome any great resistance.[3]

Bell was confident regarding the acoustic effect, but the electrical transmission, whether from the vibrations of one reed or from a series selected out of a multitude of reeds, he had no confidence in. The current must be too feeble to be effective, he thought, and there was no experience to justify any other conclusion.

In the development of great inventions accident has often played a most important part. In the case of the telephone there is singularly little to be attributed to accidental discovery, or

[1] *Deposition*, p. 38. [2] *Ibid.* p. 39.

[3] *The Multiple Telegraph invented by A. Graham Bell*, p. 13. Boston: Rand, Avery & Co. 1876.

unexpected results from experiments. Bell seems to have started out with knowledge and devised his instruments with an accurate mental conception of their operation. Knowledge grew with experience and one step led to another, but each was thought out beforehand. Nevertheless, fortuitous accident helped Bell to revert to the right line by showing that the electro-magnetic current generated as by a reed was capable of more effective use than he had supposed. It was on June 2, 1875, that Bell and his assistant Watson were engaged in testing receiving instruments such as those described on page 32. The tests were preparatory to an experiment having for its object the sending of three telegraphic messages simultaneously along a single circuit. Three stations (A, B, and C) were arranged, A and B being placed 'in a small room in the upper part of Mr. Williams's establishment,'[1] and C just outside in a room on the same floor. Bell was in charge of stations A and B, and Watson 'was observing the receiving instruments at station C.' At station A were three circuit-breaking transmitters, T, T^1, T^2, tuned to different pitches, and three telegraphic keys, K, K^1, K^2, for connecting the transmitters with the line wire as desired. At station B were three tuned-reed receivers, R, R^1, R^2, having the same pitches as the corresponding transmitters at station A. At station C were also three similar tuned-reed receivers, R, R^1, R^2.

Our mode of experimenting [says Bell] was as follows: I would depress one of the keys at station A—say, key K—and would look at the receivers at station B to see whether the corresponding receiver R was thrown into good vibration. At the same time Mr. Watson, at station C, would observe whether his receiver R was vibrating well. If the vibrations were not satisfactory, we would tune the proper receivers by shortening or lengthening the free part of the reed armature, according as the pitch was lower or higher than that desired. Sometimes we found that the reed armature was so closely adjusted to the pole of the electro-magnet below it that it would stick to the pole, instead of vibrating, the moment a current was passed through the electro-magnet. We would then cause the armature to be released by plucking it with the finger, and if it still stuck every time a current was passed through the coil of the electro-magnet, we would bend the reed armature so as to cause its free end to be normally a little farther away from the pole of the electro-magnet. On this second day of June 1875 we were engaged in the process of testing the receiving instruments at stations B and C. I depressed one of the keys at station A—say, key K—and observed the corresponding receiver R at station B. It seemed to vibrate well, but Mr. Watson called out that the reed of the corresponding receiver R at station C was sticking against the pole of its electro-magnet. I then told him to pluck it to release it,

[1] *Deposition*, p. 58.

and he did so. At that moment I happened to have my eye upon the corresponding receiver R at station B, and was surprised to see the reed of that instrument thrown momentarily into powerful vibration at the very time I supposed Mr. Watson to be plucking the reed of his instrument. I presume that the key K of the corresponding transmitter T at station A was raised, for I did not expect the receiver R to be thrown into vibration at that time. Under those circumstances, the vibration of the reed immediately riveted my attention, and I called out to Mr. Watson to pluck his reed again. He did so, and again the reed of receiver R at station B was momentarily thrown into vibration. I kept Mr. Watson plucking the reed at his station, C, while I placed the reeds of the receivers at station B successively against my ear. At every pluck I could hear a musical tone of similar pitch to that produced by the instrument in Mr. Watson's hands, and could even recognise the peculiar quality or *timbre* of the pluck. For a long time that day there was little done but plucking reeds and observing the effect. Changes were made in the arrangements of circuit, and Mr. Watson and I frequently changed places. To make perfectly sure that the vibration of the one reed had really been the cause of the vibration of the other, and that the effect had been electrically produced, two receivers having the same pitch were arranged as in fig 5 of my patent[1] of March 7, 1876, but without any battery upon the circuit. Upon plucking one of the reeds with the finger, the other was thrown into powerful vibration—vibration so strong as not simply to produce a musical tone, but also to be capable of operating the vibratory circuit-breaker of my patent of April 6, 1875, No. 161,739, and therefore capable of use in my system of multiple telegraphy. These experiments at once removed the doubt that had been in my mind since the summer of 1874, that magneto-electric currents generated by the vibration of an armature in front of an electro-magnet would be too feeble to produce audible effects that could be practically utilised for the purpose of multiple telegraphy and of speech transmission.[2]

It will be observed that the experiment did not reveal a new power but removed the doubts regarding the practical effects of that power. The accidental nature of the revelation is definitely recorded by Bell in the following letter to Mr. Hubbard:—

Salem, Mass., *June* 2, 1875.

DEAR MR. HUBBARD,—I have accidentally made a discovery of the very greatest importance in regard to the Transmitting Instruments. Indeed so important does it seem to me that I have written to the Organ Factory to delay the completion of the Reed arrangement until I have had the opportunity of consulting you.

I have succeeded to-day in transmitting signals *without any battery whatever !*

[1] Reproduced in fig. 15.

[2] *Deposition*, p. 58.

The musical note produced at the Receiving End was sensibly the equivalent of that at the Transmitting end in *loudness* as well as in pitch.

I shall call upon you to-morrow (Wednesday) evening as there are several matters I wish to talk over with you.

In haste, Yours respectfully,
A. GRAHAM BELL.

GARDINER G. HUBBARD, ESQ.,
Brattle Street, Cambridge.[1]

This letter to Mr. Hubbard shows a great deal of self-restraint on the part of Bell. The importance of the discovery was fully realised. He was so much impressed with it that 'for a long time that day there was little done but plucking reeds and observing the effect.' The experiment was being undertaken for the purpose of developing multiple telegraphy, and for this it had its value; but to Bell's trained intelligence it meant a great deal more. A transmission of sound in which *timbre* was produced as well as pitch was a practical demonstration of what he held to be possible in theory. The human voice he knew to be only sound with a frequently changing pitch and quality. Here was encouragement to press on to his goal. And encouragement was needed, for from his associates he received nothing but discouragement in any efforts towards the transmission of speech. Telegraphy was a marketable commodity, the transmission of speech an idle dream—to his financial associates. The environment is evidently responsible for that self-restraint which permitted Bell to write of his accidental discovery without a word about the telephone, when his mind must have been full of the idea that he had reached a most important stage in the development of that invention. Only a week before (May 24, 1875) he had written to his parents: 'Every moment of my time is devoted to study of electricity and to experiments. The subject broadens. I think that the transmission of the human voice is much more nearly at hand than I had supposed.' The experiment of June 2 added to his confidence and encouraged his further energies. Six weeks later (August 14, 1875) he writes to Mr. Hubbard:—

On glancing back over the line of electrical experiments, I recognise that the discovery of the magneto-electric current generated by the vibration of the armature of an electro-magnet in front of one of the poles is the most important point yet reached. I believe that it is the key to still greater things. . . . When we can create a pulsatory action of the current which is the *exact equivalent* of the aerial impulses, we shall certainly obtain exactly similar results.

[1] *Deposition*, p. 56.

Any number of sounds can travel through the same air without confusion, and any number should pass along the same wire.

It should even be possible for a number of spoken messages to traverse the same circuit simultaneously, for an attentive ear can distinguish one voice from another, although a number are speaking together.[1]

The last quoted paragraph is another example of Bell's accurate perception and of the thoroughness with which he attacked his problem. *The wire as the medium of transmission must carry the same kind or form of electrical vibrations as the aerial vibrations of the originating sound.* These are undulatory. Hence the electrical current must be undulatory. He had not yet used that term, but the paragraph shows that he meant it, and the descriptive name was to follow. The stage reached at this date was indeed important. The undulatory current was not only realised as necessary, but experiment had demonstrated that it could be created, transmitted, and its effects reproduced. The creation was mechanical—the plucking of the reed; the rest was electrical. There is a wide difference between the plucking of a reed and the creation of electrical currents by the voice. There was much ground yet to cover. Confidence was needed, and that confidence could only be obtained by Bell's own knowledge of the underlying principles and his belief that suitable instrumentalities could be devised. By study and experiment he had made much progress in the last few months in the electrical part of his work, encouraged thereto by the interest and advice of Professor Henry.

The interview between the aged professor and the youthful inventor is described by Bell in a letter to his parents dated March 18, 1875:—

We appointed noon next day for the experiment. I set the instrument working, and he sat at a table for a long time with the empty coil of wire against his ear listening to the sound. I felt so much encouraged by his interest that I determined to ask his advice about the apparatus I have designed for the transmission of the human voice by telegraph. I explained the idea, and said, 'What would you advise me to do, publish it and let others work it out, or attempt to solve the problem myself?' He said he thought it was 'the germ of a great invention,' and advised me to work at it myself instead of publishing. I said that I recognised the fact that there were mechanical difficulties in the way that rendered the plan impracticable at the present time. I added that I felt that I had not the electrical knowledge necessary to overcome the difficulties. His laconic answer was, 'GET IT.'

[1] *Deposition*, p. 73.

I cannot tell you how much these two words have encouraged me. I live too much in an atmosphere of discouragement for scientific pursuits. Good —— is unfortunately one of the *cui bono* people, and is too much in the habit of looking on the dark side of things. Such a chimerical idea as telegraphing *vocal sounds* would indeed to *most minds* seem scarcely feasible enough to spend time in working over. I believe, however, that it is feasible, and that I have got the cue to the solution of the problem.[1]

[1] *Deposition*, p. 48.

CHAPTER V

THE SOLUTION OF THE PROBLEM

THE experiments of June 2, having demonstrated the efficacy of very weak currents, became the starting-point for further practical trials on more defined lines. Bell says :—

The discovery that the vibration of a steel reed in front of the pole of an electro-magnet generated magneto-electric currents of sufficient power to produce audible effects from a receiver in the

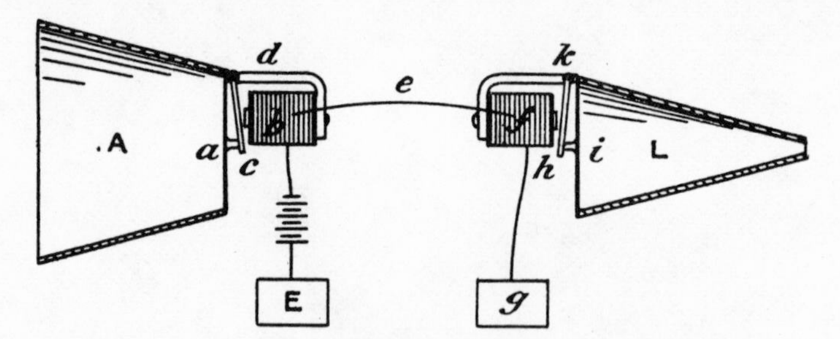

FIG. 4.—Seventh figure of Bell's patent, No. 174,465.

same circuit, convinced me in a moment that the membrane-speaking telephone I had designed in the summer of 1874 would prove a practical working instrument. Before the second day of June 1875, I believed that the instrument was a theoretically perfect speaking telephone, but I had the idea that the currents generated by the action of the voice would be too weak to produce distinctly audible effects from the receiving instrument on a real line. The experiments made on June 2 convinced me that this idea was a mistake, and I immediately gave instructions to Mr. Watson to have two membrane telephones constructed substantially similar to those shown in fig. 7 of my patent of March 7, 1876 [reproduced above as fig. 4].[1]

[1] *Deposition*, p. 60.

The apparatus here illustrated is thus described in the patent specification :—

The armature *c*, fig. [4], is fastened loosely by one extremity to the uncovered leg *d* of the electro-magnet *b*, and its other extremity is attached to the centre of a stretched membrane, *a*. A cone, A, is used to converge sound-vibrations upon the membrane. When a sound is uttered in the cone the membrane *a* is set in vibration, the armature *c* is forced to partake of the motion, and thus electrical undulations are created upon the circuit E, *b*, *e*, *f*, *g*. These undulations are similar in sound to the air vibrations caused by the sound—that is, they are represented graphically by similar curves. The undulatory current passing through the electro-magnet *f* influences its armature *h* to copy the motion of the armature *c*.

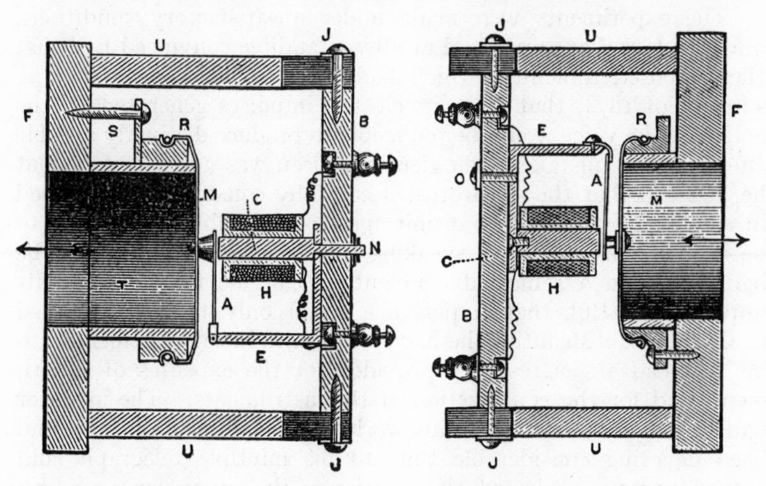

FIGS. 5 AND 6.—Telephones experimented with in July 1875.

A similar sound to that uttered into A is then heard to proceed from L.[1]

The first instruments that were made after the discovery of June 2 were constructed with armatures too heavy and membranes too light. They gave a great deal of trouble, and instructions were given to have the instrument remodelled with a lighter armature and a stouter membrane. Shortly after July 1, 1875, trial was made with the instruments illustrated in figs. 5 and 6, where M is a membrane closing tube T, and capable of being stretched drumhead fashion by the screws S, of which there were three, though only one appears in the illustration. Attached to the centre of the membrane is the armature A. Experiments were made with these

[1] U.S. specification, No. 174,465.

instruments connected together in metallic circuit. Bell 'spoke and shouted and sang' into one instrument upstairs while Watson listened to the other downstairs. Watson rushed upstairs in great excitement to say that he could hear Bell's voice quite plainly and could almost make out what was said. Bell was less successful as an auditor. He says :—

I do not remember the details of these experiments, nor exactly the results obtained, excepting that speech sounds were unmistakably produced from the receiver, and were almost intelligible, and that Mr. Watson appeared to hear a good deal more than I was able to do.[1]

The experiments were made under unsatisfactory conditions, and the fact that any sound at all was audible convinced Bell that the supposed difficulty, which had been in his mind since the summer of 1874, that magneto-electric impulses generated by the action of the voice would be too feeble to produce distinctly audible effects, was a mistake. The demonstration was not complete, but he believed that the apparatus, if carefully constructed, and tried in a quiet place, would transmit speech intelligibly, and prove to be a practically operative speaking telephone. Bell knew that he had conceived a remarkable invention, and he fully realised its importance. But that importance could only be demonstrated in the future. Meantime he had to live, and his arrangement with his financial associates only provided for the expenses of experiments and for the construction of the instruments. The inventor had to maintain himself by his work in other directions. He had been devoting considerable time to his multiple telegraph, and proportionately neglected the exercise of the profession to which he had to look for his daily bread. The 'magneto-electric current generated by the vibration of the armature of an electro-magnet in front of one of the poles' was applicable to the multiple telegraph; the 'undulatory current' was also applicable to the multiple telegraph; but these two discoveries were clearly recognised by Bell as the basis of an invention far transcending in importance any telegraphic application. The transmission of actual speech was Bell's goal, but his associates were practical men ready to supply funds for the development of a marketable commodity like an improved telegraph, but shy of countenancing adventures into unknown, and therefore presumably unprofitable, regions.

To proceed with experiments and adaptations so as to produce apparatus which would effectively demonstrate that the problem was solved would have been the course adopted by most inventors,

[1] *Deposition*, p. 70.

and would probably have been adopted by Bell himself but for the series of circumstances that effectually prevented it. Conscious that he had conceived an important invention, that his discoveries included all the essential elements for carrying it out, that his experiments sufficiently demonstrated in practice the soundness of the principles upon which he was working, he decided to proceed at once with the preparation of a specification for a patent. The specification was commenced in September 1875 and substantially completed by the middle of the following month.[1]

The delay which arose between the completion of the specification in October and its filing in February further illustrates the difficulties with which Bell had to contend. Fully conscious of what he had achieved, Bell was unwilling to limit his patent protection to the United States alone. He wished to have patent protection in other countries also. Neither Mr. Hubbard nor Mr. Sanders was prepared to embark in foreign patents. They limited their outlook to the United States. Bell consequently had to seek support from other sources to enable him to take out foreign patents. For this purpose an agreement was made with the Hon. George Brown of Toronto and his brother Mr. Gordon Brown, who undertook not only to finance the foreign patents, but also to pay the expenses of special rooms where Bell's experimental apparatus could be kept private, as he was at this time troubled by rumours which had reached him regarding visitors to Mr. Williams's workshop 'examining his apparatus with curious eyes.'[2]

The specification was handed to Mr. Brown on the understanding that the application should not be filed in the United States Patent Office until telegraphic instructions could be received from him. He sailed for Europe about January 25, 1876. The American application was sworn to in Boston on January 20, and was sent to Washington 'ready to be filed the moment word should be received from the Hon. George Brown that we might go ahead without interfering with his interests.'[3] But no word was received from Mr. Brown, and it was due to Mr. Hubbard's impatience that the patent solicitors filed the application on February 14, 1876.

With some slight changes this specification became the United States patent No. 174,465 of March 7, 1876, which for the first time revealed to the world the principles of speech transmission, and effectually protected for the inventor the method and means described.

At the time the specification was placed before the Patent Office the invention was—in its full application—a mental conception rather than an accomplished mechanical fact. The two membrane telephones previously described transmitted sounds:

[1] *Deposition*, p. 340. [2] *Ibid.* p. 81. [3] *Ibid.* p. 90.

The voice of the speaker was clearly perceptible from the receiving telephone, the intonations being heard. Articulate sounds were also unmistakably transmitted, but it was difficult to make out what was said. Mr. Watson seemed to be able to hear a good deal better than I could, and my impression is that he stated that he could almost make out what was said.[1]

But the evidence available would seem to show that no really convincing demonstration of the transmission of audible and recognisable speech was made before the deposit of the specification. The circumstances which prevented the active prosecution of experiments and demonstrations have already been referred to. But by Bell such demonstrations were not needed. He had a mental conception of the requirements which was so complete that actual demonstration seemed unnecessary. His appreciation of the value of the invention was equally keen, so that he wished to avoid the delay which might arise from the development of mechanisms to carry out in practice the principles which he had outlined. In his own words :—

I was so satisfied in my own mind that I had solved the problem of the transmission of articulate speech, that I ventured to describe and claim my method and apparatus in a United States patent, without waiting for better results ; in full confidence that the problem had been solved, and that my instruments would turn out to be operative speaking telephones. I was more concerned about taking out a caveat or patent than about further experiments. I believed I had experimented sufficiently to entitle me to a patent.[2]

Bell's anxieties at this time must have been considerable. Harassed regarding his means of livelihood, worried by fears that his experiments might be disclosed to his rivals in invention, it is easy to see that he was additionally troubled by the lack of means to convert his ideas into machinery and the lack of sympathy on the part of his associates with his ultimate aims. In these trying times Bell stood firmly upon his own judgment, was satisfied with his own knowledge of what he had accomplished, and claimed the protection of the patent laws of his adopted country. His application for a patent was allowed, and after fierce fights in later years the highest tribunal of that country decreed that the problem of the transmission of articulate speech was effectually and for the first time solved in the method disclosed in Bell's patent of March 7, 1876. How complete was the solution it is hoped will be shown more in detail in the following pages. Some scientists, patent lawyers, financiers, and others more immediately interested, have

[1] *Deposition*, p. 338. [2] *Ibid.* p. 339.

known that Bell's patent covered all practical methods of speech transmission, but the general public, influenced by the names of subsequent inventors of improvements in details, have regarded Bell as the inventor of the magneto telephone, ignorant of the fact that the specification which covered the employment of undulatory currents obtained by inductive action also covered the method of producing such undulations 'by gradually increasing and diminishing the resistance of the circuit, or by gradually increasing and diminishing the power of the battery,' and, in fact, the electrical transmission and reproduction of spoken words by such undulations, howsoever produced.

CHAPTER VI

DEVELOPMENT AND DEMONSTRATION

RECALLING the illustration and description of fig. 7 of the patent (fig. 4), we may see that when the instrument is used as a transmitter the diaphragm operates upon the magnet through the medium of a suspended armature, the same operation being gone through in reverse order when used as a receiver. Other models followed on the same general lines but with improvements in details, like those illustrated in figs. 5 and 6. One of the principal modifications in the fig. 6 pattern is the method adopted for supporting the armature. Fig. 5 followed the method of fig. 4 with a metallic hinge; in fig. 6 a leather hinge was substituted, suggesting what seems obvious now that the metallic support or suspension was not sufficiently flexible to respond to the movements required. Looking at the subject in the light of after-knowledge, it seems a little surprising that a method of control involving friction or retardation should have been adopted. The records available do not help materially to disclose Bell's reasons for this form, but they may readily be inferred. Bell had really to go through a process of education by trial and experiment before he could realise how small a power sufficed to generate an effective telephonic current. He had early rejected the inductive method, believing that the power would be insufficient. Happy accident demonstrated the contrary, but he probably still supposed that an armature of substantial size and weight was needed, and such an armature would require mechanical support, for his diaphragm was a delicate fabric. Between the diaphragm quickly responsive to sound waves and a heavy armature sluggish in its movements there was a contradictory condition which had to be overcome. It could be overcome only by education—the education of the inventor in the realisation of the fact that the current required was so insignificant. And he could only be his own teacher. To seek any other school would be fatal.

The surprise expressed by the leading scientists of the day when the accomplishment was proved suffices to show that appeals to authorities would have been fruitless. What was needed is now clear enough. The armature must respond to the vibrations of the diaphragm with greater precision than could be expected of detached mechanism, but to do so its dimensions must be diminutive indeed. Confidence in the sufficiency of feeble currents must be gradually attained. The patent was granted on March 7 although, as we have seen, it was prepared some months earlier. On May 10, 1876, Bell presented his paper, 'Researches in Telephony,' to the American Academy of Arts and Sciences. In section 12 of this paper he says :—

Two single pole electro-magnets, each having a resistance of ten ohms, were arranged upon a circuit with a battery of five carbon elements. The total resistance of the circuit, exclusive of the battery, was about twenty-five ohms. A drumhead of gold-beater's skin, seven centimetres in diameter, was placed in front of each

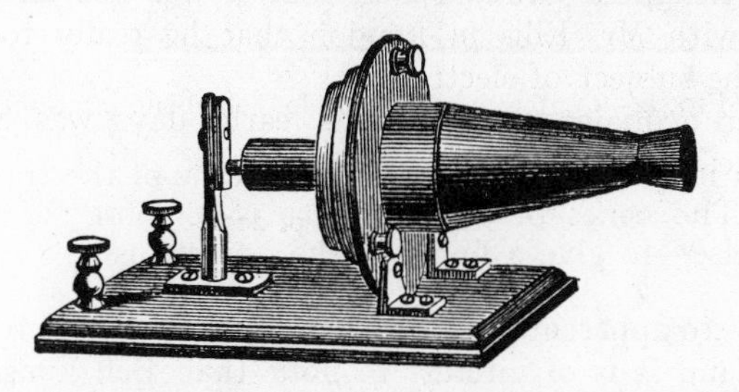

FIG. 7.—Centennial Single Pole Telephone (perspective).

electro-magnet, and a circular piece of clock-spring, one centimetre in diameter, was glued to the middle of each membrane.[1]

Detailed information of the line of reasoning or the series of experiments which led to this result is lacking, but in the instruments described to the American Academy in May, and exhibited at the Centennial in June, we see that the hinged armature had disappeared and that its place was taken by a small iron disc directly attached to the centre of the membrane diaphragm. In another instrument exhibited at the Centennial the diaphragm itself was metallic.

The instruments exhibited at Philadelphia[2] (besides a special form of transmitter which will be reserved for later reference) were :—

A Single-pole Magneto Telephone illustrated in perspective in fig. 7 and section in fig. 8.

A Double-Pole Magneto Telephone, figs. 9 and 10, and

An Iron Box Receiver, figs. 11 and 12.

The illustrations show the instruments with sufficient clearness

[1] *Proceedings of the American Academy of Arts and Sciences*, xii. 7.
[2] *Deposition*, p. 96.

to make a detailed description unnecessary. A curious feature in these instruments is the long cones or mouth-pieces provided. Asked whether he made use of 'the japanned tin cones,' Bell said :—

I tried them and satisfied myself that they would work ; but in my own experiments and tests I preferred to omit them, and simply

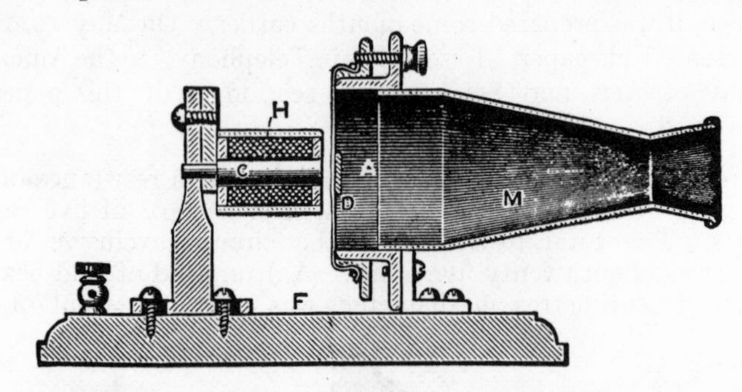

Fig. 8.—Centennial Single Pole Telephone (section).

speak with my mouth as near the membrane of the transmitter as possible. The cones or mouth-pieces were more for show than anything else—to give a finish to the apparatus.[1]

They were apparently a tribute to convention which was not to last long, and it is of interest to note that Bell himself did not regard the cones as essential or even advantageous.

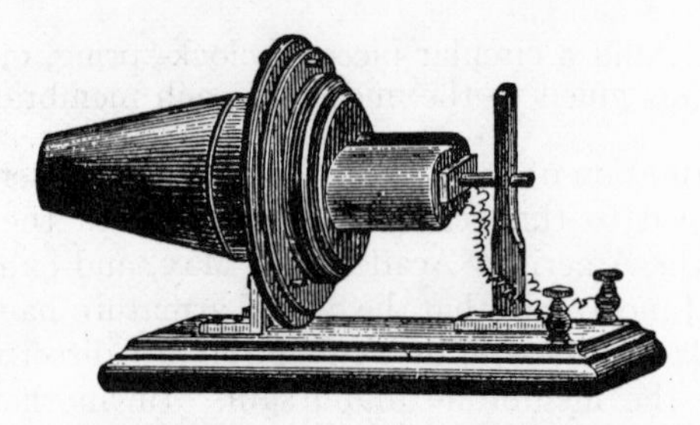

Fig. 9.—Centennial Double Pole Telephone (perspective).

Sunday, June 25, 1876, is a notable day in telephonic history. The Centennial was an International Exhibition, and the judges were of many nations also. Amongst them were the two leading scientists of Great Britain and the United States, Sir William Thomson and Professor Henry. The latter was acquainted with Bell's aims, having been consulted by him, as previously recorded. Sir William Thomson—honoured in his life by being raised to the

[1] *Deposition*, p. 102.

peerage under the title of Lord Kelvin, and at his death by being given a resting-place in Westminster Abbey by the side of his illustrious predecessor, Newton—whilst covering a wide field in his

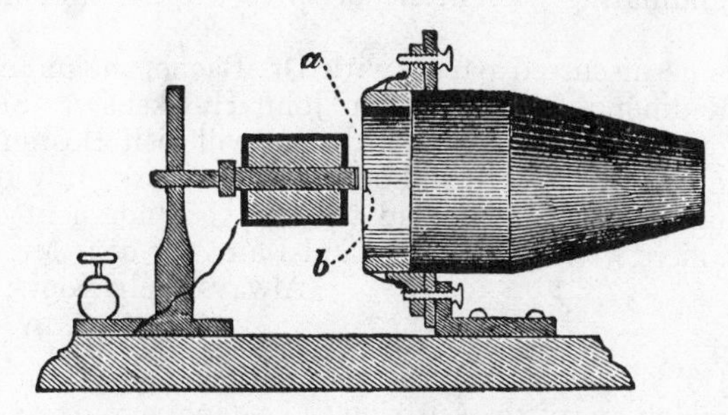

FIG. 10.—Centennial Double Pole Telephone (section).

scientific activities was a specialist in electricity and its applications. But he was a stranger to Bell's efforts and a surprised witness of their results.

The impression produced upon Sir William Thomson was recorded on the same day by Professor Hunt in the following letter :—

Continental Hotel, Philadelphia,
June 25, 1876.

DEAR MR. BELL,—I am informed that you leave tonight for Boston, so I take this way of congratulating you on your success

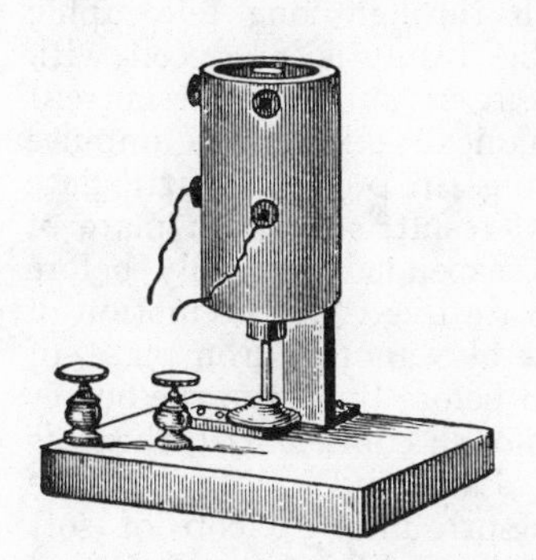

FIG. 11.—Centennial Iron Box Receiver (perspective.)

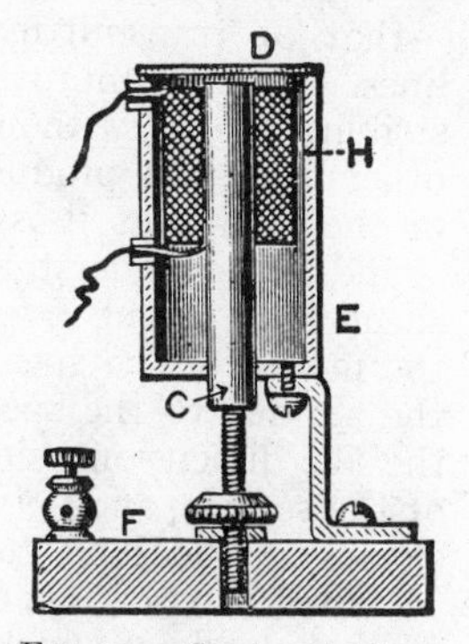

FIG. 12.—Centennial Iron Box Receiver (section).

today. I returned to my hotel with Sir William Thomson, and dined with him. He speaks with much enthusiasm of your achievement. What yesterday he would have declared impossible he has today seen realised, and he declares it the most wonderful thing he has seen in America. You speak of it as an embryo invention, but to him it seems already complete, and he declares that, before long, friends will whisper their secrets over the electric wire.

Your undulating current he declares a great and happy conception.

All this he discussed partly with Dr. Bache, and more at length with Sir Redmond Barry and Sir John Hawkshaw. Sir William leaves here on Friday for Montreal, and will visit Boston for a day or two before sailing, which will be from New York, July 19.

Thinking you would be glad to hear the judgment of one so eminent, I have written you this, and I am, my dear Mr. Bell,

Always truly yours,

T. STERRY HUNT.

P.S.—Do you know anything of Brücher's system of visible speech, of which one of the Austrian judges spoke to-day? It seems very like your father's.

GRAHAM BELL, Esq.[1]

Whether by any rights of office or merely by the tacit recognition of their colleagues that their qualifications were supreme is not recorded, but Henry and Thomson were delegated to draft the judges' reports. It was they who realised the scientific achievement and the practical utility. The General Report of the Judges was written by Henry. After reference to the multiple telegraph he says:—

The telephone of Mr. Bell aims at a still more remarkable result—that of transmitting audible speech through long telegraphic lines. In the improved instrument the result is produced with striking effect, without the employment of an electrical current other than that produced by the mechanical action of the impulse of the breath as it issues from the lungs in producing articulate sounds. To understand this wonderful result, suppose a plate of sheet iron, about five inches square, suspended vertically before the mouth of the speaker so as to vibrate freely by the motion of the air due to the speech, and suppose also another iron plate, of the like dimensions, similarly suspended before the ear of the hearer of the sounds, and between these, but not in contact with them, is stretched the long telegraphic wire. Each end of this wire is attached to two coils of insulated wire surrounding a core of soft iron, the ends of which are placed near the middle of the plate, but not in contact with it. These four cores are kept in a magnetic condition by being attached at each end of the line to the two poles of a permanent magnet. Now it is evident that in this arrangement any disturbance of the magnetism of one of the permanent magnets, increasing or diminishing it, will induce electrical currents, which,

[1] *Deposition*, p. 101.

traversing the long wire, will produce a similar disturbance of the magnetism of the arrangement at the other end of the wire. Such a disturbance will be produced by the vibration of the plate of soft iron due to the words of the speaker, and the current thus produced, changing the magnetism of the soft iron cores, will by reaction produce corresponding vibrations in the iron plate suspended before the ear of the hearer. The vibrations of the second plate being similar to those of the first will reproduce the same sounds. Audible speech has, in this way, been transmitted to a distance of three hundred miles, perfectly intelligible to those who have become accustomed to the peculiarities of certain of the sounds. All parts of a tune are transmitted with great distinctness and with magical effect.

This telephone was exhibited in operation at the Centennial Exhibition, and was considered by the judges the greatest marvel hitherto achieved by the telegraph. The invention is yet in its infancy and is susceptible of great improvements.[1]

The Report on Awards was written by Thomson. This also refers to the multiple telegraph, and then proceeds :—

In addition to his electro-phonetic multiple telegraph, Mr. Graham Bell exhibits apparatus by which he has achieved a result of transcendent scientific interest—the transmission of spoken words by electric currents through a telegraph wire. To obtain this result, or even to make a first step towards it—the transmission of different qualities of sounds, such as the vowel sounds —Mr. Bell perceived that he must produce a variation of strength of current in the telegraph wire as nearly as may be in exact proportion to the velocity of a particle of air moved by the sound ; and he invented a method of doing so, a piece of iron attached to a membrane, and thus moved to and fro in the neighbourhood of an electro-magnet, which has proved perfectly successful. The battery and the wire of this electro-magnet are in circuit with the telegraph wire and the wire of another electro-magnet at the receiving station. This second electro-magnet has a solid bar of iron for core, which is connected at one end, by a thick disc of iron, to an iron tube surrounding the coil and bar. The free circular end of the tube constitutes one pole of the electro-magnet, and the adjacent free end of the bar-core the other. A thin circular iron disc, held pressed against the end of the tube by the electro-magnetic attraction, and free to vibrate through a very small space without touching the central pole, constitutes the sounder by which the electric effect is reconverted into sound. With my ear pressed against this disc, I heard it speak distinctly several sentences, first of simple mono-syllables, 'To be or not to be' (marvellously distinct) ; afterwards sentences from a newspaper, 'S. S. Cox has arrived' (I failed to hear the 'S. S. Cox,' but the 'has arrived' I heard with perfect

[1] *General Report of Judges, Centennial Exhibition, Group XXV*. p. 20.

distinctness); then 'City of New York,' 'Senator Morton,' 'The Senate has passed a resolution to print one thousand extra copies,' 'The Americans of London have made arrangements to celebrate the fourth of July.' I need scarcely say I was astonished and delighted; so were the others, including some judges of our group who witnessed the experiments and verified with their own ears the electric transmission of speech. This, perhaps the greatest marvel hitherto achieved by the electric telegraph, has been obtained by appliances of quite a homespun and rudimentary character. With somewhat more advanced plans and more powerful apparatus, we may confidently expect that Mr. Bell will give us the means of making voice and spoken words audible through the electric wire to an ear hundreds of miles distant.[1]

What could have been more appropriate than that the leading scientist of America, who at an earlier stage had given much needed encouragement to persevere by expressing the opinion that the experiments described by Bell contained 'the germ of a great invention,'[2] and the leader of British scientists, should be the first to express appreciation of the development of that germ, to put the seal of science upon that great invention? Popular applause, however great, can never convey a gratification equal to the appreciation of the well-informed. Alexander Graham Bell experienced this gratification in full measure.

Two months later Sir William Thomson presided over the meeting of the Mathematical and Physical Section of the British Association for the Advancement of Science at Glasgow, and in the course of his presidential address narrated his experience at the Centennial Exhibition, repeating much of his judges' report, and adding:—

This, the greatest by far of all the marvels of the electric telegraph, is due to a young countryman of our own, Mr. Graham Bell, of Edinburgh and Montreal and Boston, now becoming a naturalised citizen of the United States. Who can but admire the hardihood of invention which devised such very slight means to realise the mathematical conception that, if electricity is to convey all the delicacies of quality which distinguish articulate speech, the strength of its current must vary continuously and as nearly as may be in simple proportion to the velocity of a particle of air engaged in constituting the sound.[3]

The first impressions, briefly recorded by Professor Hunt, are enlarged but unaltered. One of the greatest mathematicians of his time was impressed with the 'mathematical conception,' the

[1] *Reports on Awards, Centennial Exhibition, Group XXV*. p. 130.
[2] Page 40.
[3] *Nature*, Sept. 14, 1876, xiv. 427.

greatest electrician of his age recognised the perfection of the means for realising that conception. There was an element of good fortune in the fact that Sir William Thomson was one of the judges at the Centennial. No one had as yet dared to contemplate the possible field of usefulness for the invention, the financial associates were still pressing for multiple telegraph instruments, the inventor even now stood alone in his faith in the commercial developments of the telephone, and the words of Sir William Thomson were not only encouraging to him but were also helpful in their justification for pushing along further development. They informed the world on authority which permitted no carping doubts that speech had been transmitted, and that this result had been reached by the application of true scientific principles. At the achievement the great scientist marvelled, and the world marvelled with him.

CHAPTER VII

THE PRODUCTION OF A COMMERCIAL INSTRUMENT

ON July 12, 1876, in the presence of Sir William Thomson, experiments were made between neighbouring rooms in the Equitable Buildings at Boston over a circuit extending to New York and back. The condition of the line rendered the experiment unsatisfactory, but on short circuiting the New York loop the articulation became audible. After the conclusion of the experiment Bell presented the instruments to Sir William Thomson as a memento

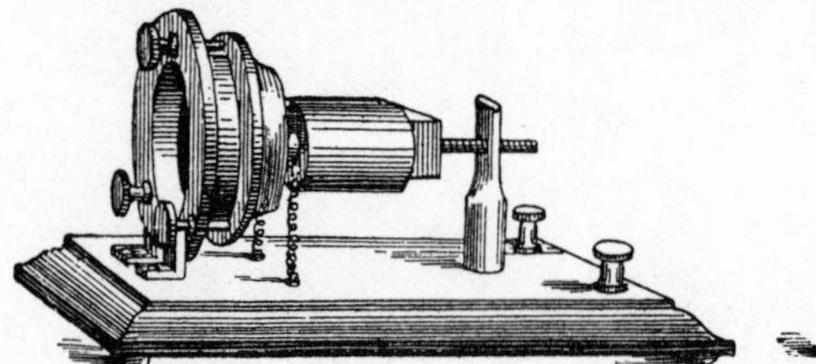

FIG. 13.—Double Pole Instrument. (July 1876).

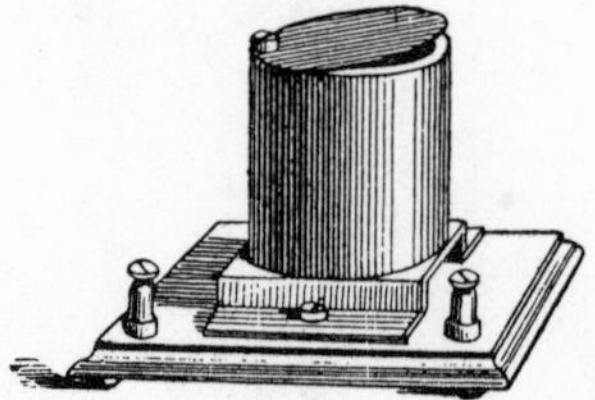

FIG. 14.—Iron Box Receiver (as exhibited at Glasgow).

of the occasion, and they were exhibited by him to the members of the British Association at Glasgow.

These instruments followed closely the design of those exhibited at the Centennial but varied in some respects. The electro-magnet of the double pole instrument (fig. 13) was larger and longer than the Centennial and was wound with finer wire. The iron box receiver (fig. 14) was of larger diameter than the Centennial and, as will be noted from the illustration, was differently mounted upon its base. It had one other difference, as exhibited at Glasgow, which is responsible for much mistaken history. The difference is related by Bell :—

The iron diaphragm or lid has a small screw at one side fastening it to the edge of the iron box. The Centennial iron box receiver had no such screw. The instrument used on July 12, 1876, had this feature. The screw was not employed during the course of experiments, but when the instrument was not in use, the diaphragm was screwed to the edge of the iron box so that it should not be lost. At the conclusion of the experiments on July 12, we had no time to pack the telephones for transportation to England. I simply gave them to Sir William, and he said he would put them in his trunk. The diaphragm of the iron box receiver, not being protected by a packing case, was probably injured on the voyage, so that upon arrival in England the iron lid was somewhat distorted, and did not lie down flat on the top of the iron box as it should have done.[1]

Sir William Thomson's address at Glasgow was reported in *Engineering* of September 15, 1876, and the instruments exhibited were illustrated in the issue of that journal for December 22, 1876. Sir William was evidently unaware that the screw was not intended to remain when the instrument was in use, or that the tilt in the diaphragm was accidental. He gave evidence in the case of the United Telephone Co. Ltd. *v.* Alexander Maclean tried before Lord M'Laren in the High Court at Edinburgh in January 1882. Speaking of the Philadelphia experiment he said: 'The only way it could be heard was by the ear being pressed upon the disc. If pressed too heavily it would kill the sound, and if too little[2] the same result ensued. The great difficulty was adjusting the ear.'[3] From the general tenor of his evidence in this case it seems clear that Sir William Thomson failed to remember that the disc was intended to touch the tube throughout its circumference. The drawing in *Engineering* accurately represented the instrument as exhibited at Glasgow. The British Association meeting was its introduction to the scientific world, and so it happens that the illustrations of the Glasgow instruments became incorporated in history. The drawing with the tilted disc went back to the United States and was included in Prescott's 'Speaking Telephone,' 1878 and 1884; it appears in Du Moncel's 'The Telephone, the Microphone, and the Phonograph,' 1879, and all other histories and text-books. It so appears even in the report of Bell's lecture of 1877 in the *Journal of the Society of Telegraph Engineers*, vol. vi. p. 408, whence the illustration (fig. 14) is taken.

The circumstance was explained by Sir William in his evidence in the Telephone Case of 1882, but in telephonic literature the error was not publicly corrected until the third edition of Mr. Kempster Miller's 'American Telephone Practice,' published in 1900, which contained a letter from Mr. T. D. Lockwood reciting the facts.

[1] *Deposition*, p. 116. [2] ? lightly.
[3] *Electrical Review*, London, Jan. 28, 1882, x. 70.

As was stated in the Edinburgh case, these Glasgow instruments could not be made to work. The reason is to be found in the illustration. It was presumed that the diaphragm was bent intentionally instead of accidentally, that the screw fulfilled a a permanent, instead of temporary, purpose. In operation the screw should have been removed and the diaphragm permitted to rest evenly on the box throughout its circumference, as shown in the illustration of its predecessor (fig. 12). The tubular form of magnet was only reached by stages. Bell had heard articulate speech produced by a tuned reed, fig. 15 (fig. 5 of patent, March, 7, 1876), when the reed was sufficiently damped by close contact with his ear, and he thought that he could reduce the tendency of the reed to vibrate to its own tone by clamping it firmly at both ends and subjecting the centre of the reed to the attraction of an electro-magnet. He devised a receiver having a magnet with three parallel cores upon a single yoke. The centre core was to have a coil wound upon it, and the two

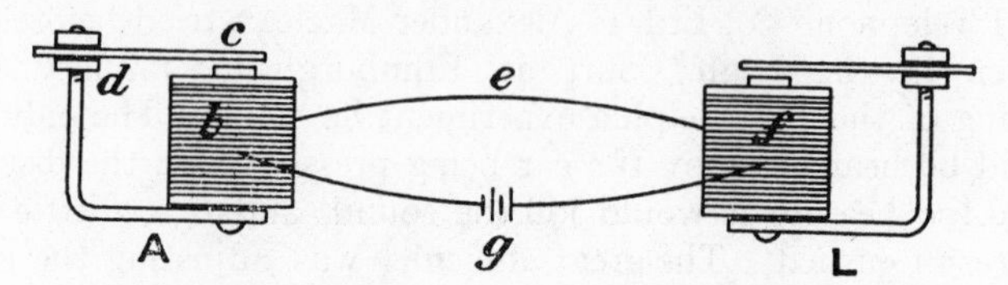

FIG. 15.—Tuned Reed (fig. 5 of patent, March 7, 1876).

outside cores were to support a bridge of steel which should pass over, without touching, the centre core. One great advantage that occurred to Bell in this connection was that the opposite poles of the electro-magnet would be very close together; the end of the covered leg constituting one pole, and the centre of the steel armature the other.[1]

From this stage he proceeded to consider the advantages of a circular plate and an increased number of limbs to the magnet, when he came across a tubular magnet in Mr. Williams's shop and realised that a tubular form would be the best. Professor Silvanus P. Thompson calls this form of magnet 'the Ironclad electro-magnet,' and says of it :—

The appropriate armature for electro-magnets of this type is a circular disc or lid of iron. It is curious how often the use of a tubular jacket to an electro-magnet has been reinvented. It dates back to about 1850, and has been variously claimed for Romershausen, Guillemin, and for Fabre. It is described in Davis's 'Magnetism,' published in Boston in 1855. About sixteen years ago Mr. Faulkner, of Manchester, revived it under the name of the

[1] *Deposition*, p. 321.

Altandæ electro-magnet. A discussion upon jacketed electro-magnets took place in 1876 at the Society of Telegraph Engineers; and in the same year Professor Graham Bell used the same form of electro-magnet in the receiver of the telephone which he exhibited at the Centennial Exhibition.[1]

It is to be borne in mind, however, that Bell's use of this form was not the ordinary use of it. In normal conditions the ironclad magnet has its centre and circumference in the same plane, and the armature makes contact with the centre as well as with the circumference. It will be seen on reference to the sectional illustration (fig. 12, p. 51) that the central core in Bell's instrument was adjustable, and the object of the adjustment was that the core should approach the diaphragm as nearly as possible without touching it. The iron box receiver is an important starting-point for subsequent developments, but as illustrated in the Glasgow pattern an erroneous idea of its operation is conveyed. The tilt of the disc gives it the appearance of a beating reed instead of a diaphragm whose circumference is in contact with the iron tube and whose centre is opposed to the core. In Bell's own words:—

By placing upon it a lid or diaphragm of iron or steel, we had an armature that was damped all around, and polarised by contact with the rim of the box. The end of the central core would then constitute one pole of the magnet, and the centre of the diaphragm or lid the opposite pole. This seemed to me an especially advantageous arrangement. This was the origin of the iron box receiver, and in order to damp the vibrations of the lid or diaphragm still further, the ear was placed closely against the lid, as it had been in former experiments against the reed of the tuned reed receiver. This was the way in which the instrument was generally employed, and it was so used at the Centennial Exhibition. The listeners were instructed to press their ears firmly against the lid of the iron box receiver.[2]

It was customary with Bell throughout his experiments to use his various instruments interchangeably as transmitters and receivers. The iron box receiver and the Centennial transmitter had both been thus reversibly operated. The best results were obtained when the membrane telephone was used as a transmitter and the iron box as a receiver.[3] The reasons were that the ear could not be applied directly to the membrane telephone when used as a receiver, and that the iron box instrument had a diaphragm 'of such small diameter and such stiff material that it was not as well adapted to be vibrated by the voice as the membrane instrument.'[4]

[1] *The Electro-magnet*, Silvanus P. Thompson, p. 52.
[2] *Deposition*, p. 103. [3] *Ibid.* p. 104. [4] *Ibid.* p. 62.

The diameter of the Centennial iron box diaphragm was $1\frac{3}{4}$ inch, the Glasgow pattern, fig. 14, p. 56, was larger, whilst the diameter of the membrane telephone diaphragm was 3 inches.

The greater flexibility of the membrane and the readiness with which its tension could be adjusted seemed to render it more suitable than a metallic plate. Yet the metallic diaphragm had been used for the receiver, and the success with which speech had been transmitted proved that the flexibility of the iron diaphragm was equal to the requirements so far as the reproduction of speech was concerned. It is probable that the realisation of the minute current which would do the work was not yet complete, and that the membrane remained the favourite for transmitters because of the greater excursions that could be expected from it. But the superiority of the metallic plate was gradually being discovered.

Bell had used voltaic batteries in circuit with his instruments in his earlier work, but soon came to realise that the function of the battery was merely to energise the electro-magnet, and that a permanent magnet might be substituted for the latter, thus rendering the battery superfluous. In July 1876, therefore, he had an instrument constructed similar in pattern to the Centennial double pole telephone, but employing permanent magnets instead of electro-magnets with battery, and after a little experimenting the battery was discarded and the permanent magnet relied upon. The diaphragm, however, continued to be of a membrane form with a metallic patch. But repeated experimental demonstrations gave confidence in the metallic diaphragm, and on January 15, 1877, an application for a second patent was filed, which contains the following paragraph :—

In my patent No. 174,465, dated March 7, 1876, I have shown as one form of transmitting instrument a stretched membrane, to which the armature of an electro-magnet is attached, whereby motion can be imparted to the armature by the human voice, or by means of a musical instrument, or by sounds produced in any way. In accordance with my present invention I substitute for the membrane and armature shown in the transmitting and receiving instruments alluded to above, a plate of iron or steel capable of being thrown into vibration by sounds made in its neighbourhood.[1]

The same specification introduces the permanent magnet :—

All the effects noted above may be produced by the same

[1] U.S. specification, No. 186,787, January 30, 1877 (application filed January 15, 1877).

instruments without a battery by rendering the central bar FH [fig. 16] permanently magnetic. Another form of telephone for use without a battery is shown in fig. [17], in which O is a compound permanent magnet, to the poles of which are affixed pole pieces of soft iron, P Q, surrounded by helices of insulated wire, R S.[1]

The eighth claim combines the two features of permanent magnet and metallic diaphragm in the following words :—

In a system of electric telephony, the combination of a permanent magnet with a plate of iron or steel, or other material capable of inductive action, with coils upon the end or ends of said magnet nearest the plate, substantially as set forth.[1]

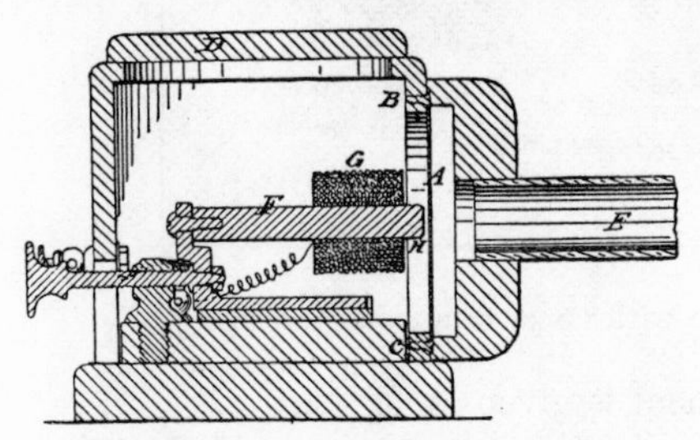

FIG. 16.—Box Telephone with Bar Magnet (1877 patent).

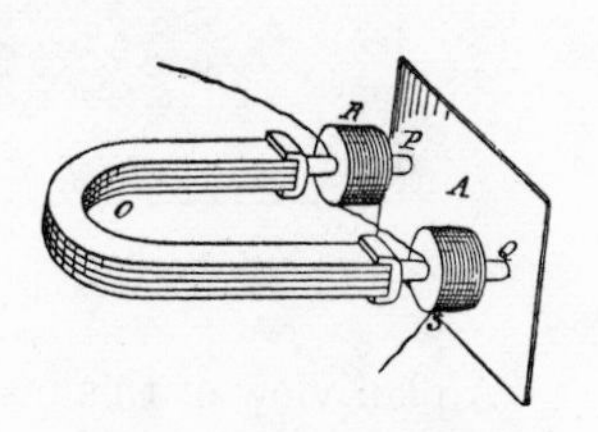

FIG. 17.—Double Pole Type (1877 patent).

This second patent is numbered 186,787, and was granted January 30, 1877.

The drawings in this specification show the continuance of a tendency to rely on boxes and long mouth-pieces, but when the combination of the permanent magnet and iron diaphragm had been reached, it was clearly but a matter of arrangement to dispose these elements in the most suitable way, and Bell devoted his attention to perfecting the form.

The double purpose served by the single instrument is thus definitely stated :—

The sender of the message will use an instrument in every particular identical in construction and operation with that employed by the receiver, so that the same instrument can be used alternately as a receiver and a transmitter.[1]

The Centennial Exhibition instruments, however, differed according to their intended use as transmitter or receiver, and this difference in form was continued, although the principles of operation were the same.

The transmitting instrument was called the Box Telephone, and is illustrated in fig. 18 with the cover removed.

[1] U.S. specification, No. 186,787, Jan. 30, 1877 (app. filed Jan. 15, 1877).

In a pamphlet[1] from which this illustration is taken the date of June 1877 is assigned to it, though Prescott[2] says that the instrument used in talking between Boston and Somerville in April 1877 was essentially like it.

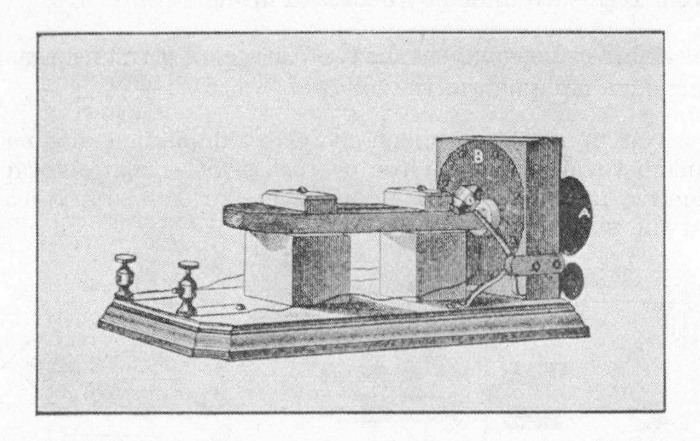

FIG. 18.—Box Telephone with cover removed (June 1877).

A plan view of this instrument is given in fig. 19.

The growing commercial use of the invention would doubtless soon demonstrate the somewhat inconvenient form of this box telephone, so that within two or three months a new model was

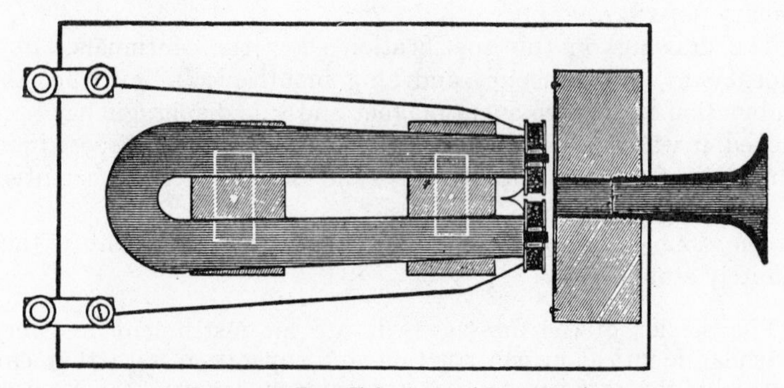

FIG. 19.—Box Telephone (plan view).

evolved, the only change being a modification in the arrangement of the parts.

When the diaphragm and mouth-piece were placed at the end of the magnet and in line with it, as in the earlier model, a long box

[1] *Alexander Graham Bell, the inventor of the Electric Speaking Telephone*, undated, but probably issued in 1881.

[2] *Bell's Electric Speaking Telephone*, Prescott, 1884, p. 440.

was necessary. Attached to a wall it would project unduly. The inconvenience of this form in commercial use was doubtless promptly recognised, for by August 1877 the model illustrated in fig. 20 was brought out. By placing the pole pieces and diaphragm at right angles to, instead of in line with, the magnet the instrument was well adapted to attach to a wall, and this constituted the transmitter.

The first hand telephone followed the box telephone in being of the double pole type. A **U**-shaped permanent magnet was placed inside a cylindrical wooden case of small enough diameter to be easily grasped by the hand, and thus form a handle by which the telephone could be lifted to the mouth or ear as required.[1] An instrument of this form was publicly exhibited at a meeting of the Society of Arts in Boston in May 1877, and though the first to be completed, it was one of four ordered at the same time, 'two to contain **U**-shaped permanent magnets, and two to contain straight-bar permanent magnets,' and all were to be of the same general shape. There are no illustrations of the double pole hand instrument, but by May 1877 the straight bar model had assumed the form shown in fig. 21:

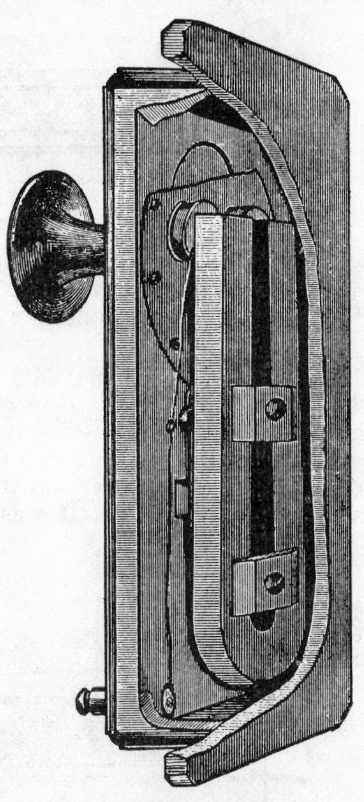

FIG. 20.—Box Telephone (August 1877).

Another model (fig. 22) was issued in June, the change being apparently limited to the exterior of the case.

But by this time scientific interest had been aroused, and numerous people were investigating causes and suggesting designs.

Amongst others were a number of professors of Brown University at Providence, who conducted experiments in the Physical Laboratory of that institution, and kept Bell advised of their progress. Of the work of these gentlemen Bell thus spoke in his London lecture to the Society of Telegraph Engineers :—

[1] *Deposition*, p. 155.

And here I wish to express my indebtedness to several scientific friends in America for their co-operation and assistance. I would specially mention Professor Peirce and Professor Blake, of Brown

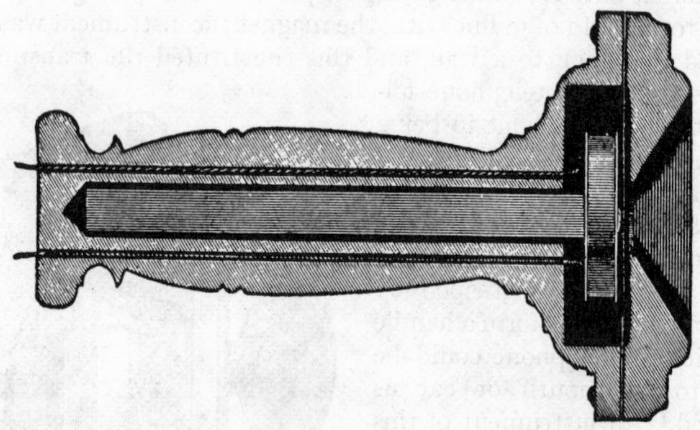

FIG. 21.—Hand Telephone (May 1877).

University, Dr. Channing, Mr. Clarke, and Mr. Jones. In Providence, Rhode Island, these gentlemen have been carrying on together experiments seeking to perfect the form of apparatus required, and I am happy to record the fact that they communicated to me each new discovery as it was made, and every new step in their investigations. It was, of course, inevitable that these

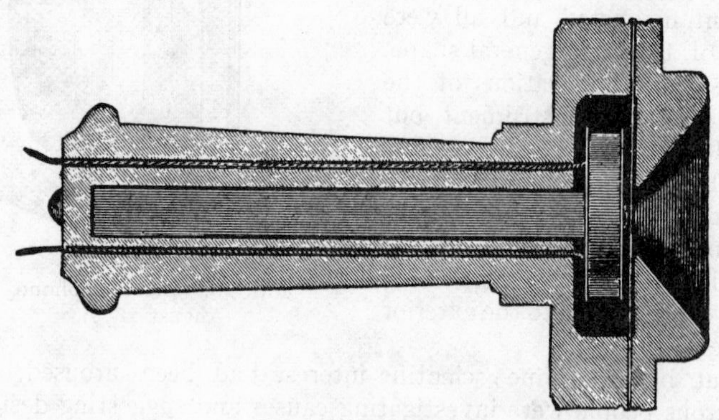

FIG. 22.—Hand Telephone (June 1877).

gentlemen should retrace much of the ground that had been gone over by me, and so it has happened that many of their discoveries had been anticipated by my own researches; still, the very honourable way in which they from time to time placed before me the results of their discoveries entitles them to my warmest thanks, and to my highest esteem. It was always my belief that a certain ratio would be found between the several parts of a telephone, and that the size of the instrument was immaterial; but Professor Peirce was the first to demonstrate the extreme smallness of the magnets which might be employed. And here, in order to show the parallel lines in which we were working, I may mention the fact that two or three days after I had constructed a telephone of the portable form, containing the magnet inside the handle, Dr. Channing was kind enough to send me a pair of telephones of a similar pattern, which had been invented by the Providence experimenters. The convenient form of mouth-piece shown in fig. [22]. now adopted by me, was invented solely by my friend Professor Peirce.[1]

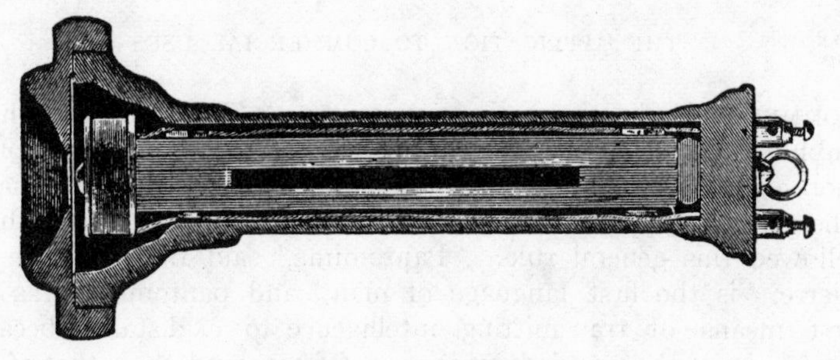

FIG. 23.—Hand Telephone (December 1877).

By April 1877 the telephone had 'got into the factory.' Some time previous to April 1, 1877, Bell and his associates made an arrangement with Charles Williams, junior, of Boston, for the manufacture of his telephones with metallic diaphragms for general commercial use.

After other slight changes in shape in the meantime, the hand telephone assumed its more definite form (fig. 23) in December 1877, when wood as a material for the handle was given up and hard rubber adopted instead. For the magnet, compound bars made up of several layers of magnetised steel were used.

This was by no means the last change to be made in the form of the telephone as a receiver, but we cannot now follow these changes further. The instrument had reached a commercial form; it was to be put to commercial uses. The direction or extent of those uses few, if any, could foresee.

[1] *Journal of the Society of Telegraph Engineers*, vi. 411.

CHAPTER VIII

THE APPLICATION TO COMMERCIAL USES

COMMERCE is, in the main, conservative. The introduction to public attention of an adaptation of some existing condition is more promising of success than the attempt to create a new demand. The practical applications of the transmission of intelligence have followed this general rule. 'Pantomime,' said Bernardin de St. Pierre, 'is the first language of man,' and pantomime was the first means of transmitting intelligence to a distance because the power of the unaided eye is more far-reaching than that of the unaided ear. So man communicated ideas to distant man by the waving of arms. When Chappé was impressed with the possibility of improving the means of communication his conception took the form of replacing the arms of the individual with arms on an elevated post. The number of links in the chain of communication was reduced, and thereby economy in time and money was effected. A code was established which put into language the wavings of semaphore arms. So when Wheatstone and his co-pioneers essayed the transmission of intelligence electrically, they employed electricity to move needles. The method of transmission was new, but the mental operations were along familiar lines.

What more natural than that the proprietors of the telephone should seek to obtain public support by grafting it upon an existing stock rather than awaiting the development of an independent root! The telegraph existed, but trained operators were necessary for its use. To supersede the telegraph instruments by telephones would not only be economical on some existing lines, but the substitution of the spoken word for a cumbrous code would materially enlarge the field of utility. The name first given to the telephone by the public was the 'talking telegraph'; the first public use was expected to be that of a talking telegraph.

So soon as a practical commercial instrument had been completed, the proprietors of the patents went into business. They lost no time in issuing a circular, of which the following is a copy :—

THE TELEPHONE

The proprietors of the Telephone, the invention of Alexander Graham Bell, for which the patents have been issued by the United States and Great Britain, are now prepared to furnish Telephones for the transmission of articulate speech through instruments not more than twenty miles apart. Conversation can be easily carried on after slight practice and with the occasional repetition of a word or sentence. On first listening to the Telephone, though the sound is perfectly audible, the articulation seems to be indistinct; but after a few trials the ear becomes accustomed to the peculiar sound and finds little difficulty in understanding the words.

The Telephone should be set in a quiet place, where there is no noise which would interrupt ordinary conversation.

The advantages of the Telephone over the Telegraph for local business are :

1. That no skilled operator is required, but direct communication may be had by speech without the intervention of a third person.

2. That the communication is much more rapid, the average number of words transmitted a minute by Morse Sounder being from fifteen to twenty, by Telephone from one to two hundred.

3. That no expense is required either for its operation maintainance or repair. It needs no battery, and has no complicated machinery. It is unsurpassed for economy and simplicity.

The terms for leasing two Telephones for social purposes connecting a dwelling house with any other building will be $20 a year, for business purposes $40 a year, payable semiannually in advance, with the cost of expressage from Boston, New York, Cincinnati, St. Louis, or San Francisco. The instruments will be kept in good working order by the lessors, free of expense, except from injuries resulting from great carelessness.

Several Telephones can be placed on the same line at an additional rental of $10 for each instrument ; but the use of more than two on the same line where privacy is required is not advised. Any person within ordinary hearing distance can hear the voice calling through the Telephone. If a louder call is required one can be furnished for $5.

Telegraph lines will be constructed by the proprietors if desired. The price will vary from $100 to $150 a mile ; any good mechanic can construct a line ; No. 9 wire costs 8½ cents a pound, 320 pounds to the mile ; 34 insulators at 25 cents each ; the price of poles and setting varies in every locality ; stringing wire $5 per miles ; sundries $10 per mile.

Parties leasing the Telephone incur no expense beyond the annual rental and repair of the line wire. On the following page are extracts from the Press and other sources relating to the Telephone.

GARDINER G. HUBBARD.

Cambridge, Mass., *May*, 1877.

For further information and orders address

THOS. A. WATSON, 109 Court St., Boston.[1]

The business tone of the circular is commendable. There is no attempt at exaggeration either as regards the distance to be traversed or the audibility of the speech. Another feature to be noticed is that at this early stage instruments were to be let out on hire, not sold outright.

A pictorial representation of the use of the telephone with two instruments on a direct line and with 'more than two where privacy is not required' was given in the *Scientific American* of October 6, 1877, illustrating an article entitled 'The New Bell Telephone.' Fig. 24 is a reproduction of this illustration.

One hundred and nine Court Street, Boston, whence the first circular was issued, was also the workshop where the telephone had been evolved. Thomas A. Watson, to whom communications were to be addressed, had been the close associate of the inventor during its evolution. Charles Williams, an electrical manufacturer when such manufacturers were few, was the occupier of these premises. Here he conducted his business, and the first line constructed exclusively for telephone use was built between Charles Williams's suburban home at Somerville and his office in Court Street. It was completed April 4, 1877, and was noticed in the Boston newspapers of the following day.

This line was promptly followed by another connecting Williams's office with Bell's laboratory at No. 5 Exeter Street, Boston. Two other lines were constructed from Exeter Street to the offices of persons with whom Bell was associated, and in the early part of May 1877 an arrangement was made with the Cambridge Board of Waterworks to put up a line connecting their office in Cambridge with the works at Fresh Pond. This line was constructed for the practical business purposes of a customer, and the use of the telephone in connection with it was referred to in the *Boston Herald and Advertiser* of May 19, 1877.

In New York a telephone line was erected on May 18, 1877, between the house and office of H. L. Roosevelt; and on May 21, telephones were permanently placed on a line at Altoona, Pennsylvania.

By June 30, 1877, 230 telephones were in regular use; this number

[1] Copied from *The Electrical World and Engineer*, March 5, 1904 (vol. xliii. p. 447), which published a reduced facsimile from an original copy in the possession of Mr. T. C. Martin.

within one month had increased to upwards of 750; at the end of August to 1300.[1] These were mainly as substitutes for telegraph instruments for communicating between two points. But already there had been a glimmering of the possibility that might await an

FIG. 24.—Illustration to article, 'The New Bell Telephone,' in *Scientific American*, October 6, 1877.

interchangeable system. There existed in Boston, as elsewhere, a system of electric burglar alarms; doors and windows were guarded by contact making-and-breaking devices, a signal being given when a contact was broken. Buildings so fitted were connected with a central point, and the signal was given there.

The Holmes Burglar Alarm Co. had a central station at 342

[1] *Boston Electrical Handbook* (1904), p. 128.

Washington Street, Boston, whence burglar alarm lines radiated to a number of banks and stores. Arrangements were made for the use of these lines, their sub-stations and the central station as an experimental telephone exchange.

The lines of Brewster, Bassett & Co., bankers (now Estabrook & Co.), the Shoe and Leather Bank, the National Exchange Bank, and the Hide and Leather Bank, together with a new line from the office of Mr. Williams, the manufacturer, were fitted out with telephones and connected at the Holmes central station with a small switchboard made for the purpose. These lines were repeatedly interconnected, and many conversations were interchanged between their stations, the burglar alarm apparatus being employed to transmit the regular call signals. This was, in fact, the first telephone exchange.[1]

A public exhibition of the working of this system was given on May 17, 1877, and on the following day the *Boston Evening Transcript* reported that 'conversation was carried on between the several points connected with perfect ease.' On May 15 Mr. Holmes ordered from Mr. Williams a new six-plug switch, that the manipulation of the lines might be more easily accomplished than his then existing burglar alarm apparatus permitted. Additional subscribers were connected at this time, and Mr. Holmes's appreciation of the utility of the system is evidenced by his desire to obtain the central office rights, as indicated in the following correspondence :—

Boston, *July* 19, 1877.

Mr. G. G. HUBBARD,

DEAR SIR,—I understand from my conversations etc. with you that I am to have the exclusive right of the use of your telephones for the city of Boston for all Central Office purposes.

The plans most definitely mentioned so far, being a system of running a single and exclusive wire from a subscriber's business place or residence to our central office for the purpose of putting said subscriber in a direct speaking communication with any other subscriber to our central office.

Another being a plan now being done by the District Telegraph Company, that is :—our Subscribers, a number being in the same circuit or on the same wire, can telephone the central office for a messenger, or call to any express office in the city—also Railroad, Telegraph and Newspaper offices, Mercantile Agencies etc., providing a demand is made for the connections, and also any other public points that our subscribers may demand as being necessary.

The above plans, yourself and I have thought of, perhaps the most, but as I understand it I am to have the use of the telephone for all central office purposes which we may now have in mind, or which may in future show themselves to us through the experi-

[1] *Boston Electrical Handbook* (1904), p. 126.

ence and acquaintance which we may gather from the constant use of the telephone in the business which we now propose to work up. . . .

Yours very respectfully,
E. T. Holmes.

Mr. Hubbard's reply was as follows :—

Cambridge, *August* 10, 1877.

E. T. Holmes, Esq.,

Dear Sir,—It is understood that you are to have the exclusive right for the use of the telephone in the city of Boston and within a circuit of ten miles for central office purposes viz. for the use of Telephone in circuits connecting a central office with houses, offices, stores and other buildings.

The Bell Telephone Company will furnish the telephones for $10 a year payable quarterly in advance.

The several cities within your territory are to be provided with similar circuits to central offices communicating with the central office in Boston within two years. If not the Bell Telephone Company shall have the right to authorise other parties to establish such circuits, these several central offices connecting with the Boston central office in terms to be agreed upon or fixed by a referee.

The right hereby granted to you to be exclusive provided you serve the public promptly and faithfully, constructing new or extending the old circuits as the demand increases, performing the entire business to the satisfaction of the Bell Telephone Company.

The Bell Telephone Company reserves the right to purchase these lines with the goodwill of the business at an appraised valuation, but not exceeding the actual cost of the lines, or for the goodwill not exceeding $5000 if taken within three years from the date hereby given,

I am, Yours Truly,
Gardiner G. Hubbard.

The exchange system so clearly in the minds of the writers was not at once developed by Holmes. The business requirement of the public which seemed to offer the more immediate prospect of success was that of connecting subscribers to a central office to which they might transmit orders for a General Express Agency, these orders being re-transmitted from the central office. The pecuniary value of such a service was more readily demonstrable, and it was along such lines that the telephone was first developed in Boston. 'Experience and acquaintance,' which might be gathered 'from the constant use of the telephone,' was not needed by those who proposed to introduce the central office system, but rather by the public whose patronage was essential for its remunerative employment.

This experience was being gathered in other directions. Isaac D. Smith was proprietor of the Capitol Avenue Drug Store at Hartford, Connecticut, and in addition to being a chemist took great interest in philosophical and electrical discoveries. For the convenience of his business he had constructed in May 1877 a line between his drug store and the office or surgery of Dr. James Campbell, 28 Buckingham Street. At each end of this line was a bell and a push-button. Signals were exchanged by means of a prearranged code. In July 1877 a second line was constructed to Boardman's Livery Stable, 104 Main Street, where Dr. Campbell's horse was kept. When the doctor required his horse he would, for example, strike two on the bell at the Drug Store and the attendant would repeat the signal to the stable. Mr. Smith's scientific interests led to his being familiar with the fact of the introduction of the telephone, and in July 1877 he obtained the agency of the New England Telephone Company, when telephones were added to the lines, first to the doctor's house and later to the stable also.

The method of repeating the calls was at first adopted with the telephones in the same way as with the bells, but after a few days, and at the instance of Dr. Campbell, Mr. Smith devised an arrangement by means of which the lines might be connected together. The successful accomplishment of this arrangement led to other doctors being added to the system, and on August 17, 1877, the *Hartford Courant* contained the following announcement :—

> At the regular meeting of the allopathic physicians on Monday evening experiments were successsfully tried with telephones, and it was proposed to have a system of inter-communication between the doctors established by means of the new invention so that by reporting to a central office at the Capitol Avenue Drug Store they can readily exchange views between office and office.

The results were so far encouraging as to induce the chemist to enlarge the field of his electrical operations, and on October 8, 1877, he advertised in the *Hartford Courant*, under the heading ' Professor Bell's Telephone,' that he was prepared to build and equip telephone lines at moderate rates. By the month of November 1877 there were seventeen subscribers with telephonic connection, though not all connected with the central office. The line of one was extended to another, and various branches thus existed with several stations on them, but all able to inter-communicate on request, though some required to do so by various stages.

The practical advantages of Mr. Smith's central office system and the utility of the telephone as a means of prompt communication were shown on the occasion of a railway accident in the neighbourhood during the month of January 1878. .This circumstance is referred to in the following extract from a Boston newspaper of the period :—

A striking illustration of the use of Professor Bell's Telephone in a case of emergency was furnished by the recent terrible disaster to the Moody and Sankey excursion train on the Shore Line [? Connecticut Western] Road.

The first news came in the shape of a message to the Western Union Office in Hartford from Tariffville, stating that a fearful accident had happened at that place, and asking that as many surgeons as could be obtained be sent at once.

The operator at Hartford immediately telegraphed to Isaac D. Smith's Capitol Avenue drug store, and the night clerk was awakened and told what was wanted.

The drug store is connected by telephone, so that twenty-one different physicians could be summoned, which was done, and at the same time the clerk telephoned to a livery stable for an express wagon, into which on its arrival was packed bandages, morphine, chloroform, and whatever was thought would be needed.

Meanwhile the doctors, as instructed, went down to the Hartford depot.

Supt. J. T. McManus of the Hartford Providence and Fishkille Road had also been aroused, none of the C. W. people being handy, and he had at once gone to work and gotten out an engine and some wrecking cars, which he had all ready on the arrival of the physicians. Extra passenger cars were also added, the bandages, drugs, etc. were taken aboard, and about a dozen doctors started for Tariffville.[1]

This was the earliest instance of the benefits that the telephone and the exchange system were capable of rendering in a great catastrophe, and its publication in the press served to arouse public interest in the invention and the application. To the press indeed the telephone has been much indebted from its earliest days for recording its capabilities. And to a member of the press credit must also be given for a recognition of the value of the exchange service in advance of its accomplishment.

Mr. Ponton, a reporter for the *Titusville Herald*, found in the telephone a suitable subject for what is known in press circles as 'copy.' He was present at many of the earlier exhibitions and lectures given by Bell. The *Titusville Herald* of October 22, 1876, had a paragraph on the telephone. The Salem experiments of February 12 and March 5, 1877, were also described in that paper. Mr. Ponton was so much impressed with the utility of the telephone that he became one of the earliest agents of the Bell Telephone Company. On August 7, 1877, he wrote to Mr. Hubbard :—

[1] *Semi-Weekly Advertiser*, Boston, January 25, 1878.

There is no difficulty whatever in starting the central system here and spreading it all over the oil region, and I will commence active operations at once so soon as our agreement is definitely concluded.

On August 19, 1877, he wrote at length on the subject. He referred to the question of interesting capitalists, made certain estimates of costs showing

that the outlay of capital, even for towns where the wire has not to be sunk underground, is four times greater than that of the telephones themselves, and this is a very low estimate—it may in some instances be five times.

The amount estimated for the telephones was $10 each. He also prepared a scheme of rentals, which should be a source of income to the Bell Telephone Company as well as the capitalists concerned, and concluded—

All I want is to have the exclusive agency or right to use in places only where a certain amount of capital is expended in the formation of the Central System as above described, and leave it to you to designate the counties in which it shall be done.

On October 27, 1877, he forwarded a copy of a circular headed 'Ponton's Telephone Central Service,' and stating—

This service is original with the subscriber (Ponton) and was proposed by him to the members of the Bell Telephone Company before any private lines were built, and it met their hearty approval. It was too early at that time, however, to start the system, as the public required further evidence that the Telephone would do its work.

The system is extremely simple. All parties who wish to adopt it must have a separate wire from their house, office, factory, hotel, store, bank or restaurant to a central switch room, where any one wire can instantaneously be connected with any other wire. Supposing that one hundred persons adopt this system, and that the average length of each wire is half a mile, it would give each person the privilege of using fifty miles of wire at a less cost than it could be done with only one mile in the private line.

All the advantages of the telephone exchange system are clearly set forth, and the circular contains a list of occupations which might have been left to the imagination of the reader. 'In domestic life,' it states, 'the telephone can put the user in instant communication with the grocer, butcher, baker,' and one hundred and seventy-six separately stated other occupations, followed by the general phrase, 'and other places and persons too numerous to mention.'

Unhappily the work of this enterprising journalist was unproductive, for he wrote under date of January 3, 1878, that he had failed to arouse sufficient interest to establish his system.

The scepticism was not confined to the public. On August 17, 1877, the *Glasgow Herald* stated that—

In America, at Boston and Cleveland, Ohio, dwelling houses and places of business are connected with the telephone with a central office. Merchants talk with each other without the intervention of a third party, and therefore in perfect secrecy. Brown telephones his wife that he is bringing Jones and Murphy to dinner at five o'clock, and straightway Mrs. Brown directs the central office to give her the butcher, and a joint is immediately ordered by means of the telephone. It is further intended that the cities of the New World be placed in telephonic communication with a Central Office, where a merchant can buy the use of the line for ten or twenty minutes, as the case might be. For example, a merchant may write to another friend miles away ' to meet him at the telephone exchange at 12 o'clock sharp,' and next day they converse as freely as if they were in private one with the other.

There is good reason to believe that this statement was the echo of Mr. Ponton's or some other press representative's ' intelligent anticipation of events ' rather than a record of facts, so that there was some justification for the London *Telegraphic Journal* of October 15, 1877, preceding its quotation from the *Herald* with the observation :—

Like other marvellous things, the telephone seems to have established quite a literature of its own. The comic papers have employed it as a vehicle for their wit, and especially for that of a rather far-fetched description. Poets have eagerly welcomed it as a new image, and there have not been wanting preachers who have hailed it as a new symbol. It has been the theme of a great deal of amusing speculation, of which it is difficult to distinguish jest from earnest.

The idea of the exchange seems to have occurred to many men in widely separated places during the latter half of 1877. Some of them were familiar with telegraphic systems which conveyed information to the public, such as that which is known in England as the Exchange Telegraph Service and in the United States as the Gold and Stock Telegraph, or of that known in both as the District Messenger Service. The Gold and Stock telegraphed to its subscribers the premium rate for Gold and the prices for Stocks and Shares. The District Messengers were summoned by a subscriber pulling a lever which sent over a common line a series of electric impulses recorded at the Central Office and identifying that subscriber.

Such services as these had been used by a limited number of business men, and had moreover educated some other men in telegraphic work, so that they were more ready than the totally inexperienced to see advantages in any new development of telegraphic communication. The telephone was such a new development. No one name is on record as the originator of the exchange system, probably for the reason that no one is entitled to such credit. There were many to whom such an application occurred simultaneously. But Bell and his associates were fully alive to its value, offered it every encouragement, and proceeded with its organisation in a masterly way.

The telephone was already an aid to commerce, it enabled distant converse between fixed points over distances 'not exceeding twenty miles,' with intermediate listeners if desired. Easy conversation was replacing cumbrous messages, and confidence—that plant of proverbially slow growth—was replacing the attitude of incredulity with which the commercial mind regarded the telephone on its advent.

CHAPTER IX

THE TELEPHONE EXCHANGE

To the present generation familiarity in the use of the telephone has dulled the sentiment of wonder once existing that speech should be transmitted or that means should be found for diverting that speech from line to line at the speaker's desire. So promptly was science applied to commercial and social uses; so generally, after a time, was the application taken advantage of, that the wonder has evaporated and the user only become the more exacting in his requirements.

Electrical communication was the function of the telegraph. Electrical intercommunication of anything like a general character was reserved for the telephone, although efforts were made to utilise the telegraph for what are now known as exchange purposes.

The earliest suggestion so far discovered for the interconnection of telegraphic lines for public use through a central office or central offices is that of François Marcelin Aristide Dumont of Paris in the Republic of France, engineer. He took out a British patent in 1851, the year in which the then recent advances in science and manufactures were signalised by the Great Exhibition in London. Dumont's patent is numbered 13,497. He states that his invention consists in :—

First. The employment of the electric telegraph for the conveyance of intelligence in the interior of large towns, as London or Paris, by a particular combination of electric wires, and by a new system of placing and fixing such wires. . . . The means I employ for effecting a communication in towns consists in establishing, in any town to which my system may be applied, a number of stations, and connecting these stations to one another, or to the house of each subscriber, by the particular arrangement of electric wires hereinafter described.

Fig. 1, sheet 1, of my drawings represents a central station, to which each of the small stations is attached by one or several wires which only communicate with it. 'O' is the central station.

Nos. 1, 2, 3, 4, 5, 6, 7, 8 and the following are corresponding small stations. Each corresponding station is for a certain number of houses or subscribers. Each house has an electric wire connecting it to the station, and a double electric apparatus, one at the subscriber's house and the other at the neighbouring station to which it is connected. The houses are telegraphically numbered from one to any number. Thus, suppose the house number 3, connected to station No. 1, and the occupant of house No. 3 is desirous to communicate with the house 287, connected with the station 12, the subscriber of the house No. 3 signalises to station 1, the number 287. The clerk of the station No. 1 then directs the central station O to place the wire 1.O in communication with the wire O.12. When this is done the clerk signalises to station 12 the No. 287. The clerk at station 12 then connects wires O.12 with the wire 12. 287.

This accomplished, the clerk of station No. 1 finally connects the wires 3 and 1.O and a direct and intermediate communication is then established between the houses 3 and 287, and both subscribers may correspond privately.[1]

Figure 25 is a reproduction of fig. 1 of Dumont's specification, and shows quite clearly that he intended to protect the exchange system. Though he illustrated telegraph instruments in his patent, the wording of the first claim would have entitled him to regard the use of telephones as coming within it. The claim is as follows :—

First, the system or mode of *applying electricity* to the intercommunication of large towns, and the interchange of communication between the inhabitants of such towns as hereinbefore described.[1]

Dumont was born at Crest (Department of the Drôme) in 1819. Educated at the École Polytechnique he entered, in 1838, the École des Ponts et Chaussées and became engineer in chief of the second class. According to the Grand Dictionnaire Universel du 19ème siècle, vol. 6, his activities covered a wide field in engineering, for he published a number of works, including an essay on the embankment and canalisation of the Rhone (1840), 'Paris port de mer' (1863), and several works on kindred subjects; but 'La réforme Administrative et les Télégraphes électriques' (1849) is the only electrical work which is recorded. A French patent, No. 10,439, dated September 1850, intituled 'Perfectionnements et Applications de la Télégraphie électrique' has an addition of November 30, 1850, which is apparently the equivalent of the English patent of 1851. In view of this patent cordial agreement may be expressed with the remark of the editor of the Grand Dictionnaire Universel, who says: 'The writings of this engineer scientist are

[1] British specification, No. 13,497 of 1851.

remarkable for a great breadth of view and a vivid comprehension of the wants of modern civilisation.'

Dumont's patent was presumably known at the time of its

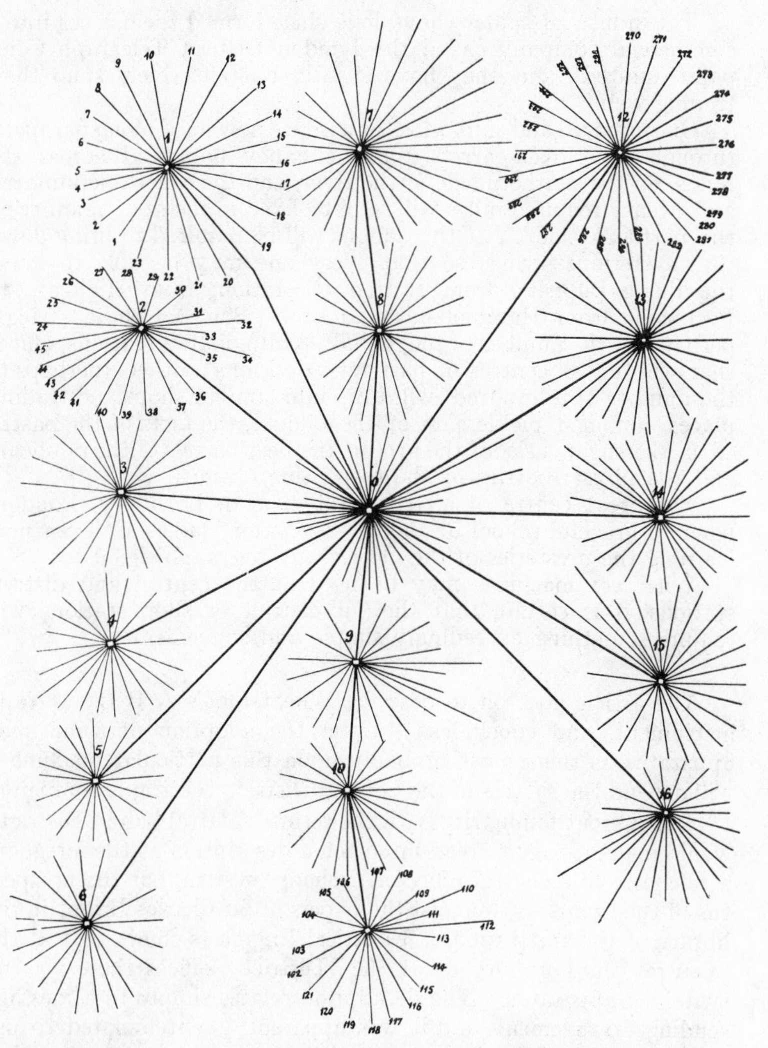

FIG. 25.—Dumont's Patent of 1851 for Telegraph Exchange (first figure).

publication to those interested in telegraphy, but I have found no reference to it in telegraphic literature, and any influence that it may have had on subsequent work in the same direction is, so far as I have been able to learn, unrecorded.

George B. Prescott in his 'History, Theory, and Practice of the Electric Telegraph,' published in 1866, quotes an article from *All the Year Round*, 1859. In the course of this article it is said :—

The industrial spiders have long since formed themselves into a commercial company called the London District Telegraph Company (limited), and they have silently but effectively spun their trading web.

One hundred and sixty miles of wire are now fixed along parapets, through trees, over garrets, round chimney pots, and across the roads, on the southern side of the river, and the other one hundred and twenty required miles will soon be fixed in the same manner on the northern side. . . . Other labour will be required to bring down the mysterious strings so that every one may be able to move the living puppets, from station to station, from Highgate to Peckham, from Hammersmith to Bow. Some of these strings, perhaps to the number of ten, will drop into district stations, offices that will act as centres of particular divisions ; others (perhaps to the number of a hundred) will drop into familiar shops and trading places, amongst pickle jars of the oilman, the tarts of the pastry cook, the sugar casks of the grocer, the beer barrels of the publican, and the physic bottles of the dispensing chemist. . . .

The great centre of all this system is in Lothbury, London, where a graceful school of about sixty young ladies are even now learning the mysteries of the old railway telegraph signals.

Whatever machines may be used at the central and district stations, it is certain that the sub-district or shop stations will require something exceedingly simple and convenient.[1]

The article goes on to describe Wheatstone's A B C telegraph instrument, and concludes : 'Upon the adoption of some such apparatus as this—most probably upon this particular machine—will depend the success of the London District Telegraph Company.'

Our present familiarity with the terms 'Central' and 'District' would make it easy to read into such a description as the foregoing a reference to a central office or exchange system, but the prospectus of the Company amongst the papers of Sir Charles Bright in the library of the Institution of Electrical Engineers shows that at the 'Central' in Lothbury or at the 'District' offices there was no switching apparatus. The description relates simply to a message sending arrangement, and it was apparently contemplated to use 'familiar shops' for the purpose of transmitting messages, so that the expense of a special staff might be saved just as post offices are

[1] *All the Year Round* (conducted by Charles Dickens), November 26, 1859, No. 31, p. 106. The article is entitled 'House-Top Telegraphs,' and appeared in the issue of the magazine which contained the final chapter of 'A Tale of Two Cities.'

combined with trading establishments at the present day. The use of overhouse wires was put forward as an important feature in the prospectus.

The *Telegraphic Journal* of May 7, 1864 (p. 218), quotes from *Chambers's Journal* an article on the District Telegraph system under the title of 'The Domestic Telegraph.' The principal office at this time appears to have been in Cannon Street. That no intercommunicating arrangement was provided for in the system is clear from the following paragraph :—

Should the message be from one out-station to another the clerk at Cannon Street copies the message, and delivers it into the hands of the person who works the particular instrument to which the office for which the message is intended is connected. The clerk there again copies the message, and delivers it to a boy, who carries it to its final destination.

Yet Dumont's patent had been published thirteen years before, and the utility of intercommunicating did not escape the writer of the article, as a later paragraph shows :—

In an upstairs room of the office in Cannon Street there is a kind of cupboard, upon opening the doors of which a mass of wires and screws, bolts and fastenings, are exposed to view. These are all lettered or numbered, and a strange mystery exists with regard to them ; they are really the junctions between various station-wires, and thus two extreme stations might at once be placed in connection with each other. We will suppose that B represents the termination of the wires that lead from Blackheath to Cannon Street, by which wires the messages are sent, and C the termination of the Camden Town wires. Now B and C might be a yard apart in the cupboard, and it would merely be necessary to join these two by means of a wire, in order to allow Camden Town to talk to Blackheath, or *vice versa*. [To 'talk' or to 'speak' was telegraphese for sending messages.]

Incidentally it may be remarked that the district system would appear to have been the pioneer of cheap telegraphy. Reference is made to the fact that the public could now procure one hundred of the company's stamps for one pound. A stamp placed on a written message, which might be enclosed in an envelope and sent to the nearest station, would ensure the transmission of the message ; thus 'the actual price of a message of fifteen words is rather under twopence-halfpenny.'

According to the report of the company for the half year ending June 1864, 152,795 messages were sent and £4,802 10*s*. received for them, so that the average price per message was approximately

7½*d*. The receipts fell short of the expenditure by £60 12*s*. for the six months, an improvement upon previous years which gave the shareholders some hopes; but there was considerable disappointment that the public did not offer greater support to the undertaking. This was attributed by one of the shareholders to the fact that the company had but fulfilled one half of what they undertook to do, and what they hoped still to do. He found that in many cases there was such a delay in the delivery of messages that they could be as speedily delivered by messengers. Many complaints had also been made of the inaccuracy of the messages, and some respecting the charges.

The experiences of the London District Telegraph Co. were not encouraging. In spite of 'publicity methods' which a later age might envy, the directors were, after four years' efforts, unable to develop a paying traffic in local telegrams over the London area, though they apparently served a useful purpose as collectors for the long distance companies, since in addition to the amounts already mentioned they received £3,202, which was paid away to other companies. Whilst the local systems were much developed in later years, especially after their acquisition by the Post Office, there was never a popular local use of the telegraph. The introduction of pneumatic tubes facilitated the transmission of messages to some extent, but the message was still only a message. Local intercommunication on a large scale had to wait until the transmission of speech became practical, and the exchange system developed, for the exchange system did not originate with the telephone.

Telegraphic exchange systems were established both in Great Britain and the United States many years before the telephone was invented. The information available regarding these early telegraphic systems is very meagre, and with a view to ascertain the facts so far as possible regarding the British systems, I inquired of Mr A. M. J. Ogilvie, C.B., one of the Secretaries of the Post Office, who has kindly furnished me with information obtained from officials of the department who were personally concerned with their working. That relating to Newcastle is furnished by Mr. Colin Brodie and Mr. Shadforth, from whose recollections the following is compiled:—

The origin of the exchange system was due to the lack of postal and telegraphic facilities in the district, so that it was necessary to have a messenger attached to each colliery or works for the purpose of conveying correspondence to and from the different agents and business firms. This arrangement was naturally slow, as fairly long distances had sometimes to be travelled. It was therefore a relief to some of the larger firms when the Universal Private Telegraph

Company[1] offered to erect lines at a certain rental and agreed to dispose of the telegrams received from the subscribers for delivery or transmission. This was in 1865, and the service began with three or four renters. By 1870, when the telegraphs were taken over by the Post Office, the number had slightly increased. In 1872 Mr. Colin Brodie sought to develop the system, and succeeded so well that a switch for sixty lines was fixed and brought into operation in 1873. He invited firms who rented private lines to extend to the Post Office and thereby secure the means of speaking to each other. There were about forty renters and twenty post offices connected to the switch when, in 1881, the subscribers were invited to adopt telephones instead. Most of them agreed, but some retained the A B C sets in case of failure. These were, however, soon discarded. The substituted telephone service was started in 1882.

The rentals for the Newcastle Telegraph Exchange were calculated as follows :—

Wires.—£7 per mile, with a minimum of ½ mile for those who were renters of other lines, and 1 mile for independent connections with the exchange.

Instruments.—A B C sets, £6 each; extra bells, £1 each; switches and reversions, 10*s.*; clerk's services, £5 5*s*; apparatus at Post Office was £3 or £3 10*s.* per annum.

In Glasgow also the Universal Private Telegraph Company carried on business, but apparently had not installed any exchange system prior to their lines being taken over by the Post Office in 1870. Mr. D. Campbell, Assistant Postmaster and late Superintendent of Telegraphs at Glasgow, relates that intercommunication was set up by the Post Office on an 'Umschalter' switch. There were at one time as many as eighty renters. The work done included not only switching for intercommunication purposes but also the reception of telegrams for local delivery and outward transmission, and the delivery by private wire of inward messages. It will certainly be a matter for general surprise that this telegraphic 'exchange' still exists (January 1914), though in a very attenuated shape. It is now part of an A B C Concentrator, and of the four lines connected to it three are those of orginal renters.

In addition to those at Newcastle and Glasgow a telegraphic exchange system also existed in Bradford in the pre-telephone era. The formation of similar systems at a later date is indicative of a want of confidence in telephonic communication, but such telegraphic

[1] The Universal Private Telegraph Co. was incorporated by Act of Parliament, 24 & 25 Vict. (Royal Assent June 7, 1861). It was intended to establish private telegraphic lines using Wheatstone A B C instruments. There was no mention in its prospectus of an exchange or switching system which apparently developed from local circumstances after a number of private lines had been put into operation.

exchanges were introduced between 1878 and 1881 in Swansea, West Hartlepool, Middlesbrough, Paisley, Leeds, Darlington, Stockton-on-Tees, Hull, Sunderland, Bristol, Barnsley, and Workington.[1]

Sir John Gavey, C.B., who was in charge of the district at the time, has kindly furnished me with the following particulars of the Swansea system.

From time to time colliery proprietors and others entered into contracts with the Post Office for the provision of private wires worked with Wheatstone A B C instruments and designed for the dispatch and receipt of telegrams, &c. In 1878 arrangements were made to provide intercommunication between these lines, and this was done in the following manner :—

Each line terminated at the Post Office on a Wheatstone A B C receiver and through an Umschalter switch to earth. A pole changer or reversing switch was in circuit with each line. A call for the Post Office, which was indicated by a repetition of the code, was answered by the operator, who switched into circuit a complete A B C set and dealt with the call. A call for another subscriber was switched through to the latter on the Umschalter, the reversing switch being turned so that the two instruments should be in unison. The A B C receivers on the intercommunication switch were always in circuit, so the operator was cognisant of what passed and knew when to sever the connection. There were ten subscribers, and the service was opened on October 7, 1878.

On March 23, 1881, the A B C instruments were replaced by telephones, and the Post Office telephone exchange at Swansea started.[2]

Similar telegraphic exchange systems were in existence at New York and Philadelphia. For the following particulars regarding these systems, also compiled from evidence furnished by participators in the service, I am indebted to Mr. Lockwood.

The Gold and Stock Telegraph Company established in New York in the year 1869 a telegraphic exchange system in connection with the Clearing House, and having direct lines to various banks. The Central Office was situated at the Clearing House at the corner of Wall Street and William Street. The business of that central office was chiefly to report to each bank in the system by means of printing instruments its daily debit or credit balance, and to repeat to any bank any message received for it by wire from any other bank. But the wires of any two banks desiring direct communica-

[1] General Webber relates that the Post Office used a switch at Colchester for the ordinary business between several small towns in that district (*Journal of the Society of Arts*, April 28, 1882, xxx. 609.)

[2] Technical descriptions and illustrations of these telegraphic systems are given by Mr. Purves, M.I.E.E. in *Telegraph Switching Systems* (Alabaster, Gatehouse & Co. 1902).

tion were also connected together at a switch. This direct connecting system was not much used at first. In 1871 the Central Office was removed to the general office of the Gold and Stock Company at 61 (and later to 197) Broadway, an instrument being left at the Clearing House, which then became equivalent to a subscriber instead of being the exchange. An operator was sent daily to the Clearing House, who reported to the Central Office at 61 Broadway each bank's balance, which was then repeated to the bank by an operator and not by connecting the various bank lines with the Clearing House lines; but from 1871, by means of a suitable battery and the regular switching devices, any bank was on demand connected with any other bank or with the Clearing House. All the wires came to the switch (a Jones Lock switch) through a relay, and these relays, one or both, were left in circuit, and of course rattled and clicked all the time the two banks were working together by their printing instruments. When the subscribers required to be disconnected they made three long pulsations as a signal, but this signal was frequently forgotten to be given, so that when the relays were silent for a considerable time they were disconnected without signal. About twenty-five banks and two private lines were connected to this system, which continued in existence until 1880.[1]

A telegraphic exchange of a similar character was established at Philadelphia, the following particulars of which are taken from a letter written by Mr. Bentley, the proprietor, to Mr. Theo. N. Vail on December 12, 1881:—

I first established a system of connecting private telegraph wires through a central office in this city in the year 1867 (in August), as I find by my records. These lines were Morse telegraph wires, and the subscribers to my plan of intercommunication included the largest houses in the city.

In September 1871 I arranged a central office for intercommunication between some twenty odd of the banks and bankers of this city, having said central office in a room adjoining the clearing house. . . . Within the five years following my establishment of a central office for private wires my subscribers increased to somewhere about fifty.

The first telephones we ever worked, so far as I find any record, were four, which were sent to myself personally by Mr. G. Hubbard, or by his direction. My books show that I received them by express August 31, 1877, worked them around on various short wires to test them, and on September 11 following we connected the office of H. D. Davis, 3rd and Walnut Streets, also in the mean time the

[1] From statement of Mr. G. F. Wiley, an officer of the Gold and Stock Telegraph Company, in the possession of Mr. T. D. Lockwood.

Stock Exchange and my office. These were of course experimental tests. I give this to you for what it is worth. Finally, Mr. Cornish was directed to get the Phones from me and that ended that matter.

The first telephones we rented for pay from this office was on February 14, 1878.

Another system of the same character was that established in New York by the Law Telegraph Company.

The origin of this system was related to me some years ago by its originator, Mr. William A. Childs.

In the spring of 1874 [he said] I proposed to establish for lawyers in New York City a circuit wire system with tickers or printing telegraph instruments—a service similar to that given brokers and bankers by the Gold and Stock Telegraph. My idea was to connect the lawyers with the various courts so that they could get each day the calendars for the next day, the decisions of the Judges, and Court news generally. I went among the lawyers and solicited them with a subscription paper, but they said, 'That will be no use to us.' After canvassing for a week or more with no encouragement a lawyer one day said to me, 'There is a great deal of intercommunication amongst lawyers. What you propose is no use; but if you had a system of telegraphs by which I could telegraph to any other lawyer who had one, that is something I would pay for.' Another lawyer said my plan was of no use. He could get all that I proposed in the morning or evening papers, but a means of direct telegraphic communication would be of value. I then conceived the idea of the establishment of a central office with a system of signals whereby if Smith wanted to talk with Jones he could signal the Central Office and have his wire switched on to Jones's wire, and so on. Then I considered the possibilities of getting the wires up and what instruments to use. We adopted the dial instrument similar to Wheatstone's A B C made by Chester, of New York. Then I printed a new prospectus setting forth the new plan and commenced to work a few weeks after. The plan went 'like hot cakes,' subscribers putting their names down readily. There was one central office with a single wire to each subscriber and an individual annunciator on each line. The annunciator was a bell. The number required was indicated by the strokes on the bell of the subscriber giving the call as - - -   - for 31. Sixty bells were placed in a space of, say, four feet long by three feet high. In front of these bells stood about three operators. On the hammer of the bell was a tag containing the subscriber's number. When a subscriber called the Central Office the key which he used for the purpose automatically shunted a large amount of resistance normally in the circuit and thereby gave the battery located at the Central Office sufficient power to move the Central Office bell which otherwise it did not have. A signal to disconnect was made with the same key as the call, hence resistance was taken out and the Central Office bell rang. In each line there was normally connected twelve cells of gravity battery at the Central Office. When we coupled two subscribers together there were thus twenty-four cells in the combined circuit, and in coupling them we necessarily had to reverse the direction of twelve of those cells which required a switch properly constructed for the purpose.

The bell described by Mr. Childs as the first annunciator was later superseded by a Morse Sounder upon a separate line—a call wire.

References are made to the Law Telegraph system in 'Telegrapher,' April 10, 1875, and May 1, 1875. The following is taken from a contemporary notice in a New York paper :—

Electricity has now become identified with the interests of the whole world, from the mightiest ramifications to the demand of local business necessities, and supplies a means of direct and private communication between lawyer and client, merchants, manufacturers, &c., and their connections, indeed, throughout the city, country, and world at large. Most especially is this effectually secured by the Law Telegraph Company of 145 Fulton Street and 261 Broadway, New York, which was organised in the year 1874, and was established for the special purpose of members of the legal profession having direct and immediate communication with their associates, clients, and personal offices, thus entirely dispensing with the uncertain and tardy system of conveying information by messengers.

After a general description of the 'instruments' and 'central office,' it is stated that 'Arrangements are now being made for direct connection with the Western Union Telegraph Company whereby messages can be sent to any part of the world by subscribers' own instruments,' a statement which clearly must not be taken too literally. We can form an idea of the subscription rates as well as learn that the Company did not stand alone by what follows :—

The subscription price of the instrument, compared with that of other companies, is in itself an important recommendation ; for instance, brokers pay such companies at the rate of $25 per month, cotton merchants $30 per month, and parties having private lines are charged from $300 to $1,200 per year, while the charges of the Law Telegraph Company are from 20 per cent. to 30 per cent cheaper. . . . Since the establishment of the Company the number of subscribers has been doubled, and some one hundred and twenty instruments are now in use.

The names are given of 'a few of the prominent houses and firms who have 'long used them,' and the article concludes with the following :—

> Setting the average cost at $5 per week for an ordinary subscription, all parties must perceive how entirely inadequate it is to the general value derived from the use of the machine, and the attention of the business and professional community should be closely interested in the system, indeed no extensive establishments should fail to avail themselves of it.

As we have seen, telegraphic intercommunication was suggested in 1851 by Dumont, whose patent is quoted above. It existed in Newcastle-on-Tyne in 1865 and in Philadelphia in 1867 ; amongst the bankers of New York in 1869, and the lawyers of the same city in 1874.

But so long as words had to be transmitted in dots and dashes or spelt out by letters upon a dial, there was little likelihood of the extensive use of an exchange system. With the introduction of an instrument that talked, the situation was completely changed. There was no need then for dependence on skilled workers. Communications were no longer restricted to mere messages, but were expanded to the very different condition of conversations. The enormous gap that was thus bridged is not recognised so generally as it should be. Question and answer may be sent by telegraph, but they must always lack the spontaneity of a conversation such as the telephone affords. This fact alone suffices to explain the little use made of the exchange idea in conjunction with the telegraph, and to indicate how the invention of a talking instrument changed the situation. The inventor of that instrument was amongst the first to recognise the advantages which the exchange system would confer. In the 'Boston Electrical Handbook' it is related that at the New York lectures on the evenings of May 17, 18, and 19, 1877, Bell outlined and eloquently advocated the proposed use of the telephone exchange yet to be developed.[1] Mr. T. D. Lockwood was the writer of this portion of the Handbook. He was present at the lectures and records his own recollections.

On August 4, 1877, Bell sailed for Europe and did not return until November 10, 1878.[2] The development of the business in the United States consequently devolved upon his associates, but he delivered lectures and gave demonstrations of the use of the telephone at the Society of Arts and the Society of Telegraph Engineers in London, and before the British Association at Plymouth.

[1] *Boston Electrical Handbook*, p. 126. [2] *Deposition*, p. 178.

Negotiations were in progress for the acquisition of the English patents, and in connection with these negotiations Bell prepared a statement, printed below. In a letter to E. J. Hall dated June 8, 1903, enclosing copy of the statement, Bell says :—

I may briefly recite the circumstances under which this paper was written :—

I was married on July 11, 1877, and went abroad very shortly afterwards and remained continuously abroad for about one year. Colonel Reynolds, of Providence, R.I., undertook to organise in England a telephone company to work my invention there. He was at work upon this in the winter of 1877 and spring of 1878, and early in 1878 Colonel Reynolds requested me to write an address to the capitalists he had succeeded in interesting, setting forth the advantages of the telephone. This I did in a paper dated from Kensington, March 25, 1878, addressed to 'The Capitalists of the Electric Telephone Company.' I remember that this paper was printed as a kind of prospectus, and a copy given to the gentlemen concerned and others interested. I beg to enclose a copy of this document. It expresses, of course, the ideas of the development of the telephone I had in mind in the autumn of 1877 when I went to England.

Kensington, *March* 25, 1878.

To the Capitalists of the Electric Telephone Company.

GENTLEMEN,—It has been suggested that at this, our first meeting, I should lay before you a few ideas concerning the future of the Electric Telephone, together with any suggestions that occur to me in regard to the best mode of introducing the instrument to the public.

The telephone may be briefly described as an electrical contrivance for reproducing in distant places the tones and articulation of a speaker's voice, so that conversation can be carried on by word of mouth, between persons in different rooms, in different streets, or in different towns.

The great advantage it possesses over every other form of electrical apparatus consists in the fact that it requires no skill to operate the instrument. All other telegraph machines produce signals which require to be translated by experts, and such instruments are therefore extremely limited in their application, but the telephone actually speaks, and for this reason it can be utilised for nearly every purpose for which speech is employed.

The chief obstacle to the universal use of electricity as a means of communication between distant points has been the skill required to operate telegraphic instruments. The invention of automatic printing, telegraphic dial instruments, &c., has materially reduced the amount of skill required, but has introduced a new element of difficulty in the shape of increased expense. Simplicity of operation has been obtained by complication of the parts of the machine —so that such instruments are much more expensive than those usually employed by skilled electricians. The simple and inexpensive nature of the telephone, on the other hand, renders it possible to connect every man's house, office, or manufactory with a central station, so as to give him the benefit of direct telephonic communication with his neighbours, at a cost not greater than that incurred by gas or water.

At the present time we have a perfect network of gas-pipes and water-pipes throughout our large cities. We have main pipes laid under the streets communicating by side pipes with the various dwellings, enabling the members to draw their supplies of gas and water from a common source.

In a similar manner, it is conceivable that cables of telephone wires could be laid underground, or suspended overhead, communicating by branch wires with private dwellings, country houses, shops, manufactories, &c., &c., uniting them through the main cable with a central office where the wire could be connected as desired, establishing direct communication between any two places in the city. Such a plan as this, though impracticable at the present moment, will, I firmly believe, be the outcome of the introduction of the telephone to the public. Not only so, but I believe in the future wires will unite the head offices of the Telephone Company in different cities, and a man in one part of the country may communicate by word of mouth with another in a different place.

I am aware that such ideas may appear to you Utopian and out of place, for we are met together for the purpose of discussing not the future of the telephone, but its present.

Believing however as I do that such a scheme will be the ultimate result of the telephone to the public, I will impress upon you all the advisability of keeping this end in view, that all present arrangements of the telephone may be eventually realised in this grand system.

The plan usually presented in regard to private telegraphs is to lease such lines to private individuals, or to companies at a fixed annual rental. This plan should be adopted by you, but instead of erecting a line directly from the one to another, I would advise you to bring the wires from the two points to the office of the Company and there connect them together; if this plan be followed a large number of wires would soon be centred in the telephone offices, where they would be easily accessible for testing purposes. In places remote from the office of the Company, simple testing boxes could be erected for the telephone wires of that neighbourhood, and these testing places could at any time be converted into central offices when the lessees of the telephone wires desire intercommunication.

In regard to other present uses for the telephone, the instrument can be supplied so cheaply as to compete on favourable terms with speaking tubes, bells, and annunciators, as a means of communication between different parts of the house. This seems to be a very favourable application of the telephone, not only on account of the large number of telephones that would be wanted, but because it would lead eventually to the plan of intercommunication referred to above ; I would therefore recommend that special arrangements should be made for the introduction of the telephone into hotels and private buildings in place of the speaking tubes and annunciators at present employed. Telephones sold for this purpose should be stamped or numbered in such a way as to distinguish them from those employed for business purposes, and an agreement should be signed by the purchaser that the telephones should become forfeited to the Company if used for other purposes than those specified by the agreement.

It is probable that such a use of the telephone would speedily become popular, and that as the public became accustomed to the telephone in their houses, they would recognise the advantage of a system of intercommunication. When this time arrives, I would advise the company to place telephones free of charge for a specified period in a few of the principal shops so as to offer to those householders who work with the central office, the additional advantages of oral communication with their tradespeople. The central office system once inaugurated in this manner would inevitably grow to enormous proportions, for those shopkeepers would thus be induced to employ the telephone, and as such connections with the central office increased in number, so would the advantages to householders become more apparent and the number of subscribers increased.

Should this plan be adopted, the company should employ a man in each central office for the purpose of connecting wires as desired. A fixed annual rental could be charged for the use of wires, or a toll could be levied. As all connections would necessarily be made at the central office, it would be easy to note the time during which any wires were connected and to make a charge accordingly—bills could be sent in periodically. However small the rate of charges might be, the revenue would probably be something enormous.

In conclusion, I would say that it seems to me that the telephone should immediately be brought prominently before the public, as a means of communication between bankers, merchants, manufacturers, wholesale and retail dealers, dock companies, water companies, police offices, fire stations, newspaper offices, hospitals and public buildings, and for use in railway offices, in mines and [diving] operations.

Agreements should also be speedily concluded for the use of the telephone in the Army and Navy and by the Postal Telegraph Department.

Although there is a great field for the telephone in the immediate present, I believe there is still greater in the future.

By bearing in mind the great object to be ultimately achieved, I believe that the Telephone Company can not only secure for itself a business of the most remunerative kind, but also benefit the public in a way that has never previously been attempted.

I am, gentlemen, your obedient servant,

ALEXANDER GRAHAM BELL.

The American District Telegraph system differed from the London District system described on p. 80. It was established in 1870, and was very flourishing at the time of the introduction of the telephone. According to Edward A. Calahan[1] it owed its origin to the perpetration of a series of burglaries at Englewood, New Jersey, including the house of E. W. Andrews, first president of the Gold and Stock Telegraph Co., who had resigned his office in order to introduce the 'stock ticker' into London. In his absence Mr. Calahan devised a system, and on Mr. Andrews' return the American District Telegraph Co. was formed. Work was commenced at Brooklyn Heights, and 100 subscribers were secured within a week. Whilst still retaining the protective features of 'police,' 'fire,' and similar calls, the principal feature of the Company's operations was to provide messengers on a request being conveyed telegraphically. The messengers were held in readiness at a central office serving a particular district. Lines radiated from this central office, and each line was 'looped in' at the offices of several subscribers. A call box was placed in each subscriber's premises. To call the central office the subscriber pulled down a lever which was held up by a spring. In course of the return of the lever to its normal position a series of electric impulses were sent over the line to the central office, the impulses varying with each subscriber according to the arrangement of the call box. These impulses were received at the central office and recorded on a Morse register. The attendant there reading the number of the subscriber dispatched a messenger to the address indicated. The messenger was then at the service of the subscriber to carry a message, letter, or parcel, as might be required, payment being made according to the time occupied.

The circuit and apparatus of the District Telegraph system may be observed from fig. 26, p. 97, by disregarding the telephone apparatus included in that illustration.

The telephone was probably first applied as an auxiliary to the district call box to communicate only with the central office. 'Instructions to Agents,' dated November 15, 1877, signed by Hubbard and Sanders, provide that a discount of twenty per cent. may be made 'For District Telegraph purposes.' Thomas B.

[1] *Electrical World*, N.Y., March 16, 1901, xxxvii. 438.

Doolittle, who filed an application for a patent on April 10, 1878, refers to the 'District Telephone System,' and says :—

In the use of the telephone system as it exists at present, where more than two stations are connected with a line wire, there is no practical means, that I am aware of, of preventing a message between any two stations from being heard or picked up at any or all of the other stations. This condition of things is a great disadvantage in the use of the telephone in a district system, where notice is sent to a main office to respond to an inquiry, or a command given to put two stations in communication.[1]

Mr. Doolittle's invention contemplates cutting out intermediate telephones on the same line, showing that it was customary to add telephones in the manner adopted for district call boxes, a number being in series on a single circuit, and it would imply that circuits were coupled in response to a demand to put two subscribers in communication.

A development from this condition was the addition of a direct line to which the telephone was attached. Mr. C. E. Scribner states :—

In Chicago the first exchange was established and the central office equipment placed in a small room in the back of an American District Telegraph Office in La Salle Street. The few subscribers to the system had the lines over which they telephoned connected with a small switchboard, and a single operator responded to their calls and made the connections they required. Each subscriber had, in addition to his own telephone line, a connection with the American District Telegraph system, and was provided at his office with the ordinary district call box, by means of which he transmitted a signal to indicate his wants. A register at the central office recorded telegraphically upon a tape the signals as received, and it was in response to these recorded signals that a connection with the subscriber's individual telephone line was made.[2]

Mr. Scribner was giving evidence at first hand but after a considerable interval of time. A contemporary record is to be found in an article by Mr. Haskins in 'La Lumière Électrique' of 1880, of which the following is a translation in part :—

There exist in America several systems for the establishment of telephone communications between private people, and the most important of these is the one known under the name of 'American

[1] U.S. specification, No. 209,115, October 22, 1878.

[2] Western Electric Company *v.* Capital Telephone and Telegraph Company. Circuit Court of the United States, 1896. *Supplemental Brief for Complainant*, p. 18.

district system,' which has, until now, been employed by the Edison-Gray Company. The number of telephones in use in this Company's service is equal, to-day, to approximately 1600, and the subscribers' list increases every day.[1] The total number of connections established every day amounts to several thousands, and our lines are capable of giving telephonic communications at a distance of seven and even nine miles.

Signal lines are wires which pass through district boxes and end at the subscribers' residences or offices in order to enable them to ask either for messenger boy, a telephone, a cab, a policeman, or for a fireman. The circuit of this wire is closed or 'metallic,' and the signals are obtained, as in the systems of the same kind, by the cutting-out of the current. It is only after the wire has gone through a relay that it goes into the central exchange, this relay actuating a second circuit in which is interposed a Morse apparatus.

The district box is composed of a cylindrical case bearing a dial upon which moves an index, and the portion of the circumference through which this index is moved shows division marks corresponding to the different kinds of service orders which may be given. At the back part of the case is adapted a pawl or lever retained by the teeth of a ratchet wheel which permits, when pulled from the upper part towards its lower part, to wind up the spring of a clock mechanism disposed in such a manner that, through the register, it provides for closings of the current in a number more or less great according to the part of the arc followed by the index, and which may thus determine the inscription of the signal. In order to obtain a given signal it is sufficient to push down the level until the index comes in front of the order which it is desired to be transmitted. After releasing the lever the clock mechanism, while moving, will determine 3, 4, 5 or 6 closings of the current, according to the order to be transmitted. This mechanism could even be disposed in such a way as to simplify the combinations in order to lengthen or to shorten the closings of the current.

All the wires are run into the operators' room, each wire ending at a small brass plate 1 inch long, ½ inch wide, and ¼ inch thick.

Each brass strip is provided with two holes disposed to receive plugs. Each one of the boards is provided with a telephone having one end of its wire connected to ground and the other end connected with a flexible conductor ending in a brass plug adapted to fit the holes of the brass strips. Let us assume now that Mr. A. has sent a signal to the central exchange, this signal having the number 26 and corresponding to 'telephone.' The operator in charge of the signals immediately consults his register book and finds that the

[1] There are in Chicago two telephone companies: the Bell Company and the Edison-Gray Company. More than 3000 subscribers are divided about equally among these two companies. [Note in original.]

signal 26 which has been sent corresponds to the board No. 984. He takes a ticket printed as follows :—

	CONNECT	Time
	- · · · · · - 984	
	DISCONNECT	Time

Underneath the word *connect* he writes the number of the calling subscriber, i.e. No. 984, and before this number he reproduces the signal which has been sent. This ticket is given to the operator of the nearest board, who immediately inserts the plug of its telephone flexible conductor in the plate No. 984 ; then he depresses the key of his manipulator in order to connect the battery to the line, and he advises Mr. A. that his signal has been received. Then Mr. A. says that he desires to speak to Mr. C. The number of Mr. C. is 516 ; then the operator writes down his signature on the ticket below the signature of the first operator, and he writes the No. 516 below the No. 984, and at the same time he writes down the hour. The ticket is then given to another operator whose apparatus has ordinarily no connection to the ground, but which possesses two keys, one for the left, one for the right. These keys have their connections established in the following manner : The person who calls being always at the right, the operator speaks through a tube to the operator of the main switchboard and asks for the numbers 984 and 516. The wires which establish the communication between the telephone of these numbers and the main switchboard are connected to the two plates which correspond to them, and the operator depresses the left key of the manipulator and listens, while effecting at the same time two or three notations. In consequence of the depression of the key, the left wire is put in connection with the telephone and the battery is brought into circuit. At the moment he has received the answer of the left correspondent, he operates the right key and calls the other correspondent ; after this, the two correspondents may speak to one another. Generally the operator does not speak, he only listens for a sufficient time to ascertain that the two called correspondents are actually talking together, and when he is sure of this he calls the operator of the main switchboard to ask him to connect together the two correspondents and to disconnect his apparatus from the circuit in order to be free to establish new communications.

When the correspondents have terminated their conversation they both place the index of their boxes in front of the signal No. 2, which means 'clear out the connection,' and the wires are immediately separated from one another. It takes much longer to describe all these operations than to do them, and yet, however rapid the system is, it is far from being quick enough on the days when there are no end of dispatches to be exchanged.[1]

In their paper read before the American Electrical Society in December 1879, Haskins and Wilson regarded the American District Telegraph system as then

too well known to require a detailed description. It is only necessary to state briefly, that in Chicago the larger portion of all the calls from telephone subscribers are received at the central office by that system over a separate wire. . . . For this purpose the American District Telegraph Co. of Chicago have used their already established wires with the apparatus formerly employed exclusively for A.D.T. calls. . . . To adapt this apparatus to the telephone exchange it was only necessary to add two signals to those already in use—'Telephone use' and 'Telephone through,' the former signifying that the central office is wanted, the latter that the subscriber desires to be disconnected.[2]

Figure 26, which illustrates the circuit and apparatus of the American District Telegraph system and of the telephones used therewith, is taken from Firman's U.S. patent No. 328,305, dated October 13, 1885 (application filed January 16, 1880).

Whilst the telephone exchange was suggested by Bell and his associates, it apparently developed in practice in various places by means of the American District Telegraph Service, and the application of telephones to the Law Telegraph system, as well as from the pioneer work on Holmes's Boston Burglar Alarm service and the Hartford Drug Store connections of Isaac D. Smith. The telegraphic exchanges such as have been described as existing in Newcastle, Philadelphia, New York, and elsewhere doubtless had some influence, but Dumont's proposals were apparently forgotten, his patent being only brought to light by research due to litigation in after years.

Thus during Bell's absence from the United States the exchange system was rapidly progressing, the existence of numerous district telegraph systems facilitating this progress. The first fully equipped commercial telephone exchange established for public or general service was opened for business on January 28, 1878, at New Haven, Connecticut. This example was speedily followed by other cities, such as Bridgeport, New York, and Philadelphia, until it became evident to others than the Bell associates that the telephone was an invention to be reckoned with.

The Western Union Telegraph Company was at that time the

[1] *La Lumière Électrique*, ii. 155.

[2] *Journal of the American Electrical Society*, 1880, pp. 50–51.

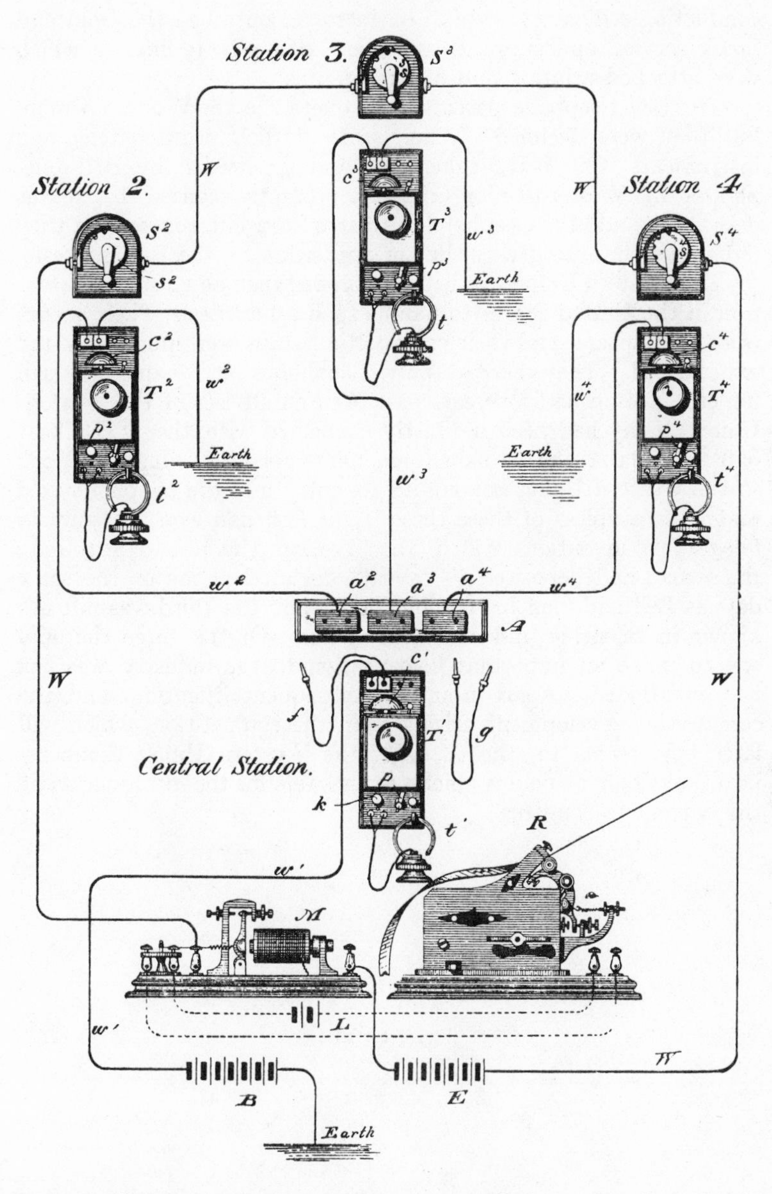

FIG. 26.—Telephone Exchange System as applied to American district telegraph by Firman.

largest electrical organisation in the United States. Besides conducting ordinary telegraph business, it controlled the Gold and Stock service, and supplied customers with private lines to which were attached printing and dial telegraphs.

The Bell telephone patents were offered to the Western Union, but they were declined. When some of their printing telegraph instruments were being taken out and superseded by Bell telephones, the Western Union Company promptly realised that it was desirable to add the telephone to their telegraph business, and they did so, not in accord with, but in opposition to, the Bell interests.

The Western Union was better equipped than any other organisation in the United States to handle such a business. Their agents were everywhere, and their construction staffs were numerous and experienced. They needed only telephones and names to sail under. Edison had for years past been an adviser of the Western Union; Gray had been indirectly connected with them; Dolbear had claimed that the permanent magnet telephone was his invention: so the Western Union announced its entry into the telephone field under the auspices of these three. The first had won his spurs in telegraphic inventions which the Western Union had acquired; the second had deposited a 'caveat' for a telephone on the same day as Bell had filed his specification; and the third was not unknown in scientific and technical circles. Of the three the only one to make an important contribution to the industry was the first mentioned. Apart from the subsequent litigation, and the commercial development arising from the competition, which will later be referred to, the entry of the Western Union Company into the telephone field is chiefly remarkable for the introduction of the battery transmitter.

CHAPTER X

THE BATTERY OR VARIABLE RESISTANCE TRANSMITTER

In describing the apparatus exhibited at the Centennial Exhibition at Philadelphia in Chapter VI, the transmitter was reserved for

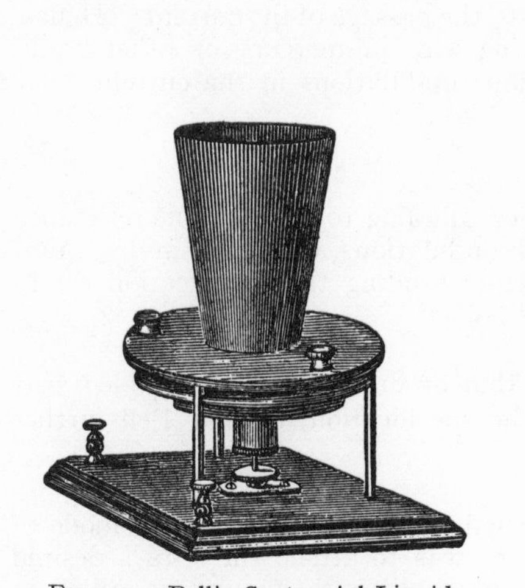
Fig. 27.—Bell's Centennial Liquid Transmitter (perspective).

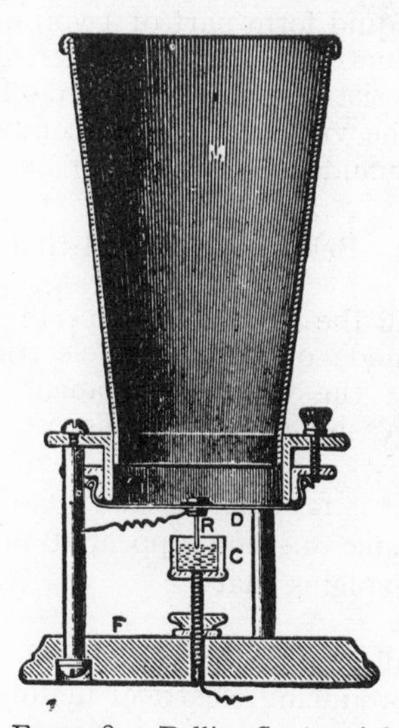

Fig. 28.—Bell's Centennial Liquid Transmitter (section).

later reference. The liquid transmitter there exhibited is illustrated in perspective by fig. 27 and in section by fig. 28. It is thus described by Bell :—

D is the diaphragm ; R is the rod attached to the centre of the diaphragm, and dipping into the liquid in the cup C. F, supporting frame, and M the mouth-piece. This also has a straining ring with adjusting screws like that used in the single and double pole membrane telephones already described. Indeed, they were all three made from the same casting. The cup is supported upon a screw, by which it can be adjusted vertically to vary the depth of immersion of the lower end of the rod R.[1]

It will be seen that this Centennial liquid transmitter illustrates in a practical form one of the methods other than inductive action of obtaining undulations in the circuit suggested by Bell in his patent, from which the following is quoted :—

Electrical undulations may also be caused by alternately increasing and diminishing the resistance of the circuit or by alternately increasing and diminishing the power of the battery. The internal resistance of a battery is diminished by bringing the voltaic elements nearer together, and increased by placing them farther apart. The reciprocal vibration of the elements of a battery, therefore, occasions an undulatory action of a voltaic current. The external resistance may also be varied. For instance, let mercury or some other liquid form part of a voltaic circuit, then the more deeply the conducting wire is immersed in the mercury or other liquid, the less resistance does the liquid offer to the passage of the current. Hence, the vibration of the conducting wire in mercury or other liquid included in the circuit occasions undulations in the current.[2]

Bell has explained that

all the parts of the specification alluding to the variable resistance mode of producing electrical undulations were put in by [him] at the very last moment, before sending the specification off to Washington to be engrossed.[3]

It is for this reason probably that no drawing of the variable resistance method appeared in the specification, though Bell further explains that

all that was necessary, it seemed to me, to illustrate this mode of producing electrical undulations was to attach the wire I desired to vibrate to the various instrumentalities shown and described in my specification and illustrated in figs. 5 and 7 [thereof]. That is, the water would take the place of the electro-magnet in the transmitting instrument, and if it was desired to produce undulations for the purpose of multiple telegraphy, the wire that dipped into the water would be attached to the free end of the reed illustrated in fig. 5 ; and if it was desired to transmit vocal or other sounds telegraphically —including articulate speech—the wire to be vibrated should be attached to the centre of the stretched membrane shown in fig. 7,

[1] *Deposition*, p. 99.
[2] U.S. specification, No. 174,465, March 7, 1876 (application filed February 14, 1876).
[3] *Deposition*, p. 88.

which would then have to be substantially horizontal to admit of the vertical vibration of the wire.[1]

The American Academy paper was presented May 10, 1876, between the date of the patent and the opening of the Centennial Exhibition. In this paper Bell said:

Electrical undulations can be produced directly in the voltaic current by vibrating the conducting wire in a liquid of high resistance included in the circuit. . . . A platinum wire attached to a stretched membrane completed a voltaic circuit by dipping into water. Upon speaking to the membrane articulate sounds proceeded from the telephone in the distant room. The sounds produced by the telephone became louder when dilute sulphuric acid, or a saturated solution of salt, was substituted for the water. Audible effects were also produced by the vibration of plumbago in mercury, in a solution of bichromate of potash, in salt and water, in dilute sulphuric acid, and in pure water.[2]

The liquid transmitter was developed from a device designed by Bell some time previously to obviate the difficulties from induced currents in telegraphic instruments. The object and the means are thus explained in a specification which was prepared but not proceeded with:—

When a Morse Sounder or other telegraphic or electrical apparatus is placed in a voltaic circuit, the induction of the current upon itself in the coils of the electro-magnet or other instrument gives rise to an induced current of high tension.

When the circuit is broken at any point, as by the lifting of the key, the induced current leaps across the break in the circuit in the form of a bright spark. When a powerful battery is employed, or when the circuit is very rapidly made and broken, the spark becomes so intense as to burn or oxidise the points between which it appears, and thus prevent the effective working of the instruments upon the circuit.

This defect has hitherto been overcome by the employment of *condensers*, by means of which the spark has been materially lessened or destroyed.

Now, I have discovered that the same effect may be produced by introducing between the points where the circuit is broken an *imperfect conductor of electricity*, which shall offer a very great resistance to the voltaic current, but afford a free passage for the induced current which occasions the spark.

Such a substance is *water*, especially when slightly acidulated.

[1] *Deposition*, p. 87. Fig. 5 of the patent is illustrated in fig. 15, p. 58, and fig. 7 of the patent in fig. 4, p. 42.

[2] *Proceedings of the American Academy of Arts and Sciences*, vol. xii. 'Researches in Telephony,' par. 13.

Retort carbon, animal and vegetable tissues, and other substances offering a high resistance answer the purpose ; but I prefer to employ a liquid (like water) which can be decomposed by the passage of the current.[1]

When including the variable resistance method in his telephone patent it occurred to him that water was not a good illustrative substance to be specified in this connection on account of this very fact of its decomposability by the action of the current. He therefore proposed to use as a typical example a liquid that could not be thus electrolytically decomposed, and specified mercury as the best example of such a liquid known to him.[2]

The caveat deposited in the Patent Office at Washington by Elisha Gray on the same day as, though several hours later than, the specification of Bell also contemplated the use of a liquid transmitter. The caveat is quoted in 'The Speaking Telephone' (Prescott), 1878, p. 202, accompanied by an illustration. After a general description of the apparatus Gray says :—

Owing to this construction, the resistance varies constantly in response to the vibrations of the diaphragm, which, although irregular, not only in their amplitude but in rapidity, are nevertheless transmitted through a single rod, which could not be done with a positive make and break of the circuit employed, or where contact points are used.

I contemplate, however, the use of a series of diaphragms in a common vocalising chamber, each diaphragm carrying an independent rod, and responding to a vibration of different rapidity and intensity, in which case contact points mounted on other diaphragms may be employed.

The vibrations thus imparted are transmitted through an electric circuit to the receiving station, in which circuit is included an electro-magnet of ordinary construction, acting upon a diaphragm to which is attached a piece of soft iron, and which diaphragm is stretched across a receiving vocalising chamber, A.

The diaphragm at the receiving end of the line is thus thrown into vibrations corresponding with those at the transmitting end, and audible sounds or words are produced.[3]

It is not surprising that in an electrical atmosphere of such high tension as may be produced by strenuous commercial and legal contentions charges of plagiarism should have been made. Let it be understood that neither Bell nor Gray had at the time of filing their respective specification and caveat put a liquid transmitter into operation. Nor had Gray constructed or designed an electro-

[1] *Deposition*, p. 85. [2] *Ibid.* p. 87.

[3] *The Speaking Telephone*, Prescott, 1878, p. 204.

magnetic receiver such as he outlined. Bell had transmitted and received vocal sounds through his magneto instruments, and had thereby been satisfied that his conception of the undulatory current theory was right. His aim was to protect without delay the undulatory current and the means for obtaining that current. One means was inductive action, another was variable resistance. The liquid transmitter was an illustrative instrumentality of the latter type.

Gray's proposition to elaborate his transmitter by increasing the number of diaphragms, each responding to a different note, is an indication that he had not reached the stage which Bell had in recognising that the single diaphragm would serve to translate the aerial vibrations into corresponding electrical currents. The idea in Gray's mind was apparently analogous to the harp telephone suggested, but not constructed, by Bell in the early stages of his investigations. Such an idea was but a stepping stone to Bell. By its aid and that of other stepping stones already detailed he reached, as his patent disclosed, the simple method by which a single diaphragm transmitted not merely a simple vibration but the resultant of complex vibrations.

A caveat is not intended to be a finished or complete specification. Its having been deposited is proof of date, and that is its main purpose. Gray's document was a caveat, the document which Bell had deposited some hours earlier was a complete specification.

But Gray's caveat was accompanied by a drawing of his liquid transmitter and an illustration of an individual talking into it. Bell's liquid transmitter was only described, and there was no such pictorial representation of talking people as Gray depicted. These were no doubt contributory causes to the claims set up for Gray and for the persistence with which his friends maintained that he invented the telephone. The Western Union Company had abandoned Gray's claims in 1879, but they were revived by others in later suits.

Gray's work in this field was well known to his telegraphic contemporaries, whilst Bell was comparatively unknown. It is not surprising that the surface indications of Gray's caveat, together with his recognised standing in the electrical world, should have produced an impression that served to influence the judgment of his friends. But going beneath the surface and examining the specification of Bell and the caveat of Gray, it is easy to see that Gray showed the externals without the essentials, and Bell exactly and specifically defined the essentials, but was content with a very general indication of some of the means.

Gray's words indicate that makes and breaks will not produce the required result, but that a single rod will permit the transmission of the vibrations of the diaphragm, which is suggestive of a recognition of the need of an undulatory current. But the practical feature of this suggestion is destroyed when he goes on to contemplate the use of several diaphragms, which were apparently intended to respond to different vibrations, and by co-operation were to send to the line a composite result.

Bell, on the contrary, emphasised the fact that in order to transmit speech it was necessary to put on to the line a current of a character that had not hitherto been used. He defined those which had been used and the kind of current that needed to be employed. He said in effect: You must have an undulatory current. If you

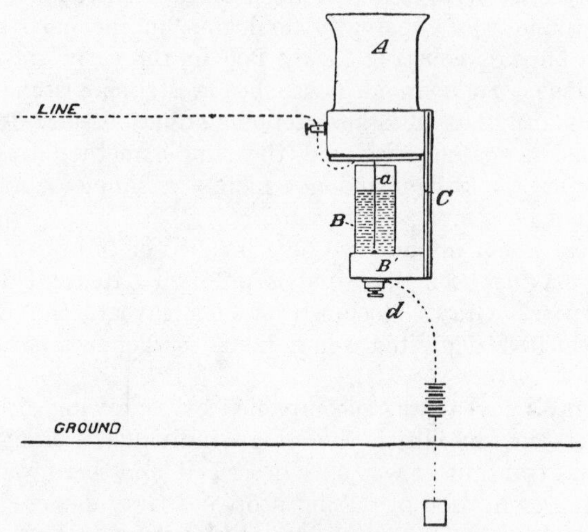

FIG. 29.—Gray's Liquid Transmitter.

talk to a diaphragm it will be set in motion by the sound waves, which are undulating. The undulatory current was the central and essential feature of Bell's patent, together, of course, with the means for obtaining it. These means were in effect three. The one which was illustrated, and therefore received most attention, was the use of a diaphragm as an electric generator, which by inductive action should produce a current varying in accordance with the vibrations imparted to the diaphragm by the air waves.

But, in effect said Bell, this absolutely essential undulating current may also be obtained by using the vibrations of the diaphragm to control a current otherwise obtained. You may go to the source and vary the position of the elements of a cell, or you may interpose an instrument in the external circuit which shall vary the current in accordance with the motion of the diaphragm.

The inclusion of such a bold feature as the variation of the source is an indication of Bell's intention to control all available methods, and a further proof of his perception of the essential feature for the transmission of speech. With this in his mind it became unnecessary to define with too great exactness, or even to illustrate by a drawing, any particular means of varying the external resistance. He contented himself with the statement of the fact that it can be so varied, and described such a method as his 'mercury or other liquid' transmitter with a preliminary 'for instance.'

Notwithstanding the inclusion of this description in Bell's patent, it was even suggested in court that his Centennial liquid transmitter was copied from that shown in Gray's caveat. Such a suggestion was indignantly combated by Bell's counsel; but it is only necessary to examine both designs to see how groundless such suggestions were. Fig. 29 represents Gray's Liquid Transmitter and fig. 30 that of Bell.

Gray's description is as follows :—

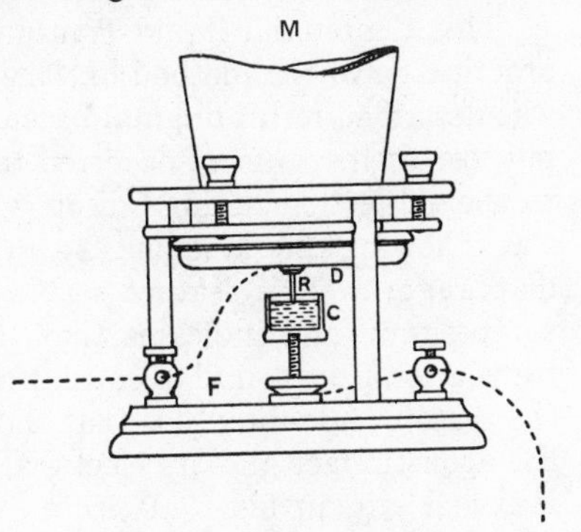

FIG. 30.—Bell's Liquid Transmitter.

A box or chamber, A, across the outer end of which is stretched a diaphragm, . . . of some thin substance, such as parchment or gold beaters' skin, capable of responding to all the vibrations of the human voice, whether simple or complex. Attached to this diaphragm is a light metal rod, [a], or other suitable conductor of electricity, which extends into a vessel, B, made of glass or other insulating material, having its lower end closed by a plug, which may be of metal, or through which passes a conductor, . . . forming part of the circuit.

This vessel is filled with some liquid possessing high resistance, such, for instance, as water, so that the vibrations of the plunger or rod, [a], which does not quite touch the conductor, . . . will cause variations in resistance, and, consequently, in the potential of the current passing through rod [a].[1]

Bell's description is as follows :—

D is the diaphragm, R is the rod attached to the centre of the diaphragm, and dipping into the liquid of the cup C. F, supporting frame, and M the mouthpiece. . . . The cup is supported upon

[1] *The Speaking Telephone*, Prescott, 1878, pp. 204, 216.

a screw, by which it can be adjusted vertically to vary the depth of immersion of the lower end of the rod R.[1]

While necessarily very similar in appearance, there is an obvious difference in design and operation. Gray's vessel was to be of glass or other insulating substance. The liquid was to have a high resistance, and its function was to complete the circuit between the point of the rod vibrated by the diaphragm and the plug or point at the base of the vessel. The resistance varied with the distance between the point of the rod and the metal plug or conductor at the base of the vessel.

In Bell's design the variations in the current were obtained by the greater or lesser surface contact of the rod, according to the extent of its immersion in the liquid.

This Centennial liquid transmitter of Bell's simply put into a practical form the method of varying the external resistance which was described in his original patent. The use of a liquid for such a purpose seems to have occurred to all who devoted their attention to the subject, including Edison, who also in his early experiments was following the same line as Gray in seeking utility from multiple diaphragms. This feature serves to distinguish Bell from all his competitors, and indicates that they had no conception that the translation from sound to current could be so simply attained. And the reason undoubtedly is that they had not mastered, as Bell had, the acoustic fact and its electrical equivalent. As Professor Clerk Maxwell said in his Rede lecture delivered at Cambridge in May 1878, Professor Graham Bell, the inventor of the telephone, was 'not an electrician who had found out how to make a tin plate speak, but a speaker who, to gain his private ends, had become an electrician.'[2]

It will be seen how far behind were those experimenters who sought in multiple diaphragms some aid in translating compound sounds into electrical currents. This is the more surprising when Reis was content with one and when the phonautograph of Leon Scott gave graphic representations of speech waves with but a single diaphragm. In truth, Bell's contemporaries were not only far behind Bell but in this regard also behind Reis, an account of whose work is, for greater convenience, reserved for a later chapter. Gray had illustrated a single diaphragm but was evidently not confident that it would suffice. A similar lack of confidence was shown by Edison in his early efforts. But the first commercial forms of variable resistance transmitter were contributed not by either Bell or Gray, but by Edison in the early stages, and by modifications of the discoveries of Hughes later.

[1] *Deposition*, p. 99.
[2] *The Scientific Papers of James Clerk Maxwell*, ii. 742. *Nature*, vol. xviii.

As we have already seen, when the Western Union Telegraph Company decided to go into the telephone business, Bell's magneto telephone was being used extensively for private line purposes. It was this practical development which sufficiently demonstrated that the telephone was no toy, but 'a business proposition.' Equally clear would it be to them that the feeble results should be capable of improvement. The very simplicity of the magneto telephone, the independence of battery attained by the use of a permanent magnet, were in the nature of temptations to continue the inductive method. The very fact that telephones talked, and were intended to be placed in the hands of the general public and not of expert electricians, would tend to justify the retention of simple forms. Only with the growth in its use, only when greater demands were made upon it, did the limitations of the magneto instrument as a transmitter become so apparent as to lead to an organised attempt at improvement. And during this eventful period of its history the Western Union Telegraph Company decided to become interested in it, and Bell himself was abroad.

Thomas Alva Edison was an expert upon whom the Western Union were justified in placing reliance. In various branches of telegraphic work he had proved his ability. He was commissioned to make improvements in the telephone, and he did.

His achievements in the telephone art are thus recounted by Messrs. Dyer and Martin :—

With his carbon transmitter he gave the valuable principle of varying the resistance of the transmitting circuit with changes in the pressure, as well as the vital practice of using the induction coil as a means of increasing the effective length of the talking circuit.[1]

In the application of the induction coil or transformer in connection with speaking telephones Edison was original,[2] but in the application of the variable resistance he introduced no new principle, but simply ascertained by experiment a suitable method of applying the principle fully described by Bell in his first patent.

In the United States patent system applications for patents which show the same object are 'put into interference' with a view

[1] *Edison, His Life and Inventions*, p. 180, by Frank Lewis Dyer and Thomas Commerford Martin. Harper & Bros. MCMX.

[2] An induction coil was applied by Dr. Wright to the Reis transmitter, a condenser being used as a receiver: 'The line current is made to pass through the primary of a small induction coil. In the secondary circuit he places two sheets of paper, silvered on one side, back to back so as to act as a condenser. Each current that comes from the sending apparatus produces a current in the secondary circuit, which charges and discharges the condenser, each discharge being accompanied by a sound like the sharp tap of a small hammer. The musical notes are rendered by these electric discharges and are loud enough to be heard in a large hall.' *Electricity*, Ferguson, 1866, p. 258.

to determine by evidence and examination the priority of invention. The claims of Bell, Gray, Edison, Dolbear and others were 'in interference,' and were adjudicated upon by 'the Examiner of Interferences,' who reported on July 21, 1883. The following are extracts from his report :—

The first conversation said to have taken place between Orton and Edison relative to the subject of transmitting speech was in July 1875, but the first conversation relative to transmitters is not said to have taken place until in March or April 1876. In November or December 1875, Edison first heard of Bell as a person working in the same line as Gray on harmonic telegraphy, hence in the same line as he, Edison, had been engaged by Orton to work up for the Western Union. Bell's patent of March 7, 1876, was undoubtedly granted prior to the interview with Orton in which the transmitter is said to have been first mentioned, and it is but fair to assume, though the record is silent on the point, that Edison was at the time familiar with Bell's specification, in which various modes of producing undulatory currents are mentioned. Following closely on the grant of this patent, and contemporaneous with the construction of the alleged first successful transmitter for articulate speech, was the description given by Johnson to Edison of Bell's Centennial exhibition of the telephone. This exhibition took place June 25, 1876, and was not only preceded by Bell's patent of March 7, 1876, but also by his lecture read before the American Society of Arts and Sciences, May 10, 1876, and published shortly thereafter.

A knowledge of these transactions and publications is not brought home to Edison, for he is not interrogated with respect thereto; but being public, and Edison interested as he was in matters pertaining to electricity, more especially that branch upon which he knew Bell was engaged, it may be reasonably inferred from the coincidence in time and other circumstances that if not already fully informed as to Bell's claims and inventions, as seems most probable, he began at once to investigate the subject when he learned from Johnson that articulate speech had actually been transmitted and reproduced. Unquestionably Edison was well acquainted with the abstract principle that a rise and fall in tension could be produced by varying the resistance, as indicated in his previous experiments and inventions; but when he heard of Bell's success at the Centennial, it would be only natural that he should look to Bell's prior patents for suggestions.

It was in July 1876 that Edison learned of Bell's Centennial exhibition of the articulating telephone, and in the same month we find the first evidence of a distinct effort being made on the part of Edison to construct an instrument designed solely for the transmission of speech. . . . Commencing with July 1876, the record indicates the active prosecution of the subject of the speaking telephone. Two lines of experiments seem to have been followed, the first currently designated the 'water telephone,' wherein the tension of the current in a closed circuit was to be varied by the movements of electrodes with reference to an interposed liquid, and the second wherein the vibration of a diaphragm or diaphragms imparted motion to several contact points and cut in and out resistance.

It will be found upon inspection of Bell's patent of March, and his lecture of May 1876, that the methods upon which work was actually begun and prosecuted under Edison's supervision are all suggested as equivalent or alternate modes of inducing electrical undulations, and it is a coincidence that these experiments should have begun so soon after public notice had been directed to Bell's patent by the exhibition of his telephone, provided Edison had received no information or suggestion therefrom.[1]

From the same source we learn :—

On the next day, March 22, 1877, an agreement was entered into between Edison and the Western Union Telegraph Company, whereby the former agreed to assign all the inventions and improvements capable of being used on telegraph lines he should make during the period covered by the agreement, the latter to remain in force for five years unless sooner terminated, and to take effect March 1, 1877. On March 23, 1877, the day following the signing of the agreement, directions were given for the preparation of Case 130, the first application filed by Edison referring to the electrical transmission of articulate speech.[2]

As an expert of the Western Union Company Edison entered the telephone field. 'The work that Edison did was, as usual, marked by infinite variety of method as well as by the power to seize on the one needed element of practical success.'[3]

As we have seen, the principle upon which Edison worked had been defined previously by Bell. It was the principle of varying the resistance of the circuit in correspondence with the excursions of the diaphragm. It remained for Edison to select a tension regulator' which should be effective and adapted to commercial use. A material subject to evaporation like water, as suggested by Gray and devised by Bell, was obviously ill adapted for commercial use. There was probably no one so well equipped as Edison to discover the best material and method for obtaining the required variations in resistance. Long experience in the telegraphic field gave him the starting point of knowledge. His experimental laboratories and skilled assistants enabled him to proceed at once with likely experiments and carry them to a conclusion. Perhaps also his characteristic method of reaching an

[1] *The Speaking Telephone Interferences*, pp. 180–2. [2] *Ibid.* p. 203.
[3] *Edison, His Life and Inventions*, p. 178.

end by the process of exhaustion of means was equally helpful in this case.

On July 20, 1877, he filed an application which is a distinctly bad start from the acoustic standpoint, since the apparatus consists of

> a resonant case with several tympans, adjusted to different degrees of tension or delicacy, and these are all so connected with contact points in the metallic line circuit that the electric pulsations will be sent over the line from one or more of these tympans, and operate upon an electro-magnet and receiving tympan of a resonator.[1]

It seems somewhat surprising that at this date Edison should have sought assistance from multiple diaphragms and 'resonant boxes,' but the specification is of interest from the casual mention that the circuit closing springs shall be 'preferably with carbon or plumbago points.'

It was in the direction of the use of carbon or plumbago that Edison was to contribute to the art. A few months later (December 13, 1877) he applied for a patent[2] in which the variable resistance first described was the familiar rheostat composed of convolutions of wire capable of being more or less short circuited by a metallic contact operated by the diaphragm. But alternatives were suggested, and amongst them 'a semi-conductor such as plumbago,' which, however, was intended to operate in the same way as the rheostat, for 'the forward movement of the diaphragm causes more and more platina to come in contact with the plumbago, thus allowing the greater part of the current to pass through the platina, according to the amplitude of the diaphragm vibrations.' This patent is also noticeable for its inclusion of the induction coil and in that a single diaphragm was shown.

But the practical variable resistance had not yet been reached. The material was there but the disposition was unsuitable. Yet years before Edison had devised a rheostat following the lines of Clerac, in which carbon was inserted directly in the circuit. Harking back on this experience, Edison commenced to experiment on more definite lines. He sought to obtain variations in resistance by the operation of the diaphragm on some substance directly in the circuit. His carbon rheostat varied its resistance by pressure. A suitable material should respond to the action of the diaphragm and transmit more or less current, in accordance with the greater or less excursions of the diaphragm. Silk fibres (or 'fluff,' as it was described) covered with plumbago, and various other applications of the same idea, finally led to the filing of an application on

[1] U.S. specification, No. 203,014, April 30, 1878.
[2] U.S. specification, No. 203,013, April 30, 1878.

March 7, 1878, in which it is explained that the carbon heretofore employed in connection with a diaphragm is of too great resistance to be adapted for use in the primary circuit of an induction coil. In this specification is also recorded the discovery that lamp black obtained from the combustion of very light hydrocarbons, such as gasoline or naphtha, can be used.

I select from lamp black thus made only the very blackest portions and then place the same in a mold, and subject it to a very powerful pressure, sufficient to consolidate the same and place it in a correspondingly shaped [i.e. to the mould] cavity contiguous to the diaphragm, with a piece of cork or a piece of rubber

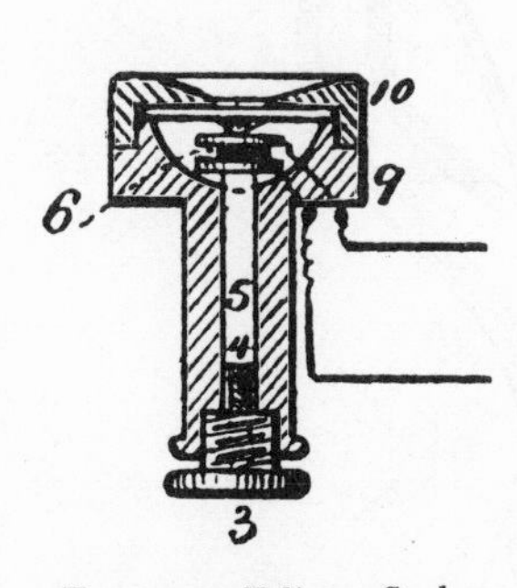

FIG. 31.—Edison Carbon Transmitter, as illustrated in patent.

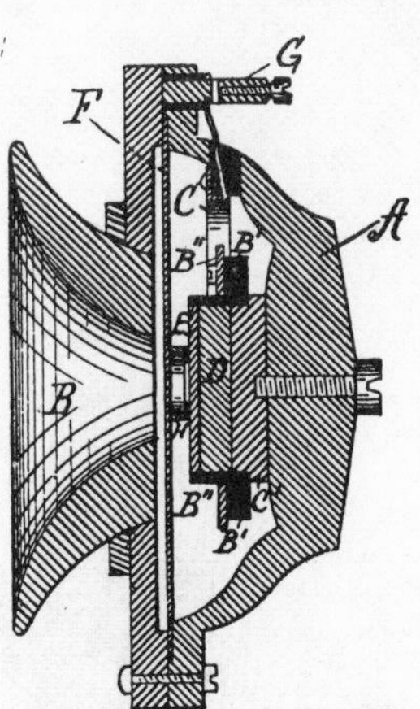

FIG. 32. — Edison Carbon Transmitter (Clarke's illustration).

intervening between the same and the diaphragm, and connect the discs of platina foil that are used at each side of the carbon in the primary circuit of the induction coil, and obtain from the pressure resulting from the motion of the diaphragm the necessary rise and fall of tension without the great resistance heretofore inseparable from the carbon in said circuit.[1]

The illustrations to this specification from which fig. 31 is obtained show the influence of the commercial form of Bell's telephone since the carbon arrangement is mounted in an instrument having the external form of the Bell hand telephone (fig. 31). This was not well adapted to the purpose, and before the Edison transmitter was commercially issued it underwent some modifications in minor points of detail at the hands of George D. Clarke [2] (fig. 32)

[1] U.S. specification, No. 203,016, April 30, 1878.
[2] U.S. specification, No. 217,773, July 22, 1879 (app. filed Sept. 16, 1878).

and George M. Phelps [1] (fig. 33), the illustrations to whose patents will serve to indicate the developments.

The theory upon which Edison worked, as stated by himself, was that the electrical resistance of carbon became materially lessened under *pressure*, and that the variations in the pressure upon the carbon mass accounted for and produced the necessary variations in the current. The theory, as we shall see, was to be open to question, but of more importance sometimes than theory is

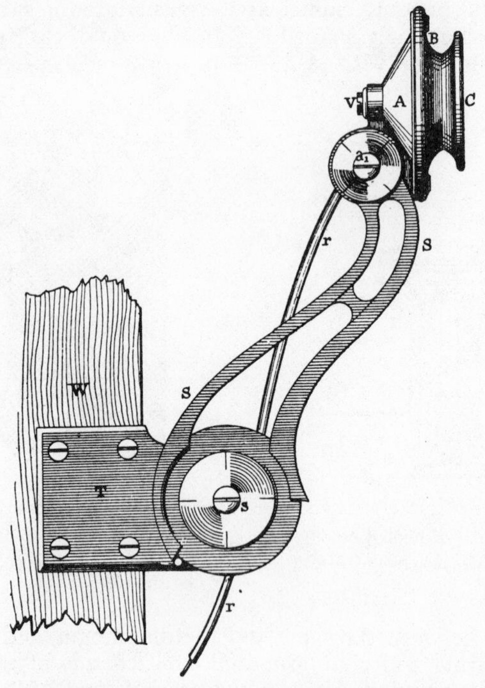

FIG. 33.—Edison Carbon Transmitter (Phelps's illustration).

practice. In practice Edison had placed his patrons, the Western Union Company, in possession of an instrument which so much increased the power and extended the range of telephonic speech as to place them in a commanding position at this eventful period of telephone exchange development.

The transmitter, operating in the manner which he had defined, was to have a short life, but the use of the induction coil was permanent, and carbon, though in a form differing from that which he had selected, was not to be removed from its pride of place.

[1] U.S. specification, No. 214,840, April 29, 1879 (app. filed Dec. 6, 1878).

CHAPTER XI

THE MICROPHONE

OF more than passing interest is the international character of the scientific discoveries leading up to the invention of the telephone. Faraday, Henry, Volta, Oersted, Ampere, and others all wrested secrets from nature contributing material with which to build; Wheatstone, Helmholtz, Page, and Reis served, each in his way, to indicate some appropriate method of using those materials. But it is of special interest to note the conjunction of Great Britain and America in the nationality of the two men who are more particularly associated with the scientific and practical development of the telephone. Bell was not a citizen of the United States when Sir William Thomson saw and heard the telephone at the Philadelphia Exhibition. But the telephone was invented in Boston by a native of Edinburgh who was to become a citizen of the United States, and the microphone was discovered in London by David Hughes, who, though a Londoner born, had spent the greater part of his life in the United States. When he was seven years of age (1838) he was taken with the family to settle in Virginia. Of Welsh origin, he inherited the national and family taste for music, and he obtained the position of professor of music at the College of Bardstown, Kentucky, subsequently occupying the chair of Natural Philosophy in the same college. Here he set out to devise a telegraphic instrument which should print the received message in Roman characters. He succeeded, and obtained its adoption not only in America but also throughout Europe. In the evening of his life he continued to take an interest as a scientist in discoveries which he had previously pursued as an inventor.

Hughes was not the first to note that a variation in the current followed from varying degrees in the intimacy of contact between two sections of an electrical conductor.[1] Berliner had not only

[1] A. M. Tanner in the *Electrical Review*, London, November 21, 1890, p. 612, recounts prior observers, Du Moncel (1856), Mousson (1861), Buff (1865), etc.

noticed it, but had claimed its use in a telephone transmitter. But Hughes was the first to study the facts and explain the causes to the satisfaction of the scientific world. His investigations were communicated to the Royal Society by Professor Huxley on May 8, 1878.

The telephone, then so recently invented, made the investigations possible, and he pays his tribute thereto in his opening words :—

> The introduction of the telephone has tended to develop our knowledge of acoustics with great rapidity. It offers to us an instrument of great delicacy for further research into the mysteries of acoustic phenomena. It detects the presence of currents of electricity that have hitherto only been suspected, and it shows variations in the strength of currents which no other instrument has ever indicated.[1]

Hughes's investigations were originally undertaken in order to observe the effect of sonorous vibrations upon the electrical behaviour of matter. Sir William Thomson and others had shown that the resistance to the passage of currents offered by wires was affected by their being placed under strains, and Hughes believed that the wire would vary in its resistance when it was used to convey sound[2] so he commenced with a study of the influence on the current of a wire under stress, but found no change indicated in the telephone until the wire broke. He then sought to imitate the condition of the wire at the moment of rupture by replacing the broken ends and pressing them together with a constant and varying force by the application of weights.

It was soon found that it was not at all necessary to join two wires endwise together to reproduce sound, but that any portion of an electric conductor would do so, even when fastened to a board or to a table, and no matter how complicated the structure upon this board, or the materials used as a conductor, provided one or more portions of the electrical conductor were separated and only brought into contact by a slight but constant pressure. Thus, if the ends of the wire terminate in two common French nails laid side by side, and are separated from each other by a slight space, were electrically connected by laying a similar nail between them, sound could

[1] *Proceedings of the Royal Society*, May 9, 1878, xxvii. 362.

[2] Bell had the same impression. In a letter to Mr. Hubbard dated May 4, 1875, he says :—'Another experiment has occurred to me which, if successful, will pave the way for still greater results than any yet obtained. The strings of a musical instrument in vibrating undergo great changes of *molecular tension*. In fact the vibration represents the struggle between the tension of the string and the moving force impressed upon it. I have read somewhere that the resistance offered by a wire to the passage of an electrical current is affected by the *tension of the wire*. If this is so, a *continuous current of electricity* passed through a vibrating wire should meet with a varying resistance, and hence a pulsatory action should be induced in the current.' (*Deposition*, p. 53.)

be reproduced. The effect was improved by building up the nails log-hut fashion into a square configuration, using ten or twenty nails. A piece of steel watch chain acted well.[1]

He relates the various substances which had been tried, and proceeds to describe the instrument which he called a *microphone* because it was suitable for magnifying weak sounds. The name was given by Wheatstone to a sort of duplex stethoscope, regarding which Wheatstone says :—

The microphone is calculated only for hearing sounds when it is in immediate contact with sonorous bodies : when they are diffused by transmission through the air, this instrument will not afford the slightest assistance.[2]

Although the members of the Royal Society attending the reading of Hughes's paper assumed that they heard the tramp of a fly upon the table, it must be recognised that what they heard was the sound created in the telephone by the mechanical disturbance of the conductor rather than the reproduction of any actual sound, however minute, made by the fly. The 'microphone' must for our purposes be regarded as an effect described in the paper rather than as a magnifier of sound. It is quite evident, says Hughes—

that these effects are due to a difference of pressure at the different points of contact, and that they are dependent for the perfection of action upon the number of these points of contact.[3]

Hughes related the experiments which he had made in the transmission of vocal sounds on the lines of this discovery, and he called any of the preparations of finely divided metals or charcoal 'confined in a glass tube or a box and provided with wires for insertion in a circuit' a *transmitter*.

Edison's carbon telephone was called a transmitter and it was the most effective transmitter available. It is perhaps partly due to the use of this word by Hughes that Edison immediately launched an accusation of piracy against Hughes and Preece, to whose appreciation of the importance of the facts and kind counsel in the preparation of the paper the professor had expressed his indebtedness. The controversy waged furiously, but Sir William Thomson (Lord Kelvin) intervened, and in a public letter summed up the facts, concluding by saying :—

[1] *Proceedings of the Royal Society*, May 9, 1878, xxvii. 362.
[2] *Quarterly Journal of Science*, 1827, Part II. Wheatstone's *Scientific Papers*, p. 33.
[3] *Proceedings of the Royal Society*, May 9, 1878, xxvii. 366.

I cannot but think that Mr. Edison will see that he has let himself be hurried into an injustice, and that he will therefore not rest until he retracts his accusations of bad faith publicly and amply as he made them.[1]

The following contemporary comment on the controversy represents the generally accepted view of the difference between the conceptions of Edison and of Hughes :—

The merit in Professor Hughes' discovery mainly consists in this—that it is a property of the *contact* of two conductors, and that it is not confined to any one, but to all conductors. Edison's claim has been until now, and can only be stated as follows :—'I discovered about two years ago that carbon of various forms, such as plumbago, graphite, gas retort, carbon, and lampblack, when moulded in buttons, decreased the resistance to the passage of the electrical current by pressure' (we give his own words). This was an important discovery, but it had been anticipated by M. Clerac, of the French Telegraphic Administration, Paris, who, as long ago as 1866, constructed tubes containing powdered carbon, the electric resistance of which could be regulated by increasing or diminishing the pressure upon it by means of an adjusting screw. Mr. Edison applied it to his carbon telephone in a manner well known. Now it appears to us, from all accounts, that Mr. Edison, though very nearly discovering the full significance of the microphone, did not do so. He explained the real nature of the property which he had found to exist in carbon, and, let us say, the vague but all-comprehensive patent phrase, 'finely divided metals,' to be a diminution of resistance in the *mass* of the material. The discovery of the microphone limits it to an effect of the *surfaces of contact*; but it is plain that Edison did not find this out, else we have no hesitation in saying so great an inventor would not have confined himself to buttons of carbon placed between platinum discs. Thus it is that Professor Hughes' discovery explains the true action of Edison's telephone, which is indeed a form, but a disguised form, of the microphone. It may enable the former to rid itself of its mask, and in addition it opens up a wide field of investigation in which other inventions and discoveries now lie hid.[2]

It is somewhat curious that throughout this controversy no reference was made to the fact that, in the provisional specification of his British patent No. 2909 of 1877, Edison uses the words 'intimacy of contact,' which were subsequently so generally used to describe microphonic action. In the first paragraph of this provisional specification he says :—

The vibrations of the atmosphere which result from the human

[1] *Nature*, xviii. 356.
[2] *Telegraphic Journal and Electrical Review*, vi. 266.

voice or from any musical instrument or otherwise, are made to act in increasing or lessening the electric force upon a line by opening or closing the circuit *or increasing or lessening the intimacy of contact between conducting surfaces* placed in the circuit. At the receiving station the electric action in one or more electro-magnets causes a vibration in a tympan or other instrument similar to a drum and produces a sound, but this sound is greatly augmented by mechanical action.[1]

In the complete specification he states (apparently in contradiction of the provisional) :—

I find that it is not practicable to open and close the line circuit in instruments for transmitting the human voice ; the circuit to the line must be always closed and the transmission be produced by a rise and fall of electric tension resulting from more or less resistance in the line. This resistance may be produced in several ways. I have shown several which will hereafter be named, but I find the most delicate to be small bunches or tufts or disks of semi-conducting elastic fiber, such as particles of silk and an intermediate conducting or semi-conducting material ; this device I call an electric tension regulator ; it is more or less compressed according to the vibrations of the diaphragm or tympan, and the electric current rises in tension as it is compressed or lessens as the fiber expands.[2]

In his United States specification dealing with the fibre device Edison says :—

I have discovered that if any fibrous material—such as silk, asbestos, cotton, wool, sponge or feathers—be coated by rubbing or otherwise with a semi-conducting substance, such as plumbago, carbon in its conducting form, metallic oxides, and other conducting material, and such fiber be gathered into a tuft and placed in a circuit, it is very sensitive to the slightest movement. I am enabled not only to obtain the regulation by the greater or less pressure, but also to increase or decrease *the extent of surface-contact* between the particles of conducting or semi-conducting material that is associated with the fiber.[3]

Again, in the British specification he says :—

In some cases I make use of a variable resistance resulting from greater or less intimacy of surface contact such as would result from a disk covered with plumbago placed adjacent to a diaphragm, also covered with plumbago or other semi-conducting material

[1] British specification, No. 2909, 1877, p. 1, lines 8–14.
[2] *Ibid.* p. 5, line 20.
[3] U.S. specification, No. 203,015, dated April 30 1878 (application filed August 28, 1877).

so that the proximity or extent of surface contact will produce rise and fall of tension, the respective parts being in the telegraphic circuit.[1]

The variations in the extent of surface contact were the means of lessening or increasing the resistance in an elastic combination of conducting or semi-conducting material which he termed a 'tension regulator,' though it varied the resistance rather than regulated the tension.

This particular form of 'tension regulator' was not commercially used. Superior results were obtained from 'the best quality of lampblack retained within a case.'[2] The lampblack was compressed into a button or lozenge, and it was this form of 'tension regulator' which was practically used, and consequently became the basis of comparison. The compressed carbon button was in some respects a departure from the theoretical design of the invention because it was not sufficiently elastic, and whilst its efficacy was attributed to the variations of conductivity under varying degrees of compression, the researches of Hughes demonstrated that the results were obtained by the variations produced upon the surfaces in contact and not by the mass. In this way, however, its operation would accord with that of the disc and diaphragm above referred to in Edison's patent, which in his own words obtained its results 'from greater or less intimacy of surface contact.' The drawing illustrated in fig. 10 of this patent and reproduced in fig. 34 is very suggestive of a double microphone, but the detailed description removes it from this category.

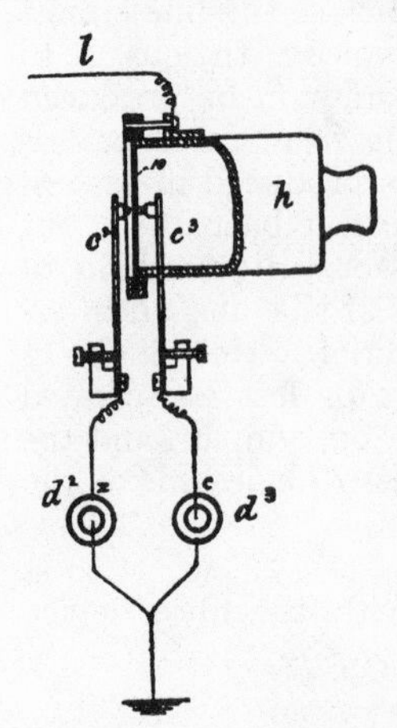

FIG. 34. — Edison's Design for Transmitter with Varying Surface Contact.

This design was intended to utilise opposed batteries, the purpose being thus described :—

The battery d^2 has zinc to the line or spring c^2, and the battery d^3 has copper to the line or spring c^3. When the springs c^2 and c^3 are adjusted to make contact with the diaphragm equally no current passes to the line, but when the diaphragm is vibrated its movement to one side, say c^2, causes a greater pressure upon the plumbago on that spring and a lessening of the pressure on the plumbago on c^3, hence the balance of the batteries c^2 and c^3 will be destroyed,

[1] British specification, No. 2909, 1877, p. 6, line 5. [2] *Ibid.* p. 6, line 32.

c^2 having the advantage will send a negative current to line ; upon the return of the diaphragm the battery currents will again neutralise each other. The vibration of the diaphragm to the other side causes the pressure to be reversed and the battery d^3 will send a positive current to the line.[1]

The diagram with the above description alone might well suggest microphonic action, which suggestion is, however, removed when the description of the points of the springs c^2 and c^3 are considered. These points are to be

made of compressed plumbago mixed preferably with gum rubber; but any substance not liable to rapid decomposition, or the elastic or fibrous tension regulator aforesaid, may be used. These points face each other on opposite sides of the diaphragm and make contact with platina foil disks secured to the diaphragm.[2]

The variation in the current is clearly expected to be obtained through the varied resistance of the more or less elastic substance of the points, and not by their varied degrees of contact with the platinum foil discs secured to the diaphragm.

The verdict of history needs not to be changed. Edison did not realise that the variation of resistance between points in contact would suffice for the transmission of speech, but he did realise that changes of surface contact in the particles composing a tension regulator were of value, and the attention which he gave this subject perhaps renders his interposition in the controversy referred to more reasonable than appeared at the time, especially when it is remembered that Hughes's claims reached him in an incomplete form.

But whilst Edison with all his telegraphic and electrical experience generally did not contemplate the possibility of a transmitter operating by simple variations in contact, Berliner, who obtained his information as to imperfect contacts at second hand, invented such a transmitter two or three months before the date of Edison's patent, and nearly a year before Hughes communicated his paper to the Royal Society.

It was on April 14, 1877, that Emile Berliner deposited a caveat, and on June 4, 1877, an application for a patent, on a transmitter consisting of a diaphragm in loose contact with an electrode. Emile Berliner was employed in Washington at the time of the Centennial Exhibition, and although he had not seen the Bell exhibit he had heard of the transmission of speech, and had become so far interested in the subject as to conduct experiments. He had formed

[1] British specification, No. 2909, 1877, p. 7, lines 23–32.
[2] *Ibid.* p. 7, lines 17–21.

the impression that the transmission might be more satisfactorily accomplished by a battery instrument than by the magneto form of Bell. Berliner was acquainted with the chief operator of the fire alarm telegraph office—a gentleman named Richard—and from him Berliner learned that in order to send an effective telegraphic current it was necessary to make a firm contact with the Morse key. Having ascertained in this way that in practical telegraphy there was a variation in conductivity due to a greater or less pressure between contacts in an electrical circuit, Berliner made two instruments, each consisting of an iron diaphragm and a steel ball. He connected two of them, one upstairs, the other downstairs in a three-storey building. He had a friend talk into the instrument upstairs, and he himself listened carefully downstairs and could plainly understand what was said.[1] He applied for a patent, and, the application becoming known to the Bell Company, that Company purchased the invention and engaged the inventor in their service.

The deposit of Berliner's application in 1877 gives him the priority in the discovery of the utility of the variable contact in telephone transmitters. The success of his experiment was perhaps in the nature of a happy accident. He did not develop or propound the microphone theory. But there is no disputing the fact of his priority of invention for a contact telephone.

Reverting to the commercial situation as it existed between the Western Union and the Bell interests, it was noted that the introduction of the Edison transmitter gave the Western Union a considerable commercial advantage, and, moreover, demonstrated that battery instruments of some form must replace the magneto instrument as a transmitter.

In the summer of 1878 Mr. Francis Blake, junior, became associated with the Bell Company. Mr. Blake had some scientific training, and was connected with the United States Coast Survey. Having studied the published accounts of the work of Hughes, he designed a transmitter on the microphonic principle, and conceived the idea that the electrodes should be attached to separate springs converging together and jointly impinging upon the diaphragm. The principles of the design were excellent, and of the details remaining the selection of suitable carbon was the most important. The manufacture of special carbons then in progress for electric lighting purposes provided the raw material, and a process of hardening said to have been devised by Berliner completed the requirements. The Blake transmitter was now available for public use. It was issued to subscribers, and by

[1] *Telephone News*, Philadelphia, Febuary 1, 1911.

its merits held control until the conditions of service had radically changed.

The merits of the Blake transmitter have been eloquently defined by Mr. Lockwood in his 'Practical Information for Telephonists,' 1882, p. 55, and in an article on the telephone in the 'Boston Electrical Handbook' (1904), which, though unsigned, is evidently by the same hand.

Professor Hughes described the effects obtained in his experiments as being produced 'simply and solely by the direct effect of the sonorous vibrations,' the diaphragm having been altogether discarded. In the numerous applications of the microphone to the transmission of speech a diaphragm has always been used. The carbon pencil type adopted by Ader, Crossley, Gower and others had a diaphragm of wood placed nearly horizontal, to which the carbon pencils were attached; the Blake, Hunnings, and the numerous successors of the latter retaining the circular diaphragm adopted by Edison. There is room for doubt whether Hughes gave a sufficiently wide meaning to the word 'diaphragm' in its application to telephones. As an archdeacon was defined as one who undertook archidiaconal functions, so it may be reasonable to argue that anything which performs diaphragmatic functions is a diaphragm. An expert witness, Professor Silvanus Thompson, in the English courts maintained that 'a diaphragm is something which separates something from something else.'[1] Edison's original British patent contained thirty claims, but it was sustained only after it had been reduced by amendments to the single claim for the combination of a diaphragm and a tension regulator.

The fixing of the diaphragm in its case involved considerable difficulty. It needed to be held firmly, yet not rigidly. Its movements should follow the movements of the air and be independent of any vibration through mechanical contact. Packing was resorted to in the form of springs or elastic substances on either face, but in such cases the edge remained unprotected. Edward F. Wilson, of Boston, conceived the idea of protecting edge and both faces with a single rubber band—

the internal diameter of which when not stretched is less than that of the diaphragm. This band is stretched over the edge of the diaphragm so as to form what may be termed a 'binding' for it, covering the edge proper, and extending upon both faces from the edge towards the centre for the distance of about a quarter of an inch. The band should be small enough to exert in a small degree a pressure toward the centre of the diaphragm, so as to render the diaphragm slightly concavo-convex. It will be found that by the application of an elastic band to the diaphragm certain troublesome

[1] *Electrical Review*, London, x. 638.

overtones are avoided which are present when the edge proper is left bare.[1]

Wilson's patent, from which the above is quoted, is an example of many very simple applications of great practical value. The rubber ring around the diaphragm first applied to the Blake transmitter is a feature which has been retained.

In the United States the Bell patent controlled the electrical transmission of speech however obtained, so that Edison's was quite subsidiary. The Blake and other carbon transmitters employing carbon as a contact material were, however, generally regarded as being subordinate to the Edison invention of the carbon transmitter and of the contact transmitter of Berliner.

When it is considered that the essentials of a transmitter were accurately recorded by Edison as two—the diaphragm and the 'tension regulator'—and when it is considered also that the diaphragm does not lend itself to many changes, it is matter for wonder that so many variations of the 'tension regulator' could be made as to have permitted almost every new entrant in the telephone field an opportunity for a new christening. Some, of course, had detail merit, but most were modifications covering no new principle and effecting no improvement in results. In consequence they need no mention, and at this stage one other only will be referred to.

The Rev. Henry Hunnings, of Bolton Percy in the county of York, a clergyman of the Church of England who combined with theology a taste for natural philosophy, devoted some attention to the telephone, and as a result of his experiments, which, he said, 'were made in total ignorance of any experiments having been made by any other worker,'[2] he invented a transmitter which is thus described :---

A front vibrating diaphragm, composed wholly or in part of suitable metal, such as, preferably, platinum, silver, ferrotype-iron, tinned iron, *et cetera*. In close proximity to the aforesaid vibrating diaphragm is fixed a disc of brass or other suitable metal, the intervening space being filled with carbon in the form of powder to the depth of about one-sixteenth of an inch. The aforesaid vibrating diaphragm and the fixed disc of brass are connected, respectively with the opposite poles of a voltaic battery. The whole may be secured in a box of suitable non-conducting material, with a mouth-piece, if desired.[3]

He remarks that

the details can be indefinitely varied, the great feature being in the

[1] U.S. specification, No. 250,616, December 6, 1881 (application filed May 21, 1879).

[2] *Electrical Review*, London, x. 350.

[3] British specification, No. 3647, September 16, 1878. (U.S. equivalents, No. 246,512, August 30, 1881, and No. 250,250, November 29, 1881.)

use of carbon used in a state of fine loose powder, not in any way compressed or consolidated, as I find the loose particles of conducting matter to be most delicately sensitive to sonorous vibrations.[1]

Fig. 35 is a reproduction of the drawing forming part of the patent specification.

Hunnings, or at any rate his patent agent, was familiar with the Edison transmitter and the Hughes microphone, but, so far as the patent specification gives any indication, does not appear to have been acquainted with the whole of Hughes's paper, or of his reference to the utility of carbon or metals in a finely divided state.

The Edison, Blake, and Hunnings transmitters, together with Hughes's discovery of microphonic effects, all influenced subsequent developments referred to in later chapters. The Hunnings instrument played but little part at the period to which we are now referring. An attempt was made by a rival Company in London to establish its independence of the Edison patent—an attempt which succeeded on a technical point in the court of first instance only to fail on appeal. The utility of the Hunnings loose carbon was to be developed later. But the Blake played a part of immediate importance. It gave the Bell Company an instrument which, though not as loud as the Edison, was its superior in clearness of articulation, in reliability, and in durability. The following paragraphs are taken from a circular entitled 'Directions for setting up the Edison Transmitter Boxes':—

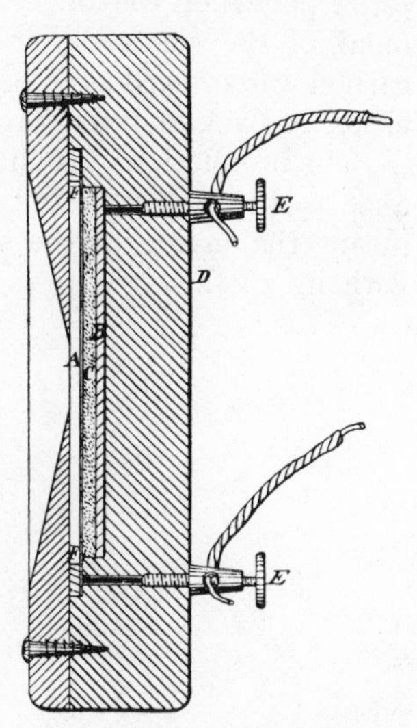

Fig. 35.—Hunning's Transmitter, as illustrated in patent.

Battery (either carbon or gravity) is needed at each station, and is used alternately as local and main. It is used as local when talking, and as main to call with.

The battery at all the stations is left on the main circuit for calling, except when conversation is being carried on. The switch changes the battery from the local circuit to the main line. The battery of one station remaining in circuit does not interfere with talking between any other two stations.

. . . .

For short lines one cell of carbon or two cells of gravity battery

[1] British specification, No. 3647, September 16, 1878. (U.S. equivalents, No. 246,512, August 30, 1881, and No. 250,250, November 29, 1881.)

are needed at each station. These are enough for speaking purposes for any length of line, but for long lines more cells are needed for calling. These may be placed at any point in the main circuit.

The carbon button of the Edison transmitter is delicate and liable to disarrangement from a slight jar. The instrument should therefore be transported with great care, and the box be fastened to the wall before the transmitter is put in place.

The adjustment of the carbon button was also a work of difficulty and delicacy. The Blake suffered much less in these respects. The electrodes were mounted on the short arms of a lever. A screw played on the longer arm permitting the most delicate adjustment of the electrodes in their relation to the diaphragm. The carbon was so hard as to be free from danger, and there was no need either to pack or transport the transmitters as if they were eggs.

The production of such an instrument was of the utmost importance to the industry, but to the Bell Company its timely arrival meant the saving of the situation in the commercial competition with its rival.

CHAPTER XII

PHILIPP REIS AND HIS WORK

At that period of Philipp Reis's career which is of especial interest to us, he occupied a post in Garnier's Institute at Frankfort-on-Main. A teacher by profession, he was in the primary sense of the word an amateur of science. He was interested in new discoveries and developments, and acoustics had a special attraction for him.

Chronologically his work should have been referred to at an earlier stage, but since it had no direct influence upon the development of Bell's invention, its consideration has been reserved for this chapter, because such consideration is facilitated by the preceding chapters.

So far as can be judged by his Telephone Memoir, he seems to have presented his subjects in a practical manner, avoiding exaggeration, yet not oblivious to a proper recognition of any contributions he might make to scientific progress. Brevity, though generally commendable, was in his case perhaps something of a misfortune, for he desired to be understood, and if he could have foreseen the diverse interpretations which were to be placed in later years upon his aims, his explanations, and his results, he would undoubtedly have elaborated his illustrations and chosen his words with such care that it would not have been possible to misunderstand.

Reis invented his apparatus which he called The Telephone, he exhibited it in operation and explained the theory before a scientific society to which he belonged, and the apparatus was described—together with the degree of success attained with it—in scientific and other publications. In 1863 the Reis telephone was exhibited before the British Association by Mr. Ladd. He obtained it from the authorised manufacturer, and received an autograph letter from Reis describing its construction and the method of its operation. Its exhibition was recorded in the 'Transactions' of the Association without any reference being made to speech. It is curious to note also that the name of the inventor was not given. Dr. Ferguson in his book 'Electricity,' published in 1866 (p. 257) describes the mechanism with precision, explains the operation of the transmitter as being that of a circuit breaker, the sound in the receiver as being due to the 'magnetic tick,' and the combined result as the transmission of pitch only. Regarding the reproduction he says: 'The note is weak, and in quality resembles the sound of a toy trumpet.'

It is natural that any developments in science or invention should revive interest in previous work on similar lines. When therefore Bell succeeded in transmitting speech, interest in Reis's work, which had been dormant, was revived; but it was not until Bell's telephone had been in existence long enough to demonstrate its practical and financial value that it was alleged Reis had previously transmitted speech.

Lack of novelty is fatal to a patent, and the effort to prove prior invention is therefore an almost invariable feature of the defence in a patent action. There were several such actions about the telephone, and the claims put forward for Reis as a prior inventor figured in most of them.

These claims were investigated by Federal Courts in Boston, New York, New Jersey, Philadelphia, at Pittsburg, New Orleans, and by the Supreme Court of the United States at Washington, who one and all decided that Reis was not the inventor of a telephone in the sense of inventing an instrument which would transmit articulate speech. English courts, in the case of the United Telephone Co. *v.* Harrison, Cox-Walker & Co.,[1] also decided that the Reis instruments and publications did not anticipate Bell.

But the work of Reis, nevertheless, was meritorious, and in giving a brief account of it the opportunity may be taken to explain the misunderstanding of that work which underlay the claims persistently submitted to legal judgment and as persistently rejected.

Like the more modern telephones which have been described, Reis's apparatus consisted of two parts—a transmitter and a receiver. The receiver had to reproduce the sounds, and whatever results were expected must necessarily be limited to the effects which the apparatus was known to be capable of producing, or that the inventor contemplated it was capable of producing.

The receiving apparatus adopted by Reis was that used to illustrate the magnetic sounds known since 1837 as the 'Page effect.' Dr. Ferguson in his 'Electricity' gives an explanation of this effect as follows:—

Magnetic Tick.—When an iron rod is made to rest on a sounding board, such as the body of a fiddle, and placed in the centre of a powerful coil, each time the current is broken a distinct tick is

[1] *Telegraphic Journal*, x. 381.

heard from the rod. If a file be placed in the circuit, so that a wire when it slides along will alternately close and open the circuit, the rasping noise of the wire sliding along the file will be distinctly rendered by the rod, each interruption giving rise to a tick; the series of ticks being in the same order exactly as the series of noises at the file.[1]

In his own record of the experiment as originally carried out, Page in *Silliman's Journal*, July 1837, said that 'when the contact is made the sound emitted is very feeble; when broken, it may be heard at two or three feet distance.'

It will be seen that Ferguson refers only to the broken current as producing the tick. Reis adopts the theory that 'at every closing of the circuit the atoms of the iron needle are pushed asunder from one another,' and 'at the interruption of the current the atoms again attempt to regain their position of equilibrium.' The 'interruption' here is clearly the *breaking* of the circuit.

Reis makes no claim to have discovered in the receiver any new qualities, nor beyond the hypothesis referred to on p. 139 does he advance any new theory in the method of its operation. It had produced sounds before by repeated ticks. Reis gives us no ground for assuming otherwise than that he not only expected it to operate by repeated ticks, but also that the ticks would be produced by breaking the circuit.

While the receiver was adopted as a ready-made appliance, the transmitter was designed by Reis himself. Both are illustrated in fig. 36.

Before Reis's time the knitting needle contrivance illustrating the Page effect had only been operated by a mechanical circuit breaker of some sort, of which the wire and file mentioned by Ferguson is one example. The transmitter of Reis permitted the operation of the circuit breaker directly by air vibrations.

Its construction was ingenious. The metal patch in the centre of the diaphragm was one electrode. The other electrode (in the standard model) was one foot of a tripod resting normally on the centre of the diaphragm in contact with the lower electrode. The other two feet of the tripod rested on the box in suitable 'cups' or recesses. In one of the cups there was placed 'a little drop of quicksilver,' evidently for the purpose of ensuring conductivity under movement. The movement, its object, and its result are very clearly described. When the diaphragm is set in vibration,

at the first condensation the movable electrode will be pushed back. At the succeeding rarefaction it cannot follow the return vibration of the membrane, and the current . . . remains interrupted . . .

[1] *Electricity*, Ferguson, 1866, p. 182.

until the membrane, driven by a new condensation, presses the lower electrode against the upper electrode once more. In this way each sound wave effects an opening and closing of the current.[1]

Early publications on the speaking telephone include descriptions of Reis's apparatus without, as a rule, an analysis of the operation of the various parts, and the descriptions are sometimes incorrect. Prescott (1878) (p. 147) adopts a translation of the article in Bottger's 'Polytechnical Notezblatt,' 1863, in which both electrodes are described as 'fastened' to the wood, though the 'hopping'

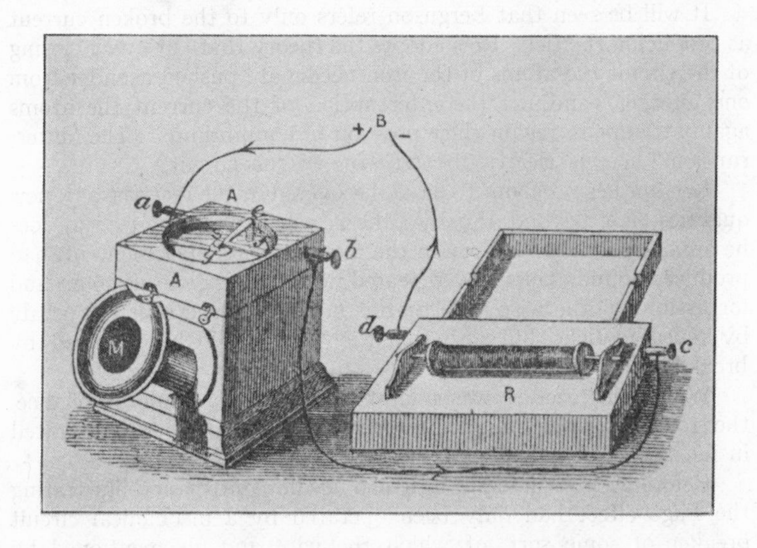

FIG. 36.—Reis Transmitter and Receiver (from 'Electricity,' Ferguson).

character of the upper electrode is referred to. Du Moncel (1879) (p. 12), after mentioning the diaphragm with its platinum disc, says that above this 'a metallic point *c* was fixed, and this together with the disc constituted the contact breaker.' 'The Modern Applications of Electricity' by Hospitalier, translated by Julius Maier (1882), says (p. 297) :—

Each time that the membrane is raised the point will touch the disc, and a current will be established. It is, on the contrary, interrupted when the membrane comes back into rest.

The description leads to the inference that the upper electrode is fixed, though the explanation of the illustration on the next page

[1] Reis, Frankfort lecture.

mentions it as a 'movable lever touching the membrane.' Dr. Maier collaborated with Sir William Preece in the authorship of 'The Telephone' (1889), in which the foregoing description from 'Modern Applications' is substantially reproduced without the explanation to the illustration.

Attention has been called to these references because it is not possible to follow Reis's ideas and appreciate his work without recognising that his upper electrode was not fixed. Dr. Ferguson's description is more detailed and more accurate—a tripod rested with two feet on the frame and one foot on the diaphragm. The circuit was normally closed by the tripod resting on the disc. When the diaphragm became agitated the circuit was interrupted in accordance with the rate of movement of the diaphragm. Thus the transmitter conformed to the requirements of the receiver in producing 'interruptions' or breaks.

It was on October 26, 1861, that Philipp Reis delivered before the Physical Society of Frankfort-on-Main his lecture 'On Telephony by the Galvanic Current'; on November 16 he gave 'An Explanation of a New Theory concerning the Perception of Chords and of Timbre as a Continuation and Supplement of the Memoir on the Telephone.' These lectures were combined and printed in the records of the Society[1] under date of December 1861. Though recorded as one paper it was, as we have seen, delivered in two parts.

In our quest for Reis's meaning it would have been a great help to know the exact portions of his paper which were delivered on the respective dates. The first was certainly a lecture on the telephone, and may be presumed to have included so much of the description and the theory as was necessary to understand it. The second was an exposition of a new theory on the perception of chords and of *timbre*, to which a table of curves presented with the paper seems to be a natural illustration.

There are other curves intended specifically to illustrate vowel tones, and these, it may be presumed, were a part of the first (or telephone) paper. What Reis meant by the vowel curves is fully explained by analogy, so that there is no need for conjecture. The more extensive illustrations of compound curves which, it may also be presumed, were a part of the second (or perception of chords) paper, are not so clearly described.

It is these last-mentioned curves which underlay the claims made for Reis as the inventor of the telephone. They have been applied to speech with a meaning that was not prevalent in Reis's

[1] The record of the lecture is given in full in *Philipp Reis, Inventor of the Telephone*, by Silvanus P. Thompson (London: E. & F. N. Spon, 1883), together with other contemporary documents.

day. Reis made no reference to these curves in connection with speech. They were used as musical illustrations, analysed to the extent of four notes, and it is possible that they were adopted by Reis without any such perception as has been ascribed to him.

It may be recalled that the theory of compounded tones, an elaborated exposition of sympathetic reinforcement, and the analysis of compounded tones by resonators were enunciated in Helmholtz's lecture at Bonn in the winter of 1857, 'On the Physiological Causes of Harmony in Music.'[1]

Where two writers are dealing with the same subject it is inevitable that there must be some points of similarity, and a common origin is not necessarily to be inferred from such similarity. But Helmholtz was known to be an original investigator in the field of physics, a great scientist and a popular lecturer, whilst Reis was an interested student as well as a capable teacher, one who (in his own words) had, some nine years previously, 'a great *penchant* for what was new but with only too imperfect knowledge of physics.'[2]

In the preface to the English translation of Helmholtz's 'Popular Scientific Lectures' (1873) it is said that the Bonn lecture had not previously appeared, so that it is not possible for Reis to have read this lecture as a whole. Whether he heard it or whether any partial publication of it may have been made, can only be matter of conjecture.

Reis was evidently not a man who would put forward as his own the idea of another, but it is to be noted that the explanation of the new theory which the curves served to illustrate is not stated to be his own theory, though some support might be found for such a suggestion in his remark that the correctness of 'my views with respect to the curves representing combinations of tones may perhaps be determined by the aid of the new phonautograph.' It is only from internal evidence, and perhaps not unreasonable inference, that it can be suggested that Reis was probably familiar with some portions of Helmholtz's Bonn lecture.

Helmholtz remarked that the theory which he enunciated in this lecture in 1857 'will perhaps seem new and singular.' The title of Reis's second lecture was 'An Explanation of a New Theory concerning the Perception of Chords and of Timbre as a Continuation and Supplement of the Memoir on the Telephone.'

Helmholtz told his audience they must 'conceive the air of a concert hall or ball-room traversed in every direction, . . . by a variegated crowd of intersecting wave-systems,' and, after describing his sound curves and other illustrations, asked, 'Now, what does the

[1] *Popular Scientific Lectures*, Helmholtz, p. 61.
[2] Reis, Frankfort lecture.

ear do? Does it analyse this compound wave? Or does it grasp it as a whole?' Developing his theory of Corti's arches he suggested the probability that it was analysed into its constituent parts by the individual fibres. He further drew a distinction between the audible sensation and the intellectual conception. 'We have, as it were, to distinguish between the material ear of the body and the spiritual ear of the mind.' Reis said:—

How does *our ear* take cognizance of the total vibrations of all the simultaneously operant organs of speech? Or, to put it more generally: How do we perceive the vibrations of several bodies emitting sounds simultaneously? . . . The function of the organs of hearing, therefore, is to impart faithfully to the auditory nerve every condensation and rarefaction occurring in the surrounding medium. The function of the auditory nerve is to bring to our consciousness the vibrations of matter resulting at the given time, both according to their number and their magnitude.[1]

In this Bonn lecture Helmholtz illustrated two simple sounds by their appropriate curves and their combination in a compound curve. Reis adopted a similar method in his Frankfort lecture, but carried the illustrations further. There is some importance in the consideration of the question whether Reis from independent thought created his curves or whether he adopted them from Helmholtz. It is without doubt that Helmholtz publicly used the illustration four years before Reis's lecture.

Our interest lies in realising, if possible, Reis's meaning of the curve, and in this endeavour to discover what was in Reis's mind we must consider the whole of his paper, the theory which he propounded, and the apparatus by which he proposed to convert it into practice.

In this consideration either of two methods may be followed. One may start with the description and illustrations which are capable of varied interpretation, according to the extent of knowledge assumed, and then proceed to consider the apparatus with a view to seeing how far it may meet those interpretations. Or one may argue backwards from the apparatus. The latter affords the more ready and probably the more accurate method of realising Reis's conception.

When he comes to define the transmission Reis says:—

As soon, therefore, as it shall be possible at any place and in any prescribed manner to set up vibrations whose curves are like those of any given tone or combination of tones, we shall receive the same impression as that tone or combination of tones produced upon us.[1]

[1] Reis, Frankfort Lecture.

He anticipates that the results at which he aims will be secured if he transmits number and strength. This is clearly expressed in his explanation of the reproduction of the vowels. He alludes to 'The researches of Willis, Helmholtz, and others,' showing that vowel sounds can be artificially produced by causing the vibrations of one body to reinforce those of another periodically.' He explains this operation by the use of an elastic spring set in vibration by the thrust of a tooth of a cog-wheel:

The first swing is the greatest, and each of the others is less than the preceding one. After several vibrations of this sort (without the spring coming to rest) let another thrust be given to the tooth; the next swing will again be a maximum one, and so on. The height or depth of the sound produced in this fashion

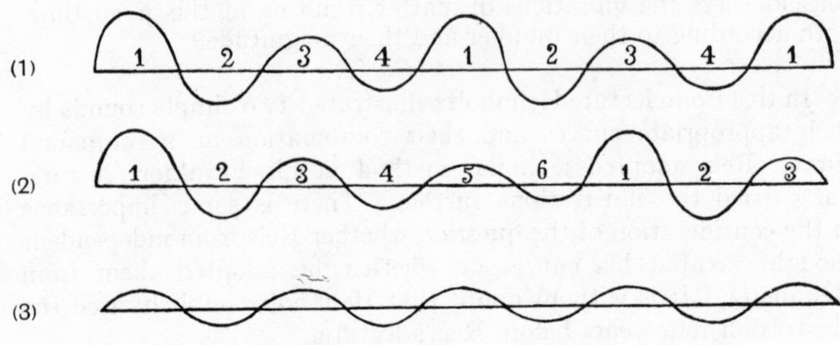

FIG. 37.—Vowel Curves from Reis Lecture.

depends upon the number of vibrations made in a given time; but the quality of the note depends upon the number of variations of amplitude occurring in the same time.[1]

He uses curves in illustration. These curves are expressly applied to speech, and it is from these that we must form our impression of what Reis conceived to be necessary. He says:—

Two vowels of equal pitch may be distinguished from each other somewhat after the manner represented by the curves (1), (2); while the same tone devoid of any vowel quality is represented by the curve (3).

Our organs of speech create the vowels probably in the same manner by a combined action of the upper and lower vocal chords, or of the latter and of the cavity of the mouth.[1]

The two different vowels illustrated by Reis are assumed to be obtained by vibrations of the same period with a reinforcement in

[1] Reis, Frankfort lecture.

the first case after every fourth beat, and in the second case after every sixth beat, there being a gradual loss of power in either case between each reinforcement.

This being Reis's conception of the differences in vowel tones, he expected his apparatus to perform the function which he had outlined for the ear :—

The greater the condensation of the sound conducting medium at any given moment, the greater will be the amplitude of vibration of the membrane and of the 'hammer,' and the more powerful therefore the blow on the 'anvil' and the concussion of the nerves through the intermediary action of the fluid.[1]

Since the parts referred to as constituting the mechanism of the ear were reproduced in his transmitter he expected to obtain corresponding results. It was to meet this requirement that he devised the peculiar form of circuit breaker in his transmitter.

The ingenuity of the arrangement may be better appreciated by considering the alternative. The first idea would naturally be to follow the vague suggestion of Bourseul and provide a vibrating diaphragm which should make and break contact with a fixed electrode. Such a plan could not effectually allow for the variations in the amplitude of the vibrations which Reis had clearly in mind. He reversed the order, and instead of arranging for the making or breaking of contact by the extended diaphragm operating as a sensitive key upon a separate stud he provided a means of breaking contact when the agitation of the diaphragm threw upwards the movable electrode normally resting on it. The upper electrode was a point and the lower electrode a disc. In a drawing the former would be regarded as a hammer and the latter as an anvil. The reversal of the ordinary operation which Reis carried out may be comprehended readily by assuming that instead of a hammer striking an anvil, the anvil by a rapid movement threw up the hammer. This would occur with a condensation, and in the subsequent rarefaction the diaphragm would retreat more quickly than gravity could operate upon the upper electrode, thus effecting the alternate separation and contact. When, in any compounded sounds, the vibrations coincided, the condensation would be of greater strength, the 'blow' would be more powerful. This is explained when he says :—

Moreover, the strength of this tone [i.e. this tick] is proportional to the original tone [i.e. original vibration], for the stronger this is, the greater will be the movement of the drum-skin, the greater therefore the movement of the little hammer, the greater finally the length of time during which the circuit remains open,—[1]

[1] Reis, Frankfort lecture.

We may interrupt the quotation here to note Reis's thoroughness in his desire to follow the process through. Since it is required to reproduce the sounds, he has to explain how these operations in the transmitter are reflected in the receiver. He continues :—

> and consequently the greater, up to a certain limit, the movement of the atoms in the reproducing wire (the knitting needle), which we perceive as a stronger vibration, just as we should have perceived the original wave.[1]

At the beginning and at the end of the lecture the same idea is expressed. The function of the auditory nerve was defined in his introduction as the bringing of the vibrations to the consciousness 'both according to their number and their magnitude,' and in his conclusion he says that his apparatus permits the transmission of 'the number and the strength' of the vibrations. Consequently he expected to produce varied qualities of sound by transmitting detached currents which should vary not only with the number of vibrations in a given time (producing pitch), but which should also vary with the force of individual vibrations, by which he expected to reproduce *timbre* because that was his conception of the audible perception of *timbre*. Whether he obtained on the line and in his receiver the expected varied effects from varied amplitudes is open to question. But it suffices to indicate that such effects were involved in his theory and that the production of such effects was considered to be provided for in his apparatus. In the ear he assumed the operation of number and magnitude. In his apparatus he thought he had provided for number and strength. The imperfections in operation which he recognised he attributed to defects in the transmission or reproduction. What he expected to attain and conceived to be sufficient may be graphically represented by circles of differing size. Fig. 38 indicates in this form the two vowels and one simple sound illustrated by Reis's curves in fig. 37. The sizes of the circles represent the varying strength of the blows assumed to be effected by each vibration. The detached currents transmitted to line were in the reverse order—the strongest blow was expected to convey the shortest current. In the reproduction of the vibrations in the receiver the variations in the degrees of loudness were assumed to be effected by the longer or shorter period afforded for the readjustment of the molecules in the knitting needle.

If it be assumed that Reis's complex curve was intended to be a graphic representation of several separate vibrations it will be in accord with the knowledge of the period, and we may read

[1] Reis, Frankfort lecture.

his paper without doing violence to any of his words. If it be assumed that he conceived the necessity of transmitting the whole of his complex curve with all its sinuosities in the manner which after 1876 was expressed by the term 'undulatory current,' we must not only do violence to his language but we must assume also that he had made important discoveries in the operation of both transmitter and receiver without making any reference thereto.

It is no compliment to Reis's memory to assume that he omitted to record any new discovery which he may have made, or that there was any looseness of expression in his use of common terms. When lecturing on a new theory regarding the perception of chords it is hardly to be expected that he would refrain from describing a new discovery of his own if he had made one. The discovery

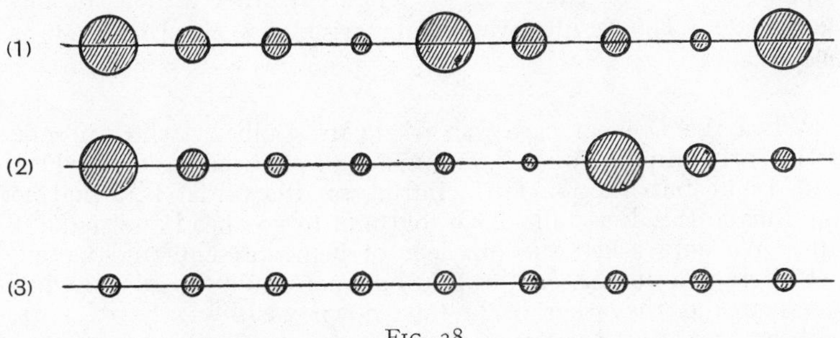

FIG. 38.

of the microphonic effect must not be confounded with a general knowledge that current was subject to some variation according to the degree of pressure on a Morse key or similar contact maker. Such an effect was on record though not generally known. The discovery that such an effect would serve to modify a current with the almost infinite variety required for the transmission of speech vibrations is a very different matter.

Throughout his papers there is not only no indication that Reis knew anything about the effect of a variation in resistance between electrodes in imperfect contact, but there are very definite statements that the contacts of the transmitter must make and break. Moreover, his receiver was a contrivance which had been used, and only used, to produce sound resulting from the molecular disturbances in the core of an electro-magnet as the result of repeated breaks in an electrical circuit. If Reis had discovered any new method of tone production from it, we may infer that he would have been glad to say so. He does not say so, but he gives an exposition of the magnetic theory of its operation, the basis of which he quotes from Muller-Pouillet,

'Lehrbuch der Physik.'[1] There is in his description of such theory no indication that he had discovered and sought to use a new property not previously known or used. Reis clearly expected to obtain from his needle a series of ticks such as had been previously obtained, and did not regard it as possible to obtain sound from it in any other form.

Scientists learned with expressed surprise many years later, and after the electrical transmission of speech had been effected by other means, that variations in contact pressure would serve to transmit articulate speech. At an earlier stage they were still more surprised at the demonstration that the variations in a magnet, operating upon a diaphragm armature, would reproduce articulate speech.

The contemporary records of Reis's demonstrations have been alluded to. In the course of the litigation other demonstrations were made. The results were summarised by Mr. Dickerson as follows :—

When the Dolbear case was tried, Mr. Dolbear, who was one of the early admirers of Bell, and who had published a book which lauded him (but who joined the infringers afterwards), said that he could make this Reis talk. We told him to go ahead and make it talk. We only asked the privilege of being present, and having a stenographer at each end, so that one man could take down what was said and the other could take down what was heard. Mr. Dolbear and Mr. Buck were the operators. They had several instruments—the box kind and cone kind. Their box set they had gone to Germany for and thought it the best they had. They had been experimenting with it for months. They met us and tried to make it talk. They talked over a thousand words into it on two different days. They used familiar phrases, such as 'Mary had a little lamb,' 'Hello,' 'How do you do ? ' 'How is this for high ? ' and all that sort of thing. This apparatus will give you the general rhythm of the sound, and from it a man can guess once in a while what he hears. Out of more than a thousand words they guessed at fifteen or twenty, and they guessed more than half wrong and less than half right. . . . Edison said, in one of his depositions about a telephone which he had thrown away as worthless, 'When you knew what it was the man was saying it sounded awful like.' . . . If you get a rhythm, identifying it with a phrase you are accustomed to, you may guess right. Suppose you are at the instrument, and the phrase 'Mary had a little lamb' was sent to you, and your assistant had used it many times, and it had been sent many times, you might be able to think you recognised it by the length of the words and the rhythm. And so in that way they guessed at fifteen words out of the thousand ; but of those they

[1] Vol. ii. p. 304, 5th edition.

guessed, eight were wrong and seven right. That is the best that has ever been done.[1]

These are the words of an advocate. The facts which he refers to were in evidence before the Court. The conclusions he forms from those facts are his own. The material interests which he was defending were great. Those interests have long since lapsed; the scientific interest alone remains. But independent effort to discover Reis's mental conception, and careful consideration of the contemporary evidence available as to the operation of his instruments, fail to modify the contention of the advocate. Whether the Reis instruments ever conveyed from talker to listener a sound which could accurately be called speech can never with certainty be known. The probabilities against it are so great that overwhelming evidence is necessary to support the contention that speech was transmitted. And the contemporary evidence does not indicate more than that sounds suggestive of words were sent. All the recorders were friendly recorders and were disposed to make the best of the apparatus. Legat had formed the highest opinion of its importance, was the most enthusiastic, and was probably the most competent demonstrator after Reis himself. Legat's report shows that chords and melodies 'were transmitted with marvellous fidelity,' single words 'were perceptible more indistinctly' but here also modulations and exclamations 'attained distinct expression.' This accords with what might be expected from an examination of the theory and the instruments. In cases where the ground tone formed the prominent characteristic of a word the word was suggested, but in any case where overtones became an essential to identification the characteristic feature was lacking and the identity was lost. Such a conclusion is not to be modified by the assumption that the Reis transmitter occasionally operated as a microphone. The possibility—even the probability—that it did so act momentarily and accidentally must be recognised, but if the transmitter did by accident send an undulatory current out upon the line it could not be converted into an audible sound with the receiving mechanism provided. At the receiving end also we must not be misled by the possibilities of a knitting needle and a coil of wire as known since March, 1876. We must remember that Reis was not looking for a still small voice. The theory upon which he founded his experiments and the circumstances under which he carried out his demonstrations indicate that his only hope of sound from the knitting needle was the tick arising from the broken circuit. Sounds produced by aggregate ticks

[1] Mr. Dickenson's argument, *American Bell Telephone Co.* v. *National Improved Telephone Co.*, Circuit Court, U.S., March 11, 1886, p. 113.

were loud, so loud that at the first public demonstration he was in a position to make them audible 'to the members of a numerous assembly.' The volume of sound necessary for this purpose shows that it was obtained by ticks, and the expectations of such volume would prevent the discovery of the extremely feeble sound (if any) resulting from undulatory currents.

The scientific world adopted the view of Reis's work, which is expressed by Dr. Ferguson.[1] The practical results had not justified any other view, and it was perhaps too readily assumed by some that his aims had been no higher than his results. In what may be called the controversial period much was made of the phrase 'reproduction of tones,' it being assumed that Reis had no higher aim than melodic reproduction. It was also sometimes claimed that because Reis provided his instruments with a Morse key and sounder he never intended to transmit speech. That Reis started out with the intention of transmitting speech is more than doubtful, but that he essayed the transmission of speech is clear from his own words. That he failed is equally clear from his own records.

Reis's work had no direct effect on the invention of the speaking telephone, for Bell fortunately went on entirely independent lines and without any reference to the prior work of Reis. It is on record[2] that Legat's report of Reis's apparatus was submitted to Edison in or about the month of July 1875, and formed the starting point of his telephonic experiments. It is perhaps on that account that Edison in his 1877 specification relates as a new discovery 'that it is not practicable to open and close the line circuit in instruments for transmitting the human voice.'[3] Bell started with the realisation of that necessary fact.

The difference in the results of Bell and Reis is plain enough, but the difference in the methods is plainer still. Between an originating sound and an attempted reproduced sound Reis placed an electrical apparatus to obtain a broken circuit, and Bell placed an electrical apparatus incapable of producing a broken circuit, but adapted to the production of an electrical current of an undulating character, following precisely the curve of the originating sound. The methods differed because the conceptions differed.

It is not because the world was unprepared for a 'phonic telegraph,' nor because Reis was a comparatively obscure experimentalist, that his apparatus came down to a later generation as a musical telephone. It is because the vital principle of a talking instrument was lacking.

[1] P. 126.
[2] *Bell's Electric Speaking Telephone*, Prescott, 1884, p. 110.
[3] Chapter XI. p. 117.

In reputation Reis has suffered somewhat from the claims made on his behalf. Those who saw in him the inventor of the speaking telephone claimed his transmitter as a microphone. Those who opposed such claim were content to demonstrate that he produced, and only intended to produce, make-and-break apparatus. Between these two extremes he has failed as yet to receive the credit to which he is really entitled. The individual breaks—' hand-claps,' Clerk Maxwell described them—have been assumed to be of equal value. But Reis contemplated a differing value in the individual units—hand-claps still, but of differing degrees of force. As already remarked, the capacity of his apparatus to carry his theory into practice is open to question, and it may be that the historians have correctly described what the apparatus did, but they failed to record what Reis hoped that it would do. His theory on the acoustic side, though incorrect, was not unreasonable with the information then available. On the electrical side his explanation of the operation of the receiver needle adopted from Muller-Pouillet was only an electro-magnetic reason for its performing the same simple function as the spring in Savart's wheel, and thus by responding to variations in *number* producing pitch. An extension of this explanation to account for the variations in *strength* was based on the hypothesis that a longer period of rest in the needle gave greater power to the subsequent tick. Some explanation of this sort was necessary to make the assumed electrical operation conform to the acoustic theory. Its accuracy is doubtful, but whether accurate or otherwise is of no practical importance in the consideration of speech transmission. The simple purpose of Reis's explanations being overlooked, more abstruse reasons suggested themselves and contributed materially to the misunderstanding upon which the Reis claims were based. The application of the same acoustic theory is observable in modifications of the apparatus which were exhibited by Legat. That Reis received little encouragement so far as speech is concerned may be taken for granted, but there was no lack of recognition for that which he actually accomplished. Fourteen years later the real inventor of the telephone, as we have seen, had to hide his hopes and bide his time because he realised that neither the scientific nor the financial world would regard the expectation of the electrical transmission of speech as other than a dream. That Reis failed does not detract from the credit due to him for his efforts in a direction which no successor had the temerity to pursue until Bell, with a courage prompted by wider knowledge, independently attacked the problem, achieving results which were immediately recognised, and which eventually revolutionised the methods of communication throughout the world.

CHAPTER XIII

CALL BELLS

WHILE the consideration of Reis's work was felt to be most conveniently undertaken in the preceding chapter, we have to recognise a certain disadvantage in the interruption thereby effected in the record of practical work, and in the recital of the means adopted for carrying on an effective service. Returning, then, to the practical stage we may recall that before the production of battery transmitters or microphones, the telephone had been applied to a new service which called for the development of accessory apparatus of new types. With the introduction of these battery transmitters the new service became capable of great expansion. But before considering the developments in this direction some attention must be given to the more important of the accessories—call bells and switchboards—which contributed to make the new service possible. Only the earlier forms of call bells will be noted in this chapter and of switchboards in the next.

Transmitters and receivers were now in existence by whose means speech could be transmitted and received with clearness over long lines, but the ability to talk carried with it the need to send a calling signal over equal distances.

Bell's original circular had intimated that the telephone itself would serve for the call, since 'any person within ordinary hearing distance can hear the voice calling through the telephone'; but it also anticipated occasions when that still small voice might be insufficient, for it was added :—'If a louder call is required, one can be furnished for $5.' The apparatus by which the louder call was to be effected was that for which Thos. A. Watson submitted on December 5, 1877, an application for a patent, in which he says :—

In using a system of electric telephones it is necessary to provide some means for producing a sound at the distant telephone station loud enough to attract the attention of persons at a distance from the telephone.

My present invention provides one means for doing this by causing an intermittent current of electricity of high intensity to pass through the line wire and the distant telephone. For producing such current I make use of an ordinary induction coil combined with a galvanic battery and a rheotome, for rapidly interrupting the current.[1]

In this case the telephone was still used as the source of calling sound at the receiving station.

F. A. Gower combined a 'musical instrument' with a telephone,[2] which was introduced into Europe, and is thus described by Du Moncel :—

The instrument can itself give a very loud call by only breathing into it instead of speaking.

For this purpose a small oblong opening is made in the diaphragm at a half diameter from its centre, and behind this the reed of an harmonium is applied to a square copper plate fixed on the diaphragm itself. On using the bellows the expelled air passes through this little hole and, on reaching the reed, sets it into vibration, and produces a sound of which the acuteness depends on the condition of the vibrating plate.[3]

Gower's telephone had a flexible tube to speak into, and this was considered to be an advantage when using the vibrator call. In describing the Gower telephone before the Society of Telegraph Engineers (I.E.E.) in London, April 23, 1879, Mr. Scott said :—

When the user wishes to call attention at the other end of the line he does exactly as he would do with an ordinary speaking tube, with the use of which every one is familiar. He blows into the flexible tube, and the air pressing through the orifice in the diaphragm vibrates the reed, and then escapes by a vent made in the side of the box.[4]

Mr. Scott also referred to a method of using the telephone for making its own call patented by Mr. A. F. St. George of the India Rubber Company :—

Outside the coil of the telephone another coil of thick wire is wound, an intermittent current from a battery being passed through this outer primary coil, corresponding currents are induced in the inner coil, which as usual is permanently connected with the line, and cause vibrations of the diaphragm at the distant telephone audible

[1] U.S. specification, No. 199,007, January 8, 1878.

[2] U.S. specification, No. 217,278, July 8, 1879 (application filed October 24, 1878).

[3] *The Telephone, the Microphone, and the Phonograph,* Du Moncel, p. 359.

[4] *Journal of the Society of Telegraph Engineers,* viii. 335.

at a considerable distance. The vibrations are rendered intermittent by means of an armature vibrated by an electro-magnet.[1]

And to the method of

Mr. Siemens, of Berlin, [who] uses a reed temporarily held in the mouth-piece of the telephone, by which not only is the sound produced by the vibration used, but it is made to act on a small ball which produces a series of blows upon the diaphragm or disc, the result being an audible signal at the further end of the line. This invention requires the telephone when not in use to be kept in an upright position.[1]

The magneto call bell had made such progress that these methods of utilising the telephone itself were but very little used in the United States, Gower's for example was never commercially used there, but a method analogous to that of Siemens was suggested by Edison in an application filed March 4, 1878. He says:—

The invention consists of a stand for the receiving instrument and a swinging metal lever, the end of which comes into contact with the diaphragm, so that it is thrown from it violently when a strong wave or current passes over the line or through the magnet of the receiving instrument. This lever, in returning, strikes the diaphragm a blow, and produces a sharp penetrating sound like that of a Morse sounder, and this may be heard in all parts of a large room.

I have heretofore shown, as in Case No. 146, an induction coil in connection with a telephone. I arrange a switch between the local and main line circuits, in such a manner as to vary the electric tension on the line by moving such switch, and thereby operating the call at the distant station; and I prefer to employ a peculiarly constructed induction apparatus, in which there is a fine wire wound helically round a larger wire, and then the two are wound to form a helix. The larger wire is in the local circuit, and the induced current is set up in the finer helix.[2]

This probably never went into operation. The Edison instruments were fitted with a local call bell operated through a relay by batteries.

All these devices were merely tentative, and quite early the patent specifications indicate the striving for a more effective means of obtaining a call. The Edison interests held to the battery possibly by reason of the Western Union telegraphic groove. The same reason probably also accounts for Morse sounders being attached—instead of bells—to the earlier instruments manufactured. The Bell interests, on the contrary, directed their attention

[1] *Journal of the Society of Telegraph Engineers*, viii. 335.
[2] U.S. specification, No. 203,017, April 30, 1878.

to alternating current magneto machines and polarised bells, perhaps in the first instance in order to obtain a continuance of that freedom from the troubles of batteries which Bell's permanent magnet telephone permitted. So early as October 11, 1877, we find Watson applying for a patent in which he frankly abandons one feature of the circular issued in May by saying :—

In operating magneto-electric telephones such as are described in the said Letters Patent, difficulty has been experienced in calling the attention of the operator at a distant station and it has been found advisable and necessary to combine with such a telephone instruments specially adapted for the purposes of signalling.

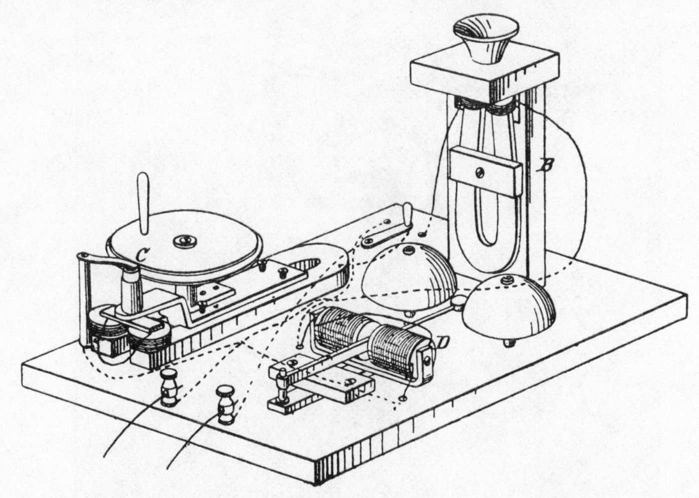

FIG. 39.—Watson's Magneto-electric-inductor Call.

The object of the present invention is to produce an audible signal at a distant station of sufficient loudness to attract the attention of the operator at a considerable distance from the instrument. To this end I combine with a telephonic circuit a magneto-electric inductor of ordinary or suitable construction. Electrical currents of high tension may thus be induced upon the line by the rotation of an armature, and these currents, on being passed through the coils of a distant telephone, produce a sound of considerable loudness. I prefer, however, to combine with the telephones at the receiving end of a circuit a bell or other contrivance specially adapted for calling attention.[1]

It will be seen from the illustration (fig. 39) that Mr. Watson was utilising the best known form of magneto electric machine, and

[1] U.S. specification, No. 202,495, April 16, 1878.

it may be considered of sufficient interest to reproduce an early illustration of such a type taken from 'The Illustrated Handbook of the Royal Panopticon of Science and Art; an Institution for Scientific Exhibitions, and for Promoting Discoveries in Arts and Manufactures, 1854.' The Royal Panopticon was situated in Leicester Square, London, and the building which it occupied is

FIG. 40.—'Clarke's Magnetic Electrical Machine.'

now known as the Alhambra Theatre. Fig. 40 represents 'Clarke's Magnetic Electrical Machine' there exhibited.

Another example of the same calling device may be seen in Roosevelt's application of May 29, 1878. Mr. Roosevelt was evidently no believer in economic waste, for 'the object of [his] invention is to make a combination telephone, which shall consist of a telephone and a magneto machine, and to use a single permanent magnet both for the telephone and the magneto machine.' [1]

[1] U.S. specification, No. 218,775, August 19, 1879.

This device was never applied to commercial uses. It was otherwise with the push-button type of magneto of which Anders' U.S. patent, No. 228,586, June 8, 1880 (applied for July 21, 1879), may be taken as an example. The patentee says :—

The second part of the invention consists in the combination, with the telephonic apparatus at each station, of a magneto induction apparatus operated by the depression of a push knob for generating the current which operates the annunciator or signalling apparatus at the central office.

The illustration (fig. 41) shows the complete outfit and indicates that its introduction followed the Blake transmitter, which is enclosed in the same case as the magneto inductor. Though put into practical use it was not reliable, and its manufacture was discontinued after one or two years.

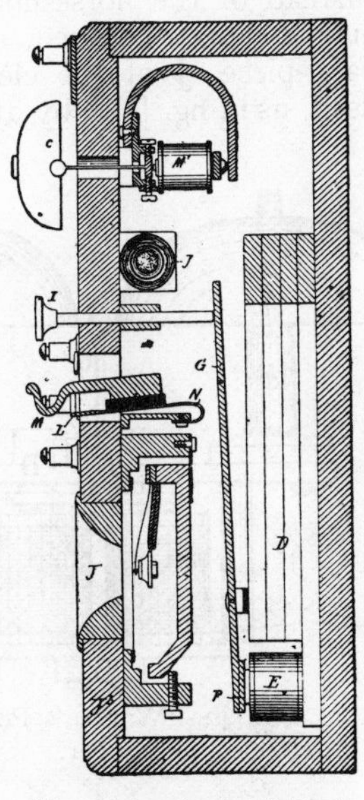

FIG. 41.—Anders' Push-button Magneto.

Another form of magneto made by the Western Electric Manufacturing Co. is illustrated in Prescott's 'Speaking Telephone' of 1878, p. 24 and following pages, with diagrams of circuits. The generator was of the Clarke form with revolving bobbins.

After a very short experience of such types as these the superiority of the Siemens armature pattern was recognised and all the makers adopted it.

Watson, who had done so much in assisting Bell with his experiments leading to the invention of the telephone, and who devoted his attention to developing the calling apparatus by adopting magneto generators as above mentioned, has also left his mark on the art in connection with the 'ringer' of the magneto-bell. On August 1, 1878, he applied for a United States patent on an 'Improvement in polarised armatures for electric bells.' The improvement relates to 'that class of electric bell strikers having their armatures and electro-magnets polarised by proximity to, or contact with, the poles of a permanent magnet, and which are operated by alternately reversed currents of electricity.' In describing his apparatus he says :—

Figure [42] represents a front, and fig. [43] a side, elevation of my improved bell ; and fig. [44], a modification in which a single bell is used.

A is one pole of the permanent magnet, and B the other. C is the armature, pivoted at the point *d* of the metal piece D, so that when one end is against the electro-magnet the other is away from it.

E E are the supports for the bells F F. These are attached to the metal piece D by screws passing through slotted holes, and the bells can thereby be adjusted in their relation to the hammer or striker G.

H H is an ordinary electro-magnet fastened to the pole B. Instead of the horseshoe magnet A B, one or more straight-bar magnets may be used, extending from the metal piece D to the back-piece J of the electro-magnet. A single bell may also be used, as in fig. [44], by attaching a hammer or striker, G G, to each

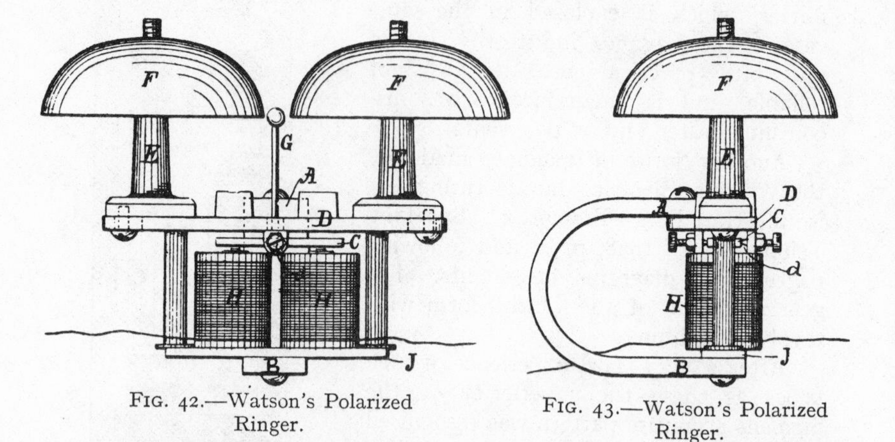

FIG. 42.—Watson's Polarized Ringer.

FIG. 43.—Watson's Polarized Ringer.

end of the armature C, and fastening the bell-support to the metal piece D directly over the centre of the armature. The hammers can strike either on the inside or the outside of the bells.

When a current passes in one direction through the coils of the electro-magnet, the armature is attracted at one end and repelled at the other. When a current passes in the opposite direction this action is reversed.[1]

This centrally pivoted polarised armature bell of Watson's at once went into general use, has survived all changes in other apparatus, and is still almost exclusively employed.

The plan of using one gong suggested in fig. 44 has not been practically used, probably because it lacks the provision for adjustment, which is so simple and so effective with the two-gong type.

At the first meeting of the National Telephone Exchange Association, representative of the Telephone Companies of the

[1] U.S. specification, No. 210,886, December 17, 1878.

United States, in September 1880 a committee was appointed who examined, as carefully as the time afforded them would permit, the different appliances in the way of call bells for telephonic use, and begged leave to report as follows :—

The bells presented are from the factories of Charles Williams, Jr., of Boston ; Post & Co., of Cincinnati ; Davis & Watts, of Baltimore ; The Gilliland Manufacturing Co., of Indianapolis ; and the Electric Merchandising Co., of Chicago.

In general it may be said that the bells are excellent, the enterprise and progressiveness of some of the manufacturers are most remarkable. We think the device of the Gilliland Company, which dispenses with the press button on the ringing of the bell, deserves attention. Also, the various improvements of Post & Co., Charles Williams, Jr., and others.

We earnestly recommend to the manufacturers the adoption of the interchangeability of parts.[1]

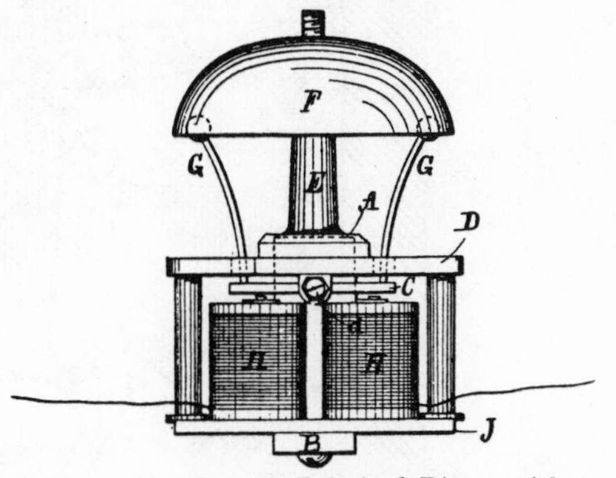

FIG. 44.—Watson's Polarized Ringer with single gong.

The design of these various magneto-electric call-bells (a name subsequently shortened by general usage to 'Magneto' simply) varied with each maker. The Magneto contained not only the generator and receiving bell (termed the 'ringer'), but also the switches, cut outs, and lightning arresters; everything, in short, except the talking instruments and the battery. The driving gear was one of the important differences in each make. In the Post bell the pinion wheel attached to the armature was driven by means of a rubber band, in the Gilliland friction wheels were used, and in the Williams friction wheels also, but of another type.

'The device of the Gilliland Company, which dispenses with the press button on the ringing of the bell,' doubtless referred to the automatic cut out. Since the generator coil offered considerable resistance a short circuiting shunt was provided and was normally completed, but this shunt required to be broken when a ringing current was to be sent out on the line. In the first instruments this was done by pressing a button with one hand whilst turning the handle with the other. Automatic devices of varying kinds were used

[1] *National Telephone Exchange Association Report*, 1880, p. 186.

to overcome this difficulty. In the Gilliland a lever with a spring attachment was connected to the handle. The revolution of the latter removed the lever from a contact. In the Williams the shaft of the generator was normally in contact with the short circuit. A V-shaped slot in a collar on the shaft altered the position of the latter on its revolution and thus broke the short circuit, and in the Post magneto a weighted spring was thrown off at a tangent when the armature was revolved. These were different methods for performing the same result and all tending towards simplifying the work the subscriber was called upon to do.

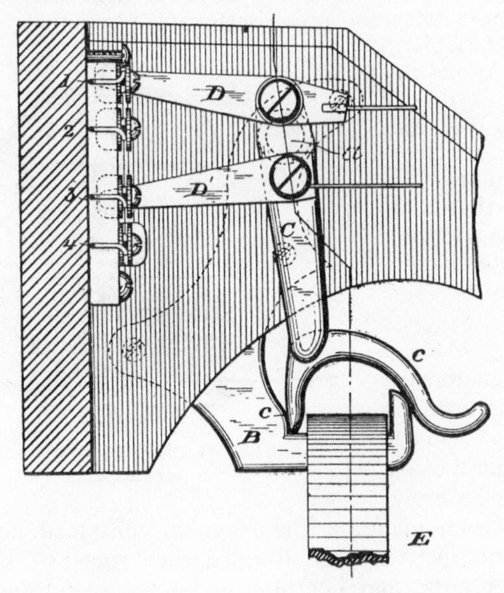

FIG. 45.—Phelps' Automatic Switch for pony-crown receiver.

In the same direction was the utilisation of the receiver to vary the position of a hook, and thus actuate springs which changed the circuit so as to cut out the ringing apparatus and cut in the talking apparatus, or *vice versa*. To effect this change by the weight of the telephone was the subject of a patent to H. L. Roosevelt of New York for a device termed 'the gravity switch.' The existence of this patent and the royalty claimed under it led to the production of an alternative automatic switch adapted to the 'pony crown' receiver. In the United States 'pony' is a prefix denoting diminutive. The crown receiver was designed with the idea that a number of magnets would add to the power of the telephone. The magnets were circular, and the instrument placed diaphragm downwards had very much the appearance of a crown.

It was soon found that any superiority which this instrument might be considered to possess did not arise from its numerous magnets. The size was reduced and only one magnet remained. The switch was so arranged that the act of removing the instrument from or placing it upon an appropriate bracket changed sliding contacts and effected the same purpose as the gravity switch. Figs. 45 and 46 show this switch as illustrated in the patent granted to G. M. Phelps.[1]

The magneto was so important an adjunct of the telephone that much attention was given to its development, but experience was needed fully to impress manufacturers and telephone companies

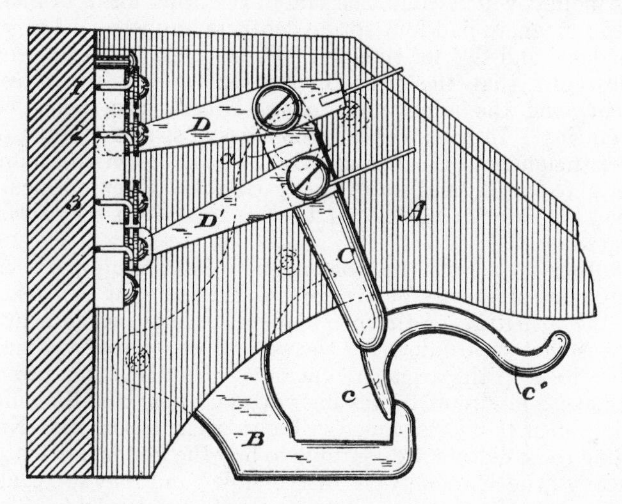

FIG. 46.—Phelps' Automatic Switch for pony-crown receiver.

with the need of sound design and reliable workmanship. The results attained during the first few years of the existence of exchange service were not satisfactory, as may be gathered from a paper by Mr. Sunny presented to the Switchboard Conference in 1887. The paper referred to the field for improvement in subscribers' apparatus generally,[2] but some of Mr. Sunny's remarks on the magneto bell may be more conveniently given in abstract here.

The magneto as constructed to-day is a cheap looking affair, except the new Gilliland, and they are all more or less unreliable, while after ten years' experience we ought to have an instrument that would look in keeping with the furnishings of the finest residence or office, and that would be free from electrical defects.

[1] U.S. specification, No. 222,201, December 2, 1879 (application filed September 10, 1879).

[2] Chap. xxiv. p. 328.

Chicago's experience with magnetos has been unsatisfactory, and this is no doubt true of other exchanges.

He records the various types of magnetos used, and continues—

We changed to the Gilliland iron magneto, which by this time was a year old, and had had some of the original defects in construction eradicated. I think that the thanks of every telephone man are due to Mr. Gilliland for the great effort he made to give us a magneto that would be perfect in every respect; and while electrically it has fallen short of our expectations, its appearance impresses the subscriber with the idea that he is getting something for his money, which cannot be said of the other forms of magnetos. The iron magneto had four spring contacts of phosphor-bronze, two for the local and two for the line. These contacts became so caked up with dirt that the trips of the repairmen were exceedingly frequent, and the annoyance to subscribers from being cut off, exasperating. In 1886 we had 2100 cases of switch trouble, chiefly in these magnetos. To remedy this, we cut down the number of contacts to three, and substituted German silver springs. This lessened the number of cases of trouble to 1600 in 1887 (December estimated).

Another fault with the iron magneto is the automatic cut out. The number of these troubles for the current year will reach 500. With the armature of the magneto in circuit we find in many instances that we cannot ring the subscriber, so that we make it a practice to keep the armature cut out. We detect the presence of an armature in circuit by testing with an apparatus that indicates the line open if its resistance is higher than 400 ohms. Notwithstanding these defects, we continue to buy the iron magnetos. They are heavy; the machinery is inaccessible; and they are constantly working loose from the wall because of their weight. The subscribers like them, however, and when we send a man to a new subscribers with a repaired Davis & Watts in a new box with shining bells, and all in apple-pie order, he comes back with the load on his back, and says that that man wants a machine like his neighbour's, which is a new Gilliland, or none at all. We rarely have to change a new Gilliland magneto for a new one, while the D. & W. are being crowded out of the service by the people who will not renew their contracts until they get a new Gilliland.

These defects in the magneto are not peculiar to Chicago. It is impossible that we should be forced to take out magnetos in Chicago because of defects, and that the same make of instrument should be a complete success five hundred miles away. Yet it is a fact that magnetos that are undeniably defective are furnished in one exchange long after the defects have been proven by another exchange. We have three makes of instruments to select from. The Gilliland iron magneto, rich in construction but poor in operation; the Chicago magneto, and the Post. The last two gotten up with a view to the strictest economy. It would be better, it would seem, if there was but one magneto, since we cannot have more than three, and that that one be a carefully constructed, well tested instrument electrically, and in appearance something that will impress the subscriber as favourably as the iron magneto.

One result of the discussion arising out of Mr. Sunny's paper was the use on magnetos of hinges having a spiral wire whose ends were soldered to the respective sides of the hinge, a device which materially reduced magneto 'troubles' by preserving the continuity of the conductor.

The development of the magneto call bell was carefully and constantly continued. The educative influence both to users and manufacturers was important. The 'interchangeability of parts' recommended by the Association became a necessity of its manufacture, and the application of labour saving devices and improvements in factory methods which has since become so pronounced a feature in the production of telephonic apparatus generally received its first impetus in the making of magneto bells. Much ingenuity was applied to their design, and the manufacture attained a very high standard in later years;[1] but at the commencement of the industry, which is the period now being dealt with, there were necessarily crudities both in design and manufacture. Despite these drawbacks a practical instrument was available for the use of the pioneers in Telephone Exchange service. And they were prompt to avail themselves of all the facilities at hand.

[1] Chap. xxix. p. 454.

CHAPTER XIV

THE TELEPHONE SWITCHBOARD

AT the time that the telephone was invented the telegraph had assumed such proportions that offices existed into which numerous lines entered. Some of the circuits were operated at that office; others went through. The changing of the circuits was accomplished by means of switches, and the switches were mounted on a board. Hence the name which has been retained to describe the most extensive and elaborate apparatus in electrical engineering.

Although the switchboards used in telegraph offices exercised an influence on the development of the telephone switchboard, the latter served a purpose not previously effected, and the influence of the telegraph board was limited to a detail of construction. Speaking generally and subject to the exceptions related in Chapter IX, a telegraph switchboard was not operated at the desire of an employee at a distant station to be connected with some other distant station, but by an employee at the office where the switchboard was placed, and usually for some purpose connected with the instruments therein.

A telephone switchboard, on the contrary, is for the purpose of interconnecting lines at the desire of subscribers at the extremities of those lines. And this purpose, though not entirely new, was practically new to all concerned in the telephone business. Telegraphic exchanges were so little used as to be generally unknown, and the switchboards used in connection with them were not described in the literature of the art. Dumont, who patented a telegraphic exchange system[1] in 1851, describes in his specification[2] a switchboard for effecting the desired results. But Dumont and his proposals were unknown, and telephone switchboards were developed on new lines to meet new needs.

Dumont's switchboard was submitted in later years to critical analysis, and it was claimed that as a practical device it was inopera-

[1] Chap. ix. p. 77.

[2] British specification, No. 13,497, 1851.

tive because no provision was made for reversing the polarity of the battery, and also having regard to the type of telegraphic instrument with which it was designed to be used. But any minor difficulty of this sort would certainly have been overcome on being put into operation. The intention to provide intercommunication between all the lines is clear enough, and as the earliest known example of a switchboard designed for the purpose of such intercommunication

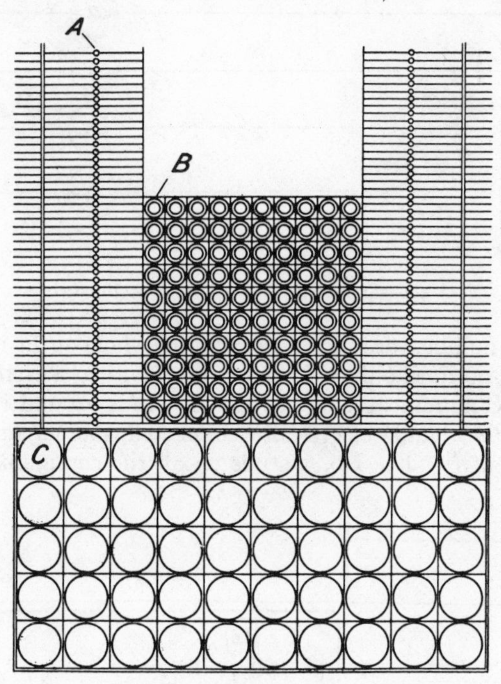

FIG. 47.—Dumont's Telegraph Switchboard. Simplified diagram from patent of 1851.

at the request of subscribers, the operation of the mechanism proposed must be briefly described.

The drawing of the switchboard forming part of the specification is partly diagrammatic, but somewhat too large for identical reproduction. A simplified representation on a reduced scale is given in fig. 47.

The parts A, B, and C in fig. 47 are reproduced in figs. 48, 49, and 50 respectively from the patent drawing sheet 2.

The specification has probably suffered somewhat in clearness in the course of its translation from the French to the English language, but it was apparently intended that each line entering the central office should pass through a two-point switch A on its way to a dial telegraph instrument B and earth. A call for intercommunication having been received, the switch A was moved so as to cut out the instrument B and earth and put the line in connection with the communicators C. Fifty communicators are illustrated, and

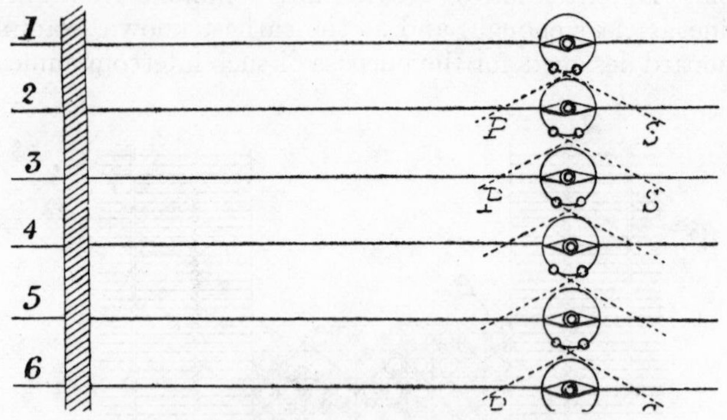

FIG. 48.—Dumont's Telegraph Switchboard (A).

although 200 numbers are shown on each, the whole outfit is intended to illustrate 100 lines only. Limiting ourselves to this 100 we will regard each communicator as having 100 studs around its circumference and two hands electrically connected, moving on the same axis at its centre. The No. 1 studs on all the communicators are

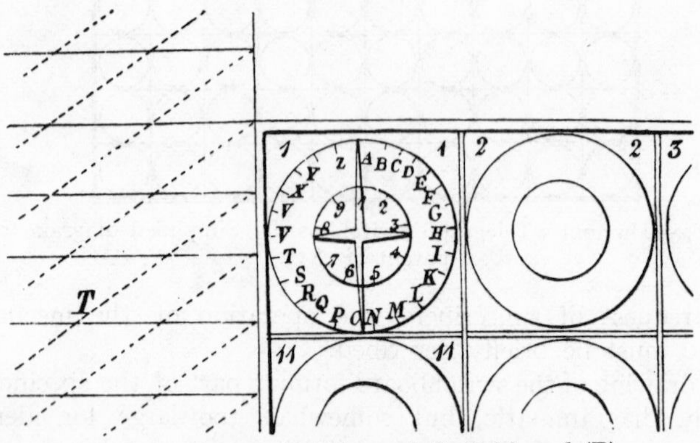

FIG. 49.—Dumont's Telegraph Switchboard (B).

connected together, and so with all the other numbers. It follows that at any one of the fifty communicators any two subscribers' lines may be connected together by placing one hand on one stud representing the calling line and the other hand upon another stud representing the called line, the switch A of the latter having been operated so as to cut out the instrument B attached thereto. The provision of fifty communicators for 100 lines enabled all the subscribers to be connected at any one time, which we know now to be in excess of requirements. Had Dumont supplied ten communicators instead of fifty, he would have made provision equal to the ten pairs of cords per 100 subscribers generally allowed in early telephone switchboards.

But this suggestion for a telegraphic exchange unhappily came too soon to be applied to public use or reward the patentee.

The ' dial ' form of switch has frequently been used for diverting a line to one of several other lines as required. Perhaps the more general form is that with one hand or arm which, pivoted at one

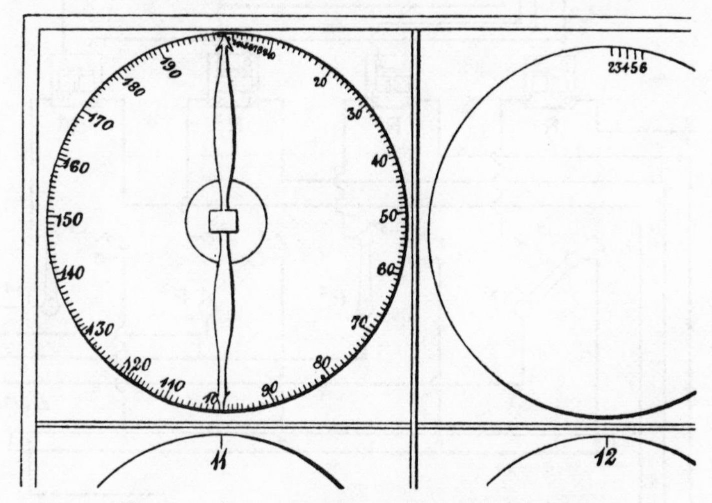

FIG. 50.—Dumont's Telegraph Switchboard (C).

end, makes contact at the other end with a stud, the line being connected with the arm. The arm revolving like the hand of a clock can make contact with any stud arranged at equal distances from the pivot centre like the figures on a clock face. To permit interchange amongst a number of lines, such dials may be multiplied according to the number of lines and the studs of each of the numerous dials connected together at the back of the board. Continuing the analogy of the clock face and assuming the studs to equal the hours in number, then all the 1 o'clocks would be connected together, and all the other studs respectively, up to twelve. Line A having its switch arm on, say, No. 3 of its dial, and line B having its arm on stud No. 3 of its own dial, then the circuit would be made up of line A through the switchboard and out at line B. Mr. Lockwood states in ' Practical Information for Telephonists ' (p. 79) that a switchboard of the dial type was put in operation by Mr. Murray Fairchild, in connection with the telegraph office at New Haven, Connecticut ' away back in the fifties.'

The first telephone switchboard actually installed for regular business communication was situated in Chapel Street, New Haven,

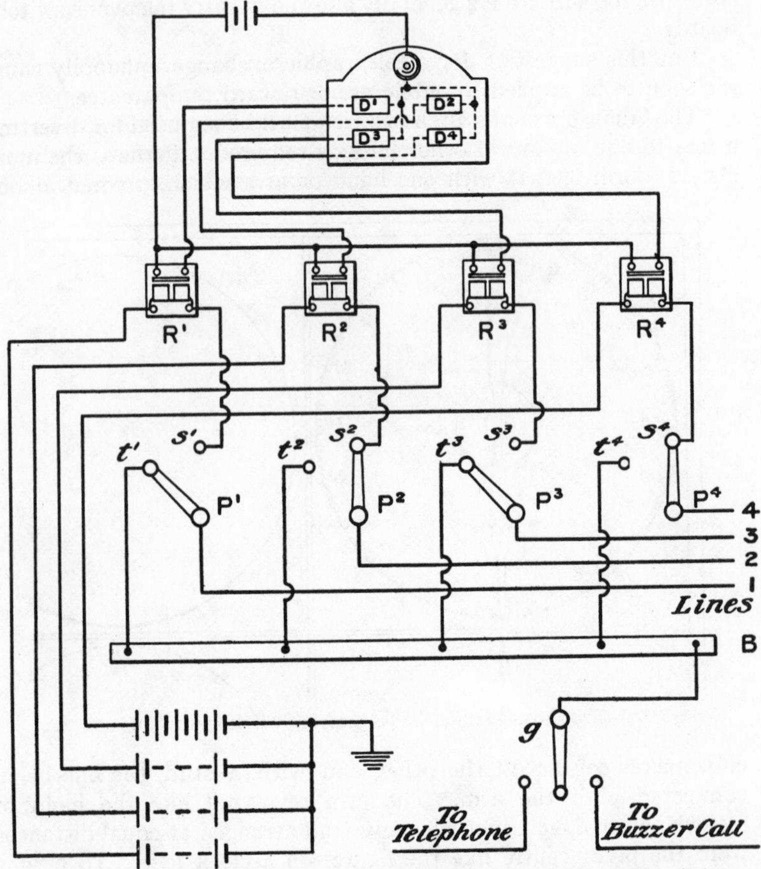

FIG. 51.—Switchboard used at New Haven, January 1878 (circuits).

in January 1878. The apparatus at that office is illustrated in the diagram,[1] fig. 51.

Each line was connected to a lever pivoted at P, resting normally on stud *s*, thence through a relay, R, and battery to ground. In the local circuit of the relay was an indicator drop D for each line, the bell of the indicator being common to all the drops. To send a call the subscriber opened the circuit, causing the relay to fall back and close the indicator circuit. On the fall of the drop the

[1] From drawing in possession of Mr. T. D. Lockwood.

FIG. 52.—Switchboard used at Meriden, Connecticut, February 1878 (front).

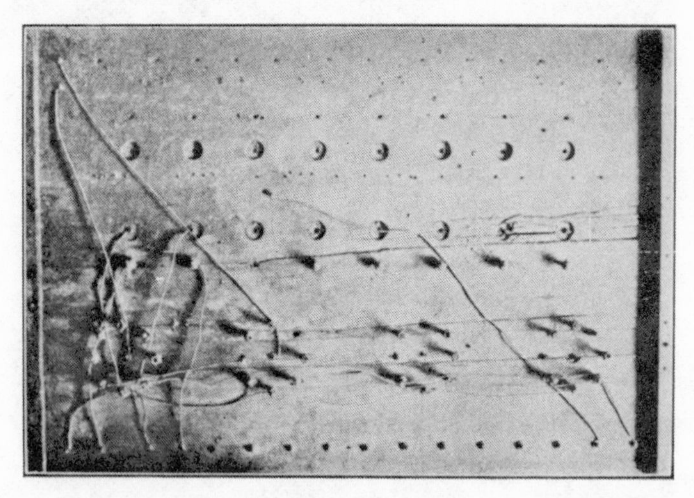

FIG. 53.—Switchboard used at Meriden, Connecticut, February 1878 (back).

central office operator moved the lever of the calling line from its stud *s* to stud *t*, thus cutting out the relay and indicator and connecting the line through the stud *t* with the bar B. At the same time he turned the lever *g* to its left-hand stud, thus connecting the line with the telephone and receiving his orders. The called line would then have its lever diverted to its stud *t*, and through the common bar B the two lines would be connected. To send a signal out to line, the lever *g* would be turned to the right, and would then be in connection with a buzzer circuit. This produced a loud buzzing sound in the subscriber's telephone.

The diagram shows lines 1 and 3 connected.

Since there is no provision for indicating the close of conversation, it must be presumed that this was effected by the operator listening in at intervals as described in the later model.

This switchboard was thus a combination of relays, indicators, and two way lever switches. It was only in use about two months, being then replaced by one of the dial type, which those in telegraphic circles in the locality were probably familiar with through Mr. Fairchild's work. It would seem, however, that the dial pattern of board must have been used in the first instance at Meriden, Connecticut.

The front and back of the Meriden board are illustrated in figs. 52 and 53 respectively. Fig. 54 is a sketch of the connections which was made in after years by Ellis B. Baker, who installed the board, and who states that it was put into use on February 1, 1878.

The dials on this Meriden board, it will be seen, are of the one arm pattern, and two dials were consequently needed to be operated to make a connection. Simultaneous connection could thus be given to four out of the eight subscribers on the board.

The operation of the board is thus described by Mr. Baker :—

The circuit started from the earth, passing through a line battery, pair of annunciator coils, thence to binding post marked 1 Out, thence to disk X^1 through switch lever S^1 to binding post 1 In, with a leg connecting with No. 1 disc of each of the circles. From binding post 1 In, the circuit passed to the subscriber's premises, where was located an ordinary closed circuit push button. From this push button the circuit passed to the ground or to the next subscriber on the line. Circuits 2, 3, and so forth were connected in a like manner to their respective posts, discs, switches, etc. When a subscriber on circuit 1 desired to obtain the attention of the operator, the circuit was broken by pressing the push button, releasing the annunciator shutter showing the number of the circuit. The operator threw the lever S^1 to the left against a stop, and the lever of circle *a*, which normally rested on disc 8 connected with the earth, to disk 1, thus completing a circuit through disc 1, lever A, telephone H, lever B, and disc 8 to earth. Upon receiving

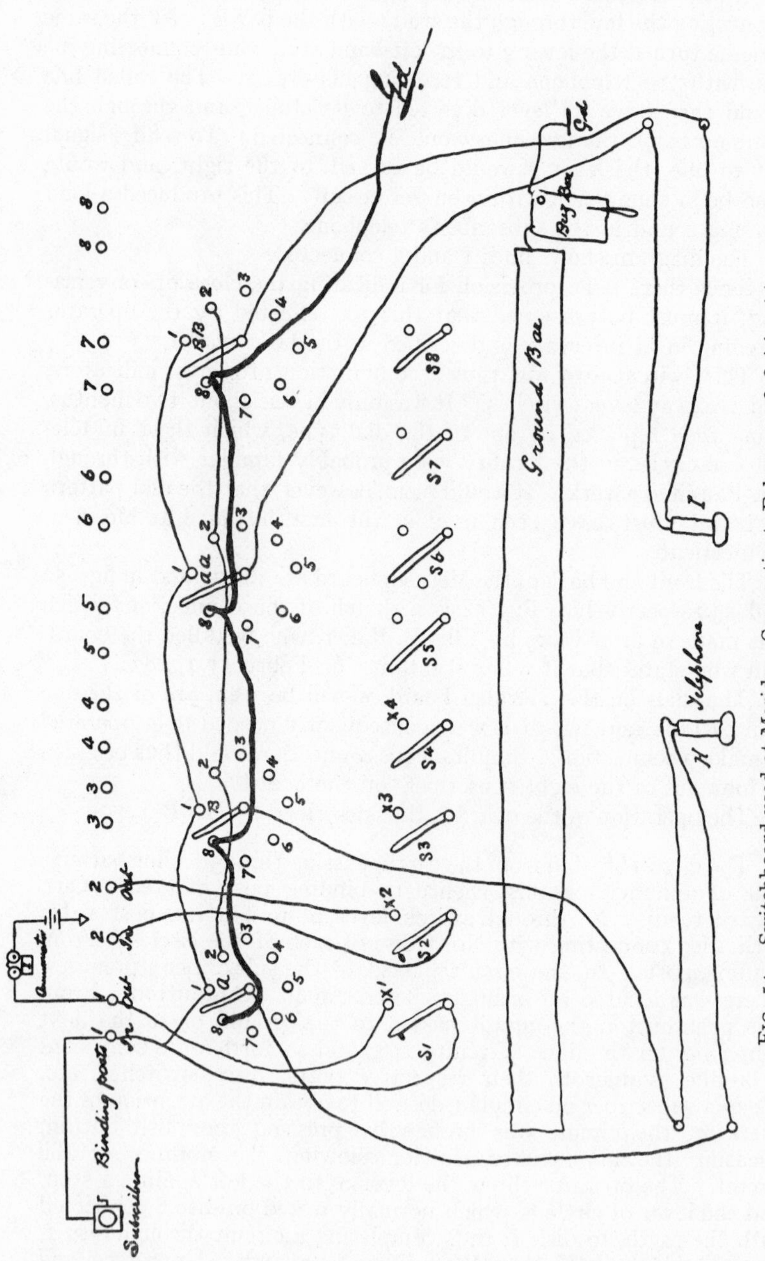

FIG. 54.—Switchboard used at Meriden, Connecticut, February 1878 (circuit).

the subscriber's call, supposing the subscriber desired to be connected with circuit 2, switch lever S^2 is thrown to the right in connection with ground or buzz-bar, and the subscriber's number sounded by vibrating the handle of the buzz-box, which was connected between the buzz-bar and the ground. After the number had been repeated two or three times switch-lever B was thrown from disc 8 or the ground to disc 2, thus connecting the two circuits together through telephone H, and if the parties were in conversation switch-lever S^2 was thrown from the buzz-bar back past the disc X^2 against the stop. This completed the connection of two subscribers or two circuits, and the same connection could be made on either pair of discs. There being no way the subscriber could signal when the line battery was thrown off, it was necessary for the operator to frequently listen in to ascertain when to disconnect the circuit.[1]

The Bell exchange in Chicago was started early in 1878, and no description is available of the switchboard first used, but about August of that year another switchboard was substituted. The following description of this switchboard is compiled and the illustration fig. 55 reproduced from evidence and sketches of employees.[2]

On a wall were placed indicators. Below and in front of them was a construction similar to the console of a three-manual organ. Upon each bank of this were placed spring jacks, not like those which have become so familiar, but simply long springs resting upon a contact. Above the console and inclined at an angle was a rack composed of metallic bars.

The line came first to the spring of the jack, through the anvil contact to the indicator and ground.

The metal rods were grouped in pairs, and between the rods of each pair was connected a clearing-out indicator.

To connect two subscribers the operator inserted in the spring jack a wedge-shaped plug with a metallic top and an insulated base, which disconnected the indicator and ground. To the flat plug was attached a cord, and at the other end of this cord a gripping plug, which has been likened to a clothes-pin. The clothes-pin plug gripped one of the rods. Another cord similarly equipped had one end grip the corresponding rod and the other connect with the spring jack of the second subscriber.

This switchboard is notable as being the first provided with clearing-out drops. A patent was applied for by Horace H. Eldred in the United States on June 9, 1880, and granted August 18, 1884

[1] Citizen's Case, Circuit Court of the United States, Western District of Michigan, 1896, *Defendants Record*, p. 213. It will be observed that the order of the respective parts is the reverse of the photographs of the board.

[2] *Ibid.*, *Complainants Record and Briefs*, pp. 177, 201, 211.

(No. 303,714). The first claim of this patent protects the combination of telephone lines, spring jacks 'for the insertion of con-

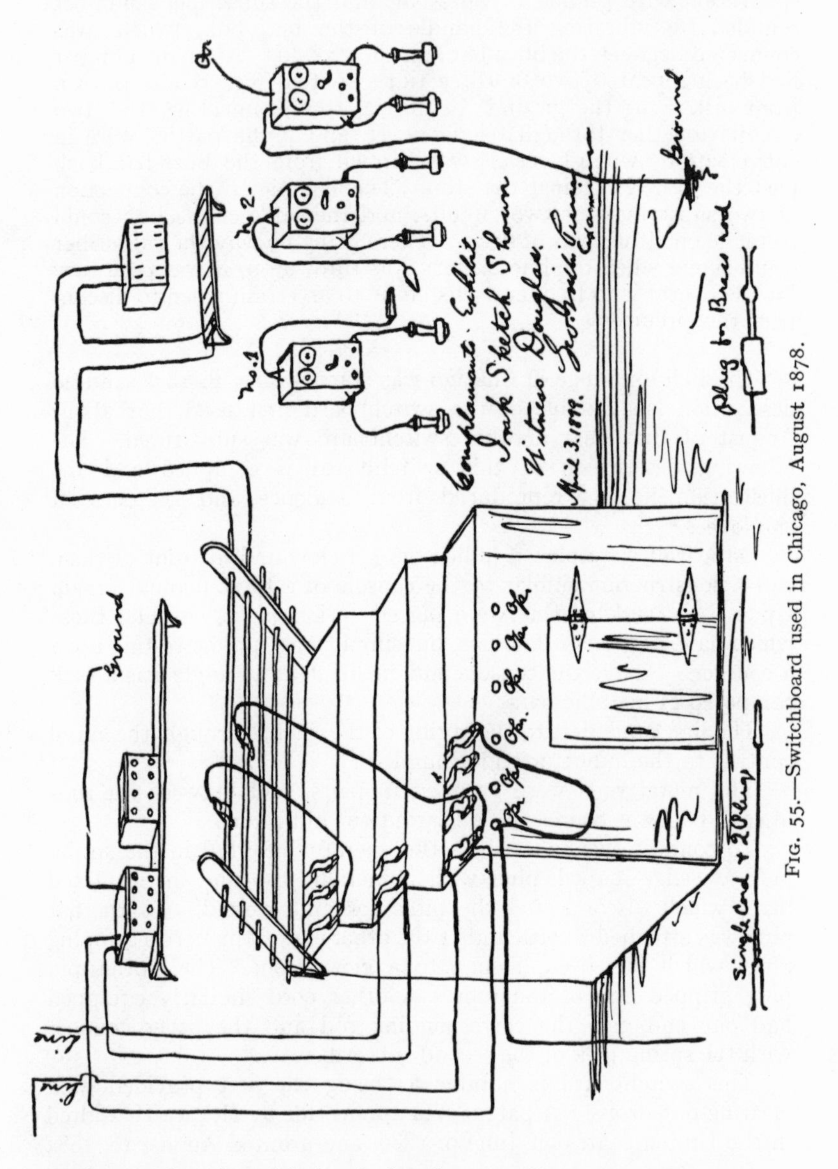

FIG. 55.—Switchboard used in Chicago, August 1878.

necting wedges,' and 'visual signals or calling annunciators . . . placed in each line at a point between its spring jack and earth, whereby the annunciators are cut out while talking.'

The second claim covers ' an auxiliary or supplemental signalling apparatus included in each of said connecting conductors, whereby the sub-stations so connected, or either of them, may notify the attendant at the central station to disconnect the said lines.'

The switchboard at New Haven, the earliest in construction and operation, was without any cords, whilst that at Chicago utilised flexible conductors between the spring jacks and connecting racks.

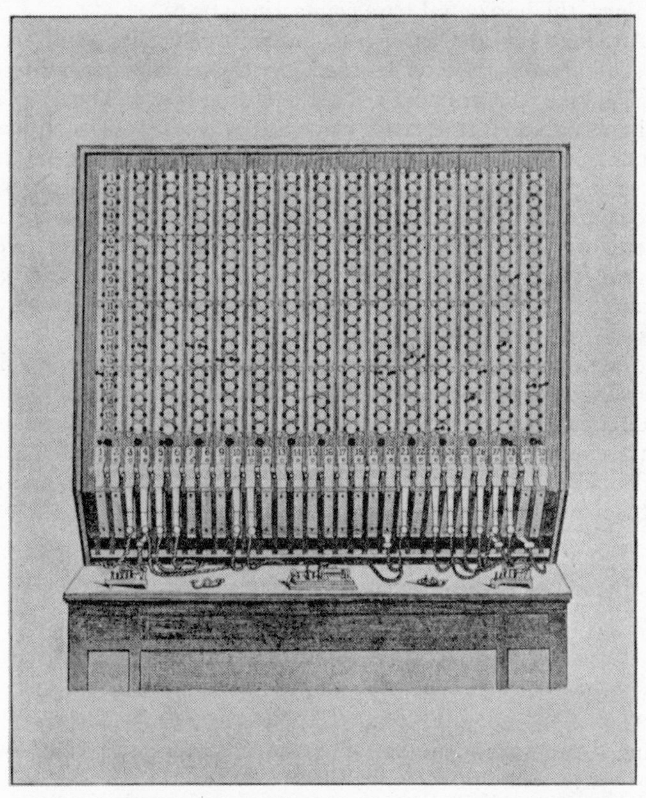

FIG. 56.—Western Union Peg Switch for Telegraph Offices.

Flexible cords were not in favour with telegraph engineers, and were the subject of much criticism in later years amongst telephone engineers. But cords could be avoided without the use of dials, which were in fact but little used for telegraph switchboards. The form generally used at the time the telephone was invented was probably that known in the United States as the Western Union peg switch (fig. 56) and in Europe as the Swiss Commutator or Universal Switch (not to be confounded with the telephone switch which was called ' Universal ' and will later be described).

Though differing in form, the same principle of operation was employed in these switches. A series of metal bars crossed the board horizontally; above them, but not in contact, was placed another series of bars arranged vertically. A horizontal bar could be placed in contact with any vertical bar by means of a metal plug. Another plug could connect the same horizontal bar with some other vertical bar. The vertical bars were continuations of the lines, the horizontal bars connecting-straps.

Although the Bell licensee manufacturers (Chap. xv. p. 184) in the switchboards with which their names are more generally identified followed the principles of this telegraph switchboard, Charles Williams, at any rate, had made other types. As early as the autumn of 1878 he supplied cord boards with annunciators and flat spring-jacks, passing connections to operators' tables where calls were controlled. But the types which came on the market in the general way as standard manufactures were thŏse having vertical bars to which the lines were connected, and horizontal bars or 'straps' for the purpose of connecting together any two vertical bars.

The examples given below are taken from circulars or catalogues issued about 1882 or 1883, and thus do not represent the first productions, but are commercial forms made after three or four years' experience.

The Williams switchboard is thus described in a circular issued by the manufacturers, the illustrations being reproduced in fascimile.

THE WILLIAMS
SPRING CENTRAL OFFICE SWITCH
Manufactured by
CHARLES WILLIAMS, JR.
109 to 115 COURT STREET, BOSTON.

The engraving, fig. [57], shows the present form of switchboard, giving a general idea of its appearance in perspective—showing a hand telephone and Blake transmitter in position.

It is composed of the upright board A and the inclined board E; upon both are arranged the connecting strips B in series of four, and designated as A, B, C, D, E, F, G, H, I, and J. Arranged between the two boards A and E are the annunciators D.

Fig. [58] is a sectional view of the switchboard, arranged to show the connections.

In this figure four connecting strips are shown on the board A and four upon the board E.

Upon the back and under sides of the board are placed the line strips. These are composed of metal springs arranged so as to normally press their free ends together as shown, and so as to form a continuous connection. The spring-jack R and wedge W shown on the front edge of board are used to loop in an operator's telephone and transmitter, and also for signalling a subscriber,

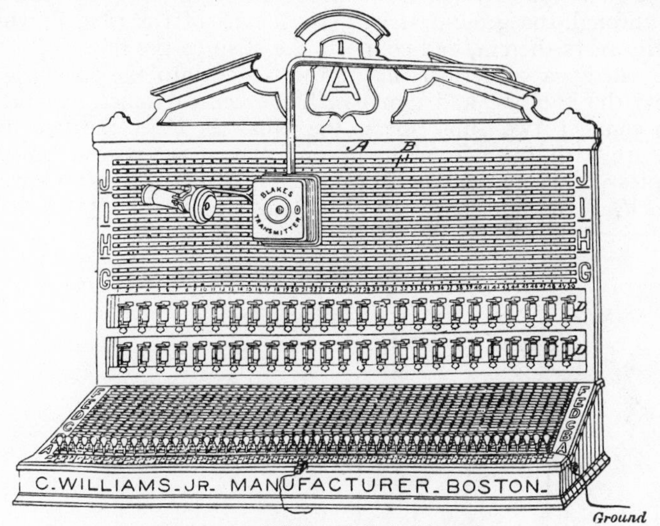

FIG. 57.—Williams' Switchboard.

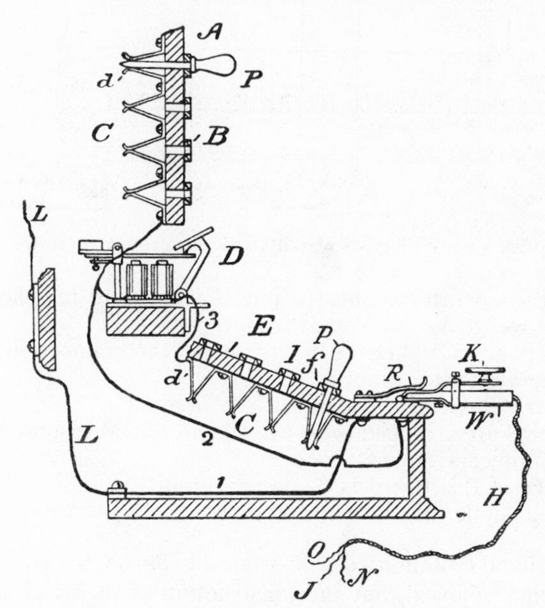

FIG. 58.—Williams' Switchboard (section).

the ends of flexible cord, O, N, being connected to the telephone and J to a magneto generator.

The line circuit enters at L, connecting with spring-jack R, through the shoe to line strip C, on the board A, a branch running through an annunciator D to the line strips on the table E, reaching earth through the ground strip *f*, by means of the plug P, which normally rests therein, and connects the strip to the line.

To call up a subscriber, insert the wedge into the spring-jack. Depress the knob K, and signal with the central office generator.

To connect two subscribers, the plugs P, P of the two lines which are withdrawn from the ground plate *f* are passed through the holes of a common connecting strip B, pressing between the springs *d*, of the line strip C, and making electrical contacts therewith.

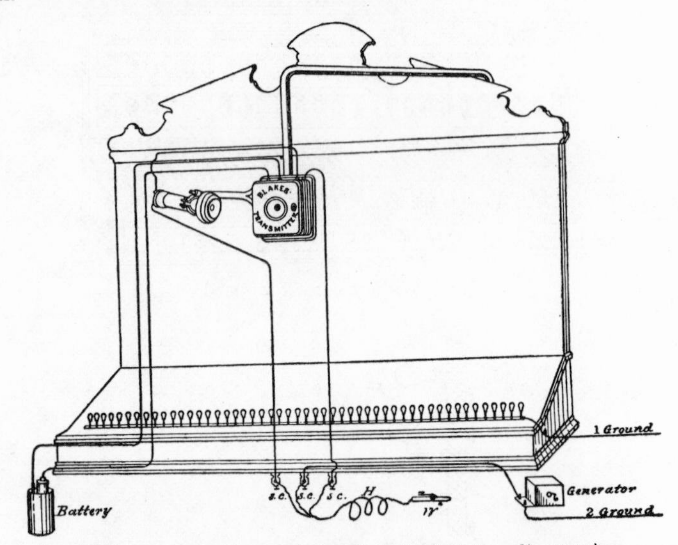

FIG. 59.—Williams' Switchboard (skeleton diagram).

When the operator wishes to put the listening telephone into circuit, the wedge W is inserted in spring-jack R.

Fig. [59] shows a skeleton board, with the connection to telephone, generator, and ground.

The figure illustrates a board of fifty lines.

The boards are arranged to be placed end to end to accommodate any central office system.

The price of this board is $3.00 per circuit, including lightning arrester, telephone and transmitter holder and patent wedge.

The Gilliland switchboard is illustrated in fig. 60, the same general features being followed, but the construction of the strips and plugs differed from other makes, being very simple, correspondingly cheap, and lacking strength and durability.

Post & Co. of Cincinnati (who later became the Standard

Electrical Works) manufactured a board of which fig. 61 is an illustration.

From the catalogue is taken the following description of the

Manner of Setting Up and Operating.

The line of posts on top of table are to connect subscribers. The connections for engine, telephone, microphone and night service

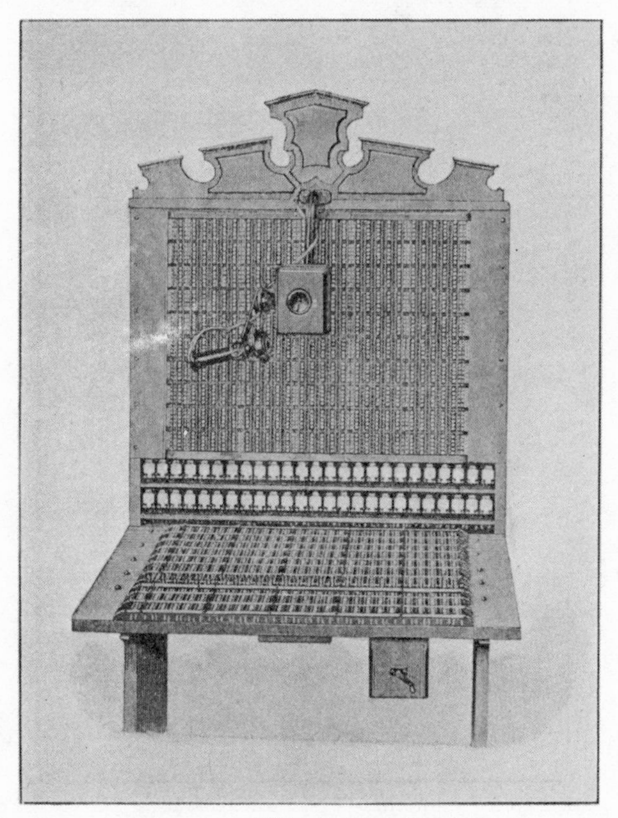

Fig. 60.—Gilliland Switchboard.

bell are all properly Tagged. The night bell, No. 6, can be thrown out of circuit at any time desired. *The operation* of the table is as follows :—A call comes in and drop No. 4 will fall, showing party desiring to talk. You answer by taking out plug No. 1 on his line and insert in second strip on board No. 2. Then move treadle, which answers subscriber by a ring. After answering, push the large button No. 7, on the lower right hand corner, when you are thrown in circuit with microphone and telephone. After answering and asking what is wanted, take the same plug No. 1, that you first used, and insert in any of the ten strips No. 3 which is not in use. Call up the party that is called for in the same manner above described. When answered by a ring, take the plug No. 1 on his line and place it in the same strip that you inserted the first plug in, which throws them in circuit. If at any time you desire to listen to subscribers conversing, push on small button No. 8, placed at right hand side of connection strips.

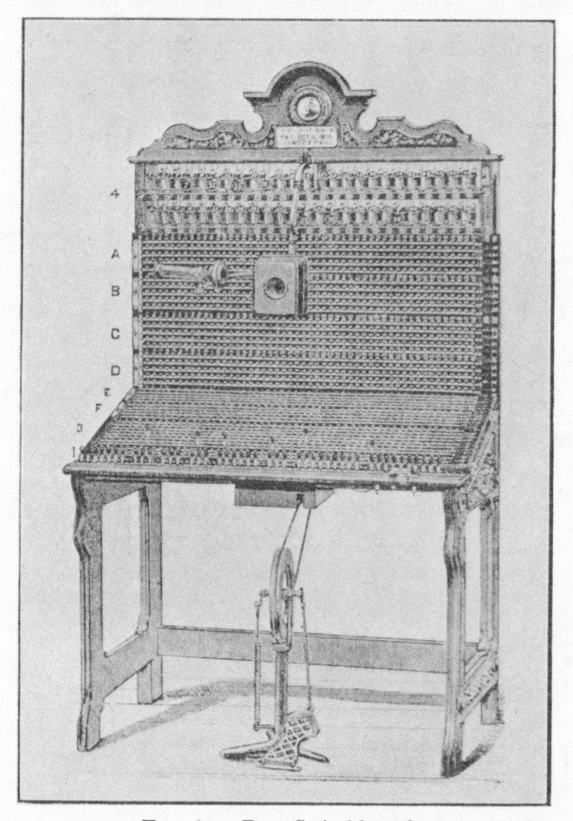

Fig. 61.—Post Switchboard.

Trunk Lines, A, B, C, D, E, and F.—Tables having trunk lines are operated in the same manner as described, with this *exception*: That when parties are wanted on any other tables, connect in strips marked A to F on any table desired. The 50 line table has connections for 6 additional tables, 5 lines each. We can increase this number, if desired, at a very slight addition to our regular rates.

We use no cords about the table, depending on *rubbing points entirely, as our long experience in the matter has fully demonstrated the fact of that being the best manner of connecting.*

Subscribers should be particularly notified to *ring off.* Use *two cells* of *Gravity Battery on the microphone* circuit of switchboard, as it is always in circuit. The night service bell, No. 6, rings continuously when call comes in until drops are set by operator.

Treadle for Generator is always off dead centre, so that the least motion of the foot starts it. Telephone and microphone stands have an universal adjustment. We furnish with each table a *microphone mouth-piece,* which Exchanges will find very convenient in conveying sound to microphone. The engine requires a drop of oil now and then to prevent working hard and to generate strongly. Tables—black walnut, highly ornamented and veneered.

The competition of rival manufacturers and the lack of capital on the part of the purchasers probably combined to produce a lightness of construction leading to cheapness in price, especially on the part of those manufacturers like Post & Co. and Gilliland, who were distant from the Bell headquarters. Williams' switchboards were more solidly made.

The switchboards already referred to are those introduced by the manufacturers licensed by the Bell Co. But the Western Union Telegraph Co. were running telephone exchanges in opposition to the Bell, and the manufacturing interest identified with the Western Union Co. was the Western Electric Manufacturing Co. The illustration of the Western Union Telegraph switch given in fig. 56 is from the Manufacturing Co.'s catalogue, and it is noticeable that whilst other telephone manufacturers took the telegraph type of switch for their model, the makers of that switch themselves started on new lines. The types of the Bell licensees were early abandoned, whilst that of the Western Electric Manufacturing Co. became the starting-point for most of the later developments.

The relations of the Western Electric Manufacturing Co. with the Western Union Telegraph Co. were close. Consequently, when the Western Union Co. entered the telephone field the Manufacturing Co. became actively engaged in the production of telephones and accessories. The Edison transmitter was made by them. The receiver adopted by the Western Union and also made by the Manufacturing Co. was the Bell invention, pure and simple, though modified in form by George M. Phelps, Senior.

But in the case of telephone switchboards, an article of new manufacture, the company were not indebted to any outside source for ideas, and even avoided following, as other makers had done, their own patterns of telegraphic switchboards. In Scribner's British patent, specification No. 4903, of November 29, 1879, is an illustration (fig. 62) of what was known as the 'Universal Switch,' from which the standard switchboard was developed.

In describing the invention Scribner says:—

I cause each subscriber's wire to terminate in a metal block, and these blocks are arranged upon a board, so that there are as many blocks upon the board as there are subscribers. Each block carries a blade similar to the folding blade of a pocket knife, and, as in such a knife, the blade folds into a groove or recess in the block, the block corresponding to the handle of the knife. A spring also tends to keep the blade closed in its recess.[1]

Here it will be observed the knife-like block is the means of access to the line instead of the vertical bars of the telegraphic form of switch, and this knife-like block is an electro-mechanical contrivance, its purpose and its operation being thus described :—

In this state of affairs when the subscriber sends a call current through the wire it excites the electro-magnet, the magnet attracts its armature, and allows an indicating shutter to fall; this shows to the attendant that this subscriber calls for his attention. Each block (corresponding to the knife handle) has two holes formed through it, and into one of these holes the attendant when called sticks a metal peg which displaces the knife blade, partially opening the knife. In this way the line wire is disconnected from the indicator. The peg inserted into the hole in the block is attached by a flexible conductor to a telephone, by the aid of which the attendant receives the subscriber's order to couple his wire with that of some other subscriber with whom he may desire to converse. Now for making these connections a series of insulated bars are provided upon the lower part of the board, and the attendant, selecting one that is out of use, connects the caller's line wire to it by a flexible conductor terminating in pegs at its ends. One peg he sticks into a hole in the subscriber's block, and the other into a hole in the coupling bar.[1]

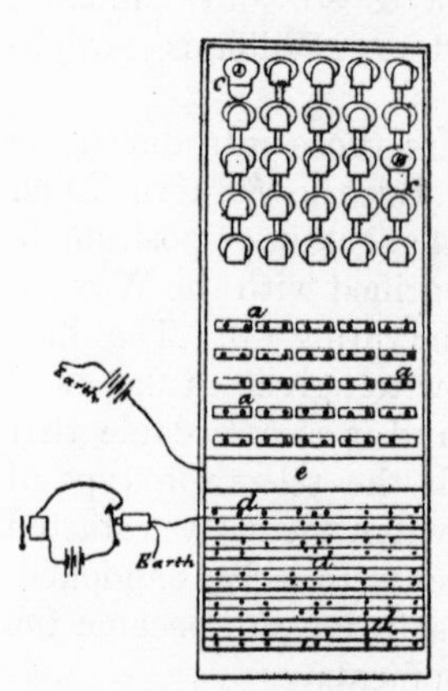

FIG. 62.—Universal Switch (from patent).

The succeeding requirement in telephone service is that so soon as the subscribers have finished their conversation they should be promptly disconnected by the operator. This requirement is met as follows :—

In connection with each of the coupling bars there is an indicator to inform the attendant when the conversation between the two subscribers lines, which the bar serves to couple, is complete.[1]

[1] British specification, No. 4903, November 29, 1879.

This specification contains further important features which render it a classic in switchboard patents, but in it the word 'switchboard' itself does not appear. The figure illustrated is described as 'a front view of a coupling board arranged according to this invention.'

The illustration fig. 62 from the 4903 1879 British patent is

FIG. 63.—Universal Switch (from Catalogue).

diagrammatic only. The details are more clearly designated in fig. 63.[1]

For the purpose of giving prices and dimensions this Universal Switch is divided into three parts:—The upper portion is called a '25 Number Shutter Annunciator (12 by 13½ inches) $50'; the middle section is '25 Number "Jack Knife" Switchboard (12 by 8½ inches), $20,' and the lower portion is '15 strap Switchboard (12 by 10½ inches), $9.' The illustration shows only a 10 strap board.

[1] Reproduced from the Western Electric Manufacturing Co.'s catalogue issued in 1878 or 1879, p. 56.

The combination, however, is called both a 'section' and a 'switchboard.'

The dimensions of each section, when mounted, are 12 inches in width and 32½ inches in height. The clearing-out annunciators and relays are set up by themselves apart from the switchboard.[1]

The 'clearing-out annunciator and relays' are illustrated in fig. 64.

The general description of the switchboard is as follows:—

The Universal Switch is the standard telephone exchange system of the Gold and Stock Telegraph Company.

Each wire comes into the exchange through a lightning arrester to one of the bolts on the jack-knife switchboard (shown just below

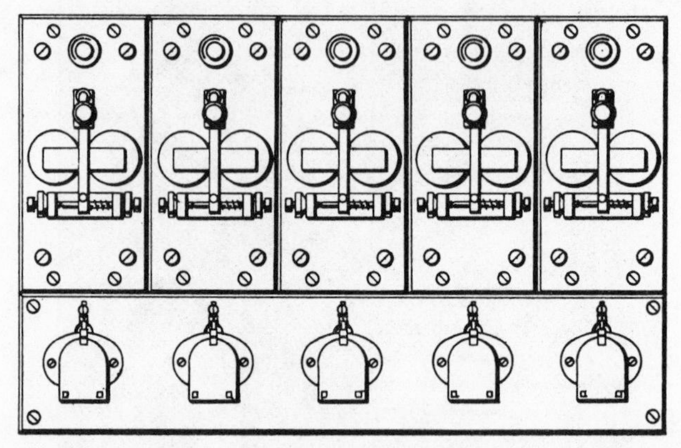

FIG. 64.—Clearing-out Annunciators and Relays of Universal Switch.

the annunciator in fig. [63]). From this bolt it passes to the corresponding annunciator magnet, and thence to the ground.

When there is no plug in the jack-knife bolt the circuit passes straight through to the annunciator magnet and ground; but when a plug is inserted in the hole in the bolt, this plug takes the line circuit, and the circuit through the annunciator to ground is opened. If, therefore, a plug is placed in bolt No. 1, and another in bolt No. 2, and the two plugs connected by a cord, lines No. 1 and No. 2 are connected for talking.

In practice the two lines are brought together by connecting them to the same strap on the 'Strap Switchboard' [i.e. the lower part of fig. 63]. Each strap is connected to ground through a 'Clearing-out Relay' of 150 ohms resistance, so that the talking is done over this ground connection. When two subscribers are through talking, the one who called first signals out in the same way as he called. This signal works the clearing-out relay, which

[1] Catalogue, p. 57.

closes a local battery, and drops a shutter on the 5-number annunciator [fig. 64]. This is a signal to disconnect at the office.

The 'shutter annunciator' is very delicate, and may be used with either a battery or a magneto call.[1]

This board, it will be observed, contained the important features that the insertion of a plug removed the calling indicator and earth (a provision also met with in Eldred's patent), but in addition brought the line into connection with the operator's telephone; both calling indicators were out of circuit during the conversation and a special clearing-out indicator 'teed on' in a branch to earth.[2] This method of connecting a clearing-out drop was a very important application, whose value was to be more fully appreciated in after years. For some time yet the clearing-out drop was placed directly in the circuit, after the manner of Eldred, but in a way far more convenient for operating.

The connecting bar across the board is a survival from previous types, but used in a different way. A cord with plugs at either end served to connect one subscriber's block with the connecting bar, a similar cord with plugs connected the same bar with another subscriber's block. Clearly, then, the only purpose of the horizontal bar was to complete the circuit of the two cords. That purpose could be served by connecting the cords together as permanent pairs. The point at which this connection was made afforded a suitable place for inserting keys by whose means the circuit might be diverted to telephone or generator as required. This modification is found in the next catalogue issued. The description and illustration are as follows:—

The Telephone Exchange Switch

Described in this circular is believed to be the simplest and most expeditious in manipulation of any yet devised. The movements necessary to connect and disconnect subscribers are reduced to a minimum.

Fig. [65] shows a section for fifty subscribers' wires.

It consists of—

Fifty annunciators for subscribers' wires.

Fifty jack-knife switches for connecting and disconnecting line wires and annunciators.

Five clearing-out annunciators for signalling the discontinuance of communication between subscribers.

A shelf with five pairs of keys for connecting to calling battery

[1] Catalogue, pp. 57, 58.

[2] These features are more fully described in Scribner's U.S. specification, No. 293,198, February 5, 1884 (application filed August 23, 1879). See figs. 87 to 90.

and telephone of central office—and five pairs of cords and plugs.

The annunciators are of a new and improved construction, recently perfected by this Company. They are much more sensitive than the ordinary forms.

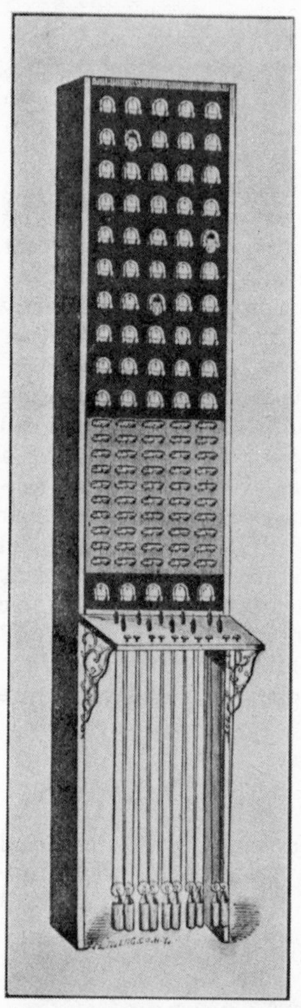

FIG. 65.—Standard Switchboard with 'jack-knife' switches.

Their magnets are wound to a resistance suitable to either battery or magneto currents.

The switch is furnished in sections of fifty wires, mounted and with connections run between the various parts, ready for setting up and connecting the line and office wires.

The sections of switch are only fourteen inches wide, and can be placed side by side, so that one operator can readily attend several sections.

Annunciators for closed circuits are furnished where desired. Part of the annunciators may be for open and part for closed circuit in the same section.

All the annunciators are provided with local circuit points, closing a local bell circuit by the fall of the drop. This is put on all the annunciators, though not always used. In small exchanges not requiring all the time of an attendant to do the switching, it is very useful. In larger exchanges it is very convenient where night service is required. A large vibrating bell may be used to awaken an attendant when he is required.

By the use of the local attachments and night bell many exchanges have been enabled to save the wages of a night operator.

The switch is intended for exchanges where either magneto or battery calls are used; and with our automatic pole changer works equally well where either or both systems are used in the same exchange.

The diagram [fig. 66][1] shows the manner in which the switch is connected.

Only five of the line annunciators and one of the clearing-out annunciators and one pair of keys are shown in the diagram.

[1] As original, but the lines should obviously read 'L 54321.'

L 1, 2, 3, 4, 5 are the line wires.
A 1, 2, 3, 4, 5 are the line annunciators.
J 1, 2, 3, 4, 5 are the jack-knife switches.
C is the clearing-out annunciator.
$K_{\prime}$ is the telephone key.
$K_{\prime\prime}$ is the battery or calling key.
$P_{\prime}$ $P_{\prime\prime}$ are plugs with flexible cords.

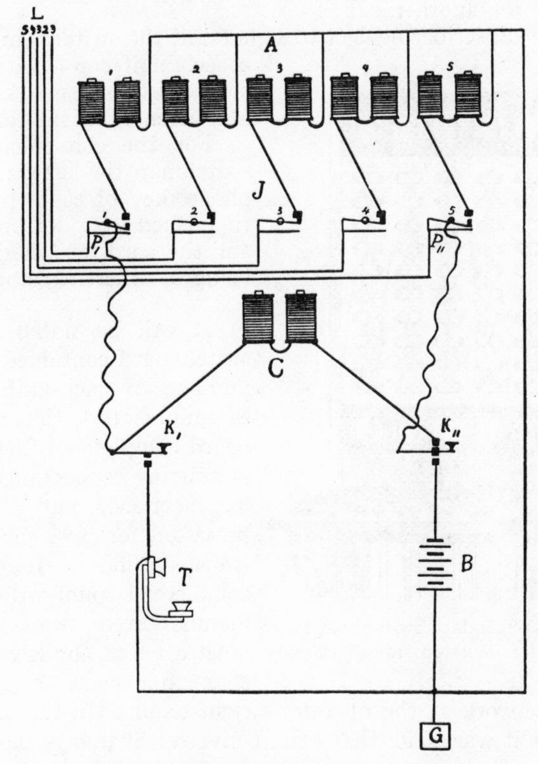

FIG. 66.—Standard Switchboard (circuits).

T is the switchman's telephone.
B is the calling battery.
The operation of the switch is as follows :—

If annunciator No. 1 indicates a call the switchman inserts plug $P_{\prime}$ into jack-knife switch 1 (disconnecting line from annunciator 1 and connecting it with clearing-out annunciator C and keys $K_{\prime}$ and $K_{\prime\prime}$), and by using telephone while pressing key $K_{\prime}$ acknowledges No. 1's call.

He then calls the subscriber desired by No. 1 (say, No. 5) by inserting plug $P_{\prime\prime}$ into jack-knife switch 5 and tapping key $K_{\prime\prime}$. He then presses key $K_{\prime}$ and listens at telephone for No. 5's acknowledgment, upon receiving which he notifies No. 5 that he is connected with No. 1, and listens until conversation begins, when he releases key K, and is at liberty to attend to another call.

Subscribers are to signify the termination of their conversation by tapping their call keys—thus dropping the shutter of clearing-out annunciator C. When the switchman observes this he pulls out the plugs from the jack-knife switches, thus restoring the lines to their proper annunciators and leaving the pair of cords and keys available for another call.

If the subscriber neglect to signal out, the switchman can easily ascertain if conversation is concluded by pressing key 1 [K,] and listening at the telephone.

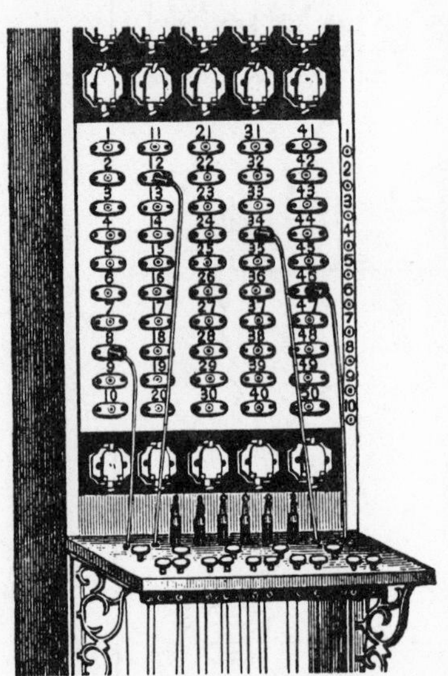

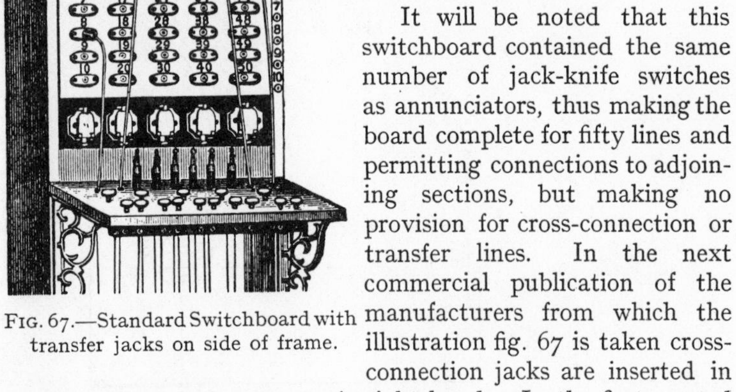

FIG. 67.—Standard Switchboard with transfer jacks on side of frame.

For the convenience of the switchman the left hand or telephone key of each pair is distinguished by a dot on the knob, and the cords are made in pairs, being of different colors.[1]

It will be noted that this switchboard contained the same number of jack-knife switches as annunciators, thus making the board complete for fifty lines and permitting connections to adjoining sections, but making no provision for cross-connection or transfer lines. In the next commercial publication of the manufacturers from which the illustration fig. 67 is taken cross-connection jacks are inserted in the frame-work at the operator's right hand. In the first quoted catalogue it was said that 'the Universal Switch is the standard telephone exchange system of the Gold and Stock Telegraph Company.' In the second the universal switch had developed into the 'Standard Switchboard,' and the name is still retained to indicate a simple as distinct from a multiple switchboard. The description given is as follows:—

This is a cord board, with the annunciator drops, and automatic switches and keys for calling and connecting telephone to subscribers' circuits. In this board the movements necessary to connect and disconnect subscribers are reduced to a minimum, and it is believed to be the simplest and most expeditious in

[1] Catalogue, pp. 5–6.

manipulation of any board yet devised. It is now in practical operation in over two hundred telephone exchanges in this country, and is used on about forty thousand telephone lines. . . .

By the use of superior stock in the construction of our cords, and with weights to take up the slack of cords, the trouble from this source is reduced to a minimum.

Our boards are 14 inches wide. The No. 1 board rises 73 inches, and the No. 2 board rises 60 inches from the floor.

In the details of construction of its various parts this board was to undergo numerous changes, but in its arrangement of calling indicators, jacks, and clearing-out indicators on the upright, with a horizontal keyboard for manipulating keys it reached at once the most convenient form, and in these essential features has remained unaltered. In the next model produced after fig. 67 the transfer jacks were placed below the line jacks and in the same panel.

An important change in detail will be noticed between the knife-like blocks in fig. 66 and the fig. 67 illustration. The jack-knife switch is superseded by the 'spring jack.' The oval form of the front of the jack in fig. 67 is due to the large shoulder with which the jack was then furnished, being fixed to the board from the front with two screws. In the next model the shoulder was abandoned and numerous illustrations of it have been published.

Electrically they are the same, the difference is in form, because it was early realised that space on the front of the board was valuable, hence the mechanism was constructed to operate from front to back instead of from side to side.

It has sometimes been assumed that the term 'spring jack' has been derived from a spring form of jack-knife switch. But this is not so. In the telegraph switchboard a spring pressing on a contact was described as a 'spring jack,' and the name was carried on from telegraphic to telephonic apparatus.

Whilst referring to questions of terminology it may be remarked that the signalling device was called an annunciator in the United States, but in Great Britain 'indicator' was more general. The difference reflects the previous practice in the use of electric bells. In the United States a device for signalling the number of the room from which a call had been transmitted was an electric bell 'annunciator,' whilst in Great Britain it was an 'indicator.' For telephonic purposes the 'Indicator' gradually became general in both the Old World and the New. They were also known as 'shutter drops,' which became shortened to 'drops.' This was especially the case with the indicators for signalling the close of a conversation, which became known as 'clearing-out drops.'

The development of a switchboard on new lines is an indication that the experts of the Manufacturing Co. recognised at an early stage that the problem of telephonic switching was a new problem and needed to be solved on different lines from that of telegraphic switching. In telegraphy the switching was an incident, the contrivance by which it was effected was only in occasional use. Facility in the operation of the mechanism and the time employed by such operations were of subsidiary importance. In telephony it was otherwise, and whilst this was eventually brought home to everybody, the Western Electric Manufacturing Co. must be credited with realising it well in advance of the lessons of actual experience. This is evidenced by the production of the Standard Switchboard, which represents not merely a collection of parts admirably adapted to serve their respective purposes, but also an arrangement of those parts in such a manner as to afford the greatest advantage in use.

The arrangement of the indicators and jacks was such as to show readily their respective relationships. A decimal arrangement was adopted, the reading being from above to below. The clearing-out drops had an obvious relationship to the cords with which they were in circuit, and the keys permitting the insertion of the generator or speaking instruments were not only original in the work they performed, but were placed upon a shelf at the most convenient position for the operators. For the first time also the equipment of the board was the result of study. The number of subscribers to which one operator could in general attend was found to be such that fifty was taken as the most convenient standard, the number of cords was determined after inquiry as to the probable maximum of simultaneous connections, and on the same principle twenty per cent. was adopted.

The framework of the switchboard was a striking departure from the elaborate ornamentation and overlapping cabinet work of its contemporaries. The simplicity, however, cannot be considered as a tribute to the æsthetic principle that surplus ornamentation is reprehensible. It was due to the recognition of the economic value of frontage space on switchboards. Ornament which increased the frontage space was something which not only cost money at the start, but, what was still more important, might entail an unnecessary expenditure in operators' wages and in rent.

The framework was consequently so arranged that a new board could be added, fitting closely to the existing board or boards so that an operator was not limited in connections to her own board, but could complete a connection on either of the adjoining boards, thus reducing the number of 'cross-connections' or transfers.

Some of the parts and circuits of the standard board are the subject of patents to Scribner, Kellogg, and Warner, but the design of the board is believed to be largely due to F. R. Welles, who appreciated the importance of the decimal principle, the readily

FIG. 68.—Williams Switchboards at Portland, Oregon, 1884.

observable relationship of the indicators with their allied jacks, the facilities which the keyboard afforded, and the utility of a compact construction permitting to the operators the common use of adjacent sections.

In Prescott's 'Speaking Telephone' of 1884[1] the Western Electric Standard Switchboard is illustrated and described[2] as the 'Gilliland Switchboard.' On p. 288 it is called 'one of Gilliland's Switchboards,' and the real Gilliland Board is illustrated on the same page, and called the 'Gilliland Standard Switchboard.' In Preece and Maier's 'The Telephone' (1889), p. 314, an illustration of the Western Electric Standard Board is described as the Williams Switchboard. Whilst many manufacturers described their switchboards as 'standard,' the word has acquired a meaning as a self-contained switchboard as distinct from a multiple board, and it is the form illustrated in figs. 65 and 67 which is so understood.

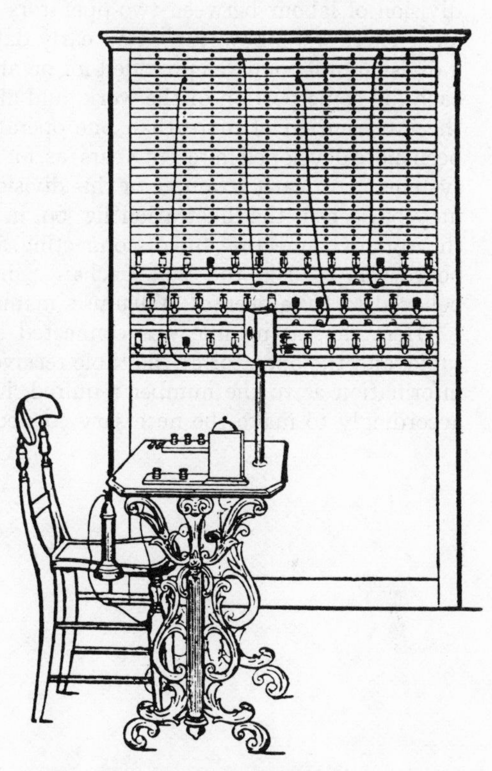

FIG. 69.—Switchboard used in Coleman Street, London, 1879.

An example of a number of Williams switchboards treated as independent units is seen in fig. 68, illustrating the central office at Portland, Oregon, reproduced from the New York *Electrical Review* of April 17, 1884.

All the switchboards above described were constructed on the principle of providing complete equipment for an operator who received a call, connected talking instruments into the line to ascertain the number required, and plugged the two subscribers together when their positions on the switchboard permitted. If the called subscriber's position on the switchboard was beyond the reach of

[1] P. 277. [2] P. 276.

the operator who received the call, it was necessary for her to obtain the assistance of a second operator. This was usually effected by means of transfer lines, to which the term 'trunks' or 'office trunks' was applied. The first operator would connect the calling subscriber's line to a 'trunk' and desire the second operator to connect the same trunk to the called subscriber. Thus there was a division of labour between two operators in some cases, but not in all. There were, however, at an early date, switchboards in which a division of labour was provided for on an organised plan, whereby each did some portion of the work, and all calls were completed by the intervention of more than one operator. Such methods have been more developed in recent years, as, for instance, in 'distributing' systems. An early example of this division of labour in switching may be seen in the illustration fig. 69, in which one operator does the work of connecting and disconnecting, another the talking. This board was used in the Bell Exchange in London in 1879, and is believed to have been of Williams's manufacture.

Here the calling line was connected by the switchman to the table, and the operator at the table received from the subscriber the information as to the number required, instructing the switchman accordingly to make the necessary connections.

CHAPTER XV

THE ORGANISATION OF THE INDUSTRY IN THE UNITED STATES.

THE first public circular issued in May 1877 [1] was signed by Hubbard and Watson, who were acting on behalf of Bell and the others interested as well as of themselves. Bell was the inventor, Hubbard and Sanders were financial supporters, whilst Watson was a technical assistant sharing with Bell the hopes and fears of experiments, and carrying out, with the aid which Williams's workshop afforded, the construction of apparatus that Bell designed.

In a few months after the issue of the first circular the developments were such as to show the need of regularising the ownership and preparing for expansion. Consequently in August (1877) the Bell Telephone Association was formed but not incorporated. This Association consisted of Bell, Hubbard, Sanders, and Watson. The patents were transferred to the Association. The shares of Bell, Hubbard, and Sanders were equal—three-tenths each; the remaining tenth was allotted to Watson. During October 1877 a contract was entered into by Hubbard 'as trustee of said patents,' but in November contracts were made by the Bell Telephone Company (i.e. the Association above mentioned) and signed by Hubbard as trustee and Sanders as treasurer.

The New England Telephone Company was incorporated on February 12, 1878, with a capital of $200,000, and was granted the exclusive right to use, license others to use, and to manufacture telephones in the New England States. The Bell Telephone Company was also incorporated on July 30, 1878, with a capital of $45,000, with the object of extending the use of the telephone throughout the United States outside of New England. These two companies were subsequently amalgamated under the name of the National Bell Telephone Company, which was incorporated March 13, 1879, with a capital of $850,000. In consequence the first New England

[1] Pp. 67–68.

Telephone Company went out of existence, but the name was revived later when a company was formed under that title to amalgamate a number of smaller original licensees and to carry on the exchange business throughout that territory.

The policy of the holders of the patents was to appoint agents in defined localities authorised to lease telephones to the users thereof at certain stated rentals, of which rentals the agent should retain a specified share. In a contract dated October 24, 1877, there is no mention of exchange systems, though the reservation to the Bell interests of the exclusive right to make contracts with 'any and all parties who desire to use telephones for the purpose of transmitting messages for hire' probably contemplates the district system.

The term 'telephone' is defined as meaning instruments made under the Bell patents, Nos. 161,739, 174,465, 178,399, and 186,787, and all patents which Bell had or thereafter might obtain for improvements, and all beneficial modifications which he should be at any time authorised to use. The term 'one set' is apparently defined rather in reference to the payments contracted for than to the practical working, since it includes either 'four small telephones, two large telephones, or one large and two small telephones.'

The agent or licensee undertakes to construct with his own capital 'in the most approved way, any and all lines which may be reasonably required' in the area specified 'by any proper person, for use in connection with telephones. Said lines to be hired or sold on reasonable terms. The rental price may be a gross sum, as rental for telephones and line, in which event the schedule rates herein shall be the sum apportioned as the rental of telephones.'

The licensee further undertakes to use his 'best efforts in all proper ways to introduce telephones to the utmost possible extent, and to procure lessees thereof for use in' the defined territory. He undertakes to 'employ at least one suitable and efficient agent whose whole time shall be devoted to the introduction and care of telephones, and will employ such further and other agents and instruments as may be necessary for said business.' The contract from which these quotations are made covered an entire State, so that 'at least one' does not seem at this date to be an extravagant requirement, but the phrase is illuminating as illustrating the caution of one of the contracting parties whatever may have been the expectations of the other. It may be remarked parenthetically that according to the official census the number of persons employed in the telephone business in the same State on December 31, 1902, was 2823 and would now be much more.

On November 15, 1877, the Bell Telephone Company issued Instructions to Agents. No. 1' which commences with the statement that

In consequence of the difficulties that have arisen in different localities for want of uniformity in price for the rental of telephones, the Bell Telephone Company has adopted the following rates for all its agencies, and prices are to be fixed in accordance herewith.

The annual rental for telephones shall be ten dollars each, payable in advance; not less than a pair of telephones must be used at each station, except as hereafter specified.

For social purposes, single telephones may be used at each station. By 'social purposes' is meant the use of telephones as a matter of convenience between private houses; between a house and private stable; a doctor's house and office, etc. etc.

For district telephone purposes a discount of twenty per cent., and for house use a discount of fifty per cent. may be made, and the use of single telephones allowed at each station.

By 'house use' is meant all places where telephones are used in one building, or group of buildings, as, for instance, several buildings in the same yard used by the party; or, in fact, where telephones substantially take the place of speaking tubes. College lines may be included in this line.

The magneto bell calls may be sold for fifteen dollars each, or rented for five dollars each per annum.

On March 8, 1878, the Bell Telephone Company entered into an agreement with the District Telegraph Company of St. Louis, who agreed 'to introduce the Bell telephone as a part of its district system, to substitute telephones for the district boxes as rapidly as possible, to construct private lines of telephones, and to prosecute the introduction and rental of telephones with all due diligence.'

On May 31, 1878, a contract was entered into with the Connecticut District Telephone Company whereby the District Company was given the exclusive license for the term of ten years to use telephones 'for district purposes' in New Haven and certain other cities.

The District Company was also licensed 'to connect said cities as above named by wire, and to transmit messages between them for hire by means of the Bell telephone, provided that such connection is made within one year from the date of this contract; but this license shall not be construed to be an exclusive license for such purposes.' In the first year 500 telephones were to be leased.

On July 3, 1878, an agreement was made between the Bell Telephone Company of New York and the Bell Telephone Company of Boston whereby the latter gave the exclusive right 'to use and rent telephones' at the annual rental of $10 'for general purposes' and $5 'for house use.'

The New York Company agreed that its capital stock should consist of 1200 shares of preferred stock and 800 shares of common stock, each of the par value of $50. The capital was to be furnished by the holders of the preferred stock. The ordinary stock was paid to the Bell Telephone Company of Boston as a consideration for the exclusive right.

The territorial system is provided for in the following clause—the 'party of the first part' being the New York Company and 'the party of the second part' the Boston or parent company :—

The party of the first part agrees that whenever the party of the second part shall be prepared to receive business in any of the towns or cities within said district for transmission to places outside of said district, the said party of the first part will turn the same over to the party of the second part, it paying a reasonable compensation for receiving or collecting same, and also turning over to the party of the first part the delivery of messages received by the party of the second part from points outside of said district to be delivered within said district at a reasonable compensation.

Nothing in this agreement shall be construed to prevent the party of the second part from establishing offices within said district for the transmission of messages to points outside of said district. And no consolidation, sale or change, or insolvency or dissolution of said party of the second part shall affect the rights or privileges granted to the party of the first part by these presents.

A contract entered into on January 29, 1879, for the first time defines the 'district use' in the following terms :—

It is further understood and agreed and made a part of the above and within agreement, that the terms 'district and exchange purposes' and 'district uses,' mentioned in the above and within agreements, refer to the use of telephones in connection with a district system to be established within the territory covered by this license, more particularly described as follows :

A central office or receiving station is established within the prescribed limits, from which point lines of wires are erected running in different directions within the prescribed territory. At some of the principal factories, stores, shops, offices, business places, dwellings, etc., along any of these said lines of wires, telephones are placed and so connected in the line that persons at any of such connected points in the line, after the proper adjustment of switches, cut-outs, instruments, etc., and exchange of proper signals, can communicate with each other on the same line, and with the said central office or receiving station, with parties located on other of said lines of wires having a similar telephone connection.

A further development in the definition of 'district or exchange purposes' is reached in a contract of singular clearness and definiteness in all its clauses. It is dated August 9, 1879, and appointed C. H. Haskins agent for the States of Wisconsin and Minnesota, and is signed Theo. N. Vail as General Manager of the National Bell Telephone Company.

The terms 'District' and 'Exchange' purposes as used in this agreement are understood to apply to a telephone business in which in any city or town one or more telephone circuits are established and connected with a general office, or general offices, to receive or execute orders or to establish connections between different lines.

The district business includes the right to transmit messages for hire over such lines, but does not include the right to transmit messages for hire between different cities or towns.

The reservation of the long distance system is also shown in this agreement by the following clause :—

This Company [i.e. the Bell Company] reserves the exclusive right of renting telephones for the purpose of communication between different towns or cities or of transmitting messages for hire, and of renting telephones to corporations or individuals whose business may be but partially carried on within the territory assigned to you, although one or more of said telephones may be used within the said territory.

The policy of appointing agents for given areas, of supplying telephones on lease, and of reserving for the company the communications between distant places was acted upon by Hubbard; and Bell, in his letter to the financial supporters in London,[1] emphasised the desirability of preventing sale outright, but the progress by the early part of 1878 was so great and the promise so much greater that Hubbard realised the importance of placing the organisation of the business in strong hands. From his own experience in the control thus far he realised to the full the need of active and far-seeing management. In July 1878 Mr. Vail became general manager, retiring from an important position in the United States Postal Service for that purpose.

Capital, hitherto supplied by Hubbard and Sanders, was now needed in quantities far beyond their slender resources. Some citizens of Boston were sufficiently impressed with the work accomplished and the future possibilities, to invest in the Bell Company. They appointed as President Col. William H. Forbes, a man of influence and strength of character, who left his impress upon the enterprise. The company was now fairly embarked upon an active career under auspices which thus provided influence and energy.

The country was mapped out, agents appointed, and agreements entered into providing for the development of a business of which all were ignorant because it had never existed before.

The essential feature of this business was the telephone, and

[1] P. 89.

provision was made for its manufacture by Charles Williams. But it was found that call bells were also necessary, and as time went on switchboards were required. Very naturally the Bell personnel took an active interest in the development of such accessory apparatus, and the company adopted the policy of licensing manufacturers in various parts of the country. From these manufacturers the exchange licensees were authorised to purchase such accessory apparatus as they might need.

The following were the manufacturers so licensed :—

Charles Williams, Jr., Boston.

Post & Co., Cincinnati.

Gilliland Electric Manufacturing Co., Indianapolis.

G. H. Bliss & Co., Chicago.

Davis & Watts, Baltimore.

Quite early in its career, therefore, the Bell Company exercised an influence in the use of apparatus outside the telephone itself, and sought such methods of control as should ensure that its licensees obtained apparatus of approved quality, whilst leaving to such licensees the selection of any of the types furnished by those various manufacturers.

The developments by 1880 were such as to need the formation of yet another company with a still larger capital, and the American Bell Telephone Company, which was organised in May of that year, succeeded the National Bell Company.

The first report of the American Bell Company, issued in March 1881, shows that the general lines of its policy were to grant licenses for local use, to retain in its own hands the long distance service, and to study all the practical questions involved for its own benefit, and the advantage of its licensees generally. The following extracts from this first report to the shareholders relate to these various features :

> The policy of making only five year contracts was adopted, in order that our company could have time to learn the best permanent basis for the relations between the company and its licensees, and to see which of them would prove satisfactory as associates. Many applications are now being made for permanent licenses, and we have begun to give such permanent contracts in places where the business is being prosecuted with energy and success in exchange for a substantial interest in the stock of the local companies. By pursuing this plan, the company will gradually acquire a large permanent interest in the telephone business throughout the country, so that you will not be dependent upon royalties for a revenue when the patents shall have expired.
>
> The business of connecting cities and towns by telephone wires has been taken up in the past year with some vigor, and the prospect

is good for a large increase in these lines. Boston, for example, is now in communication with seventy-five cities and towns, including Providence, Worcester, Springfield, Lawrence, Lowell, and other important places.

It will take some time yet to get first rate service in a large network of towns, as the practical difficulties at least equal those which were met in giving prompt connection within the limits of one city; but nothing but experience and tests of various methods are needed to enable such groups of exchanges to reach satisfactory results.

A large amount of work has been done in the electrical and experimental department, both in examining new inventions and testing telephones and apparatus, and in studying the question of overhead and underground cables, and the improvement of telephones and lines, for both short and long distance service. This work is expensive, but it is of the first importance to our company, and must be continued.

The report of the General Inspector upon this department shows that we own or control, either by purchase or by inventions made by our own electricians, 124 patents, and have applications to the Patent Office for 77 more. Among these a considerable number are of great value as a protection to our business, and from them a substantial revenue has already been received by royalties from our licensees. This source of income will be materially increased, and should eventually more than cover our experimental and electrical expenses.

The report concludes with an expression of the directors'

appreciation of the ability, fidelity, and zeal with which the general manager and his assistants have grappled with the unusually perplexing difficulties encountered in systematising our affairs.[1]

In after years the general manager, Mr. Vail, himself related in a brief preliminary statement of a legal nature, the principles underlying the organisation of the company, and remarked that

The American Bell Telephone Company early foresaw the possibilities which were offered by the development of the inventions covered by these telephone patents, and that it would be possible to establish a great system of intercommunication throughout the whole country.

. . . . . .

The ideal of the Company would not have been realised unless the users of the telephone in the territory of each of the associated companies had been enabled to communicate with the users of the telephone in the territory of every other company.[2]

[1] *American Bell Telephone Company's Report*, 1881.
[2] *Statement and Answers to Interrogatories.* Missouri, July 27, 1910.

CHAPTER XVI

COMPETITION, CONSOLIDATION, AND DEVELOPMENT

For the purpose of illustrating the principles upon which the industry was organised, the last chapter includes the period 1877 to the formation of the American Bell Telephone Company in 1880. But this period was so momentous in its events that it must again be referred to.

Brief reference has already been made to the Western Union competition. Strong in influence, with ample financial resources, with lines extending throughout the country, with agents everywhere, and employing a large number of men skilled in electrical work, it is easy to conceive the confidence with which the Western Union Company set out to oppose the Bell. It is less easy at this distance of time to realise and do justice to the resolute management of the Bell Company, whose material resources were slender, but whose faith in their patents was unshaken, and whose determination to develop the business was promptly and continuously demonstrated.

The saving of the situation by the Bell Company may be attributed to the dual factors of prosecution of patent rights and the creation of Bell exchanges wherever possible, whether in competition with Western Union exchanges or not. Apart from the fact that a business concern like the Western Union would attach importance to the business results of competition, it must be remembered that where any doubt exists, or may be alleged to exist, as to the originality of an invention, a patentee who has only a patent occupies a much less strong position in protecting that patent than one who, in addition to his patent, has a business created out of it. A patent is granted for reasons of public utility, and the existence of a business is a proof of that utility; but a patentee who is content to hold a patent without energetically working it, or arranging for its being worked, does not confer upon the public the benefits for which *prima facie* the patent was granted. The practical creation of a business on another's invention gives the creator no rights in equity, and in no way diminishes the infringement in law. The danger to the inventor is a possible bias in the interpretation of doubtful or arguable points.

It is not suggested that such ideas were in the minds of the Bell management, so far as the contemplation of the eventual results was concerned. The evidence available clearly shows that they realised the enormous public benefit which must follow from the use of the telephone, and especially from the exchange system, with corresponding financial benefit to those who controlled it. They realised also that Bell was entitled to the credit for making telephonic communication possible, and they believed that Bell's patents gave them the control. That they were undaunted in engaging in litigation with a foe much better equipped in material resources was doubtless largely due to their conviction of the strength of their case and their recognition that, even unarmed, justice is a considerable resource in conflict.

From amongst the various places in which the Western Union interests had established exchanges, Boston was selected to form the subject of a test case. The suit was technically against a Western Union agent named Dowd. The real defendants in the case were four: The Western Union, The American Speaking Telephone Company, the Gold and Stock Telegraph Company, and the Harmonic Telegraph Company.[1]

The Harmonic Telegraph Company was owned or controlled by Elisha Gray and S. S. White, 'a very rich man in Philadelphia who established dental depots all over the United States.'[1] The American Speaking Telephone Company was organised to develop the Western Union telephone interests. Gray with his partner held one-third interest in it.

Testimony was prepared, but the case never came to trial. Experts in electrical science and in law had examined all the evidence that could be got together, and advised the Western Union that it was impossible to plead anticipation or to impeach the validity of the Bell patents.

Mr. Frank L. Pope, a well-known electrical and patent expert, advised to that effect, and Mr. George Gifford, an eminent lawyer who was leading counsel for the Western Union, reached the same conclusion. In consequence, overtures were made with a view to a settlement on terms. The nature of these overtures and the result were subsequently recorded by Mr. Gifford in an affidavit, in the course of which he stated that in the years 1878–1879 he was one of the counsel for the Western Union Telegraph Company.

[1] Dickerson's Argument, *New Orleans Case*, p. 123.

At that time the Gold and Stock Telegraph Company, a company connected with the Western Union, had manufactured and were controlling the use of many thousands of telephones (there were about 10,000), and had established telephone exchanges, auxiliary to their telegraph business, in the city of New York and elsewhere. The telephones controlled by the company were composed of a receiver, generally known as the magneto receiver, and understood to be substantially the thing described in Bell's patent; but they were of a form which was claimed and had been constructed by Phelps and Gray. The transmitters were carbon microphone transmitters, constructed under the plan of Edison and Phelps, and contained the induction coil covered by the Page patent, also owned or controlled by the Western Union Telegraph Company. Among other defences set up, the European publications relating to Reis's invention were relied upon, and it was alleged that Bell's telephone, as described in his patent, was not capable of talking. Elisha Gray was set up as a prior inventor, and the inventions of Edison and Dolbear were pleaded. A very vigorous defence was (he says) made by the Western Union Company, and testimony at great length and great expense was taken in support of the answer. After the testimony was closed, or substantially closed on both sides, he (Mr. Gifford) was convinced that Bell was the first inventor of the telephone, and that the defendant Dowd had infringed Bell's patent by the use of telephones in which carbon transmitters and microphones were elements, and that none of the defences set up could prevail; Mr. Gifford advised the Western Union Company to that effect, and that the best policy for them was to make some settlement with the complainants. For the purpose of effecting such a settlement the position of the Western Union Company was (Mr. Gifford contended) very strong. In addition to the patents of Edison, Gray, and others, they owned or controlled what was known as the Page patent, which covered the induction coil used in the transmitters and was of great importance to them. Under his advice (he continues) a negotiation was opened with the Bell interests. He met Mr. Chauncey Smith, counsel for the Bell Company, by arrangement at the White Mountains, where they remained for a week in negotiation. He opened the negotiations on his part by admitting that Bell's patent was valid, and that the defendants infringed it, but he claimed that all the patents should be put together and that the Western Union should have one-half interest. This was refused and the negotiations failed. They were renewed by the principals in New York and resulted in the surrender by the Western Union Company to the Bell Telephone Company. Instead of the one-half interest claimed, the Western Union accepted one-fifth, and they agreed to go out of the business, transferring their exchanges at cost price.[1]

The one-fifth interest referred to was not the proportion of exchange receipts, but ' 20 per cent. of all rentals or royalties actually received or rated as paid in accordance with the provisions of this contract from licenses or leases for speaking telephones (exclusive of call bells, batteries, wires, and other appliances, or services furnished or performed) ' (article 1). The American Bell Telephone Company did not sell instruments to operating companies, but supplied them under specific license for their use, and the Western Union were entitled to one-fifth of the amounts received in respect of the instruments supplied. The full text of the agreement was published by the *Electrical Engineer* (N.Y.) of August 28, 1895.[2]

Thus ended the first and one of the most formidable attacks on the Bell patents and business. It was yet too early perhaps to realise that the great advantage conferred by combining existing exchanges instead of continuing them in competition was not limited to the companies concerned, but that there was a public advantage as well. The underlying principle had yet to be demonstrated in after years.

The Bell Company was now able to devote its energies with redoubled vigour to the actual prosecution of the business, with the result that the directors were in 1881 able to report :—

> The number of our instruments in the hands of our licensees in the United States :
>
> February 20, 1880, was 60,873,
> February 20, 1881, was 132,692,
>
> showing an output for the year of 71,819 instruments. This number includes 20,885 taken over from the Gold and Stock Telegraph Company [Western Union] and in use by our licensees.[3]

It is now usual to take a ' station ' as the unit for exchange statistics. A ' station ' requires both a transmitter and a receiver. It was customary in earlier years for the American Bell Company to record both transmitters and receivers as instruments. At the period in question it is probable that a number of magneto telephones were still in use as transmitters. Whilst the exact type of instrument, or the exact number of stations cannot be determined, it may be assumed that then, as now, two instruments composed a station, and therefore to facilitate subsequent comparisons

[1] Dickerson's Argument in *American Bell Telephone Co.* v. *Overland Telephone Co.*, Circuit Court of the United States, District of New Jersey, 1884, p. 65, etc.

[2] Vol. xx. p. 208.

[3] *American Bell Telephone Company's Report*, March 29, 1881, p. 1.

the above figures are transcribed in terms of 'stations' as follows :—

February 20, 1880, 30,436 stations,
February 20, 1881, 66,346 ,,

the latter including 10,442 stations taken over from the Western Union.

The agreement with the Western Union brought within the Bell Company's sphere of influence another manufacturing organisation, which had been engaged in electrical work for many years, whose facilities were considerable, and whose experts had shown, as already indicated, high appreciation of the requirements of exchange service. The Edison or Western Union business was absorbed by the Bell Company, whose personnel and influence dominated the situation, except as regards the manufacturing interest which was destined to absorb the principal licensee manufacturers of the Bell Company, and to take an important part in subsequent developments.

CHAPTER XVII

INTRODUCTION OF THE TELEPHONE IN EUROPE AND ABROAD

SCIENTIFIC men the world over evinced an immediate interest in Bell's invention, and some of them were as unstinting as Sir William Thomson in their praise of the inventor's achievement. Public attention was also early centred on the possibilities which might arise from it.

The French alone among Governments made recognition of a national character. Napoleon expressed his appreciation of the value of scientific investigation and artistic development in saying that

> the sciences which honour the human understanding, the arts which embellish life, and transmit great actions to posterity, ought to be specially patronised by an independent Government.[1]

And in 1802, when First Consul, he founded the Volta prize 'as an encouragement to him who by his experiments and discoveries, shall make in electricity and galvanism a step comparable to that which has been made in those sciences by Franklin and Volta.' This prize, originally of the value of 60,000 francs, altered to 50,000 francs when revived by Napoleon III, and so continued under the Republic, was conferred upon Bell in recognition of his invention of the telephone. The first award of the Volta prize was to Ruhmkorff in 1864 for the induction coil, the next to Bell, who was also created an officer of the Legion of Honour. The University of Heidelberg conferred upon him the honorary degree of Doctor of Medicine, in recognition of the use of the telephone in surgery. In 1902 the Society of Arts in London awarded him the Albert medal, and in 1913 the Royal Society conferred upon him the Hughes medal.[2]

[1] *Napoleon's Political Aphorisms*, p. 44.

[2] A bequest was made to the Royal Society by the late Prof. David Edward Hughes, the 'income from which was to be annually awarded, either in money or in the form of a medal, or partly one and partly the other, for

But whilst the scientific value of the telephone was highly commended, and the public had vague ideas of great potentialities, its commercial utility was not fully recognised, and there were even some hints that it might turn out to be merely a scientific toy.

The commercial developments in Europe were initiated by the American patentees with the help of a few influential and far-seeing people in each country. The importance which Bell attached to securing foreign patents, and the delay which took place in depositing the British application, resulted, as we have already seen,[1] in the United States application being deposited at a much later date than otherwise would have been the case.

Some efforts to deal with the foreign patents were made before the development of the exchange system in the United States, but the efforts were promptly renewed, and with greater prospects of success, when the great public utility of that system had been demonstrated. This demonstration was effectually shown during the competition between the Edison interests (represented by the Western Union Company) and the Bell Company. The competition was extended to foreign fields, the alliance which was effected in America not applying to the foreign organisations. The introduction of the telephone in Europe was therefore undertaken with the additional energy resulting from the rival claims of Edison and Bell, as well as some others of local origin.

Coming later into the field, Edison had more widely protected his carbon transmitter and was active in the exploitation of patents, as the manufacturing company allied with his interests was also active in its attempts to dispose of its apparatus.

Special companies were formed by the owners of the Bell patents for the development of foreign enterprises, and so soon as the National Bell Company was established and under capable management, Hubbard himself visited Europe in order to promote the foreign business.

The International Bell Telephone Company was formed in New York for the purpose of introducing the telephone exchange service on the continent of Europe, and the Tropical American Telephone Company to develop the business in South America, Central America, and the West Indies.

Colonel Reynolds of Providence, Rhode Island, came to London

the reward of original discovery in the physical sciences, particularly electricity and magnetism, or their applications.' The Hughes medal was awarded to Bell in 1913 'on the ground of his share in the invention of the Telephone and more especially the construction of the Telephone Receiver' (*Year Book of the Royal Society*, 1914, p. 175). The terms of the award are not very happily expressed, but comment thereon is reserved for Chapter xxxiii.

[1] Chap. v. p. 45.

to dispose of the Bell patents, and succeeded in interesting some important financiers, who formed 'The Telephone Company,' of which Mr. James Brand, an influential merchant, was chairman.

A circular, dated May 24, 1879, was issued by this Company. It contains a number of illustrations as 'a few examples of the applications of telephones for practical purposes.' They are all of the domestic or private line order, and the circular may be regarded as intended to develop that branch, but reference is made to the exchange system indirectly :—

The development of the telephone in England, although it has not made such rapid strides as in America, has, since its introduction, been advancing slowly and surely. A large number of instruments are now in constant use, and it has been found that the more accustomed one becomes to the telephone, the more its advantages are appreciated. But in the former country, by means of the central system, one communicates with one's tradesmen, calls cabs, transacts business of all kinds, without going out of the room.[1]

Edison's representative in London was Colonel Gouraud, at that time the resident director of the Mercantile Trust Company of New York. He formed the Edison Telephone Company of London, of which the Right Hon. E. P. Bouverie was chairman.

Both these companies started exchanges in London about the same time—the autumn of 1879. As exchanges had been growing in the United States since early in 1878, it has sometimes been suggested that Great Britain was somewhat dilatory in taking advantage of telephone exchange facilities. The reason, however, is to be found in the position of the respective companies as regards the patent situation. The Bell magneto telephone by itself was not powerful enough for general use as an exchange instrument, and The Telephone Company (Bell) could not use the Edison transmitter, whilst the Edison Company could not use the magneto receiver.

The first public exhibition of the Edison carbon transmitter in England was at the London Institution in a lecture by Professor

[1] The prices given in this circular (for a copy of which I am indebted to Mr. J. W. Ullett) are as follows :—

PRESENT PRICES OF TELEPHONES.

	Each £	s.	d.	
Telephone, Ebonite, hand pattern	1	1	0	Subject to royalty of 10*s.* 6*d.* per annum each.
Telephone, Snuffbox pattern	0	15	0	
Telephone, Wood Box pattern	1	10	0	
(*The Royalty may be commuted at any time by a payment of five years in advance.*)				
Telephone, Call Bell with Push and Automatic Switch	3	0	0	

[The policy of sale was subsequently modified and only leasing allowed.]

Barrett on December 30, 1878. A few weeks earlier one of the first long-distance experiments was made between London and Norwich over the private telegraph line of Messrs. Colman. In all these experiments a magneto receiver was used. The Bell Company drew

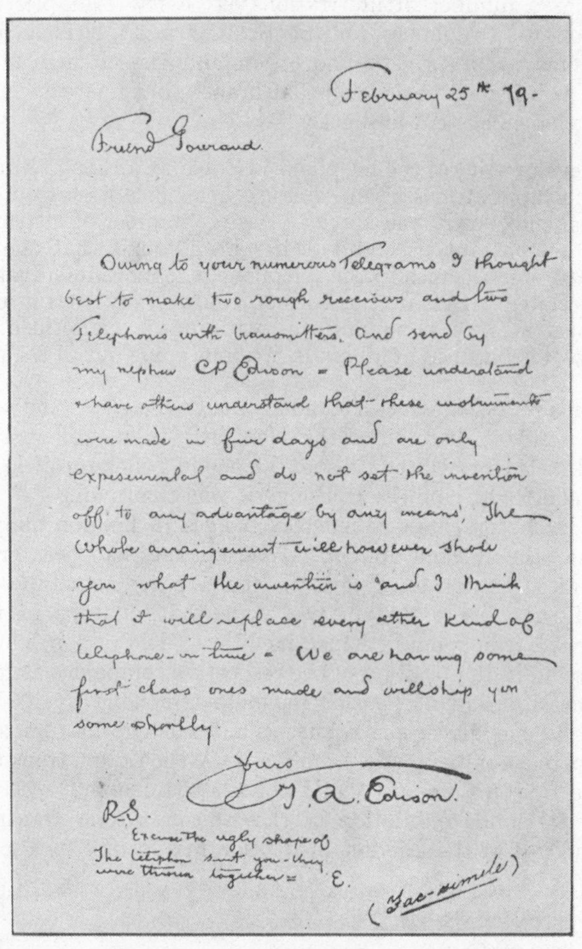

February 25th 79.

Friend Gouraud

Owing to your numerous Telegrams I thought best to make two rough receivers and two Telephones with transmitters. And send by my nephew CP Edison = Please understand & have others understand that these instruments were made in five days and are only experimental and do not set the invention off to any advantage by any means, The Whole arrangement will however show you what the invention is and I think that it will replace every other Kind of telephone - in time = We are having some first class ones made and will ship you some shortly.

Yours
T. A. Edison.

P.S.
Excuse the ugly shape of The telephon sent you - they were thrown together = E.

(Fac-simile)

Fig. 70.—Edison's Letter to Gouraud on Loud-speaking Receiver.

attention to this fact, and intimated their intention of taking proceedings to prevent the infringement of their patent. The carbon transmitter without a receiver was useless, and the commercial adoption of the magneto receiver would be sure to involve litigation. The position of affairs was reported to Edison, and he forthwith produced a receiver on an entirely different principle. The apparatus was not destined to be in use long, because it was less well adapted to the requirements of service than the magneto receiver, but its opportune production permitted the Edison Telephone Company in London to commence business free from the liability of interference by the Bell Company.

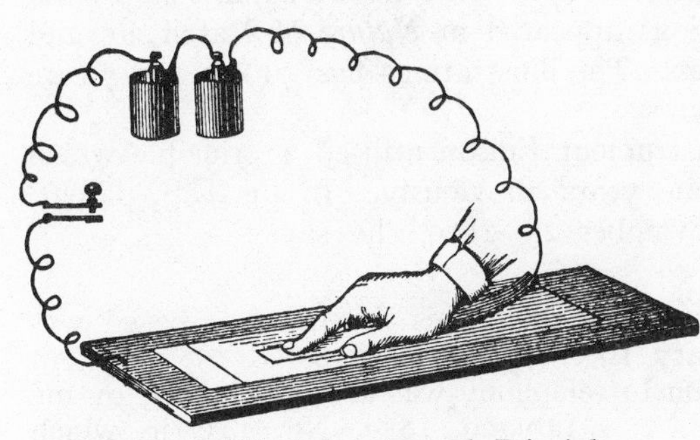

FIG. 71.—Electro-motograph Principle.

The use of this receiver was confined, in the main, if not entirely to England. Bell's United States patent controlled the transmission of undulatory currents corresponding to the aerial vibrations produced by speech. The British and other foreign patents were of less scope, and consequently, whilst Edison's new receiver was not used in the United States, it was a very valuable acquisition to the Edison Telephone Company of London.

FIG. 72.—Edison Loud-speaking Receiver. First form produced.

The letter in which Edison advised Colonel Gouraud of the despatch of the new instruments was reproduced in facsimile by means of another of Edison's inventions—the electric pen. Fig. 70 is photographed from one of the copies.

One reason for the haste with which these instruments were dispatched was that they might be used by Professor Tyndall in his forthcoming lectures on the subject of modern acoustics. They were first exhibited in operation at No. 6 Lombard Street, on March 14, 1879, and on the 17th were described in *The Times*. More detailed descriptions appeared in *Nature* of March 20, and *Engineering* of March 21. The illustrations figs. 71, 72, and 73 are from the last mentioned.

In designing this instrument Edison utilised a principle which he had discovered some years previously. In his U.S. patent, No. 221,957, dated November 25, 1879,[1] he says:—

The peculiar action upon which this invention is based was patented by me January 19, 1875, and numbered 158,787. An application of this action to telephony was also applied for by me July 20, 1877, No. 141, in which there is a band of paper moving beneath a point connected to the diaphragm. This feature therefore is not broadly claimed herein. The present application consists more particularly in devices which make the invention perfectly practicable for use in commerce and render the same reliable and effective.

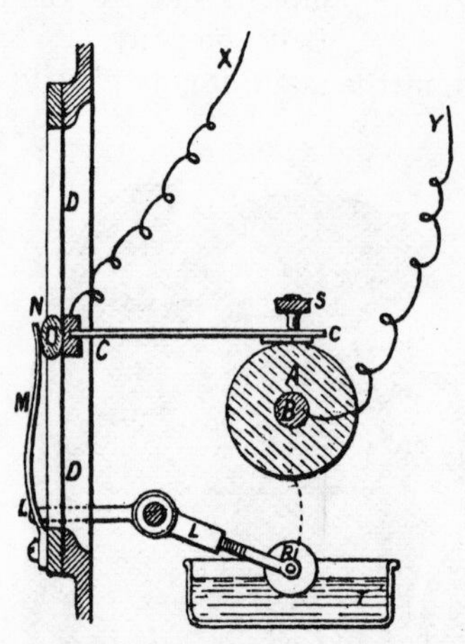

FIG. 73.—Edison Loud-speaking Receiver (principle).

The instrument which was produced at such short notice was thus rather a development than a new invention. Nearly two years previously he had applied for protection on the application of the 'peculiar action' to telephony, though only to a musical telephone with a make-and-break contact analogous to that of Reis.[2] Pressed for an independent receiver, Edison promptly set to work to complete the instrument in a practical form.

The principle illustrated in fig. 71 may be briefly described:—

A stylus connected with one pole of a battery being drawn over a strip of paper laid upon a metallic surface, Edison found to be subject to the effect of friction when the key forming part of the circuit was open and to be free from the effect of friction when the key was closed. He applied this principle of obtaining movement at a distance as an alternative to the armature and spring of a telegraphic instrument. It was useful as an invention and a patent rather than in practice, so that the electro-motograph, as he called it, was not generally known. When a telephone receiver on a new

[1] Application filed March 31, 1879.
[2] *Telegraphic Journal*, vi. 383, September 15, 1878, quoting *L'Électricité*.

principle was required, Edison's mind naturally reverted to the electro-motograph. It was originally invented to take the place of an electro-magnet and its armature. It was required now to take the place of an electro-magnet and its disc armature. Some audacity must have been required to assume that the variations in the friction would follow so closely the minute variations of a telephonic current. But a diaphragm with an arm attached to

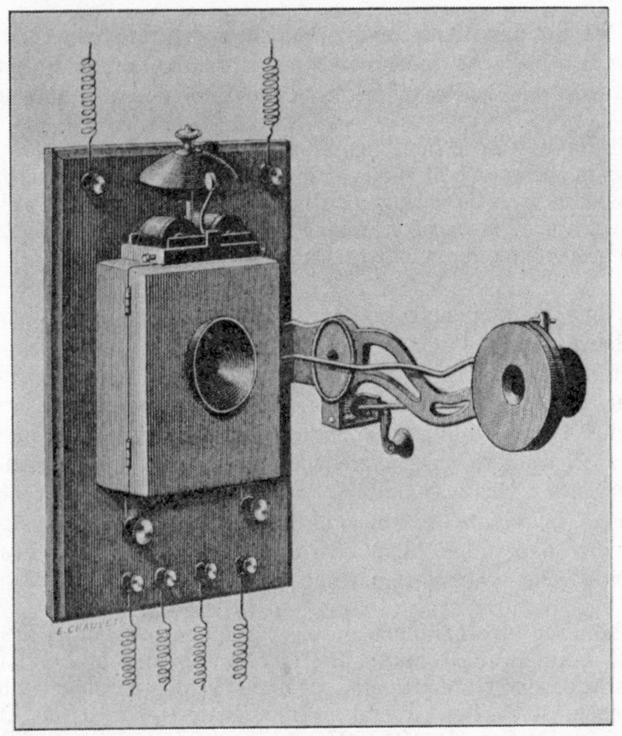

FIG. 74.—Edison Loud-speaking Receiver (commercial model).

its centre, a revolvable chalk cylinder upon which the arm pressed, and an electric circuit being 'thrown together' soon demonstrated that the instrument was a practical telephone receiver.

In its commercial form the instrument was reduced in size and attached to an arm projecting from the transmitter so as to be opposite the speaker's ear, as shown in fig. 74.

Edison's new receiver spoke loudly. This tended to increase popular wonderment, and was assumed to be a virtue. It formed one of the principal claims to the attention of capitalists. A 'private and confidential' Memorandum entitled 'Edison's Loud Speaking Telephone' commenced with the statement that—

The telephone which Mr. Edison has perfected in the course of this year differs radically from all previous instruments of the kind. Professor Graham Bell's invention, hitherto the most widely used, includes a magnet and a coil, and the sound is transmitted along the wire, losing much of its force on the way. In Mr. Edison's instrument the voice is mechanically reproduced at the end of the wire, and the speaker is heard with a volume of tone and a distinctness equal to the original utterance.

This circular was issued in the summer of 1879, and the next paragraph indicates the surprise felt in London at the growth then effected in telephone exchanges on the other side of the Atlantic :—

The extent to which telephony has developed in the United States in the course of the last few months is almost inconceivable. In towns as large as Chicago or Philadelphia, or as small as Wilmington, what are known as telephone exchanges have been started, and have been taken up on a large scale.

A line for demonstration purposes was erected between No. 6 Lombard Street (Colonel Gouraud's office) and the office of the Equitable Insurance Company of the United States, in Princes Street. Nine other lines were added and connected with a switchboard at No. 6 Lombard Street. Though necessarily of an experimental character, since there was no capacity for growth in the central office, this was the first exchange in London on the Edison system. It was in use for some time previously, but was publicly opened in September 1879. On the 6th of that month *The Times* described the system, remarking that—

telephonic intercommunication on a practical working scale has at length become an accomplished fact in the City of London. . . . The stations, or more properly speaking the private offices, which are connected with the exchange are situated—No. 1 in Copthall Buildings [Messrs. Parrish], No. 2 in Old Broad Street [Pullman Car Association], No. 3 in Suffolk Lane [Messrs. Renshaw], No. 4 in Lombard Street [Colonel Gouraud], No. 5 in Princes Street [Equitable Insurance Company of United States], No. 6 in Carey Street, Lincoln's Inn [Messrs. Waterhouse], No. 7 in Queen Victoria Street (the offices of the Company), No. 8 in George Yard, Lombard Street [Messrs. Kingsbury], No. 9 in Throgmorton Street [Messrs. Anderson], No. 10 being our own establishment.[1]

In the course of a leading article on the subject *The Times* remarks :—

We publish in another column the extraordinary new uses for

[1] *The Times*, September 6, 1879.

which this invention has been found capable. . . . There is no limit whatever to the number of points between which communication can be established, and scarcely a moment's delay in bringing them into connection with one another. . . . At the present moment there are ten favoured spots at which this privilege can be obtained; but there may just as easily be ten hundred or ten thousand, and, doubtless, before long there will be.[1]

The exchange was transferred to No. 11 Queen Victoria Street, and subscribers connected as quickly as possible. A list dated February 20, 1880, contains 172 names. In the provinces great activity was shown in starting exchanges both by the Bell and Edison companies.

The 'Telephone Company' working under the Bell patents issued a circular[2] which is not dated, but is believed to have been issued in September 1879 (a second edition is dated November 10, 1879). It states that—

A telephonic exchange has been established in the city.

Each subscriber has a wire from his own residence or office, with the necessary instruments attached, to the Telephone Company's office. A signal from the subscriber is answered by the clerk in the central office, who instantly makes a connection with the wire of any other person with whom communication is desired, and conversation can then be carried on with ease and privacy without the possibility of any third person hearing what is said.

In the third edition of the circular, dated December 24, 1879, A telephonic exchange' has developed into the plural 'Telephone exchanges have been established in the city.' About 200 names of subscribers are given in this circular. The Bell Company's first exchange and offices were at No. 36 Coleman Street, E.C.[3]

Retaining for itself the control of exchanges, the Bell Company on September 2, 1879, gave a restricted license to Messrs. Scott & Wollaston for the use of telephones for private lines and domestic purposes. Gower made a slight modification in the Bell instrument which was alleged to be of great advantage. Gower acquired Scott & Wollaston's license and formed the Gower-Bell Telephone Company. A circular issued by this Company commences with the statement :—

There are, broadly speaking, four classes of telephones :—

1. The *Original Telephone* of Professor Bell.
2. The *Electro-Chemical Telephone* of Mr. Edison.

[1] *The Times*, September 6, 1879.

[2] For which I am indebted to Mr. J. W. Ullett.

[3] The illustration on page 343 of Brault's *Histoire de la Téléphonie* shows this building with derrick on roof.

3. The *Gower-Bell Telephone,* formed on the principle of the Bell Telephone, but much more effective.
4. The *Gower-Bell Loud Speaking Telephone,* the latest and best form of instrument, being the combination of a special form of the Microphone Transmitter of Professor Hughes, F.R.S., with the Gower-Bell Telephone as a receiver.

and makes the following criticisms of its rivals :—

The *Bell Telephone,* the original and beautiful invention of Professor Graham Bell, and the foundation of the telephone system, produces very weak sounds in comparison with this company's special form of instrument, and for this reason its commercial use cannot be recommended.

The *Edison Electro-Chemical Telephone* can scarcely be considered a practical instrument. Its use has been altogether abandoned in the United States and on the Continent, and the authorities in this country find that they cannot make it work satisfactorily.

The Gower-Bell Company was subsequently absorbed by the Consolidated Telephone Construction and Maintenance Company, a manufacturing company which, seeking new markets, introduced a further competition of English origin in some foreign countries. Under the auspices of this company there was formed the River Plate Telephone and Electric Light Company, with an exchange in Buenos Aires, and in combination with the Edison interests exchanges were established in Vienna, Lisbon, and Oporto.

The competition between the Bell and Edison interests in the United States was closed by reason of the broad patent which Bell obtained there, an agreement having been come to because the Western Union Company acknowledged the breadth and validity of the original Bell patent, which really covered all practicable methods of transmitting speech electrically. The British patents were not so wide. Bell controlled the magneto telephone, except with a membrane diaphragm which was free of Bell's patent by reason of prior publication in the *English Mechanic,* and Edison controlled the carbon transmitter. The Edison Company, having carried on their introductory experimental work with a magneto receiver, became independent of the Bell patents on the production of the electro-motograph. The Bell Company, on the other hand, were incurring some risk of attack from the Edison Company by the use of the Blake transmitter. But no definite attack was made on the ground of patents. Commencing their exchange work in earnest about the same time (September 1879), they carried on their respective enterprises with all the energy which rivalry produces, and speedily demonstrated that they were supplying a service of great public utility. A common foe is a material aid to alliances, and to the British telephone companies in the year 1880 the common foe was the Government. The circumstances under which the Post Office claimed to control telephone exchanges and the result of their action thereon will be recorded in a later chapter, but it may be remarked here that the action taken by the Government facilitated the combination of the Bell and Edison interests, and no longer necessitated the separation of the carbon transmitter and magneto receiver, which would appear to be the natural complement of each other. For much as we may admire the inventive faculty which produced the electro-motograph receiver at the moment it was required, much as we may recognise the commercial ingenuity with which its loud speaking features were commended to public consideration, it is obvious that as a practical working instrument it was far inferior to the Bell form. To operate it at all a crank had to be continuously turned by hand. The loudness, in ordinary exchange use, was not merely of no value but was a definite drawback. That this was recognised by the Edison interests it would be incorrect to say, but the adoption of the Bell receiver was recommended to the directors of the United Company by the engineer of the Edison Company (E. H. Johnson) in a report dated June 3, 1880, on account of the greater simplicity in operating the apparatus. The following is an extract from this report :—

SIMPLICITY OF APPARATUS

This is a question of vital importance, for the reason that the telephone, unlike all other apparatus for communication at a distance, such as the various systems of telegraphs, is primarily so simple as to require no skill whatever in its use.

Now since the general public is notoriously incapable of grasping the simplest mechanical operations, this simplicity must be preserved. Any added complications of mechanism limits disproportionately the number of people who are able to manipulate the apparatus.

It is idle to seek to 'Educate the Public' to the observance of Rules and Regulations.' Such cannot be enforced except against employés.

In view of these facts it is, in my opinion an essential feature in effecting the necessary signalling, switching, etc., etc., that these operations should be reduced to the purely Automatic Action of taking the Telephone in the hand and laying it down (or, more accurately speaking, hanging it up). No lack of intelligence, forgetfulness, or imperfect comprehension of rules can operate to prevent the performance of so simple an act.

In order to obtain this degree of sin plicity I have been compelled to sacrifice the superior qualities of the Electro-Motograph Receiver of Mr. Edison to the superior fitness of the Magneto Receiver of

Professor Bell. This will appear more evident from a citation of a few of the operations performed automatically by the movement of a hook from which the Hand Telephone is taken when used, and upon which it is placed when the instrument is not in use.

Mr. Johnson's introductory observations are worthy of note. They record an early appreciation of the work which may suitably be put upon the subscriber, but his reference to the 'superior qualities' of the electro-motograph receiver and the 'superior fitness' as a detached instrument of the magneto receiver is less acceptable. The loyalty of the Edison entourage to their chief has been recorded by Mr. George Bernard Shaw, who for a time was employed by the Edison Telephone Company of London, being described in a list of the staff as 'Wayleave Manager.'[1] Mr. Johnson's phraseology may be excused as a retreat under cover, but then, as now, the Bell receiver had the 'superior qualities' as well as the 'superior fitness.'

The prospectus of the United Telephone Company was issued on June 8, 1880, with a capital of £500,000, of which £200,000 in shares was allotted to the Bell Company, and £115,000 in shares to the Edison Company. In the course of the prospectus it was said:—

The telephone system in this country has not hitherto been properly developed, partly in consequence of the antagonistic position of the Bell and Edison Companies. A similar state of things existed in America until these two interests became united. It is therefore to be expected that the telephone system will now make rapid strides in this country, as it has already done in America.

It is right to mention the contention of the Post Office, that their monopoly under the Telegraph Acts extends to the Telephone Exchange System, and a suit is now pending to have that question

[1] 'You must not suppose, because I am a man of letters, that I never tried to earn an honest living. I began trying to commit that sin against my nature when I was fifteen, and persevered from youthful timidity and diffidence until I was twenty-three. My last attempt was in 1879, when a company was formed in London to exploit an ingenious invention by Mr. Thomas Alva Edison—a much too ingenious invention as it proved, being nothing less than a telephone of such stentorian efficiency that it bellowed your most private communications all over the house instead of whispering them with some sort of discretion. This was not what the British stockbroker wanted, so the company was soon merged in the National [United] Telephone Company, after making a place for itself in the history of literature quite unintentionally, by providing me with a job. Whilst the Edison Telephone Company lasted it crowded the basement of a huge pile of offices in Queen Victoria Street with American artificers. . . . They adored Mr. Edison as the greatest man of all time in every possible department of science, art, and philosophy, and execrated Mr. Graham Bell, the inventor of the rival telephone, as his satanic adversary; but each of them had (or pretended to have), on the brink of completion, an improvement on the telephone, usually a new transmitter.' (*The Irrational Knot*, by George Bernard Shaw, preface, pp. ix-x. London: Archibald Constable & Co., Ltd., 1905.)

settled. The directors are advised that that contention will not succeed. The private telephone business of the company could not be touched under the Post Office Acts, and that department must pay for the right to use for profit the instruments protected by the Company's patents.

The United Telephone Company itself worked the London system and formed subsidiary companies to work provincial exchanges under license.

In France the telephone was, at an early stage in its history, the subject of much interest to scientific men. Du Moncel was one of the first in Europe to write a treatise upon it. This was translated into English and was for a time the standard work. Though published in 1879, it contains no reference to the exchange system. At a later stage when exchanges and trunk lines were making rapid progress it is to French literature that we have also to look for their earliest record. Wietlisbach in Switzerland had written upon Industrial Telephony, but mainly with a technical application. Brault in Paris about the same period (1888) published his 'Histoire de la Téléphonie et Exploitation des Téléphones en France et à l'Étranger,' in which the industrial features had the greater prominence. His work stood alone then in its record of world progress and is of much value for reference now. On the practical side Ader made a modification in the Bell instrument, and whether by reason of the modification or from any improvement in manufacture his receiver was much appreciated in European markets. He also produced a transmitter following closely the carbon pencil form of Hughes' microphone, which was one of the best of that kind. But whilst scientists devoted attention to the telephone as an instrument, in France, as elsewhere in Europe, the exchange system was developed under American auspices.

The rivalry of Bell and Edison was extended to Paris, but Gower had to some extent forestalled them, and received the first concession which applied to Paris, Lyons, Marseilles, Bordeaux, Lille, and Nantes. The Bell interests received a concession for Paris only, whilst the Edison interests were so fortunate as to cover the same cities as Gower. All these concessions were combined into one ownership in the Société Générale des Téléphones on December 10, 1880, and (except Paris and Lyons) only after that date were the exchanges proceeded with.

In Germany interest in the telephone was enhanced by the prior work of Reis. Patriotism is not a reliable aid to the investigation of scientific claims, but though unfruitful and forgotten, Reis had made experiments and constructed apparatus which he called a telephone, and these were sufficient to enable patriotic Germans to feel a proprietary interest in the invention.

On October 4, 1877, von Stephan, Director of Telegraphs, wrote to Bismarck that he had established communication between his office in Berlin and the suburb of Friederichsburg. He intimated also that he contemplated at once a practical application of the new invention in the Imperial Telegraph Service, proposing the connection by telephone of country post offices to which the telegraph service had not yet been extended. By the end of 1877 fifteen villages had been so connected with the general telegraph system. The exchange system in Berlin was not inaugurated until April 1, 1881.

The first exchange in Switzerland was that at Zurich, operated under a concession granted to a group of business men associated with the International Bell Telephone Company on July 24, 1880. During 1881 exchanges were opened in Geneva, Lausanne, and Winterthur by the Government, who also shortly after bought up the Zurich exchange. Fourteen exchanges were in operation at the end of 1883, and twice as many a year later.

The first experiments in Belgium were made in 1878. A company was formed in Brussels in 1879, and others followed. Competition was recognised as unsatisfactory, and the various companies were encouraged to amalgamate. The Compagnie Belge du Téléphone Bell was formed in 1882, this company being the Belgian subsidiary of the International Bell Telephone Company of New York.[1]

A similar company was formed in Holland under the name of the Nederlandsche Bell Telefoon Maatschappij in 1881.

In Austria the first exchange was established in Vienna in 1881 by the Vienna Private Telephone Company, but the exploitation was continued by a company under the auspices of the Edison-Gower Bell combination represented by the Consolidated Telephone Construction and Maintenance Company of London.

In Italy the International Bell Telephone Company established exchanges in Milan, Turin, and Genoa, and exchanges in a dozen of the other largest cities were started in 1881 by other interests under the auspices of a group of Paris financiers.

The International Bell Telephone Company was also responsible for the introduction of the telephone into Russia, Norway, and Sweden. In 1880 franchises were secured for

[1] Belgium was the locale of the first European factory of the American telephonic organisation, the International Bell Telephone Company and the Western Electric Company combining to form the Bell Telephone Manufacturing Company at Antwerp in 1882.

Christiania and Drammen, and in 1881 exchanges were established by the International Bell Company in Stockholm, Gothenberg, and Malmö, but in St. Petersburg (or Petrograd) and Moscow not until 1883.

Efforts to introduce the telephone into Spain were made by various interests until the Spanish concession became almost a byword amongst concession hunters. In 1885 exchanges were opened in Madrid, Barcelona, and Valencia.

In Portugal a concession was obtained and exchanges started by the Anglo-Portuguese Telephone Company under the auspices of the Edison-Gower Bell interests previously referred to. The Lisbon exchange was opened on July 2, 1881, and that in Oporto in 1883.

In India the development was undertaken by the Oriental Telephone Company, exchanges being opened in January 1882 at Calcutta, Rangoon, Madras, Bombay, and Colombo.

The Telephone Company of Egypt (a subsidiary of the Oriental Company) established exchanges in Cairo and Alexandria in 1880.

Amongst other early exchanges may be mentioned Honolulu (1880), Rio de Janeiro (1881), and Valetta, Malta (1883).

In the Argentine Republic there were three sources of telephonic enterprise. One of local origin with a Belgian instrument known as the Pan-Telephone was introduced by Mr. Fels, another by the Tropical American Telephone Company, and a third by the River Plate Telephone and Electric Light Company, formed by the Consolidated Telephone Construction and Maintenance Company. The exchanges of the local Pan-Telephone Company and the Tropical American Company were combined under the name of the United River Plate Telephone Company, and this took over later the Consolidated Company's interests, forming the United Telephone Company of the River Plate, which now exists as one of the important foreign companies under British management.

The introduction of the telephone into Australia was made under the Edison patents. Early in 1880 Mr. F. R. Welles of the Western Electric Manufacturing Company left New York for Australia, and in conjunction with a local firm—Messrs. Masters & Draper—established the Melbourne Telephone Exchange Company which made rapid progress. Efforts to obtain concessions in the other Australasian Colonies were less successful, but the attention of the local Governments having been drawn to the advantages of the exchange system, they shortly after established exchanges of their own, as did also the Government of New Zealand.

The Melbourne Exchange and those subsequently established at Ballarat and Sandhurst, having been purchased by the Government, were taken over by them on September 22, 1887. There were then 1019 subscribers to the company's exchanges, and of these 752 were at Melbourne. 'The Victorian Year Book 1887–8,'[1] from which these figures are obtained, states that 'An exchange has also been opened at Geelong, on a guarantee that not less than forty persons become subscribers.'

[1] Par. 978.

CHAPTER XVIII

PUBLIC APATHY AND APPRECIATION

THE various companies formed for establishing telephone exchanges in Europe and abroad based their expectations of success on the results attained in the United States. They were able to point to accomplished facts indicating the great utility of such a service.

In the United States the public generally were extremely sceptical of its general adoption, and even those who could foresee the great public advantage were not disposed to let their hopes run too high. The district telegraph system was an important aid to development in that it permitted demonstrations of interchangeable connections amongst a few subscribers with but little work or outlay. The lines were there with call boxes connected up. To attach telephone instruments was a comparatively simple matter. In Europe there were no district telegraph systems, the telephone exchange was adopted in its complete form, and consequently expensive construction work was needed in order to permit of the demonstration of the utility. But the pioneers on both sides of the Atlantic were undaunted, and provided the capital which was required to tide their enterprises over the initial stages and embark them upon highly successful careers.

The public apathy in the United States was as pronounced as in Europe. Appreciation came sooner, but only as a result of demonstration. Various causes have been suggested for the earlier adoption and more rapid development in the United States. Without examining these in detail or contesting any of the numerous arguments which have been based on them, a sufficient explanation may possibly be found in the confidence and commercial energy of the exploiters, together with the economic conditions prevailing. The great distances separating American cities and the consequent delay incidental to postal communication contributed to the general use of the telegraph for communications which would have been forwarded through the post in Europe. It is doubtful also if at the time of the introduction of the telephone the local telegraphic service was either so cheap or so generally used in the United States as in Great Britain. Thus the comparatively local telephone (as it was upon its introduction) supplemented the long distance telegraph and superseded the strictly local district messenger service. The high cost of labour in the United States in comparison with Europe naturally tended to the more ready adoption of any expedient which economised labour or time. The cheapness of labour and also the cheapness of the telegraph service were urged as reasons why the telephone exchange system would be likely to be less successful in Great Britain than in the United States.

In 1879 a Select Committee was appointed by the House of Commons to consider some questions relating to electric lighting. On May 2, Sir William Preece was asked by Lord Lindsay :—

As to the question of induction, you spoke of the use of the telephone in all this research that you have been making on this subject ; that is, of course, an extremely delicate instrument for testing ; but do you consider that the telephone will be an instrument of the future which will be largely adopted by the public ?—I think not.

It will not take the same position in this country as it has already done in America ?—I fancy that the descriptions we get of its use in America are a little exaggerated ; but there are conditions in America which necessitate the use of instruments of this kind more than here. Here we have a superabundance of messengers, errand boys, and things of that kind. In America they are wanted, and one of the most striking things to an Englishman there is to see how the Americans have adopted in their houses call bells and telegraphs and telephones, and all kinds of aids to their domestic arrangements, which have been forced upon them by necessity.[1]

Lord Lindsay's question was interpolated in an inquiry on another subject, and Sir William Preece's answer may therefore be considered as not having been carefully weighed, but it certainly repeats in very similar terms the remarks made by him a few days earlier (April 23) in a discussion at the Society of Telegraph Engineers on a paper by Mr. Scott on 'Recent Improvements in Professor Bell's Telephones.' The 'improvements' were those of Gower. Mr. Scott had complained of the backwardness in the adoption of the telephone in England in comparison with the United States. Regarding this, Sir William Preece said :—

The telephone had been used in this country to a large extent, but there does not appear to be the want of it in England that there is in America. One thing which strikes one in America is the enormous extent to which they apply the telegraph and the telephone for their own domestic purposes.

[1] *Report Lighting by Electricity*, June 1879, p. 69.

In Chicago, where there are from 7000 to 8000 calls daily, there is scarcely a house which has not in its hall a call bell, by which you may dispatch a message for a doctor, or a porter, or anything else you want, and the reason they are driven to that is—necessity being the mother of invention—that it acts as a substitute for servants. Here we have no difficulty in getting servants if we pay them, but the difficulty in America is to get 'buttons' at any price to run about for you as in England, and the result is the absence of servants has to a certain extent compelled the Americans to adopt this system of telegraphy for their own domestic purposes, and the telephone is to be found in almost every house as the only available substitute for the old system.

Few have worked at the telephone much more than I have. I have one in my office, but more for show, as I do not use it because I do not want it. If I want to send a message to another room, I use a sounder or employ a boy to take it; and I have no doubt that is the case with many others, and that probably is the reason why the telephone has not been more adopted here. The efficiency of the instrument in England has been seriously interfered with by those fearful inductive effects which are not felt to the same extent in America, because they have no long underground lines and they do not use that fast speed apparatus which produces such a tremendous roar with us. It is impossible with sixty wires in an underground pipe to speak through the telephone, the inductive effects being so great.[1]

These observations may therefore be taken as the expression of a then prevalent idea in England, an idea which also presumably governed official action or inaction.

Few men have done more than the late Sir William Preece to popularise the telephone, or have given expression to greater appreciation of its beauties as a scientific instrument. It was his pride that he had brought the first pair of Bell's perfected type to England. At the Royal Society, the British Association, and other meetings he had lectured on the telephone, and the researches which he and his associates had made upon or with it. To him Hughes communicated his first ideas on the microphone. Sir William Preece may well be considered to have been an enthusiast on the telephone, and he was electrician to the Department which controlled the telegraphs. The Department refused to purchase Bell's patent, and the Department has been charged with short-sightedness and English people generally with backwardness. But in thus refusing, the Government who controlled the telegraphs in England were acting in no wise different from the company which controlled the greater part of the telegraphs in the United States, though the opinion of the latter had altered before 1879.

[1] *Journal of the Society of Telegraph Engineers* (I.E.E.), viii. 337.

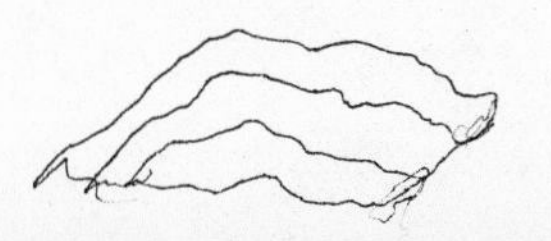

The failure in both cases to anticipate the actual results may be ascribed to a too long acquaintance with the word 'message.' Telegraphists sent messages, and they were keen on sending them as quickly as possible. It was to send messages to villages that von Stephan first saw utility in the telephone, but except for such purposes the telephone would not be likely to appeal to a telegraphist as an ideal means of transmission. There were instruments in existence which would send more words in a minute and would in addition leave a record for reference.

The saving in time which would result from having the message delivered direct to the receiver might have been expected to be apparent, but the time element in the delivery of messages had been carefully studied, and it was probably considered that this saving would not appeal to the public sufficiently to justify the expectation that they would in large numbers pay for the expense of erecting wires.

The real difference between the people who foresaw great results for the telephone and those who were disposed to weigh comparisons of costs and differences of local conditions lay in the word 'message.' The telephone made practicable what had been considered impossible before—distant conversation. Question and answer immediate, spontaneous; the meanings of words accentuated by inflexion and emphasis. The 'message' was no comparison for this. As well compare the deaf and dumb alphabet with full, free, eloquent speech. The telephone brought distant people together with all the advantages of close converse which audibility permitted. Such advantages were not to be weighed up against the cost of messengers, though time showed that even so the telephone effected an economy as well as a revolution.

Not from the light of after events comes an expression of surprise that anyone who took part in a telephonic conversation could have any doubts of the telephone's commercial success. To have conversed through the telephone before it became generally known was an experience which must be regarded as a privilege. The prevailing feeling was that of awe and wonderment, tempered by unbelief. When the circumstances left no room for doubt that the speaker was really at a distance, the clear transmission of his words, the proof by prompt reply that he had heard equally clearly the words you yourself had uttered, produced an overwhelming impression that here was an instrument fated to be of enormous value to the public, and that should produce commensurate profit to its introducers.

Public as well as official apathy gave place to appreciation wherever exchanges were established. Commencing in the United States in 1878, 'in towns as large as Chicago or Philadelphia or as small as Wilmington '—as the 'Memorandum' prospectus of the London Edison Company explained—extending to London in 1879, by 1881 or 1882 telephone exchanges were established in the principal cities of Europe, in India, South America, the ancient land of the Pharaohs, and the modern city in Australasia which takes its name from Lord Melbourne, Queen Victoria's first Prime Minister.

And these exchanges were progressing so rapidly as to tax the resources of the proprietors, and with their growth arose new problems requiring the highest scientific treatment and introducing a new branch of engineering.

CHAPTER XIX

THE MULTIPLE SWITCHBOARD

THE problems which faced the telephone engineers at the start of the exchange business were in the main solved by the light of the experience gained in telegraph engineering. Switchboards were adapted, the line construction was carried on from the older system to the newer. But as the subscribers grew in number and the transmission distances extended, the problems assumed an entirely new aspect.

Intercommunication required that lines should be brought to a central office in order that they might be connected together, and the telephone engineer had to deal with hundreds or thousands where his predecessors had to deal only with units. So with the switchboard. The board which was adopted at New Haven permitted eight subscribers to be interconnected, the Williams and the other licensee boards would work for a few hundreds, and the Standard switchboard gave a satisfactory service up to five or six hundred—sometimes more. But the subscribers increased, and with the growth in the number of lines the calls increased in greater proportion. The cost of operating was becoming a serious matter, but still more serious was the question of determining how to make the required connections at all.

The two problems of outside or line construction and inside or switchboard work were attacked concurrently, but attention will first be given to the latter.

The early switchboards, as described in Chapter XIV, had one point of access to a subscriber's line. The number of subscribers to which one operator could attend under the conditions then obtaining was usually limited to twenty-five, and rarely exceeded fifty, so that a connection between subscribers who were not both in any one operator's group would need the intervention of two operators, with some arrangement for making the connection between the two boards upon which were to be found the points of access to the two subscribers' lines. The Standard Board reduced this difficulty in part, since it was so designed that an operator could make direct connection with the board at either side, but beyond that the intervention of two operators was necessary together with transfer arrangements that became the more complicated as the numbers of subscribers increased.

The solution was found by giving to each operator a means of access to every subscriber's line—that is to say, by multiplying the points of access—and the switchboard by which this was effected was called the ' multiple ' board.

Next to the telephone itself the invention and development of the multiple switchboard must be regarded as the most potent factor in the telephone art. Without it, large exchanges would have been impossible, and the service through numerous small exchanges so slow and unsatisfactory that continued growth would have been unlikely.

The germ of the multiple switchboard is found in the United States patent of Leroy B. Firman, No. 252,576, dated January 17, 1882. The application was filed January 7, 1881, though the apparatus described was in operation nearly two years previously. Firman was the general manager of the American District Telegraph Company at Chicago. This company was introducing the Edison telephone in connection with their district system, and, finding much difficulty on the part of the operators in connecting subscribers through separate boards, Mr. Firman made some experimental designs towards the end of 1878, and about February or March 1879 he fitted up two boards at 118 La Salle Street, Chicago, upon each of which boards there were connecting points for all the subscribers to the exchange arranged in the manner illustrated in fig. 75, which reproduces fig. 1 of the patent.

The prior state of the art and the scope of Firman's invention are clearly and concisely stated in his specification as follows :—

Prior to my invention the individual lines were grouped upon a single switchboard at the central office, or grouped upon two or more boards. In the latter case trunk lines were used when it was necessary to connect a line of one board with a line of another board. A large exchange was thus divided up into a number of exchanges, which could be worked together when occasion required, as one, by means of trunk lines between the boards. When the number of subscribers increased, so that a single switchman could not do the amount of switching required, I gave the switchman an assistant. I soon found, however, that a single switchboard would not accommodate the number of attendants necessary to do the switching for an exchange of four or five hundred subscribers.

I find by the use of my new system of multiple switchboards, as hereinafter described, an exchange of a thousand or more subscribers may be successfully handled.

My invention consists in providing two or more switchboards instead of one, as heretofore, and so connecting the several lines therewith that any two lines can be connected on either of the boards, and also apparatus, whereby attendants at a given board may without delay see what lines are connected at other boards than their own.[1]

Apparatus such as that referred to in the last few lines, whereby an operator could ascertain if a given subscriber were already connected, is an essential on any switchboard having multiple

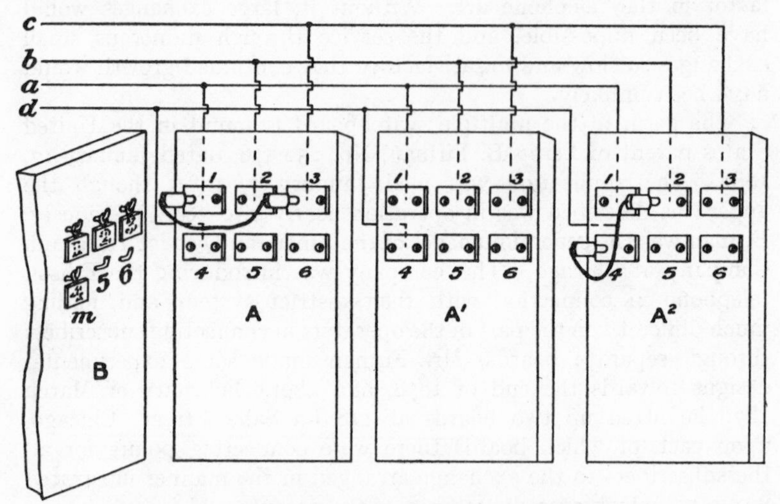

FIG. 75.—Firman's (the first) Multiple Switchboard.

connecting points. Firman's plan was of a very simple character. In the patent it is described as a 'dummy board or indicator' upon which were recorded the subscribers' numbers. This board was placed in view of all the operators. The method of operating this primitive 'engaged test' is thus described:

Suppose lines *a* and *c* are connected at multiple board A, and lines *b* and *d* at multiple board A^2, as shown by cords and plugs. The switchmen at the boards, immediately on making these connections, notify the attendant at the dummy, who thereupon hangs up the shields or targets over the figures 1 and 3 and 2 and 4; and in the same manner, when any line is connected upon either of the multiple boards, the figure which indicates its number is covered, and a switchman, by glancing at the dummy, sees what lines are con-

[1] U.S. specification, No. 252,576, January 17, 1882.

nected. For example, if the subscriber connected with plate 6 were to ask for the subscriber connected with plate 1, the attendant at board A′, before making the connection, must glance at the dummy board, and in case he should see the target over figure 1 he would know that the line wanted was in use at another board, and instead of connecting plates 6 and 1, he would notify the subscriber connected with plate 6 that the person wanted is busy.[1]

The next paragraph is of interest as illustrating the alternative methods of sending a disconnecting signal either communicating with operators by means of a circuit wire, as in the 'district' system,[2] or by the subscriber's direct line :—

> The central office may be notified when the subscribers are through talking by the American district system, or by sending a current to line at either terminal station and tripping an annunciator number in the circuit at the central office. As soon as the signal to disconnect is received the switchman pulls out the plugs from the terminal plates or switches and immediately notifies the attendant at the dummy board to remove the targets.[1]

Mr. C. H. Wilson, who was indirectly associated with the District Telegraph Company, says that the arrangement of the indicator board

> did very fair work up to a certain point of the development of the business, but as the business grew and the lines became more numerous, . . . the necessity arose for a quicker method of determining whether a line was in use or not, and the necessity of a larger number of switchboards was also apparent.[3]

The President of the District Company had some misgivings regarding the multiple principle, but Mr. Wilson reassured him on that point. He considered, however,

> that it would be necessary to develop some scheme by which an operator could determine instantly whether or not any given line was in use upon the other sections of the switchboard. I mean to say the sections other than the one at which the particular operator was at work.[3]

The necessity for the development of some such plan was also apparent to others, and Scribner was the first to devise a test system. This is shown in fig. 76 (fig. 6 of British patent, No. 4903

[1] U.S. specification, No. 252,576, January 17, 1882.
[2] Described in Chap. xxiii.
[3] Capital Telephone and Telegraph Co. Case, Circuit Court, U.S., Northern Dist., California, 1896, *Supplemental Brief*, p. 74.

of 1879), where a local circuit, distinct from the line and normally open, is completed when a connecting plug is inserted in a 'block' or spring-jack. 'The closing of the circuit by the insertion of a peg may operate indicators at each board in any well-known manner.'[1] The manner suggested in the patent was the electro-magnetic control of a valve in connection with a pneumatic tube where air under pressure should operate a piston to serve as an indicator. In this selection of a visual signal may be traced the line of development from Firman, whose 'dummy board' or 'target board' was looked at by the operator to obtain information as to the engagement or otherwise of a required line. This development is still more apparent in the United States specification, covering the same invention in greater detail.

In fig. 10 sheet 3 [says the inventor] instead of a target for each

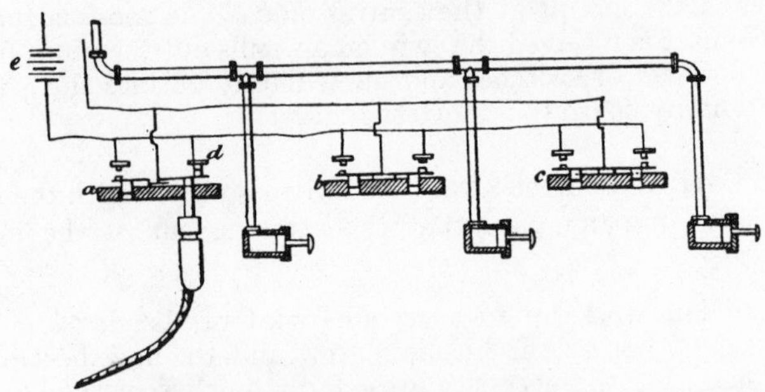

FIG. 76.—First Electrical Multiple Test System. (Scribner.)

bolt, I have placed single targets, one for each subscriber's line, or series of bolts on an annunciator board in sight of the operators of all the boards.[2]

In other words, for the manually placed target of Firman was substituted a target electrically displayed when the operators made or broke a connection. While this alternative method was suggested in the United States specification, the plan contemplated for use was that the visual signal or target indicator should be brought on to the switchboard in front of the operators, multiplied in proportion to the number of lines and sections, and actuated automatically by the insertion or removal, as the case might be, of the plug with which the connections were made.

The following description is from the British patent :—

[1] British specification, No. 4903, 1879, p. 8, line 39.

[2] U.S. specification, No. 266,320, October 24, 1882 (application filed January 7, 1881).

The insertion of a pin into either of the blocks of a series completes a local circuit, and so operates an indicator at each of the boards. In the figure *a*, *b*, and *c* are three blocks on as many different boards, forming a series belonging to one subscriber. Behind these blocks contact springs are placed in such positions that the insertion of a peg into either block deflects the spring into contact with a stud *d*. *e* is a battery; all the springs are connected with one pole of the battery, and all the studs *d* with the other pole. The closing of the circuit by the insertion of a peg may operate indicators at each board in any well-known manner. A very suitable way is to arrange an electro-magnet in the local circuit to operate on a valve and so admit air under pressure to a pipe leading to a small cylinder and piston indicators, of which one is provided in the vicinity of each block. When air is admitted to the pipe the pistons are raised to the tops of the cylinders, where they remain until, by another movement of the valve, the air is allowed to escape from the pipe, when the pistons fall by their own weight.[1]

The test of the pneumatic system was made in Chicago at the Western Union building, where compressors were in use for the pneumatic transmission of messages.

In the course of an affidavit made in 1884 Mr. Scribner says that a portion of the invention was completed and described to others by diagram as early as February or March 1879, and that a test of the pneumatic duplicate switchboard system took place in the summer of 1879.

Mr. Wilson, in conjunction with Mr. Clark C. Haskins, developed another plan, shown in the United States patent, No. 266,287, dated October 24, 1882, the application for which was filed October 23, 1879. After an introduction reciting the prior usage which has already been explained, the patentees say:—

The object of the present invention is to provide a means whereby several duplicate switchboards may be employed, to each one of which all the telephone stations are connected, and so arranged that when any one line is in use or occupied at one of the switchboards, that fact will be rendered apparent at all the other switchboards, thus preventing any confusion between the different operators at the several switchboards, and also preventing any accidental crossing of the lines.[2]

Fig. 77 follows closely the sketch of Firman (fig. 75), but instead of having an indicator board upon which targets were placed by an assistant when a line was engaged, provision is made for the

[1] British specification, No. 4903, 1879, p. 8, line 32.

[2] U.S. specification, No. 266,287, October 24, 1882 (application filed October 23, 1879).

indication of an engaged line by an electrical test. The explanations of the drawings given in the specification are as follows :—

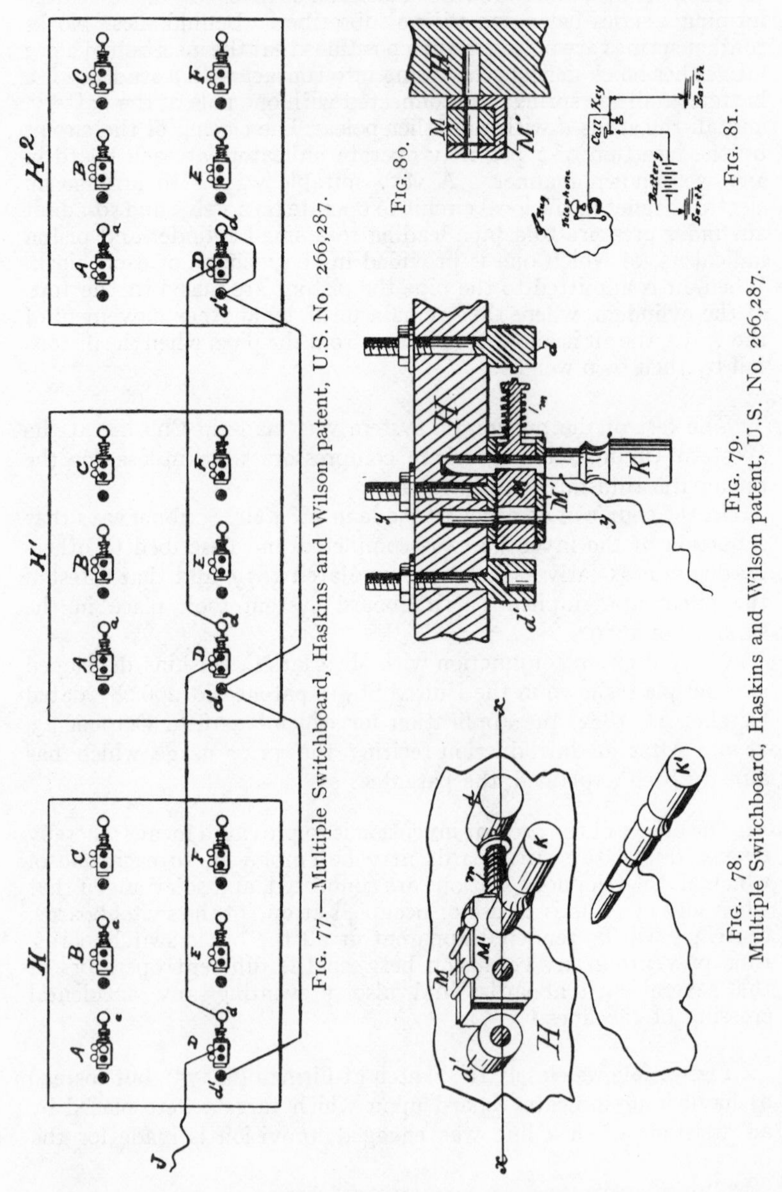

FIG. 77.—Multiple Switchboard, Haskins and Wilson patent, U.S. No. 266,287.

FIG. 78. Multiple Switchboard, Haskins and Wilson patent, U.S. No. 266,287.

FIG. 79. Multiple Switchboard, Haskins and Wilson patent, U.S. No. 266,287.

FIG. 80.

FIG. 81.

Figure [77] is a view of several duplicate switchboards, there being three in this instance, arranged according to our invention. Fig. [78]

is a perspective view of an improved switch such as we prefer to employ in working our invention. Fig. [79] is a sectional view on the line *xx* of Fig. [78]. Fig. [80] is a sectional view on the line *yy* of fig. [79]. [K′] Fig. [78] is a perspective of one of the plugs removed from the switch. Fig. [81] is a diagram of the call circuit employed in conjunction with the apparatus at the central office.[1]

It will be noted that the switch contains the essential features of, and performs the same operations as, the modern spring-jack. No insulating material being used in its construction, it is necessarily mounted on the board in three parts, *d′*, M′, and *d*. M′ consists of two separate parts, one fixed (M′), the other (M) sliding within a suitable guide piece. M has a projecting bar with collars at either end and a helical spring (*m*) over it. When no plug is inserted the spring forces the sliding piece into contact with the stud *d*. Holes are bored through M and M′ adapted to take a plug, but when the slide is in contact with the stud these holes do not exactly register, but the movement of the slide is so limited that enough of the hole remains uncovered to receive the point of the plug, and upon its further insertion the plug, acting as a wedge, overcomes the operation of the spring and forces the slide M out of contact with *d*. *d′* is called a 'try piece.' The combination is therefore comparable with a single circuit multiple jack, M′ representing the spring, *d* the contact, and *d′* the frame or testing point.

The method of wiring is clearly shown by fig. 77, where 'J is a line wire from a distant telephone station entering the central office. This wire is connected first to the metal of the switch D at the board H. This switch being closed, the circuit is through it to the contact plate *d*, and from this to the switch D on the board H′.' The operation is repeated on the board H^2, where the wire continues to 'trying post' *d′* on that board and to each 'trying post' on the preceding boards.

The circuit may thus be described in modern terms as commencing with the spring of a multiple jack through the contact on that jack, and similarly through successive jacks to the last section, where it is connected to the frame of that jack and successively to the frames of each jack of the series to the first section.

Connection between two jacks is made by two metallic plugs, K and K′, which are connected by a flexible cord, and the insertion of the plug in either jack breaking the contact as already described. It being assumed that A and D lines are connected together,

if an operator at either of the other switchboards has a call for either of the lines represented by the switches A and D, he will first proceed to ascertain if those lines are in use at either of the other switchboards, and this he will do as follows :

[1] U.S. specification, No. 266,287, October 24, 1882.

The central office is provided with a magnetic call bell and circuit (shown in fig. [81]). The call bell may be one upon a telephone, for convenience. We will suppose that the operator on the switchboard H′ has received a call for the subscriber connected with the lines of the series of switches D. He tests the question as to whether this line is in use upon either of the other switchboards by inserting a plug from the circuit at fig. [81] into the trying plate *d′* on his switchboard H′. If he gets circuit he knows that the line is unoccupied, as in this case he would be in circuit shown at fig. [77]; but if the switches A and D on the board H were connected, as previously supposed, he would not, by trying at the board H′, get the circuit, because the connection would be broken between the switch D and the contact plate *d* on the board H′; nor would he get circuit if the switch D on the board H^2 were in use, because in that case the contact plate *d* and switch D on that board would be disconnected, and there would be no circuit from the trying plate *d′* on board H′ beyond the contact plate *d* on H^2. Thus it will be seen that the circuit is always cut out behind the operator at each switchboard when any switch is in operation.[1]

It will be understood that in ‘getting circuit’ a signal is made.

In the preceding description the operation of making a connection has referred to only one hole of the switch, but, in the words of the patentees,

> it will be observed that we provide each switch with two holes. While this is not absolutely necessary, we find it convenient, for the reason that it may be desirable to connect two stations with the telephone at the central office before connecting them with each other, and two holes being provided, the plugs connecting the switches together may be inserted before the plugs belonging to the office telephone are removed.[1]

Another point of detail to which attention may be drawn is the provision made for retaining the plug when inserted in a switch. On reference to K′, fig. 78, it will be seen that a slot or notch is cut in the plug. The patentees say:—

> A plug when inserted should be turned so that the slide piece or lever M may fit into the notch of the plug. The plug is thus held securely in place.[1]

It has been stated previously that Firman was manager of the district telegraph system at Chicago, which concern was connected with the Western Union Company, and that the Western Electric Manufacturing Company was associated with the Western Union Company. Their factory was at Chicago, and various new devices which were required as the exchange progressed

[1] U.S. specification, No. 266,287, October 24, 1882.

came under their notice. Their technical staff included C. E. Scribner, who joined the Gold and Stock Telegraph Company at Chicago in November 1876, doing then occasional work for the Western Electric Manufacturing Company, and in the summer of 1877 was transferred to the latter company. To Mr. Scribner Mr. Firman described his system, and the apparatus was constructed by the Manufacturing Company. In the course of evidence in a law case, Scribner relates that he

personally had to do with this work, saw the apparatus constructed and installed, and witnessed its successful operation.[1]

From the same source there may be taken with advantage the evidence of this eye-witness as to the position of the exchange system in these early days and the problems with which the pioneers were faced.

The value of the telephone exchange came to be appreciated by the public very soon, and the growth of the telephone exchange gave to its promoters surprise following surprise. A switchboard designed and well calculated to provide for the estimated growth of years would be found of too small capacity for the needs of the time when completed—the growth of the business between the date of the order and that of completion being sufficiently great for this result. The single operator of the early exchange was first given an assistant, other operators and assistants were added, and a difficult problem soon presented itself to the pioneers of telephony. With the increasing business and the necessary employment of a number of operators, it was found that each operator must be assigned a limited number of calls.[2]

After describing the single form of switchboard, Mr. Scribner continues :—

With the earlier growth of the telephone business, and up to the time when three or four hundred subscribers were connected with the system, this plan of operation succeeded fairly well. It was apparent to those studying the question and directly dealing with the problem that with increasing business the limit of the capacity of this system would soon be reached. The problem presented to those beginners in the telephone business was to provide for the interconnection of any two lines of an extremely large number of lines ; to avoid the possible connection of three lines together at any one time ; to provide for the prompt disconnection of any two connected lines, and to give to each operator of the system the apparatus with which to accomplish this result. That is to say, each

[1] Capital Telephone Case, Circuit Court of United States, Northern District of California, 1896, *Complainants Record*, p. 20.
[2] *Ibid.* p. 19.

operator of the exchange must be provided with facilities for connecting any one of the lines assigned to her with any other line in the entire exchange. But she must be protected against making a connection with any line which was already in service, whether such line be one assigned to her or to any other operator of the exchange. The problem of interconnecting lines for communication originated with the advent of the telephone, and was an entirely new one to electricians and engineers. Many different solutions of the problem were offered and many tried. Failure followed failure, and in the first months of 1879 perfect chaos existed in the larger telephone exchanges of that time. No single switchboard could be constructed which would be large enough to provide for the switching of the exchanges at that time; increasing the number of operators increased the difficulty, for the reason that it resulted in a division of the lines of the exchange into smaller groups and increased the number of groups, thus increasing the difficulty of

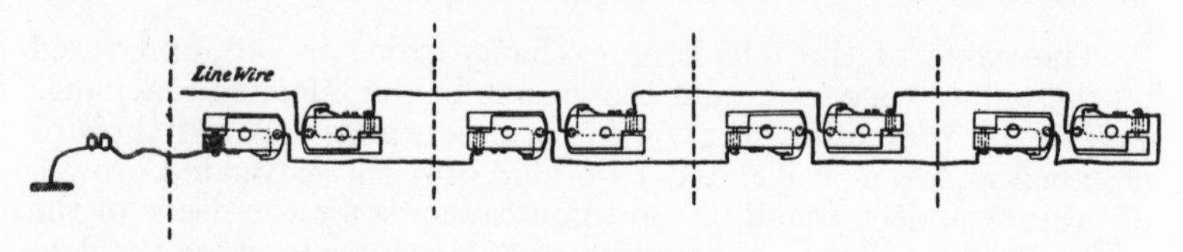

FIG. 82.—Multiple Switchboard double 'block' method of connection (1879).

establishing a connection between a member of one group and a member of any other group.[1]

In 1879 Scribner was in London for the purpose of taking out the patent already referred to, which included several features in connection with telephones which his company had improved. It was partly a communication from George D. Clark, Milo G. Kellogg, and George B. Scott. It included an improvement on the Edison transmitter, and the circuits of the Edison sub-stations as they were introduced by the London Edison Company. But the principal feature of this patent was the multiple switchboard.

It will be seen that the circuit of fig. 82[2] has a resemblance to the plan of Wilson and Haskins, but instead of passing through one switch on its way out and returning to a stud of that switch, two switches are used. The switches are of the 'jack-knife' form—a blade pivoted in a frame and resting on an insulated contact.

The purpose of the arrangement shown in fig. 82 is thus described :—

When the number of subscribers is large it becomes necessary to

[1] Capital Telephone Case, Circuit Court of United States, Northern District of California, 1896, *Complainants Record*, p. 19.

[2] U.S. equivalent, No. 321,390, June 30, 1885.

divide them into sets, so that one attendant may attend to the calls of one set, another to those of another set, and so on. I then have a separate board for each set, and on the board are the call indicators of the set, and those only; but the line wire of each subscriber is brought to the board, for a subscriber belonging to the set may desire to speak with another subscriber out of the set, and it is the duty then of his attendant to make the necessary connections. But a difficulty arises when the subscribers are thus divided into sets having different attendants, for when a subscriber is called for at one board the attendant does not know whether he may not be already engaged and coupled for conversation at another board. To avoid inconvenience arising from this cause I provide for each subscriber two blocks on each board, an upper and an under one; they are coupled in series, the upper block on the first board makes contact for the upper block on the second board, and this for the upper block on the third board, and so on. Then, at the end of the set of boards the connections are continued through the lower blocks, each in succession until last in the series, the lower block upon the subscriber's own board is arrived at where (as previously described) the connection is made with the call indicator and so to earth. Now, for the purpose of calling a subscriber the peg is inserted into the hole in the lower block, and so will not interfere with any connection which may have been previously made at either of the upper blocks; but when such a connection exists the call will not pass, the connection with the subscriber's line having been previously broken.

It is obvious that for upper and lower, right and left may be substituted. The arrangement of the connections above described to be used when multiple boards are used is clearly shown by fig. [82] of the drawings.[1]

The connections are, in fact, so clearly shown as to make it unnecessary to quote the description in further detail.

The design shown in fig. 83 is stated by Mr. E. M. Barton in a letter[2] to Mr. F. R. Welles, dated August 2, 1883, to be Kellogg's contribution to the patent. The description of this system is as follows:—

Or in place of this [the fig. 82] arrangement another requiring but one block to each subscriber on each board may be adopted. Thus, a small finger key or contact maker may be provided in connection with each subscriber's block, and the attendant when he desires to know if a subscriber is at liberty places his finger on this key or contact maker; if the subscriber be occupied the contact

[1] British specification, No. 4903 of 1879.

[2] The occasion of this letter was the then recent introduction of a multiple test system on a trunk exchange at Coleman Street, London. This was independently designed by Mr. F. B. O. Hawes and was subsequently applied to a subscribers' board at the Chancery Lane Exchange (*Journal of the Institution of Electrical Engineers*, xxv. 367).

thus made completes a circuit in which is a battery and a small bell or ' buzzer ' ; and the sounding of this bell or buzzer indicates that another connection ought not to be made. This circuit, however, only becomes complete when one of the knife-like blades, corresponding to this line wire, is pegged out on one or other of the boards, the blade being by this movement brought against a contact point and so completing the circuit ; but when by the removal of the peg the blade is permitted to spring back this local test circuit is broken, and although the finger key or contact maker may be pressed down the ' buzzer ' will not sound.[1]

The arrangements to be made when this system of local circuits is applied to ascertain when a subscriber is at liberty is illustrated by fig. [83]. Here portions of three boards are represented, and

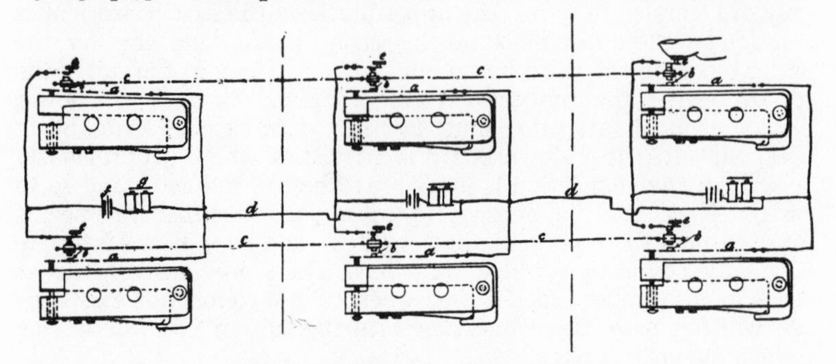

FIG. 83.—Kellogg's Multiple Test.

two blocks, belonging to different subscribers, are indicated on each board. The blocks employed are similar to those shown at figs. [85] and [89], except that in each case an insulated peg is provided upon the blade which, when the blade is lifted off its contact screw by the insertion of a peg, raises a spring *a* into contact with a metal stud *b*. The three studs *b* on the different boards appertaining to each subscriber are connected by an insulated wire *c* ; two of these wires are seen in the drawing.

The springs *a* of all the blocks on all the boards are connected to one common wire *d*. Over each block there is a finger key *e*, which when pressed down makes contact with the corresponding stud *b*. Each board has its battery *f*, and one pole of the battery is connected with all the finger keys *e* on the board, whilst the other pole of the battery is connected through the electric bell or ' buzzer ' *g* (of which also there is one to each board) to the common wire *d*. The attendant at either board, when wishing to ascertain if a subscriber is disengaged, presses down the finger key corresponding to this subscriber's block, and thereby causes the bell or ' buzzer ' *g* belonging to the board to sound if at either of the boards the subscriber's wire is already coupled, but otherwise the bell or 'buzzer ' does not respond.[1]

[1] British specification, No. 4903 of 1879.

The test system of Haskins and Wilson [1] is a circuit starting from earth at the exchange through a battery, a bell, a flexible cord ending in a plug inserted in the 'try piece' *d'*, the line (if disengaged) and earth at the subscriber's office. The circuit is thus complete and the bell sounds, indicating a *disengaged* line. In fig. 83 (Kellogg's) the bell circuit is local and distinct from the line. There is a 'buzzer' (the equivalent of Haskins and Wilson's bell) at each section, and the 'buzzer' sounds as an indication of an *engaged* line. The test circuit is completed by a finger key attached to each jack-knife switch. In the letter from Mr. E. M. Barton previously referred to it is said :—

Scribner states that Mr. Kellogg's first plan for this style of switchboard was a separate testing plug at end of a cord [similar to Haskins and Wilson], and that when he was describing the apparatus to Carpmael for him to lodge the specification he described to him this identical arrangement; but it would seem that Mr. Carpmael did not think it worth while to put in a separate description of it in the patent, and Scribner thinks that the use of the term 'contact maker' in the English specification was employed by Carpmael to make the description include such a 'contact maker' as the special cord and plug.

It seems probable, however, that Mr. Carpmael did more than cover two specific plans, that he adopted a phrase which should include any effective test by making contact and thus closing a circuit. If so it was an example of an important addition by a patent agent who had perceived the scope of the invention, for neither the finger key nor the special cord and plug survived as the means for obtaining information by the operator as to the engaged or disengaged condition of the line. This British patent was not contested throughout its life, and the words added by Mr. Carpmael were an important aid to that result. The completeness of the invention is remarkable when the early date of the industry is considered. In London, for example, a telephone exchange had been established only about three months when the specification was deposited at the Patent Office.

The modifications needed for metallic circuit lines are provided for in fig. 84, where are shown the changes in the 'knife-like blocks' which have to be made when 'it is advisable to arrange telephonic circuits with two line wires in place of completing the circuit by means of earth connections.' [2]

The jack-knife switch shows a considerable advance in design over the switch illustrated in Haskins and Wilson's specification.

[1] Figs. 78–81.

[2] The equivalent U.S. specification is Scribner's 266,319, October 24, 1882.

But it would appear that their design was not put into use, for in their paper read before the American Electrical Society in December 1879 they say :—

> A connecting plate devised by Mr. C. E. Scribner of the Western Electric Manufacturing Company of Chicago, which has proved

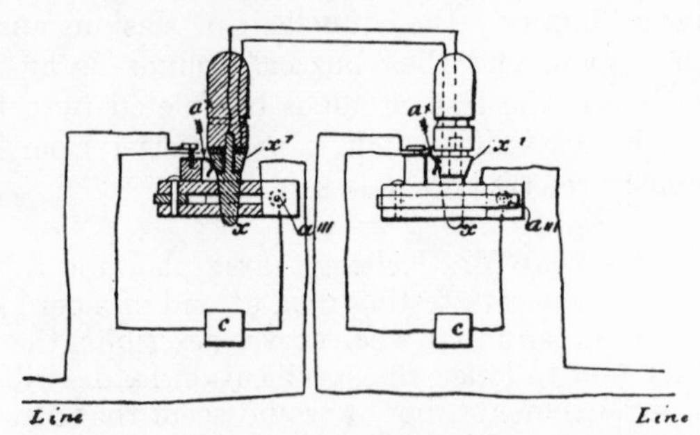

FIG. 84.—Metallic Circuit Multiple (1879 patent).

> of practical utility in this connection, has been adopted by the several companies using the multiple system. It is known as the 'jack knife.' [1]

Insulation in Haskins and Wilson's design was obtained by separation, and the three parts were individually mounted on the board. The jack-knife switch is compounded of metallic and insulating parts, and forms a structure fixed to the board by one bolt, which is also the connecting point for the line. The purpose served is the same, the line coming to the block (figs. 85 and 86) with which the blade *a′* is metallically connected and leaving by the screw *a′′′* which, though mechanically a part of the structure, is insulated by an ebonite sleeve.

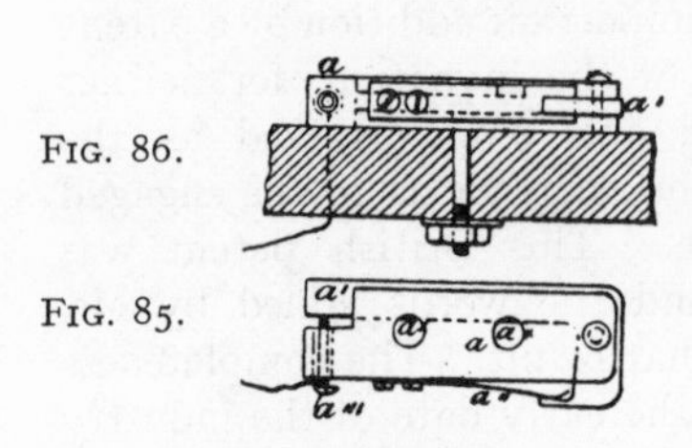

FIG. 86.

FIG. 85.

Jack-knife Switch.

The jack-knife switch is more fully described in Scribner's U.S. patent, No. 293,198 as follows :—

> Figure [87]—the original form of a part of my device—is a flat piece of metal, N, slit as shown, and provided with holes *x* and *y* and contact points C and D, the latter of which, D, is insulated, as shown, by hard-rubber bushing. The lever B, (shown by Fig. [88]

[1] *Journal of the American Electrical Society*, 1880, p. 48.

in detail,) in combination with spring I and frame or back E, and pivoted to E, as shown at *r*, Fig [89], takes the place of the single piece N. In the single piece N advantage is taken of the elasticity of the metal. Frame E is provided with two holes, *x* and *y*, and

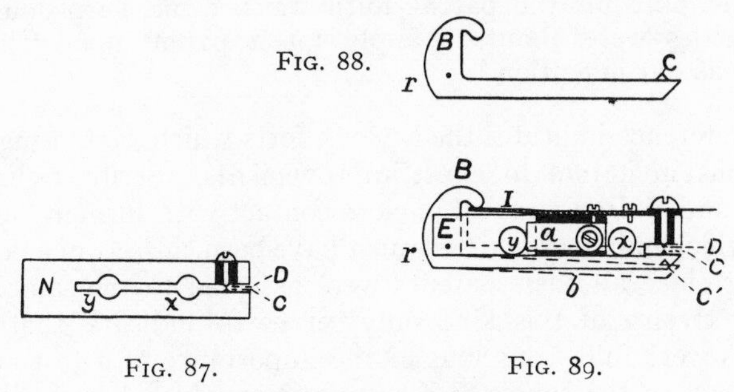

FIG. 88.

FIG. 87. FIG. 89.

Jack-knife Switch in Detail.

a slot, in which lever B rests. The inner edge of the lever B projects so as to come within the edges of the plug holes *x* and *y*, as shown by dotted line, and when a plug is inserted in either hole the lever is forced down, as shown by lower dotted line, and the points of contact D C are separated, C taking the position shown as indicated

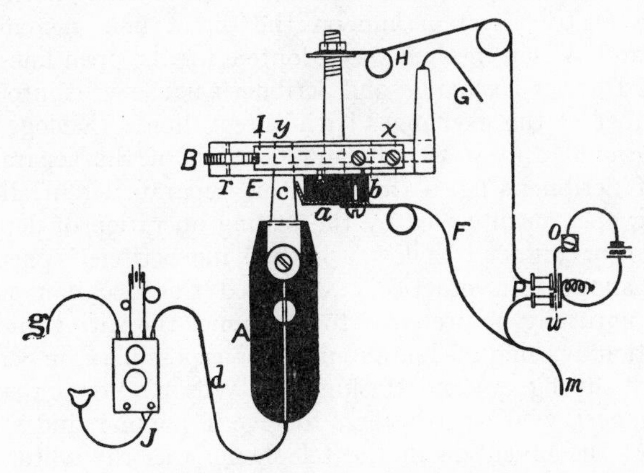

FIG. 90.—Jack-knife Switch (section and connections).

by C′. *a* is a metallic plate screwed to hard-rubber block *b*, and thereby insulated from the other portions of the switch.[1]

Giving evidence in a patent suit, Mr. Wilson said that the exact time when the invention of Haskins and himself was made could

[1] U.S. specification, No. 293,198, February 5, 1884 (application filed August 23, 1879).

not be stated, but it was some weeks, and perhaps a few months, prior to the making of the application for the patent.[1]

I remember distinctly [added Mr. Wilson] that we were in no hurry to apply for the patent until we became suspicious that other parties were about to apply for a patent upon what we regarded as our invention.[2]

The reference indicates the active efforts which were being made by various inventors to effect improvements. Scribner, Kellogg, Haskins and Wilson were all in close contact with Firman, and any rivalry which may have existed must have been entirely of a personal character since all their patents were assigned to the same firm.[3] But any rivalry of this kind only serves to indicate that these inventors were fully conscious of the importance of a test system. Mr. Kellogg, in the course of a conversation which I had with him in 1891, said that all produced their methods within a fortnight in October 1879. But whilst this may refer to Scribner's fig. 4 of 1879 (fig. 82), to Haskins and Wilson's and to Kellogg's own method, the evidence is clear that Scribner's fig. 6 (fig. 76) antedated that period by several months, and thus stands as the first multiple testing system. But it was not used in any working exchange. In Chicago at this time they were working both on the American District system and on the direct line system. The Haskins and Wilson method was adopted for the open lines in the American District Exchange, and Scribner's fig. 82 went into service in a number of the exchanges with direct lines. Kellogg's plan (fig. 83) would appear to be a modification of the separate test circuit of Scribner's fig. 6 (fig. 76), this separate circuit however remaining open until closed by the testing operation of depressing a finger key or contact. Kellogg's method, like Scribner's pneumatic, was not adopted in practice. Nor need this occasion surprise. It would naturally be preferred to economise the wiring and avoid complication by using the line circuit so far as possible as in Scribner's fig. 4 (British) (fig. 82) and Haskins and Wilson's 'try circuit.'

Du Moncel, who was the first to give a popular and scientific account of the invention of the telephone, was the editor of *La Lumière Électrique*, and, so far as Europe is concerned, it is to this publication that we must refer for contemporary descriptions of the multiple switchboard. These appeared in 1880, and abstracts of them are included in the fourth (French) edition of Du Moncel's *Le Téléphone*, published in 1882. It is to be observed that no

[1] October 23, 1879.
[2] Capital Telephone Case, Circuit Court of United States, Northern District of California, 1895, *Complainants Record*, p. 73.
[3] The Western Electric Manufacturing Company.

attempt was made to translate such technical terms as 'try circuit,' 'try plate,' or 'jack knife' which last, however, appeared as 'jack knif.' In the subsequent English edition of Du Moncel's work no reference is made to the switchboard apparatus.

The articles in *La Lumière Électrique* cover much the same ground as that in the *Journal of the American Electrical Society* (1880), which was prepared by Messrs. Haskins and Wilson at the request of the Publication Committee of that Society in order to complete the brief description given at the annual meeting held on December 10, 1879. This description was probably the first public explanation of the multiple switchboard. The record thereof in the society's journal was evidently not written by a telephone expert, but the final remarks indicate that the general idea was accurately absorbed :—

Mr. Charles H. Wilson gave the members a description of a new switchboard of his invention, to be used as a governor for wires, whether telephonic or telegraphic. He drew diagrams on a blackboard illustrating the process. One principle was the introduction of a polarized relay, to operate when a subscriber called, but would not sound when the operator responded, and a plug was so arranged that the operator had control of all the tie bolts at once. The board seems to be in every way practicable, and meets valuably a demand that has long bothered operators. The arrangement is such that an operator can readily tell when another is using the desired wire.[1]

The various types of switchboard described in Chapter XIV were gradually discontinued, except the standard switchboard, which was retained for small offices. For large offices the multiple board, the development and principles of which have been described above, was becoming general but not yet universal.

The terms 'large' and 'small' being relative, it becomes necessary to define them, and this may best be done by a quotation from a printed document of the period. In 1883 the manufacturers issued a 'Descriptive Circular of the Multiple Switchboard for Telephone Exchanges.'

The opening paragraphs of this circular are :

The two most important requisites in switchboards for large telephone exchanges are speed and economy.

It so happens that these two requisites go together ; for whatever tends to reduce the amount of work to be done also enables the operators to do it quickly, and therefore diminishes the cost of the work.

[1] *Journal of the American Electrical Society*, 1880 (Appendix).

The requisites of speed and economy are met with in a high degree in the multiple switchboard.

In this circular it is stated that :

This system is now in use, or soon will be, in the following cities : Minneapolis with 800 subscribers ; Milwaukee with 900 subscribers ; Nashville with 800 subscribers ; Kansas City with 500 subscribers ; Melbourne, Australia, with 700 subscribers ; London with 500 subscribers ; Baltimore with 1600 subscribers ; Washington with 1000 subscribers ; New Orlèans with 1500 subscribers ; Indianapolis with 1000 subscribers ; Liverpool with 1000 subscribers ; Columbus with 600 subscribers ; Toledo with 500 subscribers ; Dayton with 800 subscribers ; Memphis with 600 subscribers ; Peoria with 500 subscribers.

From this it may be gathered that an office became 'large' when it approached 1000 subscribers, and that it was then becoming generally recognised to be sound practice to instal the multiple board in the event of new switchboards being required for a central office that was expected to exceed 1000 subscribers.

The principle upon which the multiple board is based has been fully described, so that it is unnecessary to follow the circular in detail, but the method of operation and some other points are so clearly and tersely related in it that some quotation is permissible.

An operator has a certain number of subscribers whose calls she answers. When one of these subscribers has called for a connection with any other subscriber of the exchange, the operator who takes the order makes the connection of the two lines instantly and rings the bell of the subscriber wanted.

When the conversation is finished, the same operator who connected them together receives the notice by the subscribers to disconnect, and then disconnects the lines. This is all done by the operator without moving more than one step.

It is hard to imagine a system more simple or more easily manipulated. It is in consequence of its simplicity that it is so economical.

As many switchboards are required in an exchange as there are times two hundred subscribers. For instance, with one thousand subscribers, five switchboards would be necessary.

Several different arrangements of the spring-jacks and annunciators, with reference to their disposition on the board, have been tried.

The annunciators have been placed all above the spring-jacks, all at the two sides, and part above and part at the two sides.

None of these arrangements, however, seems to possess the advantages which are secured when the annunciators are placed across the boards below the spring-jacks, with the operator's keyboard below the annunciators, and the shelf for holding the cords and pulleys above the spring-jacks.

One important advantage of this general arrangement is that it allows the placing of the boards together, end to end, thus enabling an operator at any board to make connections between the spring-jacks upon her own board and those upon the adjacent board, in cases where those spring-jacks are nearer than the corresponding ones upon her own board. . . .

The accompanying cut [fig. 91] shows the jack as used on the multiple board. The frame of the jack is inserted at the back of the board and screwed fast by a lug which forms a portion of the frame.

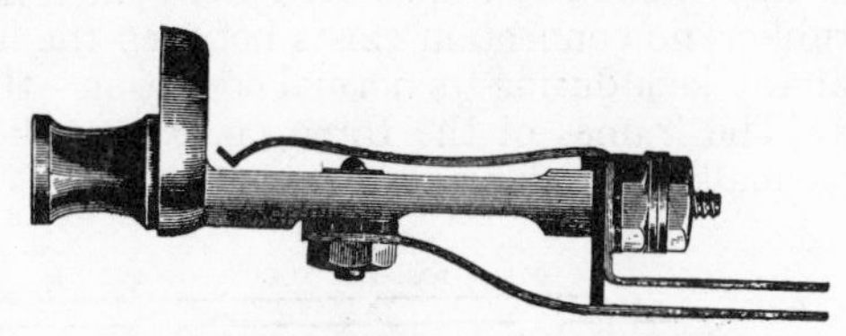

Fig. 91.—Multiple Jack (1883).

The phosphor-bronze spring, and the contact point upon which it normally rests, are insulated from the frame of the jack by hard rubber bushings and plates.

The connections of the jacks are all made at the back of the board, and, to facilitate this work, phosphor bronze connecting pieces are extended from both the insulated contact point and spring, bringing all the connecting points together in one plane surface.

When a plug is inserted into the jack [fig. 92] the phosphor-bronze

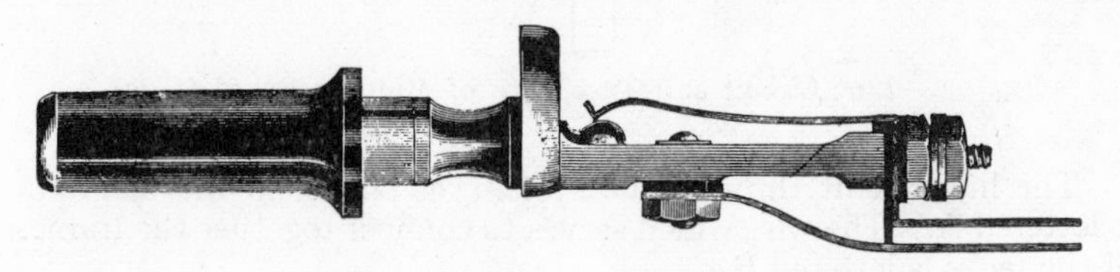

Fig. 92.—Multiple Jack and Plug (1883).

spring is lifted from the insulated contact point, breaking a connection therefrom, and a new connection is established to the tip of the inserted plug upon which the new spring now rests, and thence to the cord connecting with the plug.

A connection, or *cross*, is also established between the phosphor-bronze spring and the frame of the jack, through the medium of the plug. It is this cross as thus established (and which is more fully described hereafter) which provides for the operators' test.

The accompanying diagram of circuits [fig. 93] shows three spring-jacks of a line placed respectively on three multiple switchboards.

The circuit through these three jacks and the annunciator is from the subscriber's line through the lightning arrester to the spring of the first jack; from the contact point upon which this spring rests to the spring of the second jack; from the contact point upon which the second spring rests to the spring of the third jack; and from the contact point upon which this last spring rests the circuit is through the annunciator belonging to that line, and thence to the ground.

As is evident, the annunciator may be placed at any of the three boards, but at whichever board it is placed, its magnet must be included in the circuit between the last spring-jack and the ground.

The springs and contact points included in this circuit being, as above described, insulated from the frames of the jacks by hard rubber, no connection exists between the line circuit and the frame of any jack during its normal condition—that is when not in use.

The frames of the three jacks are all connected together, but normally to nothing else.

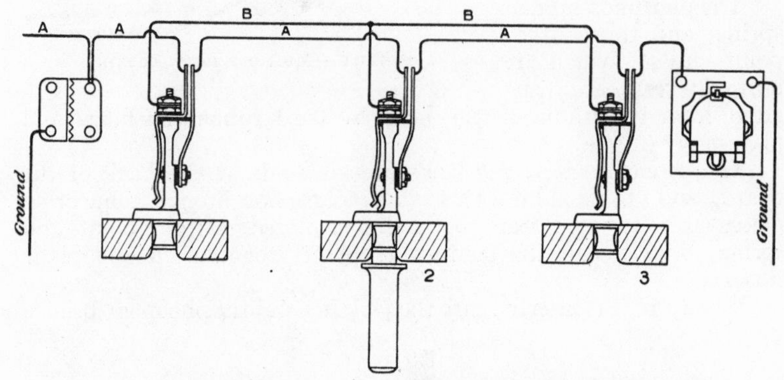

Fig. 93.—Line Circuit and Test Wire of Multiple Board (1883).

The line circuit through the system, as shown in the diagram, is lettered A. The wire which serves to connect together the frames of the jacks is lettered B.

The insertion of a plug at any of the three jacks, as shown at the second jack, lifts the phosphor-bronze spring free from the contact point upon which it normally rests, thus disconnecting the annunciator and ground, and establishing connection between the line and the cord connecting with the plug, as well as the cross-connection between the line and the frame of the jack.

The frames of all the jacks of the same number being connected together are consequently all crossed with the line.

An operator at any other board may therefore determine the fact of a cross between the jack frames and the line by touching the frame of the jack at the front of the board with the plug which she is about to insert into the jack to complete a connection.

The keyboard connections are shown in fig. 95. The first impression of those familiar only with more modern boards is

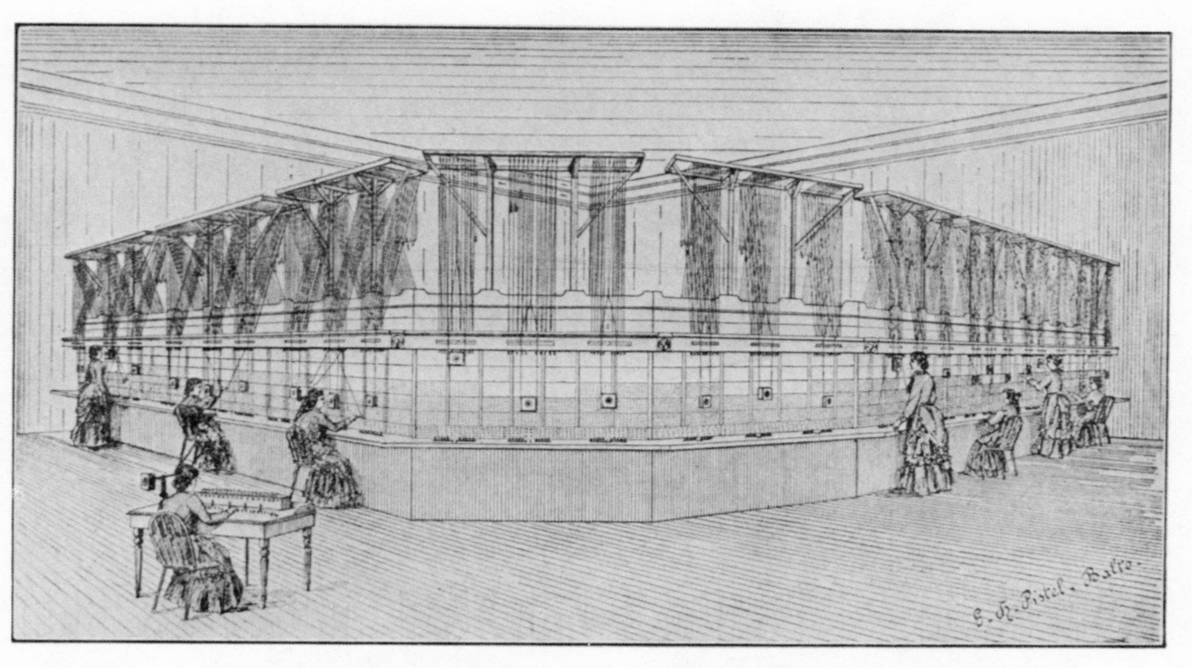

Fig. 94.—Multiple Switchboard at Baltimore. (From the *Electrical Review* (N.Y.), March 6, 1884.)

likely to be that the block has been inverted in the course of prepara tion for the press. The reader, however, may be reassured. In the early multiple switchboard the plugs and cords were not upon the keyboard, but were placed above the jacks and indicators, as will be observed more readily from subsequent illustrations.

The circuit and operation is too familiar to need quotation from the description, but an illustration of the earlier cabling methods will be of interest.

Heretofore considerable trouble has been experienced in cabling the connections from the multiple switchboard, occasioned by the wires accumulating at the sides of the board in large cables, covering over the spring-jacks and rendering them inaccessible.

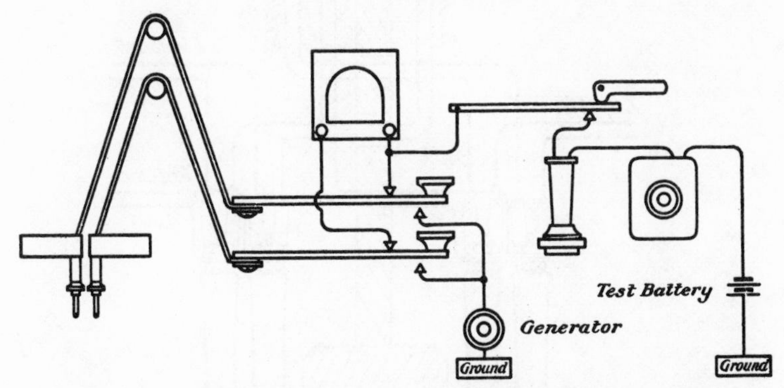

FIG. 95.—Keyboard Connections Multiple Switchboard (1883).

To obviate this difficulty spring-jacks are now arranged in divisions of twenty, leaving space for the cable to be run perpendicularly between these divisions. Perpendicular strips dividing the boards are placed in these divisions upon which to run these cables.

The boards are perforated, and the wires in cables, of no larger than twenty wires each, pass through the perforations and are supported thereby.

The wires by this method are evenly distributed over the entire surface of the board, the cables coming out from the board at regular intervals, and at no point is there an accumulation of greater than twenty wires in one cable.

All the spring-jacks of the system are accessible and can be removed in case it becomes necessary to replace them.

The accompanying diagram [fig. 96] shows a cross-section of the multiple board, the dividing strip or cable support, and the cables passing through it, as described.

The circular has an illustration of 'The Multiple Switchboard for Indianapolis' reproduced in fig. 97. It will be understood that the legs and the overhead cord structure are omitted.

Of some of the other boards mentioned in the circular as being, or about to be, in use, illustrations when fitted in the various exchanges are reproduced in figs. 94, 98, and 99 from the sources mentioned; and from the London *Electrical Review* of October 18, 1884,[1] an illustration (fig. 100) of the multiple switchboard installed at Liverpool, the first in active operation in England. The London board referred to in the descriptive circular was

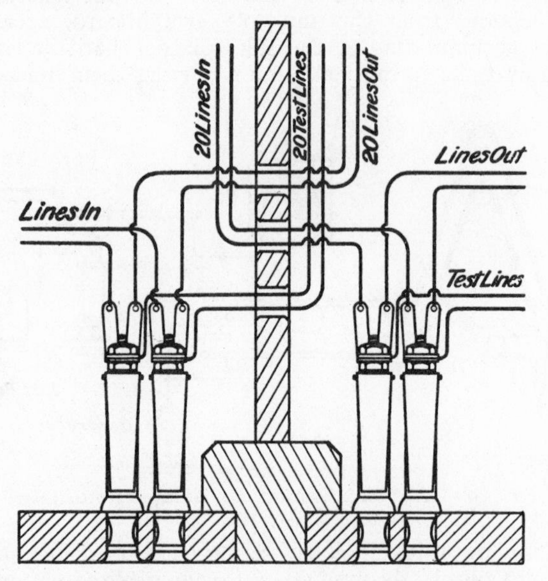

Fig. 96.—Cabling System Multiple Switchboard (1883).

supplied to the London and Globe Telephone Company, whose service was restricted by the action brought against them by the United Telephone Company for infringement of the Edison patent by the use of the Hunnings transmitter.

The reason why the plugs were placed above the boards was that in the then normal height of the keyboard from the floor there was not sufficient room to take up the slack of the cord of the length required to cover the jack field. From a new issue of the 'Descriptive Circular' in the following year (1884) it is to be inferred that there were very certain drawbacks to the overhead system. The 1884 board had other important changes, but we will first quote the paragraph relating to the cords.

[1] Vol. xv. p. 312.

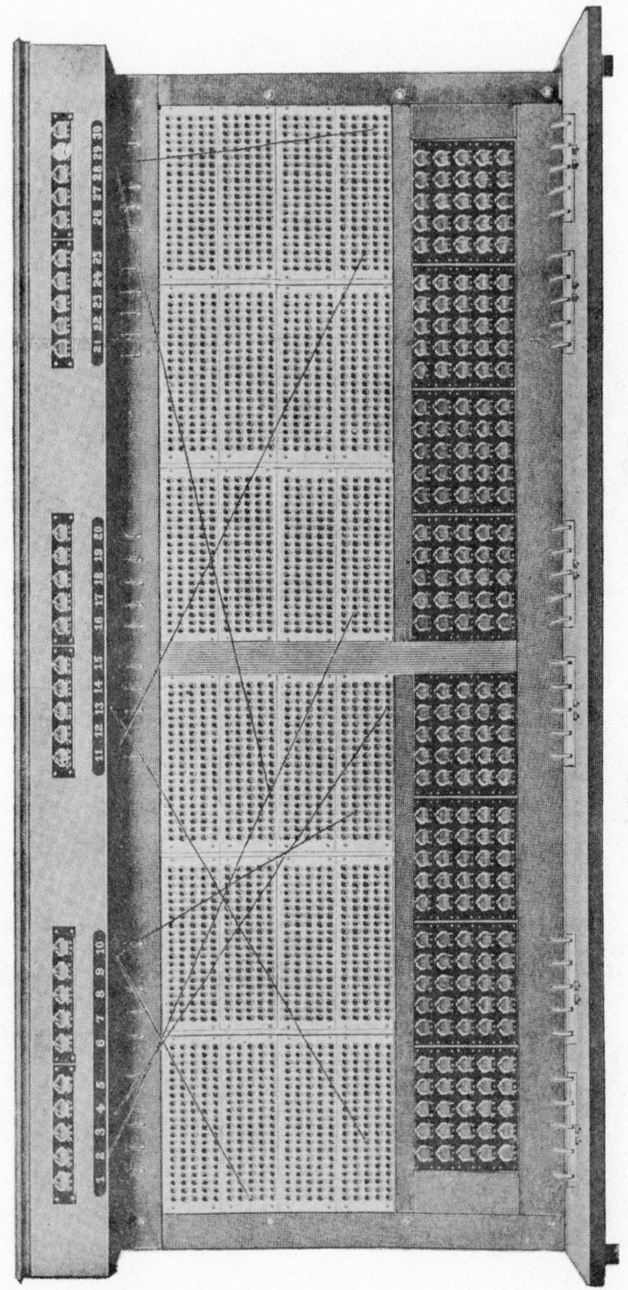

FIG. 97.—Multiple Switchboard for Indianopolis (1883).

FIG. 98.—Multiple Switchboard, Melbourne, Australia. (From the *Town and Country Journal*, September 10, 1887.)

FIG. 99.—Multiple Switchboard, Melbourne, Australia, in Operation. (From the *Town and Country Journal*, September 10, 1887.)

Fig. 100.—Multiple Switchboard, Liverpool. (From the *Electrical Review* (London), October 18, 1884.)

The cords are all below, thereby avoiding the use of several pulleys and the consequent squeaking and rattling noise caused by them. Having the cords separated also gives more room for the weights under the table, and less chance for trouble in moving them. The weights are about one-third as heavy as on the overhead cords, therefore the cords will last longer.

The illustration fig. 101 forms the frontispiece of the 1884

FIG. 101.—Multiple Switchboard (1884).

'Descriptive Circular,' and indicates important changes in details apart from the restoration of the plugs to the keyboards, for, as it will be noted, there are two shelves. The board was consequently known as a 'double-decker.' Originally the only jacks upon the board were those in the multiple field. These were arranged in blocks of one hundred. As the sections were designed for 200 subscribers, it follows that the operators answering calls at one section were in competition for the jacks in a space of two blocks. This is a drawback which would be very obvious in practical work, and it was remedied by the introduction of 'answering jacks.' The 'multiple' multiplied jacks on each section. The answering jack was an extension or refinement of the same principle, inasmuch as the jacks relating to the subscribers at a section were duplicated on that section, one appearing in the multiple field above, the other in the answering field below. Interference between operators in the multiple field was thus reduced to the selection of the jack for the called subscriber ; that of the calling subscriber was directly before the operator to whom it was allotted.

The pairs of plugs were divided, one, the answering plug, being

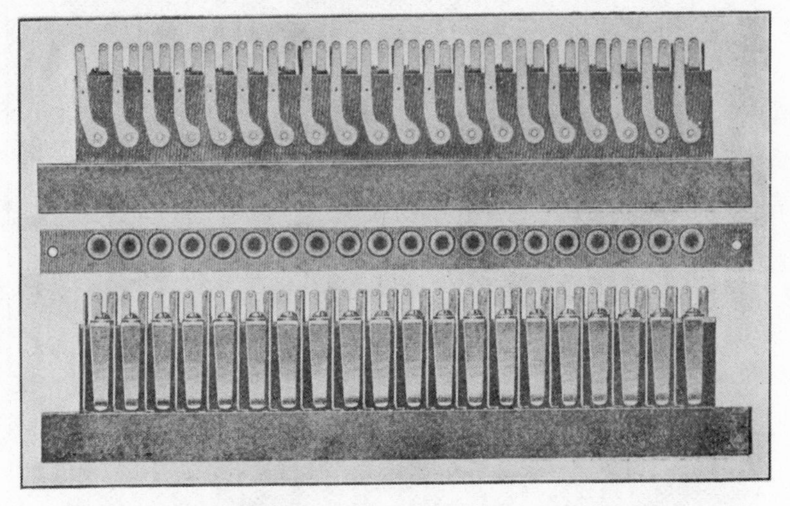

FIG. 102.—Multiple Spring-jacks (1884).

placed on the lower shelf, and its mate, the connecting plug, on the upper shelf.

There were important changes in the component parts of the board, which the circular shall describe :

The new spring-jacks are mounted upon narrow strips of hard rubber, so that very much better insulation is secured between the parts (see cut of spring-jacks) [fig. 102]. They are very much smaller in size, occupying only about one-half the space on the face of the board formerly occupied by the larger size. Five of these strips constitute a section of one hundred jacks.

The annunciators are also much condensed, and are furnished in sections of ten on a strip one inch wide by fifteen inches long.

Exchanges will grow, and the boards are made large enough to admit of any growth that is liable to occur.

Fig. 103.—Multiple Switchboard at Boston. (From the *Electrical Review* (N.Y.), September 26, 1885.)

The use of 'half sections' or dummies,' as they were sometimes called, probably preceded 1884, but they are first mentioned in this circular. After describing how the jacks at the beginning of No. 2 board are available for the use of the operator at the end of No. 1 board and so on, the circular says :—

This principle may be carried to the extent of placing a half section of board at each end of the system of multiple boards, thus giving the operators at the beginning of the first board and at the end of the last board the same advantage in this respect.

The following exchanges with the number of subscribers against each are quoted in the 1884 circular as having adopted the multiple board in addition to those previously quoted from the circular of 1883 :—

New York	2 exchanges	with	2,000	subscribers.
Buda-Pesth, Hungary	,,	,,	600	,,
Gothenburg, Sweden	,,	,,	600	,,
Boston	,,	,,	2,000	,,
Detroit	,,	,,	1,600	,,
Toronto	,,	,,	1,000	,,
Lawrence, Mass.	,,	,,	600	,,
Pittsburg	,,	,,	1,500	,,
Louisville	,,	,,	1,350	,,

The illustration fig. 103 represents the new type of board (described in the 1884 circular) as fitted in the Boston exchange, and is reproduced from the New York *Electrical Review* of September 26, 1885.

CHAPTER XX

OUTSIDE OR LINE CONSTRUCTION

At the time that the telephone was commercially introduced, telegraph engineering had reached a high stage of development. For land lines galvanised iron wire was almost universally used. The wires were strung overhead and attached to poles or fixtures by means of insulators of types which varied with localities: those of glass being very generally used in the dry climates of America, whilst porcelain ones were almost universal in Europe. Submarine cables were numerous, and the scientific principles governing their operation in the transmission of telegraphic signals were subjects of constant investigation. The conductors in these cables were of copper, and the insulation was gutta-percha or india-rubber. Land cables when used followed the submarine type of construction, but for underground conductors in congested districts gutta-percha covered wires were drawn into iron pipes in groups rather than made into cables.

The line construction methods of the telegraph system were naturally adopted by the early telephone engineers; but, as the installations grew in number and the transmission distances extended, new problems were constantly being met with which required for their solution important modifications in established practice.

For telegraph work, experience was available which permitted estimates to be made with reasonable accuracy so that telegraph lines could be planned and built with due regard to requirements and results. But there was no experience of telephone exchanges; and, if the promoters had estimated for the numbers which actually came, they would certainly have been accused of rashness and, as certainly, they would not have obtained the capital to carry out their plans. In consequence there was, in the early days, an element of haphazard in design as well as an effort at cheapness in construction which, whilst the subjects of criticism by telegraph engineers of the period, were really inevitable and, in the long run, not disadvantageous, because the quicker and cheaper construction permitted earlier use and more prompt and widespread demonstration of the advantages of the service, whilst there was less compunction about sacrificing such lines and substituting others when experience showed that it was desirable to do so.

The route of a telegraph line and the office to which it should be led were to some extent under the control of the authority constructing it, and the office might be chosen with special regard to getting the line or lines into it. But the essence of the telephone exchange business was that the new means of communication should be available at the place of business or residence of any member of the public, so that the lines must start from points which could not be chosen by the engineer but must, nevertheless, be provided for. These points were naturally more numerous in congested business and residence areas, so that the telephone engineer on the line side of his work was faced with just the same problem as on the exchange side—how to provide facilities for the growth in numbers which rapidly came.

While one terminal of the line—the user's office—was beyond the engineer's control, the other terminal—the exchange—was his own choice; and, so soon as a little experience had been gained, this was usually placed in the top floor of some high building capable of carrying a number of wires which radiated to the offices of users direct or were carried in groups to a more distant roof whence distribution commenced. There were thus two clearly defined types of line construction—'routes' and 'distribution.' The routes were composed of a number of wires carried sometimes for considerable distances, whilst the distribution was from a defined centre over a given area. For routes as well as distribution, overhouse construction was first adopted both in America and Europe. In the latter they were very generally so continued, but in the former facilities were afforded by the local authorities which permitted the routes to be transferred in large measure to pole lines in the streets, overhouse construction being retained principally for distribution.

The increasing number of lines on routes and the difficulty of providing accommodation for them, led to cables being suggested as a source of relief.

Under date of April 19, 1878, Mr. J. H. McClure, the manager of the Telephone Co., Ltd., 'sole proprietors of Bell's Telephone,' wrote a letter to Sir William Preece asking his opinion of the best form of cable for telephonic purposes, adding—

It is proposed to lay a cable from the various docks in London to a central office which, as you are aware, would be connected with various merchants' offices in the neighbourhood of such central office. I am of opinion that it would be less expensive and more convenient to lay a cable containing ten metallic circuits, than to lay a number of wires, if indeed we could obtain permission from the railway company for the larger number of wires: but I shall be glad to have your opinion on this point also.[1]

At this time Professor Bell was in London, and less than a month previously had issued the circular [2] to the Directors of the Telephone Company in which the use of cables is suggested and the connection with the docks referred to.

One of the earliest instances of the use of cables as a means of bringing in a route of lines to a telephone central office was that at San Francisco in the year 1879.[3] A forty-wire cable 75 feet in length was first run from the roof into the exchange room, and later another thousand feet, in lengths of from 75 to 200 feet, were added. In 1881 there were some 7000 feet of different lengths, varying from 60 feet to 650 feet. "In that way [said Mr. Sabin] we manage to get into the office very well."

The composition of the outside construction in New York as at May 1, 1880, is recorded in an inventory which formed the basis of the valuation of the Western Union and Bell exchanges on that date. According to this inventory the Metropolitan Telephone and Telegraph Company (which was the title of the new organisation) received from the former companies 1892 miles pole wire, 970 miles house top wire, 1117 house top frame fixtures and from 70,000 to 80,000 feet of cable averaging ten conductors each; of this 25,000 feet was aerial cable located principally over the East River Bridge to Brooklyn.[4]

The new Company

constructed large and substantial lines to connect suburban territory, with a capacity of about fifty wires but carrying an average of about eight. On the top pin is strung a number 9 galvanised iron wire which is called the induction wire, and connected with earth at suitable intervals, also with metallic cups in the cross arms in which the insulator pin is inserted.[5]

As early as 1880 it was evident to those engaged in telephone work that the difficulties of construction and maintenance would necessitate a change from overhead lines. This was clearly stated by Mr. Pope at the first meeting of the National Telephone Exchange Association in the following words:—

1 I am indebted to Mr. A. H. Preece for this letter.

2 P. 89.

3 Mr. John J. Sabin, *National Telephone Exchange Association Report*, 1881.

4 Mr. H. W. Pope, *National Telephone Exchange Association Report*, 1880, p. 154.

5 *Ibid.* p. 155.

In New York and perhaps in Chicago the matter of underground lines must be soon taken up. Before coming to this convention that was made a prominent feature and was brought to my notice, and I was requested to do what I could to get a competent committee on this subject. It should be a Committee who have had some little experience, or some knowledge that they have obtained abroad or elsewhere, in the matter of underground telegraphy. It is a subject that necessitates perhaps the establishment or the organisation of companies for this purpose, and it is a matter that needs immediate attention, and the thing should be gotten into shape.[1]

This statement was undoubtedly inspired by the controlling spirits of the business. The needs were becoming obvious, the difficulties foreseen, and the necessity for careful examination recognised. Some time was yet to elapse before an organised system of underground work was practicable. The cables then available were described and discussed. After describing one such system, Mr. E. M. Barton said :—

There is an important distinction in regard to the difficulties of laying subterranean and submarine cables and those of laying overhead cables, because there is not the difficulty of keeping out moisture in a cable suspended in the air or over buildings that there is under ground ; and while the question of making good cables is a prominent one for telephone managers to consider, the question of whether these cables shall go overhead or underground is also an important question.[2]

The question was not settled at that meeting. There was a concurrent development along both lines ; but when it is realised that to bury the wires involved the use of a suitable cable and the excavation of the ground as well, whilst the use of aerial cable saved the latter expense, it is obvious, as Mr. Barton foresaw, that the aerial cable afforded the easier method of providing the relief which was becoming more and more pressing from the congestion of the wires on routes and especially so in the neighbourhood of the central offices.

The immediate problem was attacked in the light of the existing conditions. The eventual need of undergrounding was recognised, but it was clearly seen that a general underground system was a vast undertaking which required the most careful consideration and much more experience than was then available. Inquiry and the acquisition of experience on a smaller and safer scale were diligently pursued with a view to greater developments.

[1] Mr. H. W. Pope, *National Telephone Exchange Association Report*, 1880, p. 147. [2] *Ibid.* p. 54.

There was of course no novelty in either underground work or the suspension of aerial cables.

Sir Charles Wheatstone, in his patent of October 10, 1860, contemplated the use of a large number of electrical conductors in one cable suspended over housetops or on poles. And in the circular or prospectus of the Universal Private Telegraph Co., formed in 1861, it is said :—

In place of the thick iron conducting wire in ordinary use fine copper wire (number 22 gauge) is employed, carefully insulated over its entire length with a thin coating of india-rubber by Messrs. S. W. Silver & Co.'s patented process, and further protected from abrasion by a covering of tarred tape. Thus perfectly insulated from one another, these 'electric highways' are afterwards combined into ropes of about two hundred and thirty yards in length, containing twenty, thirty, fifty or one hundred wires according to the requirements of the district through which they are intended to pass.

Mr. Lockwood stated in 1882 that :

Until an extensive network of telephone lines had been erected, very little aerial cable construction had been done in the United States. When the Gold and Stock Telegraph Company moved its offices and operating room from 61 to 195 Broadway in New York, it was requisite to retain the former place as a testing station ; and all the wires were accordingly laid from 195 to 61, before proceeding out to the brokers' offices. To accomplish this, they were carried in kerite cables of nine and ten wires each. This was in 1874. In 1875, the Law Telegraph Company was organised, and also to a certain extent adopted cable construction.[1]

The aerial cables for telegraphic use were almost invariably composed of wires separately insulated with india-rubber or some substitute therefor, such as the kerite above mentioned. The use of aerial lines, whether suspended singly or in cables, was adopted only after the experience of unsatisfactory results from underground lines, for on its introduction it was contemplated that the telegraph would be used through the medium of buried conductors.

The demonstration line laid by Sir Francis Ronalds in the garden of his house at Hammersmith in 1816 (later the Kelmscott House of William Morris) was an underground line. A trough of wood two inches square, well lined both inside and out with pitch, was laid in a trench four feet deep. In the trough were placed thick glass tubes jointed with soft wax, and the wires were run through the glass tubes.

[1] *National Telephone Exchange Association Report*, 1882, p. 90.

Mr. Lockwood pays a tribute to the prescience of Sir Francis Ronalds, who

described methods of insulating the wires, either on poles or underground, with all the details of tubes, joints and testing boxes, testing stations, linemen and inspectors, as at the present day. But the most wonderful feature of the affair is that Ronalds clearly foresaw the great trouble which, even at the present day, limits the speed of submarine and subterranean telegraphy, and apparently bars the way to successful telephony on underground lines of great length, i.e. the retardation of signals, due to the induction of the earth ; for he says in his book, which was published in 1823, 'That while over his short line the charges were apparently instantaneous, he did not contend, nor even admit, that an instantaneous discharge through a wire of unlimited extent would occur in all cases.' Further on he says: 'That objection which has seemed, to most of those with whom I have conversed on the subject, the least obvious seems to me the most important, and therefore I begin with it, viz., the probability that the electrical induction which would take place in a wire enclosed in a glass tube of many miles in length (the wire acting like the interior coating of a battery), might amount to the retention of a charge, or at least might destroy the suddenness of the discharge, or, in other words, might arrive at such a degree as to retain the charge with more or less force, even when the wire is brought into contact with the earth.' [1]

And Sir William Preece said of Ronalds: 'It is perfectly astonishing how that man's instinct saw the various troubles that were likely to be met with in the construction of long underground lines.' [2]

Messrs. Cooke and Wheatstone's 1837 patent included a plan for laying subterranean wires, and in that year they established telegraphic communication between Euston Square Station and Camden Town, the conductors being carried underground. The means adopted in this case were to place five copper wires, each separately covered with cotton and insulated by a preparation of resin, into five grooves cut longitudinally in a piece of timber. Tongues of wood were placed over the grooves, the whole covered with pitch and buried in the ground. Thus in Great Britain the first telegraph lines were laid underground, and it was intended that the first in the United States should also be so laid. Ten miles of the line from Baltimore to the Relay House were laid underground in December 1843. The line consisted of four No. 16 copper wires covered with cotton and shellac drawn into lead pipes. This experiment was a complete failure, and the wires were taken out of the tubes and placed on poles. The remainder of the line was also placed upon poles.

[1] *National Telephone Association Report*, 1882, p. 80.
[2] *Journal of the Society of Telegraph Engineers*, xvi. 427.

It may be presumed that the method adopted in 1837 had by 1839 been proved to be unsatisfactory, for in the latter year the wires on the Great Western Railway were carried as far as West Drayton through an iron tube an inch and a half in diameter, fixed about six inches above the ground, parallel to the railway and about two or three feet distant from it. *Chambers's Journal* (1840), from which this information is taken, further states that :—

Each of the four wires in Professor Wheatstone's apparatus is wrapped round with a well-rosined thread, and the whole are then tied together with a cord possessing a similar coating, so as to present the appearance of a tightly bound rope. This it is proposed to place in a small iron tube like that used for bringing gas into houses, and the tubes, united to any length, are laid below the ground, or in a wooden case on the surface, to preserve them from injury.[1]

These early efforts were succeeded by others which are concisely recorded by Mr. Fleetwood in his paper read before the Society of Telegraph Engineers (I.E.E.) in 1887,[2] to which reference may be made for details. Up to 1849 the insulation was resin ; a mixture of tar, resin, and grease ; or some similar compound. Conductors (generally four) covered with cotton or yarn and so insulated were drawn into lead tubes about half an inch diameter. Two such tubes were drawn into a three-inch cast iron socket pipe (still the standard size), thus allowing eight wires to the three inches. Lines of such a type were laid in 1846 from 345 Strand, the office of the Electric Telegraph Co., to Nine Elms Station, the then metropolitan terminus of the London and South-Western Railway. After 1848 gutta-percha took the place of the compounds previously used and the wires were drawn directly into the iron tubes.

From 1850 to 1854 there was great activity in telegraphy, and underground wires were laid not only in cities but for long-distance work such as London to Dover (six wires), London to Liverpool (six wires by one company and ten by another). But the life of these lines was short, for in 1857-8 they were condemned and lines on poles substituted. Mr. Fleetwood considers the failure to have been due to defective covering on some of the wires and still more to defects in the process of laying.

The chief cause of the rapid decay of the lines was the rapidity with which the work was carried out and the absence of supervision. The majority of the men employed were totally ignorant of the necessity for the greatest care being exercised, and it was impossible for the few able men to be everywhere along the works at the same

[1] *Chambers's Edinburgh Journal*, July 25, 1840, ix. 209.
[2] *Journal of the Society of Telegraph Engineers*, xvi. 404.

time, so that it was not surprising that these early lines had to be so soon abandoned.

Mr. Robert Sabine, reporting on telegraphs in the Paris exhibition of 1867, compares the six underground systems in that exhibition with the one only shown in the London exhibition of 1862,

> plainly indicating the increased attention which is being devoted to this [underground] branch of telegraphy, and therefore a growing necessity for it. . . . Underground lines, almost abandoned in 1862, have since that date been creeping gradually again into favour.[1]

Mr. Sabine, writing at a period nearer than that of Mr. Fleetwood, gives the same reason for the earlier failures. 'The principal difficulties,' he says,

> which underground lines have had to contend with have been the carelessness with which the wires were laid and the decay of the insulating materials. These difficulties have been met more completely in France than elsewhere.[1]

He describes some of the exhibits in which gas tar, asphalte, or bituminous compound of some sort are the insulating substances. One system (Donald Nicoll, Kilburn) had an ingenious method of jointing the conductors of the separate sections. At one end the wires were twisted in hollow coils; at the other they were straight and the straight end was pushed into the adjoining helix and soldered. A great point in favour of the proposed system in Mr. Sabine's opinion was that it required little skill or practice on the part of the workmen employed. Another exhibit (A. Holzmann, Amsterdam) was a trough system in which the wires were separated by glass supports, which Mr. Sabine considered objectionable. But of both he says:—

> These systems are not new: they have been tried repeatedly in France and in England, but failed, partly because the material employed for the insulation was too brittle, partly because insufficent care was taken to keep the wires apart. But with the benefit of all the experience and failures of other inventors, it is to be hoped that both Mr. Nicoll and Mr. Holzmann may succeed in their endeavour to give us cheap underground lines, by which they will be doing a most welcome service to telegraphy.[1]

It is evident that this hope was not realised, for the draw-in system became general. The reason may be gathered from Culley, who says that it is very much more easy to place the wires in a

[1] *Illustrated London News*, November 16, 1867, p. 550.

trough than to draw them into a pipe, but it is much more difficult to execute repairs in a trough.

The plan which has been pursued for many years in London and the larger towns is to lay down a pipe under the flagstones amply large enough for all the wires likely to be required. Oblong ' drawing-in ' boxes, 30 inches by 11 inches and 12 inches deep . . . are placed at every hundred yards if the line be straight, and nearer if it be curved.[1]

On the transfer of the British Telegraphs to the State in 1870, considerable additions to the underground work in London were made; and on January 17, 1874, when the wires were diverted to the new office in St. Martin's le Grand, the length of the pipes in the metropolitan district was about 100 miles, and the total quantity of wire in the pipes amounted to 3000 miles. Later, a new line of three-inch cast-iron pipes was laid down and 100 miles of wire drawn into it, in substitution for 126 miles of overhouse wire.[2] A route of underground lines was also laid between Liverpool and Manchester by the Post Office in 1870.

The interest of the citizens and the Government in the amenities of the capital city led to the underground system in Paris being developed to a considerable extent, and in 1864 it was well reported on by Professor Hughes. There were (he said) over 2,000 kilometres of wire laid, all in a perfect state of high insulation, which, from the constant and uniform good results, made the French system worthy to be studied by electrical engineers. There were seven distinct underground routes leading from the central station in Paris to the limits of the city, where they were joined to the air lines leading to the different provinces of France. Iron tubes 2½ yards long were laid in a trench one yard deep. A pilot wire was placed in the tubes whilst being laid. With this a rope was drawn in serving to introduce the cable. He adds :—

There are now 52 kilometres of subterranean lines in Paris, 1450 kilometres of wire enclosed in these tubes. . . . Tubes of 60 millimetres receive easily 28 wires; 100 millimetres receive 49 wires; those of 120 millimetres have 77 wires. The copper-conducting wire is formed of a strand of four wires, each having a diameter of half a millimetre. They are covered with two coverings

[1] *A Handbook of Practical Telegraphy*, fifth edition, 1871, p. 153.

[2] The process of placing the wires underground wherever possible has been considerably extended, and of the 1,745 lines of wire entering the Central [Telegraph] Station in London, not one is open. In many cases the wires are conducted underground for distances of 12 to 22 miles from this office. (*Postmaster-General's Report*, 1888, p. 9.)

of gutta-percha, giving in total a diameter of five millimetres. These are separately covered with tarred hemp.[1]

Professor Hughes gives a record of his tests, and he attributes the satisfactory results attained to the fact of the iron tubes being solid and hermetically sealed, the wires being thus protected against any mechanical injury as well as from the effects of gases which were considered to be the main cause of the deterioration of the gutta-percha. He gives also particulars of the capital outlay which will be of interest in comparison with that for existing underground circuits :—

Iron tube, 120 millimetres diameter, in which are placed sixty-three wires, contained in nine cables of seven wires each. The entire cost of the tube, trenches, placing wires, etc., complete, eight francs per yard, or twelve centimetres each wire per yard. The cables, with seven insulated wires, cost 2900 francs per kilometre. This would make 414 francs each wire per kilometre, and, adding the cost of the iron tube, with laying, etc., the total price would be 534 francs per kilometre ; and as these tubes are, after four years' use, in a most perfect state of preservation, it is but reasonable to suppose that they will be serviceable for many years to come.[1]

While in Great Britain and the United States the system adopted was determined by conditions of economy or commercial efficiency, the exigencies of military requirements had some sway on the Continent and probably account for the further efforts at the development of underground work in both France and Germany.

The later German system of underground telegraphs dates from about 1876. The cables were made on the general plan of an ordinary submarine telegraph cable. Each cable contained seven conductors. Over 2,000 miles of cable, containing 15,000 miles of wire, connected the principal cities in 1882.

In the United States, underground telegraphs were not extensively used, but when adopted the European practice of separately insulated wires with suitable protection by pipes or wooden conduits was followed.

In 1875 William Mackintosh, Superintendent of Construction for the Western Union Telegraph Company, laid a cable underground through a three-inch pipe from Broadway and Dey Street to 18 Broad Street, New York. It was composed of five conductors of No. 16 copper.

The insulation was 'best English gutta-percha.' The length of this underground line was about 1600 feet. The iron pipes were

[1] *Telegraphic Journal*, No. 33, August 13, 1864, vol. ii. p. 73.

dipped in coal tar and the joints were leaded. The cable cost $2\frac{1}{2}$ cents per foot, the iron pipe about 50 cents per lineal foot, and the digging, laying, and repaving a further 50 cents per lineal foot.

Since the laying of this cable, another of twenty-eight wires has been added, and was pulled through the pipe a distance of 800 feet by the aid of a team of horses and twenty-five men. This underground line was opened and inspected about four months ago (December 1880) and found to be in perfect condition and the gutta-percha as pliable as ever.[1]

The early underground work, in which the insulation of numerous conductors was dependent upon one outer covering, had failed and the causes of its failure are attributed in part to the materials and in part to the want of care or the impracticability of obtaining the necessary conditions to do satisfactory work in the process of laying. But efforts were made to overcome these difficulties, and it is somewhat remarkable that amongst the earliest efforts are to be found the principles of the latest successes. On August 4, 1845, William Young of Paisley, manufacturer and dyer, and Archibald McNair of the same town, merchant, obtained a British patent (No. 10,799) for ' an improved method of manufacturing electrical conductors.' The conductors were to be ' formed of one or more copper, iron, or other metallic, or mixed metallic, wires.' The preferred method of covering the wires was ' with threads in a plaited or braided form by means of a braiding machine,' stress being laid on the suitability of the covering thus applied to withstand the subsequent processes of manufacture. The wires, having been covered, were wound on reels to which was applied suitable tension. They were unwound from the reels through a lead press, first passing through a cistern of molten pitch on their way through the press. Two methods of lead-press operation are described ; ' containers ' and ' die blocks ' are fully illustrated ; the consistency and temperature of the lead defined, and suitable methods of applying heat to both the insulating material and the lead suggested. After passing through the lead press the cables were to be wound on reels, the reels mounted on wheeled carriages, and, one end of the cable being held fast, the carriage was to be driven in the desired direction, the cable unwound and delivered on the ground ' without risk of injury and with great facility.' Having been laid, the ends were to be brought into joint or test boxes provided with mercury cups through which the continuity of each conductor and of the sheath was effected.

The scope of the claim would indicate that this may be the first

[1] *National Telephone Exchange Association Report*, April 1881, pp. 49–50.

recorded suggestion of such a method of manufacture, for while the patentees say they have no intention to claim

> the adaptation of wires surrounded with non-conducting substances enclosed in tubes for electrical conductors, . . . that which [they] do claim is, the construction and manufacture of the electrical conductors by the employment of machinery having a tubular mandril or hollow rod through which wires may be drawn, whilst the leaden or other soft metal tube is forming, by pressure between a core and die, such wires being at the same time imbedded in pitch or other non-conducting material.[1]

The patent specification contemplates a multi-conductor cable, but if ever made by this process such a cable does not appear to have reached a commercial stage. A single conductor insulated with gutta-percha and enclosed in a leaden tube was exhibited by McNair in the London 1851 Exhibition.[2] According to the Jury Report there were many specimens of subaqueous wire in the Exhibition, but McNair's wire was the basis of most of them. The process of manufacture is briefly but clearly described in the Report,[3] and the opinion is expressed that the article seems to answer its purpose well.

The adoption of a similar process of covering in the United States is shown in Shaffner's statement[4] that 'Mr. Bishop has constructed extensive machinery for the covering of the insulated wire with lead of any required thickness.'

Enclosing a single conductor the hydraulically pressed lead covering was an article of commerce in 1851, but for a multi-conductor cable it is probable that there was not sufficient confidence in the lead covering so formed, for in 1869 William Alfred Marshall of Canonbury, in the County of Middlesex, telegraphic engineer, took out a patent[5] 'for improvements in the process of insulating and protecting the wires of subterranean, submarine, and other electric telegraph cables.' The wires were to be covered with cotton or other fibrous non-conducting material, which was to be previously dried and then placed in a vessel containing melted paraffin wax until the whole of the cotton or other fibres were thoroughly permeated. The wire or wires thus covered or prepared were introduced into a lead or other soft metal tube of convenient length laid or stretched horizontally in a straight line. The lead tube was to be wound on a skeleton drum to which

[1] British specification, No. 10,799, 1845.
[2] *Official Description and Illustrated Catalogue*, i. 455.
[3] *Reports of the Juries*, p. 293.
[4] *Telegraph Manual*, 1859, p. 606.
[5] No. 2587, December 11, 1869.

heat was applied. At one side of the axle of the drum was a tube extending through the bearings. The inner end of the tube terminated in a nozzle to which one end of the cable's leaden covering was attached. The outer end of the tube had also a nozzle to which was connected a flexible tube attached to a pump. The air was to be first pumped out of the cable and paraffin was then pumped in. The full description of the process, which is here abbreviated, shows that Marshall had a clear idea of what he required to attain in the course of manufacture. The reason for first expelling the air is thus stated :—

By introducing the paraffin wax into the tube [the cable covering] in which a vacuum is created as above explained, I prevent the formation of air bubbles therein, which would impair the insulating properties of the paraffin wax.

And further—

In order to provide for the contraction of the paraffin wax on cooling, I proceed to cool the cable gradually in the direction of its length, commencing at the disengaged end.[1]

Recognising that there was no novelty in the use of cotton, paraffin, or lead or in their combination, Marshall, like the patentees previously quoted, made ' no claim to the use of any of the materials mentioned,' but limited his claims to (1) the improved process described ; (2) the skeleton drum for carrying the length of cable in combination with the heating tank (or hot air chamber) ; and (3) the combination with the skeleton drum of the hollow journal, the coil and tube for the supply of paraffin wax from tank to cable during the rotation of the drum.[2]

A length of this cable was laid across Windsor Park (in a clay soil) in connection with a private telegraph line, but, after being down for some months, it failed, and on examination it is said the lead covering was found to have become completely decayed in several places, thus letting in moisture and destroying the insulation.[3] The cause of the decay in the lead pipe appears to have been attributed to the clay soil in which the cable was laid, but experience has shown this to be inaccurate. Decayed vegetable matter in the soil may, however, have been the destructive agent.

Marshall's patent appears to have attracted but little attention, and a single failure seems to have been sufficient to close its commercial career. It deserved a better fate. Even with the defects which are noticeable, more extended trial would have shown it to

[1] British specification, No. 2587, 1869, p. 6.
[2] *Ibid.* p. 9.
[3] *Electrical Review*, London, November 11, 1882, p. 368.

be reliable. It was an early attempt to transfer to a factory the combination in a completed form of conductors, insulation, and covering, and the attempt was made in a manner which (but for the slight yet important feature of the air treatment) led to eventual success.

Confidence in the outer covering was essential for the commercial success of any system in which the insulation of the interior wires was dependent upon the integrity of the envelope, and the experience with lead-covered cables was not encouraging. David Brooks of Philadelphia was of opinion that the lead pipe was liable to damage in the process of coiling and uncoiling on the drum, and he proposed the use of iron pipes with a liquid insulation. His system, patented in the United States July 13, 1875,[1] and Great Britain (No. 4824) December 19, 1877, had the merit of simplicity. Cotton-covered wires were drawn into iron pipes of which the joints were thoroughly caulked. The pipe was then filled with mineral oil under pressure, so that, in the event of any defect in joint or pipe, the tendency should be for the oil to leak out rather than for the moisture to leak in. Introduced in the United States in 1877 or 1878, the United States patents on the Brooks' system were acquired by General Anson Stager on behalf of the Western Electric Manufacturing Company in 1879.[2] Mr. Barton stated in 1880[3] that some of the cables laid at an early period were still in operation. Between twenty-five and fifty sections of different lengths had been used for telephone lines for a period of a year to a year and a half. Whilst some had failed from mechanical injury, the others had been successful. A pipe an inch and a quarter in diameter was provided for a core of fifty wires.

The system was introduced into England and Belgium. An illustrated description appeared in the London *Telegraphic Journal* of December 1, 1879, and in 1880 Mr. Brooks made a proposition to the Postal Telegraph authorities that an experimental length should be laid between Waterloo and Nine Elms on the London and South-Western Railway, on the condition that it should be paid for if it worked satisfactorily for a period of six months.[4] The proposition was accepted, the line laid, and the conditions met. In the following year (1881) the line was extended to Clapham Junction. The extension was carried out by the India Rubber, Gutta Percha and Telegraph Works Company under the direction of Mr. Brooks.[5] It was completed during the summer,

[1] U.S. specification, No. 165,135.
[2] *Journal of the Telegraph*, (New York,) May 16, 1879, p. 149.
[3] *National Telephone Exchange Association Reports*, 1880, p. 47.
[4] *Telegraphic Journal*, August 15, 1881, p. 306.
[5] Some personal recollections of the laying of this line are given by 'J.T.L.' in the *Post Office Electrical Engineers Journal* for July 1915, viii. p. 118.

and tests published in the *Telegraphic Journal*. The use of this line for telephone purposes is thus stated :—

Experiment shows that the Brooks' system may be employed with great advantage for telephone working ; a *single* wire of the cable between Waterloo and Nine Elms has been employed as a telephone circuit for some time, and, although the other twenty-nine wires are heavily worked, yet the inductive interference has not been such as to cause inconvenience. A telephone, placed on two wires of the entire length between Waterloo and Clapham did not emit any sound, although a Wheatstone automatic instrument with high battery power was being worked on one of the other wires ; the metallic loop, it may be observed, was not a twisted one, but was formed of two wires picked at random from the group of thirty. No doubt the satisfactory result obtained is in a great measure due to the fact that the relative positions of the various circuits in the pipe is different at every point throughout the length. In the cases where a number of wires in a cable are employed for the telephone working alone, it is found that the inductive interference between wire and wire diminishes in proportion to the number of wires in the pipe, a result due, no doubt, to the distribution of the effect.[1]

This experimental line continued to work well up to 1882,[2] but the eventual result is thus told by Mr. Fleetwood :—

After this line had been completed, the wires, with one or two exceptions, gave very good results as regards insulation ; but from the first there was a very great loss of oil, and this continued more or less although every effort was made to trace and repair the leaks. Probably the constant vibration of the viaduct has had something to do with the leaky condition of the joints. Owing to the viaduct being widened for a considerable length, it was necessary last year to divert the wires working through this system between Westminster Bridge Road and Queen's Road Station, and to recover the pipes and wires along the railway between the above points. This was done, and the remaining portion from Queen's Road to Clapham Junction thoroughly overhauled, one section of about 300 yards being renewed. Since last October the line has worked well, and, although the cost of maintenance has been very heavy, it does appear to me that the system is worth a further trial under more favourable conditions—for instance, along a country road, where it might form part of a through line and where the pipes would not be subjected to such frequent disturbances as in the busy thoroughfares of London.[3]

The liquid oil system of insulation was, however, found to be

[1] *Telegraphic Journal*, August 15, 1881, ix. 306.
[2] *Electrical Review*, London, September 30, 1882, xi. 262.
[3] *Journal of the Institution of Electrical Engineers*, xvi. 418.

unsatisfactory even on routes not subjected to vibration, sometimes from leakages, sometimes possibly from the absorption of moisture. It had, moreover, the defect of previous systems in that it was not complete until it was laid. It was not an article of manufacture but a combination of conduit and cable separately constructed and the combination completed *in situ*. It did not survive, but it deserves commendation and remembrance as the starting-point from which subsequent developments arose.

The Brooks cables were laid in the United States under the supervision of Mr. W. R. Patterson. His company sought to improve the existing types of multi-conductor cable, and, in common with the rest of the world, they had the knowledge from experience that liquid oil in a rigid pipe would leak out and solid wax in a flexible pipe would crack. Liquid oil in the flexible pipe or solid wax in the rigid pipe were considered to be equally impracticable.

On May 14, 1881, Mr. Patterson filed an application for a patent, in which he related the facts that paraffin, resin, beeswax and other similar substances alone or in combination had been used as an insulating substance, and it was well known they shrink on becoming cold ; thus cavities had been left within the pipe which had caused great trouble.

In order to compensate for this shrinkage [said Mr. Patterson] I charge the melted paraffin or other insulating substance with carbonic acid gas or other suitable gas or mixture of gases and force the liquid thus prepared into the heated pipe around the core. . . . The gas is held in very minute bubbles hardly visible to the naked eye. As the substance cools and contracts, these bubbles expand and the pipe is thus kept entirely full. These bubbles are isolated from each other and uniformly diffused throughout the mass of solid substance. The insulating substance is thus formed into a light, porous, homogeneous mass.[1]

Patterson's cable was Marshall's cable with one very important difference. Marshall excluded air because he conceived it would be detrimental to the insulation, and Patterson included air in the form of a suitable gas because it would render the insulating substance with which it was combined elastic instead of brittle, expansive instead of contractive.

The manufacture of this type was immediately undertaken.

At the meeting of the National Telephone Exchange Association in 1883 it was stated by Mr. Fay of Chicago :—

In 1882 a 100-conductor Patterson lead-pipe cable was placed in the Washington Street tunnel, which is still working and in good condition.

[1] U.S. specification, No. 248,209, October 11, 1881.

Some time prior to this, the system of housetop wires had grown into a forest of tall frames, cumbrous to handle and obnoxious to property owners. In the spring of 1882 it was decided to commence a gradual but sweeping change in this city by substituting aerial cables wherever the bodies of wires exceeded forty in number.

It is not improbable that, at some time, city exchanges will be operated generally upon the metallic circuit system; and in buying cables which were intended to last for some years, it seemed advisable to have them made of pairs of conductors twisted together, so that each pair could be used as a metallic circuit, if desired, in the future. We purchased some Patterson lead cables, made by the Western Electric Company, with this in view. They consisted of fifty pairs of conductors of No. 26 copper wire. One wire in each pair is insulated with white cotton and one with red.[1]

A catalogue of the Western Electric Manufacturing Company dated 1882 gives a list of Patterson cables then in practical operation (*see* table on p. 257).

The aeration of the paraffin wax resulted not only in the production of an elastic insulating material well adapted as a filling for cables, but was also of great benefit in another respect. The advantages of low capacity were well recognised by telegraph engineers, and David Brooks had included this amongst the advantages of his liquid oil system, but the Patterson system was a further advance. A descriptive circular of 1884 states this advantage as follows:—

Another and not less important function of the gas is to reduce the specific inductive capacity of the insulator. Solid paraffin possesses about the lowest static capacity of any insulator, except air and gases; an admixture of dry gas in the above-described manner reduces the specific inductive capacity at least 15 per cent. below that of pure solid paraffin.

The practical utility of this extremely low inductive capacity is that it enables us to use a smaller conductor than if such materials as gutta-percha or rubber were used as insulators. Large conductors will give better results than small ones, other things being equal.

But as the retardation is affected equally by the capacity and resistance of the wire, if we reduce the specific inductive capacity of the insulating covering, the resistance of the conductor can be increased in the same ratio without affecting the working of the circuit; or, to state it in another way, if the same size of wires is used, the cable with the smallest capacity will give service through the longest distance. Conversation can be carried on with equal loudness and clearness through 1500 ohms and 10 micro-farads as through 1000 ohms and 15 micro-farads.

The use of small conductors is advantageous in some respects provided the retardation is no greater than in large ones.

[1] *National Telephone Exchange Association Report*, 1883, p. 68.

The lead-tube type of cable was generally adopted for both underground and aerial purposes, and its manufacture was now undertaken by several firms. The first to be put into actual use

FOR TELEPHONE LINES

Subterranean and Subaqueous

Length	Number of Conductors	Size of Conductors	Situation
Ft.		No.	
2640	5	26	Evansville, Ind.
2300	50	26	Philadelphia.
1400	25	26	Norfolk, Va.
1000	25	26	Norfolk, Va.
490	3	26	McGregor, Iowa.
440	3	26	McGregor, Iowa.
400	7	26	Bridgeport, Conn.
360	30	26	Buffalo, N.Y.
320	12	26	Newburyport, Mass.
280	7	19	Ludington, Mich.
265	15	26	Green Bay, Wis.
260	25	26	Michigan City, Ind.
215	50	26	Toledo, Ohio.
200	7	19	St. Joseph, Mich.
150	50	26	Trenton, N.J.
78	15	26	——, Wis.

Aerial

Length	Number of Conductors	Size of Conductors	Situation
Ft.	Double	No.	
1100	75	26	Chicago.
3150	50	26	Chicago.
1050	50	26	London.
7720	40	26	Jersey City.

[The size of conductors above given is that of the B and S gauge.]

for telephone purposes would appear to have been the Philadelphia cable of Brooks Junior mentioned by Mr. Sargent at the meeting of the National Telephone Exchange Association in April 1881.[1]

Mr. Sargent stated that they had a cable in Philadelphia which was laid some time in May 1880 and had been in successful operation ever since. The length was about 600 feet, and there were eighty-four

1 *National Telephone Exchange Association Report*, April 1881, p. 28.

twin wires in an inch lead pipe. It was made by David Brooks Junior a son of the inventor of the Brooks cable previously described. The insulation was a combination of resin and coal oil forced in in a warm state and becoming solid when cold. In the manner of making splices and in the material employed, this cable was similar to that used by the Electric Telegraph Company between the Strand and Nine Elms Station in London in 1846. It differed by multiplying more than tenfold the number of wires within one lead covering and by transferring to a factory much of the work hitherto done upon the line, just as Young and McNair and Marshall had proposed, the only difference being in the material used as insulation.

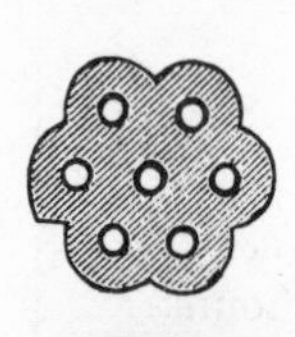

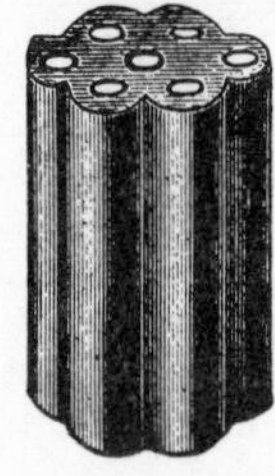

FIG. 104. — Waring Cable (Anti-Induction.)

The Patterson cable was the next of this type put into commercial use, and was followed by the Waring System of the Standard Underground Cable Company. In a descriptive circular of the Waring Underground Electric System (apparently issued in 1887) it is said that Mr. Waring's system had been developed and applied 'during the past five years.' The insulation is described as a hydro-carbon compound, inorganic in its composition, not liable to disintegration or deterioration. . . . Unlike paraffin, gutta-percha, and many other commonly used dielectrics, the Waring insulation may be subjected to a 'high degree of temperature without decentralising, carbonising, or breaking up into vapours and gases, but upon cooling returns to its normal condition, unimpaired in its insulating qualities.' The cables are classified as follows:—

1. *Anti-Induction Cables.*—Each insulated conductor being separately surrounded by a body of metal, thus destroying the effects of induction. Used for telephone purposes, where secrecy is important and distinct articulation a prime requirement; for telegraph and general electric purposes, where exactness of current, free from extraneous influences, is necessary to the proper adjustment and working of instruments. (Fig 104.)

2. *Bunched Cables.*—Where a greater number of conductors is required in a minimum space. Used for ordinary telegraphic, telephonic, and other electric purposes. (Fig 105.)

4. *Submarine Cables.*—Either anti-induction or bunched. Used for all above purposes where crossing water *en route* becomes necessary. (Fig. 106.)

Amongst the earliest manufacturers in Europe to develop the lead-pipe type of cable were Messrs. Berthoud & Borel, of Cortaillod, Switzerland. In 1879 they introduced a lead-pipe cable which had also a leaden conductor. The latter was a peculiarity and was adopted with a view to the simultaneous manufacture of core and tube with insulating material separating them.[1] The high resistance of the lead conductor and the unstable nature of the insulator (sulphur or crushed resin) prevented this cable from being successful; but in 1881 the makers introduced a modification, returning to copper as a conductor, using cotton as a separator and paraffin as an insulator, but retaining the method of putting on

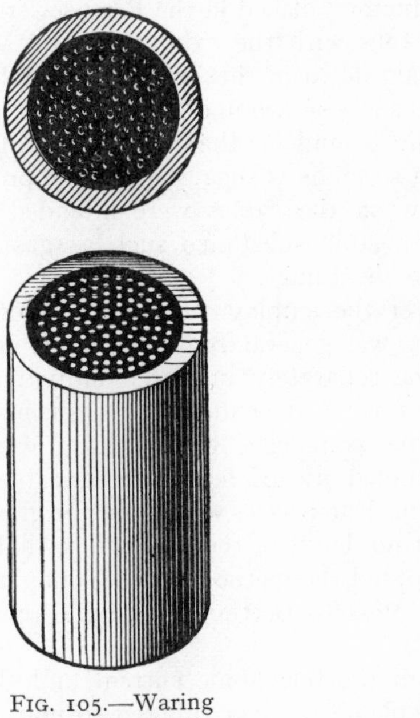

FIG. 105.—Waring Cable (Bunched).

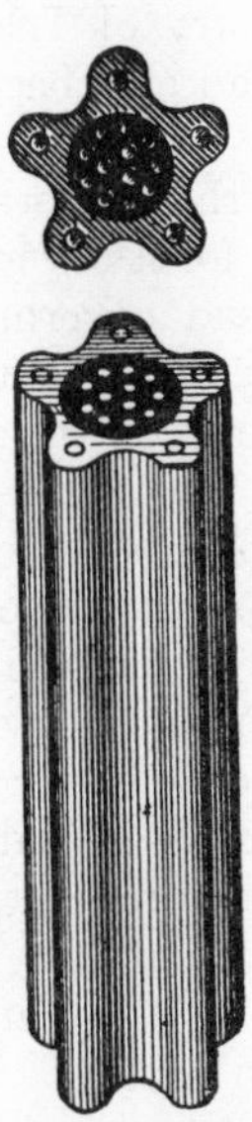

FIG. 106.—Waring Cable (Submarine).

the lead covering in the process of manufacture. In this respect Messrs. Berthoud & Borel were, so far as practice was concerned, in advance of the times though, as we have seen, the process was patented in 1845. The Standard Underground (Waring) Co. also adopted lead press covering. The early Patterson cables were made by drawing a core of suitable length into already constructed lead pipes and jointing those pipes together. To quote the Patterson circular: 'The core may be made in continuous lengths of 1500 feet and the protecting pipe is jointed over it in lengths of 75 to 100 feet.' The reason for this was the lack of confidence in the integrity of the lead pipe without a previous test. 'Any flaw

[1] *Telegraphic Journal*, April 1, 1879, p. 116.

which may exist in the pipe and any leaky joint is detected by the process of filling.'

Messrs. Berthoud & Borel, having made up the core, passed it through a lead press to receive its covering. The process is described in the London *Electrical Review* of November 11, 1882, where it is stated that nearly ten miles of these cables were employed for the transmission of power and light in different parts of the [1881 Paris] exhibition building; 5000 metres had since been laid down for the Société Jablochkoff for the lighting of the opera; a cable with several conductors, placed in the Paris sewers, connected the Ministry of Telegraphs with the exhibition, and several kilometres had also been laid down in the sewers for telephonic transmission.[1] The lead press was shown in operation at the exhibition.[2]

That there was some ground for the general lack of confidence then felt for a covering so made is suggested by the provision of a second lead covering when the cables were intended for underground use, an 'impermeable substance such as gas tar' being placed between the two sheathings.

Some few years later the employment of the hydraulic press system for lead covering was general, but, in the early stages of the efforts to supersede the separately insulated individual wires of the core by a group of wires dependent for their insulation on the mechanical envelope, confidence in the latter was of prime importance. The continued use of lead-covered cable prompted improvements in hydraulic lead presses which effected great economy in manufacture and left no doubt of the product, so that all manufacturers eventually adopted the method which Young and McNair patented in 1845 and Messrs. Berthoud & Borel commercially revived in 1879.

The characteristics of the telephonic current and the extreme sensitiveness of the telephone receiver differed so materially from the currents and the instruments used in telegraphy that a new line of experience was necessary in order to demonstrate the effect in practice of combining conductors together in cables and using those conductors for speech transmission in the form of individual messages for separate and distinct correspondents.

There was considerable experience of the conditions governing the use of telegraphic currents, and the effect in practice was known so far as the transmission of detached signals at intervals was concerned. A Joint Committee was appointed by the Lords of the Committee of Privy Council for Trade and the Atlantic Telegraph Company to inquire into the construction of submarine telegraph cables. Their report was published in 1861. To this Committee

[1] *Electrical Review*, London, xi. 368.
[2] *Mechanical World*, December 30, 1882.

Sir Charles Wheatstone submitted the results of experiments which were made by him to determine the amount of inductive discharge in wires of very considerable lengths according to variations in the lengths of the wires, their diameters, the material and thickness of the insulating substance, and in the temperature and pressure of the medium by which they were surrounded. The introductory paragraphs of Sir Charles Wheatstone's report will serve to illustrate one of the effects incident to the use of cables applied to telegraphs :—

The introduction, in late years, of submarine telegraphic cables, and the substitution of subterranean lines of considerable lengths for the aerial lines almost exclusively employed formerly, have brought into evidence certain conditions of electrical charge which so materially influence the rapidity and frequency of the signals transmitted as to make it an object of the highest importance that they should be strictly investigated, in order that it may be ascertained whether it be possible to obviate the evil effects arising therefrom, or in any degree to alleviate them.

These conditions could not fail to come under the observation of persons employed in telegraphic operations as soon as such lines were constructed; but before they were made the subject of investigation by Mr. Werner Siemens [1] and Dr. Faraday,[2] only vague notions respecting their cause prevailed.

When a metallic wire is enveloped by a coating of some insulating substance, as gutta-percha or india-rubber, and is then surrounded by water or damp earth, the system becomes exactly analogous to a Leyden jar or coated pane; the insulating covering represents the glass, the copper wire the inner metallic coating, and the water or moist earth the external coating. The electricity with which the wire is charged, by bringing the pole of an active battery in contact with it, acts by induction on the opposite electricity of the surrounding medium, which in its turn reacts on the electricity of the wire, drawing more from the source, and a considerable accumulation is thereby occasioned which is greater in proportion to the thinness of the insulating covering.

One mile of copper wire, one-sixteenth of an inch in diameter, presents a surface of 85–95 square feet, and receives the same charge from a source of the same tension as a Leyden jar having an equal number of square feet of tinfoil coating. There is, however, one material difference between the two cases. Though both are discharged in a time inappreciably minute to the senses, the discharge from the wire occupies a comparatively much longer interval than that from the coatings of the jar.

A wire on insulating supports, in the open air, when it is unconnected with the earth, receives also a charge, but very much smaller

[1] *Annales de Chimie et de Physique*, troisième série, tome xxix. 1850.

[2] *Proceedings of the Royal Institution*, January 20, 1854, 'On Electric Induction—Associated cases of current and static effects.'

in amount, the inductive action of distant surrounding bodies exerting but little influence upon it.

Although certain general conditions of this inductive action in telegraphic wires have been made the subject of experiment and are now well known, our knowledge in regard to this subject still remains very limited.[1]

The subject of inductive effects received further consideration by Professor Hughes who read a paper at the Society of Telegraph Engineers in March 1879: 'Experimental Researches into means for preventing induction upon lateral wires.' Professor Hughes said that the induction upon lateral wires had of late years been a serious question upon telegraph lines—the constantly increasing number of wires upon the same poles, added to the adoption of high speed and consequent sensitive apparatus, rendered the study of these effects of the first importance. In 1868, at the desire of the French Telegraph Administration, he undertook a series of practical experiments with a view to finding a remedy, and he at once perceived that the question was of a more complicated nature than was at first supposed.

We found then that we had to deal with the static charge of its own line and the dynamic induction of the lateral wires; that the effects of each were very different; and, whilst it was easy to deal with the comparative feeble static charge, the more powerful and rapid effects of dynamic induction could hardly be suppressed.[2]

Other duties prevented me at that date from following up this line of research. The disturbances on the lines from this cause have been on a constant increase from the adoption of more rapid, and consequently more sensitive, organs; and in the telephone we have at last arrived at an organ of rapidity and sensitiveness, which not only reveals the constant induction, but which is the main reason why the telephone has not been more largely adopted upon telegraph lines.[3]

The conclusions reached from his experiments were that,

if a telegraph line or a telephone line had a return wire upon the same poles, and absolutely equidistant from the inducing wire, we should have perfect protection, from the fact that the primary would then induce parallel currents in both wires in the same direction, but contrary to itself, and these parallel currents would, being of equal force, neutralise each other.[4]

Professor Hughes expressed the opinion that 'the cost of a double wire would probably prevent its use'[5] and more attention

[1] From *Report of Joint Committee*, Eyre & Spottiswoode, 1861; quoted in Wheatstone's *Scientific Papers*, pp. 168–9.

[2] *Journal of the Society of Telegraph Engineers*, viii. 163.

[3] *Ibid.* p. 164. [4] *Ibid.* p. 166. [5] *Ibid.* p. 167.

was perhaps given to the method of limiting inductive effects by a metallic screen. During the discussion on Professor Hughes's paper, Sir William Preece related the results of experiments he had made which demonstrated that a wrapping of lead-foil or tin-foil considerably diminished the induction between wire and wire. Before Hughes demonstrated the scientific basis for the return wire equidistant from the inducing wire, David Brooks suggested a twisted double wire in cables as a preventive against inductive disturbance. There were apparently two types of twisted wires in use according to the particulars of Brooks's cable given by members of the National Telephone Exchange Association. At the first meeting Mr. Barton said that the wires had sometimes been double, with one of the double wires connected to the ground. The object of this double wire was stated to be to avoid induction,[1] and the use of the term ' double ' would imply that the two wires were of equal gauge as in twisted pairs; but another form was described at the second meeting of the Association in 1881 by Mr. Haskins of Milwaukee, who said :—

> We have three hundred wires in six cables—fifty wires to the cable, of the Brooks form, across the rivers. . . . Each conducting wire is insulated with cotton, and then a very light copper wire, insulated with cotton, is wound spirally from one end to the other on the outside of that.[2]

From this description it must be inferred that the conductor was not twisted with a wire of corresponding diameter, but remained straight and was surrounded by a helix consisting of a wire of appreciably smaller diameter than the conductor itself. This method appears to have been efficient in practice, for Mr. Haskins continues :—

> I will say for those cables that the wires which run around them, the spiral wires which go to the ground at each end, entirely prevent induction between the wires.[2]

But this method is not the equivalent of the twisted conductor so arranged that the two limbs of the line should be equidistant from any source of disturbance. This feature, insisted upon by Hughes, is quite clearly expressed in David Brooks's patent.

> The proximity of the two telephone wires to each other is such with reference to neighboring wires, which might cause disturbance in the telephone circuit that such disturbing effect will be neutralised,

[1] *National Telephone Exchange Association Report,* 1880, p. 46.
[2] *Ibid.* 1881, p. 27.

or in other words, the inductive effects in one wire will be neutralised by the inductive effects produced in the other wire of this telephone circuit.[1]

Brooks describes and illustrates both parallel wires and wires which are 'twisted round each other.'

Bell also devoted attention to the twisted pairs, and is with Brooks quoted by Hughes in the paper referred to. In a note Hughes says that Bell had lately written him that the date of his patent was November 1877. In a specification[2] for a twisted pair cable Bell disclaims the arrangement of the wires for obviating inductive disturbance and the arrangement of the wires in twisted pairs, 'for these matters I have claimed in my application for Letters Patent for improvements in telephone circuits filed June 10, 1878, of which the present application is a division.'

In the course of correspondence on this subject Mr. T. D. Lockwood, writes me :—

I think there can be no doubt that the first suggestion of a metallic circuit in telephony is Professor Bell's British patent (which by the way has no drawing) No. 4341, November 20, 1877. Professor Bell, after getting this British patent, applied for a patent in the United States, as did also David Brooks of Philadelphia. . . . The U.S. Patent Office held that Bell should be put into interference with Brooks and was not entitled to go any further back for the date of his invention than the earliest date of his British patent. Consequently the interference was decided in favor of Brooks, who obtained U.S. patent No. 238,195, dated February 22, 1881. Bell, however, did succeed in getting from the ruins of his case a specific kind of metallic circuit cable patent, i.e. No. 244,426, July 19, 1881.

The application for a British patent and the omission to apply at the same time in the United States, is probably explained by the fact that Bell was then in England. Under date of December 20, 1877, he wrote from 57 West Cromwell Road, South Kensington, to Preece, advising the receipt of some rheostats and condensers, and adding :—

I send double wire for experimental telephones. Please take what you wish and return the remainder as I want to make it into experimental cables.[3]

In an article on 'Electric Communications,' published in

[1] U.S. specification, No. 238,195, February 22, 1881 (application filed March 4, 1878).

[2] U.S. specification, No. 244,426, July 19, 1881.

[3] I am indebted to Mr. Llewellyn Preece for this letter.

Chambers's *Papers for the People* in 1857,[1] allusion is made to the extensive development and application of telegraphs in the United States. The almost limitless breadth of territory, the author says, necessitated a proportionate extension of 'the metallic indicator of intellectual supremacy,' the lines in many instances being carried across the country, regardless of travelled thoroughfares, over tracts of sand and swamp through forests and solitudes. The writer of the article further remarks that 'Economy and rapidity of construction are prime desiderata in America.'

The telephone was following the precedent of the telegraph whether or not its promoters realised it. Not in solitudes, but in the very centre of the busiest communities, telephone exchange lines had their beginnings, but with them, as with the telegraph in the United States and elsewhere, 'Economy and rapidity of construction were prime desiderata.' Telephone exchange service was a new service. Its value had to be demonstrated. The larger the number of subscribers the greater its value to all. Hence the need of rapidity of construction. The need of economy was even greater, for, until the value had been demonstrated and a demand created, funds could not be obtained for financing the enterprises. These are ample reasons for the original adoption of single lines. Metallic circuits would have nearly doubled the cost, and in congested districts would have halved the numbers to whom service could have been afforded.

When, therefore, cables were sought as aids to conduct the wires into the central offices or to facilitate the construction along congested routes, one of the first sources of trouble was that arising from overhearing. Two principal remedies were applied: the use of double wires in the cable, of which one was used as a conductor, the other earthed; or a single conductor was wrapped with lead- or tinfoil. The latter type became known on both sides of the Atlantic as anti-induction cable and was expensive to manufacture.

To provide a remedy in a more economical way was the purpose of W. R. Patterson's United States patent No. 253,501. In the cable described in this specification the conducting wires were single, but there was, in addition to the line wires, a central wire of much larger diameter than the line wires. That larger wire was connected to earth at both ends. The theory prompting the design was thus described by the patentee :—

The establishment of a current in any conductor causes an induced current in the opposite direction in all the other circuits including the low resistance circuit. It follows from Ohm's law

[1] Vol. ix. p. 27. Partly reprinted in *Railways, Steamboats, and Telegraphs*, 1868, p. 27.

that, the current being equal to the electro-motive force divided by the resistance, a much greater tertiary current will be induced from the circuit of low resistance than from any of the others. I find that when the resistance of this circuit is very low in comparison with that of any of the others—say from one-tenth to one-fifteenth or less—this tertiary current will practically neutralise the secondary currents in the small conductors.[1]

The theory even then was questioned, but it was found in practice that, when the circuit of the low resistance wire was closed, there was no important inductive disturbance, and when it was opened the ordinary effects (overhearing) were observed,[2] and the cable was largely used.

The inclusion of this low resistance conductor in a cable was somewhat analogous to the extra wire over routes of open wires as mentioned by Mr. Pope,[3] and an earth connection contiguous to the conducting wires was early adopted for telegraph lines. One such plan is that of Mr. Highton, who proposed

to place between wire and wire a direct communication with the earth, so that any of the electricity transmitted, as it escapes from the wire, may be intercepted by this communication with the earth, and so transmitted direct to the earth, without the possibility of entering an adjoining wire.[4]

Cables of the lead-pipe pattern were growing in favour and becoming firmly established during the first ten years of the telephone's use. Whether of paraffin or hydro-carbon insulation, of central wire or anti-induction formation ; whether called Patterson, Waring, Brooks Junior, Berthoud Borel, or other names, they all relied upon lead covering and material insulation. They were yet to undergo important changes, but the later developments in cables are reserved for consideration in Chapter XXII.

[1] U.S. specification, No. 253,501, January 17, 1882 (application filed July 14, 1881).

[2] *National Telephone Exchange Report*, 1882, p. 92.

[3] P. 242.

[4] *The Electro Magnetic Telegraph*, by Laurence Turnbull, M.D. Philadelphia, 1853.

CHAPTER XXI

TEN YEARS' PROGRESS

In the preceding chapters it has been attempted to record the principal features in the invention and development of the telephone, of telephone exchanges, and accessory apparatus during the ten years which had elapsed since the first exchange was started, together with the prior work which has a relation thereto. They record mainly technical progress. It is proposed in the present chapter to consider the progress made in the telephone exchange as an industry—a public service enterprise—during the same period. Since it is not possible to deal with the various features either strictly in the order of dates or of localities, somewhat abrupt changes chronologically and geographically will be inevitable.

The rudimentary central office systems connected with E. T. Holmes's burglar alarm system in Boston [1] and with Isaac Smith's Hartford Drug Store [2] had prepared the way for the introduction of specially constructed exchanges such as that which opened for general commercial business at New Haven, Connecticut,[3] on January 28, 1878.

By December 31, 1887, Boston, with the neighbouring towns included as branch exchanges, had 5767 subscribers, Hartford 1176, and New Haven 1393.

The development in the United States was so rapid that on March 1, 1880, 138 exchanges were in operation or about to open; on February 28, 1881, the number was 408, and by the end of 1886 there were 736.

The energy put into the business, and the promptitude with which available places in the United States were offered exchange facilities, may be observed from a statement in the first Report of the American Bell Telephone Company, dated March 29, 1881, that there were then only nine cities of more than 10,000 inhabitants and one of more than 15,000 without a telephone exchange.

[1] Chapter viii. p. 69. [2] *Ibid.* p. 72. [3] P. 96.

That this covering of the ground was due to initiative and enterprise and did not arise from the public demands may be observed from the Report for 1884, which states that as a rule all the larger exchanges had a steady growth, and there seemed no reason to doubt that that would continue for some time to come. On the other hand, there was a pause in building exchanges in small places, and some seventy-eight of those already started had been for the present given up, while sixty-one new ones had been established. The view is expressed that

> the establishment of these systems in small towns was probably pushed too rapidly, in view of the stagnation in general business which followed. Many of those now abandoned will be restored upon a revival of business, and others can be put into operation under a system which is being worked out for small exchanges without a central office, and which, if successful as we hope, will carry the telephone into a large number of towns and villages where it is now impossible to place them upon a paying basis.[1]

The numbers of subscribers in the United States at successive periods were as follows :—

February 20, 1880	30,436
December 31, 1880	47,880
,, 1881	70,525
,, 1882	97,735
,, 1883	123,533
,, 1884	134,601
,, 1885	137,570
,, 1886	147,068
,, 1887	158,732

At the last-mentioned date there were 743 main and 444 branch exchanges, 127,902 miles of wire on poles, 9458 miles of wire on buildings, 8009 miles of wire underground, 363 miles of wire under water, and 6182 employees. The estimated number of exchange connections for the year 1887 was 369,203,705.

Classified in cities of specified populations, the exchanges and subscribers at December 31, 1887, were :—

	No. of Exchanges	No. of Subscribers
Cities with population exceeding 150,000 (A)	16	47,030
,, from 50,000 to 150,000 . . (B)	26	29,068
,, from 10,000 to 50,000 . . (C)	192	54,232
,, under 10,000 . . . . . (D)	505	28,382
	739	158,712

[1] *American Bell Telephone Company's Report*, 1884, p. 4.

The above figures are summarised from the Company's statement of exchange statistics as at January 1, 1888. The number of subscribers (158,712) is less by twenty than those given in the Annual Report (158,732).

In Canada there were 2082 subscribers connected with the various exchanges at the end of the year 1880, 9614 in 1885, and 11,600 in 1886.[1]

The following represents the development in the United Kingdom at the nearest available dates to the end of the year 1887:—

	Exchange Lines	Private Lines	Total
United Telephone Co. (London and district) approximately . .	4500	1250	5750
National Telephone Co. (Scotland, Midlands, and North of Ireland), June 30, 1887	6729	1125	7854
Lancashire and Cheshire Telephonic Exchange Co., June 30, 1887 .	3788	1347	5135
Western Counties and South Wales Telephone Co., December 31, 1887	1572	462	2034
South of England Telephone Co., April 30, 1888 . . .	1435	373	1808
Northern District Telephone Co., December 31, 1887 . .	562	150	712
Telephone Co. of Ireland . .	—	—	936
Sheffield Telephone Co. approximately	500	120	620
Post Office, March 31, 1888 .	—	—	1370
	19,086	4827	26,219

The establishment of exchanges in France[2] and Belgium[3] by years was as follows:—

1879 Paris (September 30).
1880 Lyon (October 15) and Marseilles (December 15).
1881 Nantes, Le Havre, and Bordeaux.
1882 Lille.
1883 Reims, Roubaix, Tourcoing, Calais, Rouen, Alger, Oran, and St. Quentin.
1884 Halluind, Troyes, Dunkerque, Elbeuf, Nancy, and St. Etienne.

[1] *American Bell Telephone Company's Reports*, 1880–5–6.
[2] Extracted from Montillot's *Téléphonie Pratique*, 1893, p. 436.
[3] Extracted from *La Téléphonie*, par Émile Piérard, 1894, p. 343.

1885 Armentières.
1886 Boulogne, Cannes, Amiens, Nice, and Caen.
1887 Fournies and Châlons.
In Belgium at the end of 1884 there were seven exchanges—Anvers, Bruxelles, Charleroi, Gand, Liège, Louvain, and Verviers.
There were no additions during 1885. In
1886 Ostende—Middlekerke—Nieuport, La Louvière, Mons, and Namur; in
1887 Courtrai—Iseghem, Malines, and Termond—Alost—Lockeren—St. Nicolas, were opened.

Wietlisbach, the Director of Telegraphs at Berne, gives [1] the following particulars of the numbers of subscribers and the approximate rates of subscription in the principal countries on the European Continent at the end of the year 1885 :—

	Number of Exchanges	Number of Subscribers	Rates of Subscription
			Francs
Germany . .	91	14,732	250
(Berlin)		(4248)	
Italy . . .	16	8346	190–250
(Rome)		(2054)	250
France . . .	20	7175	250–600
(Paris)		(4054)	600
Sweden . .	15	5705	190
(Stockholm)		(2326)	190
Russia . . .	20	5280	190–600
Switzerland . .	36	4900	170
Belgium . .	7	3365	190–250
Austria . . .	11	3032	250–375
Holland. . .	8	2493	275
Denmark . .	2	1370	250
Spain . . .	3	594	440
Portugal . .	2	350	190–275

The progress made in the various countries at the nearest dates available to the end of 1887 is indicated in the following table summarised from Brault,[2] except the figures for Belgium which are taken from Piérard.[3]

[1] *Traité de Téléphonie Industrielle*, 1888, p. 244.
[2] *Histoire de la Téléphonie et Exploitation des Téléphones en France et à l'Étranger*, par Julien Brault, 1890.
[3] *La Téléphonie*, par Émile Piérard, 1894, p. 343.

	Date	Number of Exchanges	Number of Subscribers
Germany	Oct. 1887	123	22,695
Sweden	Dec. 31, 1887	148	12,864
France	1887	—	9,571
Italy	Oct. 1887	25	9,183
Russia	Sept. 1887	37	7,585
Switzerland . . .	Jan. 1, 1886	41	4,988
Belgium	Dec. 31, 1887	—	4,876
Austria	Sept. 30, 1887	13	4,200
Holland (estimated) .	Dec. 31, 1887	11	4,000
Denmark . . .	1886	22	2,677
Spain	Oct. 1887	8	2,200
Portugal	Dec. 1887	2	895

Comparing the two tables it will be observed that, placed as the countries are in the order of numbers of subscribers, Sweden has become second, France third, and Italy fourth. The order of the other countries does not vary.

Particulars are also given by Brault of the following :—

	Date	Number of Exchanges	Number of Subscribers
Norway	July 1, 1887	21	3900
Havana	Oct. 31, 1887	—	373
New Zealand . . .	Jan. 1, 1887	—	1800
Brazil	Apl. 1, 1885	7	3335
Honolulu . . .	July 1, 1885	—	1000

A copy of the daily 'Report of Connections' in the city and suburbs of New York for December 14, 1880, came into my possession many years ago. It was probably sent by the New York Company for the information of the London Company. The note is an autograph addition by 'H. W. P.' (presumably H. W. Pope). Both note and report will be of interest as indicating the position of the business at that date (see p. 272).

The average calls per line in the city exchanges, it will be noted, were 5·31 and in the suburbs 2·60 for the day.

Throughout the United States Exchanges the average daily calls per line for the respective years were as follows :—

1883	4·85
1884	5·18
1885	5·43
1886	5·82
1887	6·37

METROPOLITAN TELEPHONE AND TELEGRAPH COMPANY—DEPARTMENT OF STATISTICS

Report of Connections for December 14, 1880

Central Offices	Vacant Drops	District Number	Number of Subscribers	No. of Connections		Total No. of Connections	No. of Connections per Subscriber		Number of Connections per Subscriber for Total Number of Connections
				Local	Trunk		Local	Trunk	
NEW YORK									
82 Nassau Street	41	1	443	657	2042	2699	1·48	4·63	6·11
97 Spring Street		2	413	594	1894	2488	1·44	4·59	6·03
923 Broadway	81	3	220	124	983	1107	0·56	4·47	5·03
37 Whitehall Street	62	4	228	800	738	1538	3.51	3·23	6·74
35 William Street	72	5	224	114	1129	1243	0·51	5·04	5·55
33 Murray Street	83	6	284	152	1669	1821	0·54	5·88	6·42
1415 Broadway	73	7	130	31	505	536	0·24	3·88	4·12
58 Broadway	45	8	499	2029	667	2696	4·07	1·34	5·41
198 Broadway	256	9	420	615	566	1181	1·46	1·35	2·81
348 Canal Street		10							
46 East 14th Street	107	11	79	33	274	307	0·42	3·47	3·89
	820		2940	5149	10467	15616	1·75	3·56	5·31
Suburban Calls.									
Brooklyn	108	1	95	54	346	400	0·57	3·64	4·21
Coney Island		1							
Yonkers		1	51	90	16	106	1·76	0·31	2·07
NEW JERSEY.									
Jersey City	17	1	59	54	51	105	0·91	0·87	1·78
Newark		1							
Elizabeth		1	82	79	2	81	0·96	0·03	0·99
New Brunswick		1	35	34	—	34	0·97	—	0·97
Paterson		1	160	470	69	539	2·94	0·43	3·37
Orange		1	103	199	56	255	1·93	0·54	2·47
			585	980	540	1520	1·68	0·92	2·60

Calls were not received from the offices where columns are left blank.

Respectfully submitted,

NOTE.—*The suburban territory makes a poor showing for the reason that we have not had time to work it up and its growth has been wholly voluntary and service neglected.*—(*Signed*) H. W. P.

In the critical ten years now under consideration, telephone service, which was daily proving itself to be of enormous public benefit, was being developed with energy and resource against difficulties of varied kinds. Wherever official or public action was taken it was repressive, calling for the exercise of additional energy on the part of the promoters to overcome the artificial resistance inserted by public authorities against the advancement of a public benefit.

Much was done, but much more might have been done if the public and official bodies had either been merely quiescent or had studied the problem more carefully. In the United States there were no fundamental restrictions as in Great Britain. It was rather on the questions of overhead wires and of rates that local or State authorities harassed the telephone companies. The New York Legislature passed an Act requiring telephone companies in New York and Brooklyn to place all their wires underground before November 1, 1885,[1] a time when no practicable system had been evolved which would permit the working of an entire telephone exchange system through underground conductors.

In April 1885 the State of Indiana passed a law restricting the price of rental of a telephone to three dollars per month. As it was impossible to conduct the service at that rate, the Central Union Company was obliged to close its exchanges in the principal cities and large towns,[2] the public being put to serious inconvenience through the act of its own government.

A Bill to repeal the law was passed by the House of Representatives in the following year, but did not pass the Senate owing to the arrest of business from other political questions. In 1888 it was unconditionally repealed by a large majority in both houses. In the meantime[3] it had 'stopped absolutely any extension of business, and checked that steady improvement which is constantly going on towards greater efficiency and convenience to the public in the conduct of the business.'

Such cases as these indicate the futility of public action in ignorance. The Telephone Company asked that the question should be studied, claimed that it was worth studying:—

The telephone situation is worthy of more careful attention than it has received at the hands of the public. Its features are peculiar and the development of telephone facilities is certainly of high importance to our people. Although in the ten years that have

[1] *American Bell Telephone Company's Report*, 1884.
[2] *Ibid.* 1885.
[3] *Ibid.* 1886.

passed since the invention became public, 156,000 miles of telephone wire have been built, over which 275,000,000 communications now pass annually, we are in reality only at the threshold of the business. It is possible already to talk with ease between Boston and Philadelphia over our experimental wire, yet the connection of our principal cities and large towns by thoroughly practical telephone systems has, in fact, only been begun.[1]

But for the confidence and persistence of the promoters the telephone system, even of the United States, advanced as that was, would have been seriously hampered by public action. But the promoters were confident in the present and even more confident in the future :—

With the improvements that are rapidly coming into use, the aim must be nothing less than to provide a complete working system throughout the United States, which will give facilities for instant conversation between all points within many hundreds of miles of each other, such as is now possible within the limits of a single exchange.[1]

The arrangements with the operating companies had been made to this end ; the telephone companies throughout the land had been held under such a general plan by the force of a Government patent. In no other way, it was urged, could even the telephone development then attained have been reached so soon. And the Telephone Company asked the public to regard the matter as one in which the public were interested. There were rights of property in patents and in security of trade ; there were advantages to the public in improved means of communication. The Company asked for no privileges ; they only pleaded for the maintenance of ordinary rights :—

To anyone who will think what it would mean if instant verbal communication could be had by any person here with every city and town in New England, and as far as Philadelphia, perhaps even Washington and Chicago, with similar facilities in all parts of the country, it must be clear that this is something well worth the risk of money, thought, and labour by the companies ; well worth encouragement by the people. The work that has been done is at least some guaranty of what will follow, if the protection promised by our patents is not interfered with. The public is ripe for the use of such a convenience, the inventive talent of the country is providing new methods for making it available ; will the legislatures interpose to prevent all progress, through misconception of the problem, or because misled by interested parties ?

[1] *American Bell Telephone Company's Report*, 1885.

What has already been accomplished has been under every discouragement that can be thrown upon a property resting on patent rights. But a small percentage of patented inventions prove valuable, and all such are usually compelled to run the gauntlet of law suits from pretenders, so long as their patents live. Capital, therefore, is timid about such investments, and nothing but the expectation of large temporary returns will bring it out for the development of a business like this.

While it is true considerable returns have been received upon our property since 1883, it is equally true that for years we were without dividend or other means from our investment; our patents have but seven more years to run, and we are still working, as always before, under a heavy fire and an expenditure for defence which nothing but a prospect of liberal returns would, for a moment, justify.

To undertake such a work as has just been outlined, large sums are required for construction. It is within the power of State legislatures to so alarm capital that might be attracted to such a business, that it would be wholly useless for a time to attempt any important work of the kind suggested. The attack upon rates is one of the most direct methods of removing all inducement to extend telephone facilities; but, even if the right to absolutely regulate this matter were conceded, the question whether the few thousand persons, who now are connected with telephone exchanges, shall have telephone service a little cheaper than at present, is of trifling importance to the country compared to the completion of a telephone system adequate to the needs of the whole people.[1]

This is a somewhat long quotation, but I make no apology for recalling it, for it gives us a picture of the period which is of great value. In spite of occasional and local difficulties the Company succeeded in accomplishing much of its object, and if reason and logic had been useful public aids might have accomplished more sooner. The telephone must triumph, even over public neglect and public resistance such as it received in part in the United States. But its difficulties there were local or ones of detail. In general the companies could go forward with their business in their own way without any restrictions from the Federal Government.

In Great Britain the Telephone Exchange service started under restrictions. The first report of the United Telephone Company to April 30, 1881, states that an agreement had been come to with the Government and records that:—

There are a great number of conditions and limitations imposed, which tend greatly to restrict the use of the telephone for ordinary business purposes. From the point of view of the authorities of the Post Office, this may no doubt be necessary and justifiable. The

[1] *American Bell Telephone Company's Report*, 1885.

Directors do not complain of it. On the contrary they have to acknowledge that so far as their negotiations with the Department have already gone, they have been met in a spirit of enlightened liberality, as could fairly be expected when it is remembered by what considerations the authorities must be hampered in their desire to protect the Post Office telegraph monopoly acquired at so great a cost. But the Board believe that experience will speedily prove the great superiority, for many purposes, of the telephone over the telegraph in facility of transmission, dispatch, and economy, and that it must be soon largely employed by the Post Office itself as a useful and indispensable auxiliary to their important business.[1]

In London, as in Boston, the great advantages of the telephone were recognised by those who had undertaken to exploit it. They knew that its use must be restricted by the conditions and limitations imposed. The principal limitation was the one of distance, and the object of the limitation was to conserve the local telegraph revenue. This condition remained in force until 1884. Its bearing may be observed from the report of the Telephone Company of Ireland, whose chairman (Mr. Dwyer Gray, M.P.) rendered important service in assisting to obtain its removal. In the report of that company to June 30, 1884, is to be found the following statement :—

That the telephone supplies a want not met by the telegraph—namely that of obtaining an instantaneous reply—is demonstrated by the fact that the gross rental of the company from the Dublin district alone is now already upwards of £8000 per annum over an area with a radius of five miles from the central Post Office, whilst the amount received by the Post Office Department from telegrams despatched from an area with a radius of twelve miles from the Post Office to addresses within the same area, as stated by the Postmaster General in the House of Commons on June 9, 1884, was only £1390 per ann.[2]

The Irish Company is the only one recording this feature of comparison, but it cannot be regarded as typical. In other localities there was probably a greater local telegraph business, but the restriction must have had a very retarding effect on the adoption of the telephone exchange service by the public. In the new license the limitation of distance was withdrawn. The effect of its removal and the expectations of the directors thereon are thus recorded in the same report :—

The new license will permit the company to open public call-offices, and also to give intercommunication between towns. The directors anticipate that these privileges and conveniences will

[1] *United Telephone Company (London) Report*, 1881.
[2] *Telephone Company of Ireland Report*, 1884.

tend greatly to facilitate business and be of considerable advantage to the company.[1]

The Governmental restrictions were not the only ones the British companies had to strive against in their efforts to introduce a public service. Local authorities sometimes helped to retard the work. The report of the Northern District Company for the year ending December 31, 1882, contained the following paragraph :—

The directors feel that this increase [in number of calls] would have been very much larger but for the difficulty and delay caused by the opposition of some members of the Council who are strongly represented on the Highways Committee. This difficulty would be no doubt very much lessened if it did not altogether disappear if the business men of the town would impress upon their representatives in the Council the necessity of giving every facility for cheap and improved telephonic communication. The action of the Corporation prevents the company connecting quickly those who are desirous of joining the exchange, which is not only injurious and inconvenient to the member who has to wait, but to all the other exchange subscribers.[2]

The improvement effected by Mr. Fawcett's removal of the four-mile limit [3] is reflected in the report of the Lancashire and Cheshire Company for the year ending June 30, 1885. This stated that the average daily calls made by subscribers still maintained its steady increase, which 'is the best evidence that the telephone service is giving satisfaction.' In several of the smaller exchanges there had recently been a marked improvement both in the number of inquiries for telephones and in the average of calls made. 'No doubt the new trunk-wire and call-office services will have largely aided in awakening in these outside towns this increased appreciation.'

A lack of knowledge on the part of the public of the value of the exchange system is indicated by a suggestion made by a shareholder at the Annual Meeting of the South of England Company (1888) that 'the directors might by means of lectures, at the cost of a few pounds make the uses of the telephone more generally known.'

But in spite of all the drawbacks placed in its way, telephone exchange service was progressing. Public appreciation was growing, subscribers were increasing. To the larger class of business people telephonic communication was becoming more of a necessity and less of a luxury. The use of the service was still within a very

[1] *Telephone Company of Ireland Report*, 1884.
[2] *Northern District Telephone Company Report*, 1882.
[3] Chapter xxxii. p. 504.

restricted field. But the field was ever extending its boundaries. Demonstrations of its usefulness to the few were becoming apparent to the many, and subscribers were increasing accordingly. Improvements in the technical features had followed the demands of the business as experience forced those demands upon attention. Co-operative efforts towards improvements were in some measure available through the exchange of experiences at the annual conferences or conventions of the National Telephone Exchange Association, but improvements mainly resulted from the inventive ingenuity of telephone engineers or the individual discoveries of scientific men in America or Europe, and of the efforts on the part of manufacturers to keep abreast of the requirements of their customers.

From the first those responsible for the initiation of telephone exchange business saw that it was to be of great public benefit, and endeavoured to impress the public with the same idea. But the service was new, and they equally well realised that they had to learn their business in the school of experience. They placed no limit ultimately on distance or development, but only asked that they should not be called upon to carry out works of construction that were physically impossible in the then state of knowledge. But the school of experience was teaching much.

Advocates of systems were being subjected to the practical test of results as well as the criticisms of their colleagues at the annual conventions previously referred to. These conventions were noticeable for the full and free information mutually exchanged. But at the end of the period to which reference is now made it had become apparent to the management of the American Bell Telephone Company that the future growth would be such that provision could only be made for it satisfactorily by a standardised system of plant and apparatus, along lines which should commend themselves to the most capable and experienced minds in the business. Accordingly, during the year 1887 the company arranged for the meeting of conferences of experts on cables and switchboards, thus inaugurating a method of organised co-operation in the study of problems and their solution, which had important results in the development of the industry.

CHAPTER XXII

THE DEVELOPMENT OF DRY CORE CABLE

(THE CABLE CONFERENCES OF 1887, 1889, AND 1891)

THE report of the 1887 Cable Conference[1] ('printed' but 'not published') covers 152 pages.

The purpose of the conference and the state of information at the time are admirably set out in Mr. E. J. Hall's opening words from the chair, as follows :—

We meet here to-day for the purpose of discussing in an entirely informal manner the subject of telephone cables. We all realise how great the need is for more information on this subject, and how few sources of information are open in electrical literature or in the current published records of telephone work.

Looking at the matter from different standpoints, we wish to add together the results of our observations and experience, to discuss the theories which explain observed phenomena, and to illustrate by practical tests the application of the formulas which we hope to have presented, for our practical use in the future.

It is, perhaps, too much to expect that we can now reach the desired result of formulating definite rules covering all the details of specifications for telephone cables for any stated use. The art of telephony is probably too new to have yet reached any such

[1] The members were :—
Edward J. Hall, General Manager, American Telephone and Telegraph Company (in the chair).
E. M. Barton, President, Western Electric Company.
W. W. Jacques, Electrician, American Bell Telephone Company.
W. R. Patterson, Electrician, Western Electric Company.
John A. Barrett, Electrician, American Telephone and Telegraph Company.
A. S. Hibbard, General Superintendent, American Telephone and Telegraph Company.
W. D. Sargent, General Manager, New York and New Jersey Telephone Company.
W. H. Eckert, General Superintendent, Metropolitan Telephone and Telegraph Company.
Joseph P. Davis, Consulting Engineer, American Bell Telephone Company.
H. B. Thayer, Manager of the Western Electric Company in New York.

state of exact knowledge. We may, however, agree on some propositions that will so far help in our work as to keep us progressing in the right direction and enable us to avoid some of the costly experiments involved in testing, as we have all been obliged to, untried hypotheses on the large scale of practical work.

If the results of our conference show that we have not yet reached any broad basis of agreement, they will serve to emphasise the necessity of continued investigation and study, and we will have in the report of this meeting a large amount of valuable data preserved in a permanent form available for future reference. To this we may hope to add from time to time the results of our own broader experience, and the mass of information which is daily being accumulated by the many able observers who are working in all parts of the telephone field.

During the discussion on Thompson's paper on 'Telephonic Investigations' at the Society of Telegraph Engineers (I.E.E.) on February 10, 1887, Preece propounded a law to determine the limiting distance of speech :—

We take the total capacity of a circuit (K), which is found by multiplying the length of line by its capacity per unit length, and the total resistance (R), which is found by multiplying the length of line by its resistance per unit ; and the result is that we get a law determining the distance to which we can speak that is simply expressed by the product K by R.[1]

Professor Thompson in his reply objected to the comparison of a telephone line to a submarine cable in which capacity and resistance were the only things to be taken into account. He, on the other hand, held that in a telephone line the important things were the mutual induction, the self-induction, and the resistance ; the capacity being negligibly small.[2] Professor Ayrton during the same discussion contributed valuable remarks on the factors to be considered on telephone lines and cables, of which the details are available for ready reference and do not therefore need quotation. It suffices to call attention to the fact that early in 1887 scientists were considering what had previously received but little attention —the conditions which controlled the transmission of the electrical equivalent of vocal vibrations in the line and the improvements to be sought.

In was in September 1887 that the conference of specialists sat in New York to inquire into the subject of cables and to advise the American companies upon the most suitable types to adopt. The first paper was by Mr. Jacques, who appears to have adopted

[1] *Journal of the Society of Telegraph Engineers*, xvi. 84. [2] *Ibid.* p. 135.

the views of Preece in part. In the course of his paper he stated that :—

No matter what may be the distance between two points good business conversation may be carried on between them, provided they be connected by a pole line or cable or both, the product of whose total resistance by its total capacity is less than $\frac{2000}{4500}$ if transmitters of the $\frac{\text{Blake}}{\text{Hunnings}}$ type be used.

He added, 'This rule is purely the result of experiment.'

Mr. Jacques' rule was subjected to some criticism, that by Mr. Patterson being selected for quotation as being the most illuminating and borne out by subsequent experience. Mr. Patterson said :—

I have been inclined to doubt the applicability of any general formula as simple as resistance multiplied by capacity as giving the correct working value of telephone circuits. First, the doctors disagree.

Mr. Preece says that the constant varies for the different materials of the circuit, and gives for an overhead line of copper a value of 15,000 ; for copper cables, 12,000 ; and for overhead iron, 10,000.

Dr. Wietlisbach gives as the result of his experiments and theorising, that if CR equals 100, the transmission is excellent ; if 300, it is good ; if 1000, it is possible ; if 2500, it is impossible ; making no variation for different transmitters.

Dr. Jacques says that for the Blake transmitter the constant is 2000, and for the Hunnings, 4500. I do not see how the transmitter can be ignored as it is by Preece and Wietlisbach, as we have every day practical demonstration of their difference. I am inclined to believe that capacity, or the factors which affect capacity, have more than a proportional share in producing retardation. If the wires are near together, not only is the capacity higher for the same material, but the dynamic or kinetic effect is increased and has a retarding effect on the current. If we connect a number of conductors in a cable in multiple arc the resistance is diminished and the capacity as measured is not increased in a corresponding ratio, but the more such conductors we get in the circuit the worse it works.

Mr. Patterson, while criticising the laws sought to be deduced, assumed that all were agreed 'that resistance and capacity are the principal obstacles to telephone working, and they should be as far as practicable removed.' Of the three ways of destroying the efficiency of telephonic transmission he had the idea 'that resistance attacks principally the loudness, and capacity and other things, which depend upon the same factors as capacity, attack the clearness.' Approaching the subject from the point of view which should be ever present to an engineer—how to obtain specified results for the lowest expenditure—Mr. Patterson continued :—

In looking at the problem to see where we can get rid of the most resistance and capacity, we see that circuits in which we approach the limits are generally made up of long pole lines and shorter cable lines, say 100 miles of pole line and 10 or 15 miles of cable. The great bulk of capacity is in the cable—of resistance in the pole line. To reduce the resistance of the cable implies that we must increase the thickness of insulation to keep a corresponding capacity, and consequently the size of the cable and conduits must be increased. If we reduce the resistance of the pole line the increase in its capacity is hardly appreciable, and there is no cost except the additional weight of copper. Then the short lines around the exchange alone will work any way. They are many more in number than the long pole line. The money to be expended in increasing the size of all these wires and in correspondingly increasing the insulation would accomplish much more if applied to increasing the size of the comparatively fewer overhead wires. Let the money which is to be expended upon cables go principally to keeping the capacity low and the wires well apart.

The economics of the situation were here well stated and prevailed.

On the exteriors of cables evidence was given by Mr. Sargent as follows :—

The first lead cables that were laid were manufactured by the Western Electric and Brooks of Philadelphia. The former have now been down almost three years, and the lead pipe shows scarcely any signs of decay. There is on the surface a hard white crust, but it is very thin and does not seem to be increasing with the lapse of time. The Brooks cable at the end of eight months was found to be badly corroded, a crust of what appeared to be carbonate of lead forming on the outside in that time one-sixteenth of an inch thick, cracking and peeling off whenever the cable was handled. Six hundred feet of this cable was drawn out on account of defective insulation and replaced by another piece. This also has corroded badly. Otherwise the whole length of about 2400 feet is still in service, but the corrosion appears to be going on with every prospect that it will continue until the lead is destroyed. As far as my information goes the only difference in the lead of these two cables is that the Brooks claims to be pure lead, while the Western Electric has a small percentage of tin mixed with it for the purpose of hardening it.

Based upon the results of experience, it has continued to be the practice of telephone companies in the United States to use the alloy of tin in cable sheaths, but the practice is not universal in

Europe, the British Post Office and many continental engineers being quite satisfied with pure lead. The durability of lead pipes is amply demonstrated by the continued existence of Roman work of this character, of which that in the city of Bath may be taken as an example. It should, however, be borne in mind that the modern 'pure lead' of commerce has all the more valuable ingredients extracted from it, which was not the case with the material used by the Roman artificers.[1] The introduction of the tin alloy serves to restore the lead more nearly to the condition of the metal of whose durability we have concrete evidence. As, however, telephone cables can hardly be laid with a view to use so far in the future as Roman work is in the past, there is still room for argument as to the money value of the alloy. But in respect to tensile strength it has been satisfactorily determined that the alloy is superior, so that to obtain equal results a thicker sheath of pure lead must be used than would be required where tin is added.

As the result of practical tests and theoretical calculations the conference recommended the use of a No. 18 Brown and Sharp gauge (·040 inch) conductor with insulation to bring the diameter up to $\frac{1}{8}$ inch, which it was calculated would enable conversation to be carried on through twenty-five miles of cable. For single circuits frequent transposition of the wires was recommended for the reduction of cross-talk. The efficacy of this method, it may be noted, was observed by the officials testing the Brooks cable in London in 1881.[2] Whilst considering this expedient it was further recommended that any companies contemplating the introduction of metallic circuits in their system within the next two or three years should at once use cables with twisted pairs, leaving one limb of the pair idle in the intervening period.

The specification subsequently prepared to conform to the conclusion of the members of the conference is given in full as follows :—

American Telephone and Telegraph Company

Specifications for telephone cables prepared on the basis of the conclusions reached by the Cable Conference, September 1887.

Size of Conductors.—Each conductor shall be of No. 18 Brown & Sharp gauge, best quality Lake copper.

Thickness of Insulation.—Each conductor shall be insulated to ·125 of an inch, with not less than two wrappings of cotton so put on that when the conductors are laid up in a cable there shall be no substantial compression of their coverings.

[1] Mr. A. P. Trotter referred to this point in the course of a discussion at the Institution of Electrical Engineers.—*Journal*, xviii. 365.

[2] Chapter xx. p. 254.

Twisted in Pairs.—The conductors shall be twisted in pairs, the twists being regular and uniform, not less than $2\frac{3}{4}$ inches, nor more than $3\frac{1}{4}$ inches in length.

Reversed Layers.—The cables shall be laid up in reversed layers, each layer being served with one covering of cotton ; the pitch of each layer shall be as long as is consistent with necessary flexibility.

The cables shall be so constructed that there shall be no substantial compression between the layers or between the outside layers and the lead pipe.

Outside Coverings.—The core shall be enclosed in a composition lead and tin pipe (for underground cables 97 per cent. lead and 3 per cent. tin) of uniform thickness and free from holes or other defects, weighing not less than :

$3\frac{1}{2}$ lb. per foot for $1\frac{3}{4}$ inch pipe.
3 ,, ,, ,, ,, $1\frac{1}{2}$,, ,,
$2\frac{3}{4}$,, ,, ,, ,, $1\frac{3}{8}$,, ,,
$2\frac{1}{2}$,, ,, ,, ,, $1\frac{1}{4}$,, ,,
$2\frac{1}{4}$,, ,, ,, ,, $1\frac{1}{8}$,, ,,
2 ,, ,, ,, ,, 1 ,, ,,
$1\frac{3}{4}$,, ,, ,, ,, $\frac{7}{8}$,, ,,
$1\frac{1}{2}$,, ,, ,, ,, $\frac{3}{4}$,, ,,
1 ,, ,, ,, ,, $\frac{5}{8}$,, ,,

Dimensions are for inside diameter of pipe.

Insulating Material.—The spaces in the core and between the core and pipe shall be filled with insulating material which shall give an electrostatic capacity not exceeding ·20 of a micro-farad per mile and an insulation of not less than 100 megohms per mile when laid and spliced and connected with terminals ready for use ; and such capacity shall not increase nor such insulation decrease beyond the limits above specified for one year after the cable has been laid and spliced, except from mechanical injury.

(Dated March 27, 1888.)

The 1889 Cable Conference[1] Report is a volume of 125 pages. Like its predecessor, it was printed but not published.

[1] The representatives were :—

Edward J. Hall, Jr., Vice-President and General Manager, American Telephone and Telegraph Company.

John A. Barrett, Electrician, American Telephone and Telegraph Company.

J. E. Crandall, Electrician, Chesapeake and Potomac Telephone Company.

William H. Eckert, General Manager, Metropolitan Telephone and Telegraph Company.

C. H. Wilson, Superintendent, Chicago Telephone Company.

George A. Hamilton, Electrician, Western Electric Company.

Hammond V. Hayes, Electrician, American Bell Telephone Company.

F. A. Pickernell, Superintendent of Equipment, American Telephone and Telegraph Company.

I. H. Farnham, Electrician, New England Telephone and Telegraph Company.

W. R. Patterson, Electrician, Western Electric Company.

In brief introductory remarks the Chairman (Mr. E. J. Hall) said :—

In September 1887 we had a conference here on the subject of telephone cables, at which we reached certain general conclusions ; the object of this meeting is to revise those conclusions in the light of the experience of the last two years, and determine whether the specifications, which were prepared on the basis of our discussions then, should be modified in any respect, and, if so, in what way.

The first paper read was by Mr. Barrett, who remarked that :—

The cheapness of lead pipe cable, added to some very important electrical advantages which it possesses, makes it beyond question the best known cable for general telephone purposes. . . .

The cheapness of the lead-pipe cable does not require discussion at this point, but low inductive capacity needs to be earnestly dwelt upon, since this is the chief burden which subterranean telephoning staggers under.

This burden is becoming more and more prominent as the lengths of the underground cables are increased and as greater distances overland are brought into connection with the city cable systems.

It is the point of chief importance to have this matter of low static capacity keenly and thoroughly appreciated early in the development of the underground cable system, so that the steps taken in this direction may not be to future disadvantage. . . .

Since the adoption of this standard by the conference of September 1887, I have, in connection with Mr. W. D. Sargent, of Brooklyn, been engaged in an effort to reduce the specific inductive capacity of the wrapping used upon the conductors so as to secure a considerably lower limit for static capacity while still using the same dimensions for the cable.

We have had an almost unlooked-for success in this direction in the employment of manilla paper in the place of cotton as the wrapping for the conductors.

Difficulty in getting our paper covering manufactured into core for our use has prevented our putting any large amount of this cable into service, or to the exhaustive tests which would enable us to determine as closely as we desire the difference in specific inductive capacity between paper and cotton ; still we have several lengths of cable of 500 feet each upon which we have at various times made tests in connection with Mr. F. A. Pickernell, of the American Telephone and Telegraph Company, Mr. Geo A. Hamilton, now of the Western Electric Company, and Mr. Chas. Matchett, of the New York and New Jersey Telephone Company.

A. S. Hibbard, General Superintendent, American Telephone and Telegraph Company.

Joseph P. Davis, Consulting Engineer, American Bell Telephone Company.

J. C. Reilly, General Superintendent, New York and New Jersey Telephone Company.

As a general deduction of these tests, without going into a detail of the figures, I have no hesitation in saying that, other things being equal, the substitution of paper for cotton in the manner in which we have employed it will give a reduction of 30 per cent. in static capacity.

Mr. Hamilton has a schedule of his tests which he may be willing to submit to consideration with his remarks upon this subject.

In respect to the mechanical properties of paper for cable purposes and the method of manufacture which has met our requirements most satisfactorily, I will say that such paper as we have used is much less hygroscopic than cotton, so is easier to prepare in expelling the latent moisture and easier to maintain in high insulation through subsequent exposure. It can be laid upon the conductors very dense and hard and in smooth round form and with the requisite degree of flexibility. It is slightly lighter per cable foot than cotton. It retains its shape well in cabling and is convenient to handle in making splices and connections. Its cost per cable foot is approximately that of cotton. It is a little cheaper per pound as material and a little more expensive in the labor of applying.

The cable from which we have secured our good results has been made with core furnished us by the Norwich Insulated Wire Company, now of this city. We have tried various methods of applying the paper to the conductors, but the product of this company alone has had the requisite hardness, smoothness, and flexibility to give us to the full degree the good results we have sought for.

In respect to the standard dimensions for cable, two inches in external diameter is the size determined upon in view of the mechanical conditions of the underground problem. There is a constant temptation and tendency to take advantage of every diminution of specific inductive capacity of material to the end of increasing the number of conductors placed in the given external dimensions and so to diminish the cost of the cable per conductor.

Thus in the standard formerly fixed, the conductor being placed at No. 18 B. and S. and a maximum limit of 20 mf. per mile named, as soon as it was discovered that a few more than the prescribed number (fifty-one or fifty-two pairs of conductors) could be squeezed into the space, this was done instead of using the advantage to the reduction in static capacity to a possible ·18 mf. or ·17 mf. per mile.

It seems to me that if this matter of the relations of static capacity in cables to the future of the telephone system were appreciated to its full extent, then every gain in reducing the specific capacity of insulating media would be received with satisfaction for its own sake alone and not for the sake of a possible cheapening in the present cost of the cable. And I believe that this being agreed to—namely, that the telephone companies can afford to employ cables in which fifty-one or fifty-two pairs go to the two inches—then this standard should be adhered to and the chief recommendations to a cable be laid upon its low static capacity. The standard of fifty-two pairs to a 2-inch cable was substantially agreed upon as the maximum limit of cost which local telephone companies could afford for the purposes of their local service. The experience of the long-distance telephone service indicates the necessity of a more liberal provision for long trunk lines. A single instance will illustrate this point, and will indicate pointedly how truly the static capacity of cables is a limiting function of telephonic transmission.

The line from New York to Buffalo is approximately 475 miles, of which about five miles consists of the Conference Standard Cable. Tests indicate that this five miles of cable constitutes about one-fourth of the whole telephonic distance to Buffalo. The practical effect of this is shown in the fact that a good Blake transmitter over 470 miles of the pole line gives excellent transmission. The same instrument through the line with five miles of cable added is almost entirely inoperative.

In this same case of pole line alone the long-distance transmitter has a good clear margin of safety with which to meet the adverse conditions of average service, while with the addition of this five miles of cable the dependence of the same service upon favourable conditions at all points is seriously enhanced.

On this account I should recommend for long-distance trunk line service a special cable standard. My suggestion for this would be lead pipe cable, No. 16 B. and S. conductor, paper wrapping and twenty pairs to the 2 inches.

This cable would probably give about ·05 mf. per mile. If upon trial it is found that twenty pairs to 2 inches would give more than ·05 mf. per mile, I should reduce the number of pairs rather than to run over that limit of capacity.

At a later period of the sitting Mr. Barrett said :—

In the matter of the paper material which I have introduced our opportunities for testing and getting results which are as full, and reliable as we would like, have been very meagre, and I should like to have that matter left more or less open, to be determined by some more tests that are in process of being applied. We are having other lengths, to a considerable extent, of paper cable now made, and within a month or two we shall have some very definite results on that point.

Mr. Hibbard suggested that the wording of the specification should be so altered as to include paper or anything else which might be found to be better than cotton.

The prevailing type of cable hitherto was one in which a covering of cotton served as the medium of separation between one wire and another. Mr. Barrett and Mr. Sargent experimented with paper with the expectation that a lower capacity would result. These experiments were limited to the substitution of paper for cotton as a separator. For insulation, paraffin wax, resin oil or other compound was still relied upon, and the tube was filled so far as possible with the insulating material. The experiments with the paper cable, though 'very meagre' as Mr. Barrett said, bore out the expectations of the designers in showing a lower capacity than the prevailing type.

Following the remarks last quoted, Mr. Patterson said:—

In regard to the use of paper, I tried some experiments some time ago and at the time came to the conclusion that the lower capacity was due to the fact that more air was contained in the pores of the paper and that the filling material did not so readily penetrate, and that just so much as the capacity was lowered its power of absorbing water was increased. This was rather a hasty conclusion, but it was enough at the time to set me off into another line of work which, of course, appears rather risky and which under careful consideration I think results in a very good cable: that is, to leave the greater part of the length of the insulation dry without any filling material at all; when the cable is brought into one of these ducts, especially when there is a large number of ducts together in one body and properly protected, there is almost an infinitesimal chance of mechanical injury in the length of the cable. About the only trouble will come in our manholes where workmen are engaged in drawing in other cable and splicing, and it may be on account of explosions in manholes. I tried some experiments in making cables in which the core was introduced into the pipe perfectly dry; then a pressure of gas was put on, say 90 lb. to the inch, and after the air was compressed in the pipe, the filling introduced from both ends simultaneously at an increase of pressure of 10 to 15 lb., so that the filling would penetrate for perhaps 25 feet from each end, while the body of the cable was left dry and without any filling. The experiments on this have not been very extensive, but have been such that I think the static capacity of cables made in that way with cotton would come down inside of ·15 of a micro-farad per mile, and possibly lower than that if jute was used; and, if paper is of really lower specific inductive capacity, there would be a still greater advantage in the use of paper. In the case of any mechanical injury to the cable between manholes, that length of cable has got to be pulled out and another put in its place. Now, if the core is dry and loose, the core can be pulled out of the pipe and used elsewhere; whilst if it is enclosed in the pipe tightly it has got to be spliced, and a sleeve splice between two lengths of cable would not be very easily pulled into the ducts. Some of this dry cable was made last year for use in New York where we crossed the route of the steam pipes.

These observations occur on page 61 of the report. The committee continued its deliberations to the extent of another sixty-four pages of print without any further reference to the omission of filling material, and finally compiled the specification abstracted below, which, following that of the 1887 conference, provided that 'the spaces in the core and the pipe shall be filled with insulating material,' though the electrostatic capacity specified was reduced from ·20 to not exceeding ·18 of a micro-farad per mile.

The attitude of this committee of practical men in thus ignoring a suggestion which lay at the root of the matter is to be explained by the fact that confidence in the protecting envelope was not yet attained. Mr. Patterson put the suggestion forward tentatively, considered that it appeared 'rather risky,' and yet indicated truly that danger from damage in the interior of the conduit was remote. Where danger existed was at the manholes, and here it was proposed that 'filling' should be provided. The dry cable had only been used in practice for a special purpose—not with a view to the reduction of capacity, but because the temperature in its neighbourhood would have melted an insulating material.

The need of a low capacity was thoroughly recognised, but the security of an insulating material was tenaciously held to by men who had ever present to their minds the serious consequences which must ensue from even a microscopic defect in the lead covering. The service of a large number of circuits would be interrupted, and continuity of service was already recognised as an essential.

The 1889 specification followed generally that of the 1887 conference given on page 283. The *size of conductors* was retained at No. 18 B. and S. gauge; but the copper was to be '98 per cent. pure with a resistance per mile not greater than 35 ohms at 60 degrees F. after the cable is laid and connected to terminals.'

The *thickness of insulation*, the *twisted pairs*, the *reversed layers*, and the *outside coverings* were unaltered.

The capacity was reduced to ·18, 'each wire being tested against all the others grounded.'

There are the following additions:—

Covering for Pipe.

1. The pipe to be thoroughly coated with asphalt.

2. A protecting jacket to be laid on outside; this jacket to consist of at least two wrappings of tape put on in reversed layers and thoroughly impregnated with asphalt.

Referring to the clause providing for wrapping the conductors with cotton, the following resolution was passed:

It is the sense of this conference that it is very important to follow up the experiments in the use of paper or other materials as a substitute for cotton, and that it is understood that it is desirable to authorise such substitution whenever the manufacturers will guarantee a substantial reduction in static capacity therefrom.

Two years later (1891) further conferences were held. The official designation was altered to that of the Cable Committee.[1] The reason for the change was indicated by the Chairman (Mr. Hall) at the first meeting as follows :—

In view of the frequent development of new ideas in the construction of cables, Mr. Hudson has thought best to have a standing committee take the place of the irregular meetings which have been held heretofore, and this meeting this morning is intended to be the first meeting of that committee.

The committee proceeded to discuss the prior specification and modifications suggested therein. The discussion on the sixth clause indicates so clearly the starting-point of confidence in the envelope and the general attitude of experts previously referred to, that it is considered desirable to give it *in extenso* :—

Mr. HALL : We will take up No. 6, Insulating Material.

6. *Insulating Material.*—'The spaces in the core and between the core and pipe shall be filled with insulating material which shall give an electro-static capacity not exceeding ·18 of a micro-farad per mile and an insulation of not less than 100 megohms per mile at 60 degrees F. when laid and spliced and connected with terminals ready for use, each wire being tested against all the others grounded ; and such capacity shall not increase, nor such insulation decrease, beyond the limits above specified for one year after the cable has been laid and spliced, except from mechanical injury.'

That brings up the question of what the requirement for static capacity should be, and I would suggest that we talk about that first, as this clause is a pretty long one and brings in three or four different subjects. The present specification has been objected to on the ground that it required that the space in the core and between the core and the pipe shall be filled with something. I understand Mr. Patterson has tried to convince some people that the filling might be dry air or anything else, but that they thought they ought to have some tangible substance for their money. I should think that you could maintain a successful law-suit on your interpretation of the specification. We all agree that air is an insulating material

[1] At the April meeting the members present were :—

Edward J. Hall, Jr., Vice-President and General Manager, American Telephone and Telegraph Company.

George A. Hamilton, Electrician, Western Electric Company.

Hammond V. Hayes, Superintendent Mechanical Department, American Bell Telephone Company.

F. A. Pickernell, Engineer, American Telephone and Telegraph Company.

W. R. Patterson, Superintendent, Western Electric Company.

J. J. Carty, Electrician, Metropolitan Telephone and Telegraph Company.

and you could probably prove that it was filled with it where it was not filled with something else.

Mr. HAYES: Of course that could be left out entirely and be left entirely optional. I should think nothing need be said about it.

Mr. HALL: I should think it would not be worth while to say anything about that now, because we have left the material, which is to constitute the core, practically to the manufacturer.

Mr. HAYES: So it seems to me.

Mr. HALL: So it perhaps might take this shape. I will read a suggestion that Mr. Patterson has made.

'*Insulation.*—Insulation resistance shall not be less than (blank) megohms per mile after cables are laid and spliced, or not less than (blank) megohms per mile measured on reel. Insulation measurements to be made on one wire against all the others and the pipe.

'*Capacity.*—Electro-static capacity shall not be greater than (blank) micro-farads per mile for dry core, and not greater than (blank) micro-farads per mile for filled cable. This capacity to be that of one wire against all the others and the pipe. The capacity of one wire against its mate shall be not greater than (blank) and (blank) micro-farads per mile respectively for dry and filled cable.'

You separate the two things, insulation and capacity, in your specification. I should think that was—I was going to say that I thought that was a wise arrangement, but I don't know that it is. Really this old specification applies, as it is commonly supposed, to a filled cable. Now you have the question before you whether you want to have a filled cable or a dry cable or a partially filled cable, and, until you have discussed that, I don't see how you can name any figure for capacity.

Mr. PATTERSON: I think all the saving that has been made in capacity since the 1887 conference has been caused by bringing in air and leaving out the other filling material. It is gradually worked down now so that everybody is willing to consider a cable that has no filling except air, and the capacity can consequently range all the way between something like ·08 and something like ·19 or ·20, if it is thoroughly and honestly filled and saturated.

Mr. HAYES: Have any measurements been made with dry core or partially dry core cables that have been down a year or more? They have been down in New York here I think a year or more.

Mr. PICKERNELL: We had a test made of a partially dry core cable—that is, the ends of the section were filled back 15 or 20 feet; a cable of five miles long that had been laid one year. This cable was in substantially the same condition as it was when laid. The insulation at that time was something like 1000 megohms per mile. The capacity was the same as it was when made.

Mr. HAYES: I have heard it suggested that it would be impossible to make a lead pipe so dense that moisture would not get through it in time, and that there must be a deterioration. Of course water is fatal to the life of a dry core cable.

Mr. Hall: Isn't it fatal to the life of any cable? If water can get through the pipe it will get through the core anyway.

Mr. Carty: I think our submarine cable experience would give us as severe a case of that as we could get.

Mr. Pickernell: We have used Patterson's submarine cable three or four years and never have had but one case of trouble with it, and that was caused by a pickaxe at the shore end. The insulation is as good as when first laid.

Mr. Hayes: If that has been the result of experience, then why shouldn't we use dry core cables?

Mr. Pickernell: I don't see any place where we could use filled cables to advantage, not when they are drawn into the conduits. In case of injury between manholes a length has got to come out, and there is no more risk in dry core cables than in filled core.

Mr. Patterson: I have been about twelve years combating the idea that there are pores in lead big enough for moisture to get through. I never saw water inside of a pipe that did not get through a hole, not through microscopic pores. The original inventor of the dry core cable is Mr. Brooks. He patented cables with dry air, keeping air under pressure. He laid a cable in that way and found moisture inside, and came to the conclusion that it sweat on the inside as he had seen pipes sweat on the outside, and he would not have anything more to do with air as an insulator.

Mr. Carty: It is my belief that there never has been a cable failure not caused by mechanical or chemical injury to the lead pipe.

Mr. Patterson: Or some original imperfection.

Mr. Carty: Mr. Pickernell has had some experience with some faults in cables in Boston, which seem to show that the filling is not of such great importance. Was it in Boston?

Mr. Pickernell: I don't know that it showed that. We had some fault in a Patterson cable; there being a pinhole in a wiped joint where it was entirely submerged in water . . . about one-half of the conductors were affected; the other half were in good condition. That was a filled cable. . . . the moisture did not penetrate very rapidly even when water was in the pipe. . . . We had rather an interesting experience with a dry core cable in Philadelphia. That cable ran at one point through an open sewer; men at work punched two or three holes through it with a pickaxe; . . . one-half of the conductors had an insulation at that time of 600 megohms to the mile, and the other half had an insulation of less than 10,000 ohms to the mile, nothing practically. That shows that the penetration of moisture in that kind of cable does not take place rapidly.

Mr. Patterson: Sisal and jute and paper can be classed about together in regard to the penetration of water, and it will go much more rapidly into cotton.

Mr. Hall: You made a point about that also, that water would not get in any faster than air could get out through any ordinary pinhole ; the exchange of air for water would be a very slow process.

Mr. PATTERSON : It depends altogether on where it is. When you get a hole on the top of the pipe and the pipe slopes, it will penetrate very rapidly, but in the case of a hole on the underside of a pipe, it goes very much more slowly.

Mr. PICKERNELL : The case I had was a case where the cable tipped down, going from one side of the manhole to the other. It was downhill, and on that account it was only on one side of the cable that was injured. The water was very slow in penetrating through a pinhole fully as large in diameter as a pin.

Mr. HALL : Now if we take a vote on the low figure for stating capacity, the effect, of course, would be to commit us absolutely to the dry cable ; and I would like, as that is an extremely important matter, the most radical step which has as yet been taken in cable manufacture, to have it considered very carefully. Perhaps you have considered it and are ready to commit yourselves. I think it would be well to make a motion, and I assume that it is made, that we recommend a figure for capacity which shall mean a dry core cable ; that without stating now just what the figures should be.

Mr. HAYES : I second the motion.

(Carried unanimously.)

Mr. HALL : That is the most important step that has been taken yet in cable making.

Mr. HAMILTON : I think our experience so far warrants us in taking the step.

Mr. PATTERSON : There are dry sections which have been in use in New York for about four years.

The conclusion thus reached from knowledge by men who recognised their responsibility has been supported by results. Dry core cable has been used ever since. It reduced the cost and increased the length available for commercial conversation, thus rendering the general introduction of underground systems possible. The principal factor was the omission of insulating material suggested by Mr. Patterson, but an important contributory was the use of paper due to the experiments of Mr. Sargent and Mr. Barrett. Paper provided covering as a separator which was much more efficacious than the somewhat solid covering of cotton previously used or even the looser braiding subsequently tried.

The methods of putting the paper covering over the wires varied. Cable specifications and cable manufactures were to undergo considerable changes yet, but these were, in the main, matters of detail in manufacture or of increase in the number of conductors in one tube, gradually increased from 50 to 300 pairs and larger numbers for smaller conductors, European organisations being perhaps in advance in this respect. The conference of 1891, however, determined the adoption of the dry core, and in essentials there is no change to record in the cable itself.

The principal changes in the 1891 specifications were :—

Conductors reduced to No. 19 B. and S. gauge (·03589 inch).

Thickness of insulation modified so that instead of the diameter of the individual insulated wire being given, the diameter of a core of 52 twisted pairs was specified to be not less than 1⅝ inch to be placed in a pipe not less than 1¾ inch internal diameter. Other sizes of core and pipe to be in the same proportions.

Twisted Pairs.—Length of twist omitted, and in place thereof 'The conductors shall be so twisted in pairs that there shall be no inductive disturbances between the circuits.'

Outside Covering.—The largest size of pipe (hitherto 1¾ inch) was increased to 2½ inch (5½ lb. per foot), the weights for 2¼ inch and 2 inch being 4¾ lb. and 4¼ lb. per foot respectively.

Electrostatic Capacity.—The Committee contemplated at the April meeting an electrostatic capacity of ·085 microfarad per mile average and ·090 maximum but, as the result of practical demonstrations available immediately thereafter, the specification issued called for ·080 average and ·085 maximum. In July there was a further reduction to ·075 average and ·080 maximum. To make allowance for the filled ends and to permit the manufacture of a firmer core the ·080 and ·085 figures respectively were restored at the December meeting.

Covering for Pipe.—The outside covering and asphalt treatment of the lead pipe retained only when laid in wooden conduits. It was previously considered necessary that this should be done generally in order to prevent disintegration of the lead by chemical action, but careful inquiry had shown by 1891 that such protection was not needed when laid in conduits other than wood.

An article in the *Engineering Magazine* of 1843[1] describing the telegraphs laid on the Blackwall, London, Leeds and Manchester, and other railways, states :—

That the minutest changes in the insulation of the wires from dampness can be detected by this valuable instrument [the 'detector'] and corrected by blowing through the pipe a draught of dry air from the reservoir.

In 1858 Captain Drayson of the Royal Artillery and Captain Burney of the Royal Engineers obtained a patent (No. 2326) for a cable consisting of a conductor varnished and silk covered

[1] Quoted by Mr. A. Watts in a paper on the evolution of underground work read before the Manchester Telephone Society, January 15, 1904.

enclosed in an elastic tube; and in 1876 a patent (No. 3099) was granted to Henry Potts Scott (a communication from Wilson Strickler of the United States) for a cable formed of a conductor with grooves and furrows and atmospheric insulation, the object being to prevent retardation. William E. Prall took out a patent in the United States in 1876 (No. 172,495) for a combination of one or more telegraph wires or cables with a continuous line of airtight enclosing pipes charged with atmospheric air under compression.

In 1882 Dolbear took out a patent in England (No. 1368) for a submarine cable consisting of a conductor surrounded by a spiral cord then covered by a waterproofed paper tube enclosed in a gutta-percha covering. The object of the construction was to obtain freedom from the retardation due to capacity, and the capacity effect was clearly related in the specification as it had been by Wheatstone many years previously.[1] An attempt was made to obtain a renewal of the English patent in 1896 on the plea that the dry core cable, which had by that time come into general use, was Dolbear's invention. The petition was heard by the Judicial Committee of the Privy Council on March 18, 1896, and failed.

Amongst others, M. Fortin-Hermann

> sought to combine the advantages of the air line with those of the underground cable. . . . The conductor is inserted in small wood cylinders touching one another by the ends, and thus forming a chaplet or chain, covered by sheet lead or inserted in a lead pipe.[2]

Glass beads were also tried in place of the wood cylinders, but neither form was sufficiently practicable to survive. The Fortin-Hermann cable, however, was an early example of the effort to attain low capacity and was used in the underground portion of the Paris-Brussels line (p. 430).

Individual invention may claim some credit on minor points of detail, but the dry core system as a whole is not the subject of patent. Its utility was recognised and its adoption was recommended by a committee of experts after experience had demonstrated that confidence could be placed in the mechanical integrity of a lead tube.

[1] Chapter xx. p. 261.

[2] *Electrician*, February 21, 1885, p. 303.

CHAPTER XXIII

EARLY EXCHANGE 'SYSTEMS'

DURING the first ten years of telephone service numerous systems were in practical operation. Some managers were responsible for the design of a complete 'system,' others were content to invent a switchboard, but most had some detail which in their judgment made their own exchange a model suitable to be copied by everybody else. Space does not permit giving particulars of all the various modifications which were put in practice, but reference will be made to the principal ones.

From the construction standpoint there were two main systems—'one wire' and 'two wire.' These terms might be understood as earth circuit and metallic return respectively, so that it is advisable to add that they are not here so used. The 'one wire' was a direct line to the exchange, through which line was conveyed instructions to the exchange operator as well as conversation between subscribers. 'Two wire' meant two lines to the exchange, one direct for conversation with other subscribers, the second (usually common to several stations) for conveying instructions to the exchange operator.

From the standpoint of the calling subscriber the choice of systems may also be taken as two :—

(*a*) To call the exchange by the mechanical operation of a magneto generator or by pressing the button of a battery circuit.

(*b*) To communicate directly with the operator through the telephone.

A combination of these two, whereby the indicator operated by magneto or battery connected the line by means of a local circuit to the listening operator's telephone, may be mentioned, but its use was not sufficiently extensive to be included in the above. The earliest device of this kind is to be found in the United States patent of John A. McCoy, No. 278,351, where the shutter of an indicator is part of the line circuit and on being operated breaks an earth contact and makes contact with the operator's telephone circuit.

In other words [as the patentee says] my invention consists in providing an automatic switch to connect the central office telephone with that of the person signalling, the said switch to be operated at the same time that the central office is signalled.[1]

A similar purpose is served in the system described in Scribner's United States patent No. 278,367.[2]

A further classification is needed for indicating the conclusion of the conversation or the desire for the disconnection of the lines :—

(*c*) Operating magneto or battery bell in the same manner as for calling.

(*d*) Operating magneto or battery so as to send a different current from that used to call.

(*e*) By communicating directly with the operator through the telephone.

Taking the service as a whole and combining the calling and clearing, (*a*) and (*c*) represent the more general magneto (or battery) system.

(*a*) and (*d*) represent the magneto (or battery) 'Ring through' system, and (*b*) and (*e*) the Law (or call wire) system.

The Magneto System

In the magneto system there is a direct line from subscriber to exchange. At the subscriber's office a magneto generator with bell to send or receive signals, and talking instruments ; at the exchange an indicator and spring jack.

The diagram fig. 107 represents the lines of two subscribers to an exchange : A is the magneto at the office of subscriber No. 1, B is the spring jack and C the indicator of No. 1 subscriber at the exchange, D is the magneto at the office of subscriber No. 2, E is the spring jack and F the indicator of subscriber No. 2 at the exchange.

The handle of the magneto generator being turned, a current is generated which drops the indicator shutter, and the operator is thereby apprised that attention is required. The operator, having ascertained the number required, completes the connection. The lines when connected may be represented by the diagram fig. 108 : where G and H are plugs and I a clearing-out indicator operated by the same current as the calling indicators C and F. A having called and desired connection with D, the operator calls D by means of devices not illustrated on the diagram and then

[1] U.S. specification, No. 278,351. Dated May 29, 1883 (application filed April 5, 1880).

[2] Dated May 29, 1883 (application filed November 6, 1880).

connects them for conversation as shown. The request to disconnect is communicated to the operator by either subscriber turning the handle of his magneto exactly as in the original call. The clearing-out drop I falls and the operator withdraws the plugs G and H, when the lines assume the condition shown in the diagram fig. 107.

To carry out this method satisfactorily, it is required that a subscriber should only send a current over the line for the purpose of calling the exchange or for disconnecting. It was with difficulty that subscribers were brought to understand that attention to such a limitation was essential. It was not unusual for two subscribers

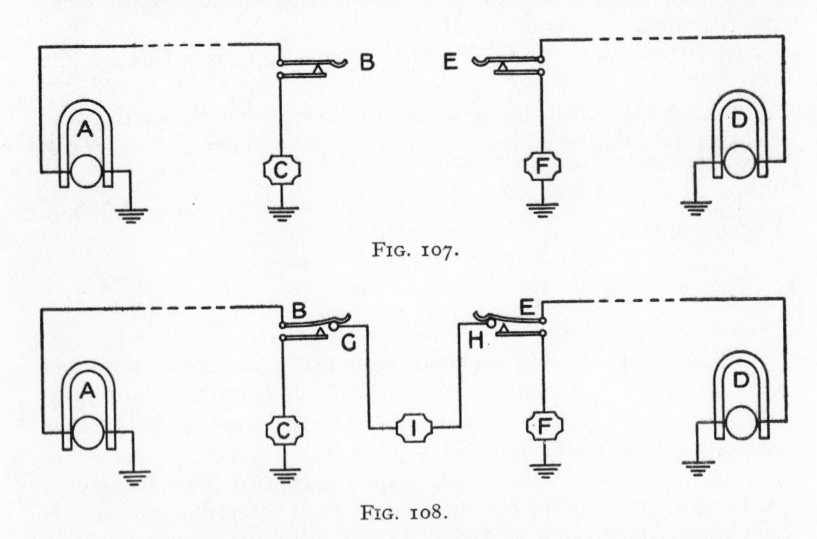

Fig. 107.

Fig. 108.

(A and B) to hold a prolonged conversation, and for A perhaps to leave the instrument to acquire information for his correspondent B. After a time B, becoming impatient, would turn his magneto handle under the impression that he would thereby bring A again to the instrument. The result, however, was of course to throw the clearing-out indicator, which was the signal for disconnection.

The uncertainty was considerable, and many authorities had not acquired sufficient information to realise the importance of insisting upon the accurate carrying out of instructions or placing upon the subscriber the penalty of any infraction. To quote one such set of instructions on this point (Melbourne, Australia, 1883): 'When connected do not again use your bell except to ring off. Should you do so you will be immediately disconnected.' In many exchanges there was not the courage to instruct the subscriber by carrying out this salutary rule. In consequence there was uncertainty, and operators were sometimes accustomed to 'listen in' and make sure before disconnecting.

The ring-through system avoided this uncertainty and in addition provided what was regarded by some as a valuable facility in permitting two subscribers to be connected for an indefinite period. The advantages of the 'ring-through' were clearly limited to a single switchboard. If more than one switchboard were concerned in the call, the first operator having connected the calling subscriber would have connected him, not to the correspondent he required, but to another operator. The second operator would connect the line to the called subscriber, and the subscriber originating the call would ring.

The encouragement to two subscribers to leave their lines connected for an unnecessary period was a mistaken policy in a single exchange. Other subscribers might call for one or both, and their continued engagement would be a disadvantage. But, in the case of two exchanges, it would be the more serious because trunk (or junction) wires would be employed. The number of junction wires between exchanges is in proportion to the traffic, and the number must be increased if the durations of the connections are prolonged.

One other respect in which the ring-through system commended itself to some telephone engineers was that it placed the responsibility for delay in answering calls upon the subscriber. A calling subscriber awaiting the answer to his call attributed to the exchange operator the responsibility for any delay. So much of it as was contributed by the subscriber was in no sense the operator's fault, but in the absence of any information to the contrary the subscriber placed the responsibility for all delay on the exchange service. To get rid of some of this responsibility was attractive, and the ring-through system consequently had its adherents.

An example of the ring-through system with batteries was that at Manchester described by the designer, Mr. Poole, in the first edition of his 'Telephone Handbook,'[1] also in the 'Manual of Telephony' by Preece and Maier. Batteries rendered the system simple, since it was only necessary to use one pole for calling and the other for clearing.

An example of the ring-through system with magneto was that in operation at Omaha, Nebraska. The magneto was commutated and an alternating or direct current could be sent at will. The calling indicator at the exchange was of the ordinary type operated by alternating current. The clearing-out indicator was specially designed so as to operate only by a direct current. It was

[1] *Practical Telephone Handbook*, Poole, 1891 edition, p. 168.

called a 'galvanometer drop,' a name which is fairly descriptive of its design. But a still earlier plan is that of C. A. Hussey,[1] in which also a commutated magneto was employed.

The Law or Call Wire System

The Law system derived its name from the fact that it was the system used by the Law Telegraph Company of New York in the operation of its dial telegraph service prior to the introduction of the telephone, and was retained by that company when the dial instruments were superseded by telephones.[2] It was one of the earliest exchange methods in use. A circuit calling wire was used in Chicago as described in detail in Chapter IX, and the multiple board of Firman [3] had no indicators. For calling and disconnecting the 'district' circuit wire was used. Though there is evidence of use by others and the suggestion by the district system is obvious, priority of publication must be given to the Law Company for the application of the idea to telephone exchange service by reason of the grant of a patent [4] to Frank Shaw of New York, Assignor to Law Telegraph Company of same place. Priority of invention was claimed by Firman, who was granted a patent (No. 328,305) on October 13, 1885 (application filed January 16, 1880), antedating that of Shaw. Shaw's specification, following the usual custom, sets forth the drawbacks of prior systems for the purpose of indicating the superiority of the method which he claimed to have invented.

The object of the invention is

> to enable any one of a large number of subscribers to what is commonly known as a 'telephone intercommunication exchange, district or central office system' to be quickly, easily, and without confusion put in private and direct communication at any time with any other subscriber.[4]

He describes as the usual method the employment of a line from the central office to the office of the subscriber, over which direct line are conveyed the signals or instructions to the operator as well as the conversations of the subscribers, and proceeds :—

> This system is open to several serious objections as follows: First, it frequently leads to confusion, because many subscribers sometimes do, and all may, at one time call or signal to the central office for connections, which cannot be properly received and executed

[1] U.S. specification, No. 247,359. Dated September 20, 1881 (application filed April 9, 1881). [2] Chapter ix. p. 86. [3] Chapter xix. p. 213.
[4] U.S. specification, No. 220,874, October 21, 1879.

simultaneously; second, it requires a large number of operators for such emergencies, who at other times stand about idle and unemployed; third, it consequently leads to frequent mistakes and delays; fourth, it requires considerable space and a large amount of apparatus.

Therefore, while this system seems simple, owing to the use of a single wire for calling the central office and communicating with a subscriber, and the same wires for communications between subscribers, it is in fact complicated in operation.

I overcome these objections by employing two wires and circuits—one for connection or disconnection calls or signals, and the other for private communication between subscribers—constructed and operated in the manner hereinafter set forth.[1]

The next paragraph in the specification has nothing to do with the system and is not referred to in the claims, but it is of interest as a contemporary record of the inconvenience resulting from the use of subscribers' names instead of numbers:—

It has been found that where there is a considerable number of subscribers, where firm-names are composed of several individual names, and especially where more than one subscriber has the same name, confusion and mistakes are likely to occur both at the central office and at subscribers' stations. I avoid these difficulties by numbering each station and printing a schedule or list of subscribers, with the name or title of the subscriber opposite the number of his station, which list or schedule is posted up near the instruments at each station and office. The calls and notices for connection are then all conducted without error by the use of numbers to designate the subscriber, which fixes absolutely the actual parties desiring to communicate or communicating.[1]

In describing his invention he says:—

From each subscriber's station I run two wires to a common central office, one of which wires I preferably run to no other station or stations, although several stations may be located thereon; and the other wire I run to as many other subscribers' stations as the amount of business done by them warrants. The first is the private, and the last is the call or signal wire. These wires are provided with a battery or other source of electricity and with a suitable switchboard and instruments at the common central office, and suitable instruments and batteries or other source of electricity at all the stations. Then any one subscriber can at any time obtain private and direct telephonic communication with any other subscriber by requesting the common central office, by means of the last-mentioned call or signal wire, to connect together the private wires of the two subscriber-stations, and afterward, having finished

[1] U.S. specification, No. 220,874, October 21, 1879.

their communications, can signal in like manner the common central station to disconnect said private lines, so that either may be in readiness to be connected at any moment with the private line of any other station.[1]

The advantages of the circuit wire system are thus set forth :—

The call or signal wire being common to a considerable number of subscribers' stations and the instruments there, the result is that a subscriber going to his instruments is able to learn whether or not any other subscriber is signalling, and to await his turn, or until the call or signal wire is not in use before attempting to signal. The result is that the subscribers, as it were, form themselves in cue, each taking his place in succession.[1]

The method of operation in the exchanges, as described in the specification, contemplates the employment of two operators, one to listen, the other to do the switching. Such a method was not limited to the Law system and was not generally employed in Law exchanges, switchboards for which were simple. Though the work was divided in the first instance, it became general for the operator on the circuit wire to carry through the actual connections as well as receive the instructions.

At the office end of the signal wire I place a receiving operator, who sits with the receiving instrument or telephone constantly at or near his ear, ready to receive the name or number of any subscriber desiring to communicate and of the correspondent with whom he desires to be placed in communication. The names or numbers so received he repeats so as to be heard by the subscriber calling through the transmitting-instrument in front of him, and at the same time by another operator at the switchboard in the central office, who instantly connects the two private wires and subscribers so indicated. As he does so, the operator at the switchboard notifies both subscribers by means of a bell in circuit on the wire. The same course is pursued when the conversation is concluded and disconnection is desired, except as to the ringing of the bell.[1]

The descriptions are as follows :—

Fig. [109] is a plan of my system for connecting a central office and its subscribers. Fig. [110] is a plan of my system for connecting the subscribers who belong to one central office with those who belong to another central office. Fig. [111] is a plan showing two subscribers' stations and central office with instruments and receiving operator in position.

In fig. [109] X represents a central office or station, and *a, b, c,* represent a given number of subscribers' stations, each of which

[1] U.S. specification, No. 220,874, October 21, 1879.

is connected with the common central office or station by two wires, the first of which is marked W, and the second a^2 b^2 c^2, respectively and also W^1 in common.[1]

The method of calling in a single exchange is described but need not be quoted. There is some interest, however, in the reference to numerous exchanges :—

> In cases where distances are great I establish two or more common central stations, converging at each the wires from all stations nearest to it and connecting all the common central stations together by as many wires as may be required.

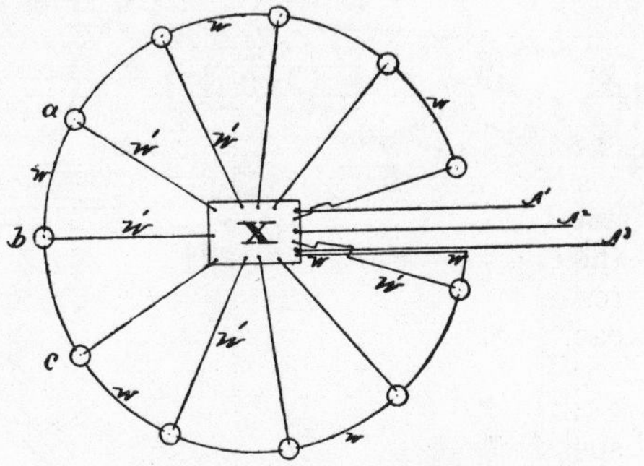

Fig. 109.—Law System Single Central Office (Shaw's patent).

> Now, should *a* wish to communicate with the subscriber of any other central office, as *d*, (fig. [110]), he makes known the fact to the operator at X by means of the wire W, who immediately connects the wire a^2 with one of the wires A^1, A^2, or A^3, and instructs the operator at the station Z by means of another of the wires A^1, A^2, or A^3, used as a call or signal wire between central stations (two or more) to connect the wire d^2 with the wire A^1, A^2, or A^3, first mentioned.

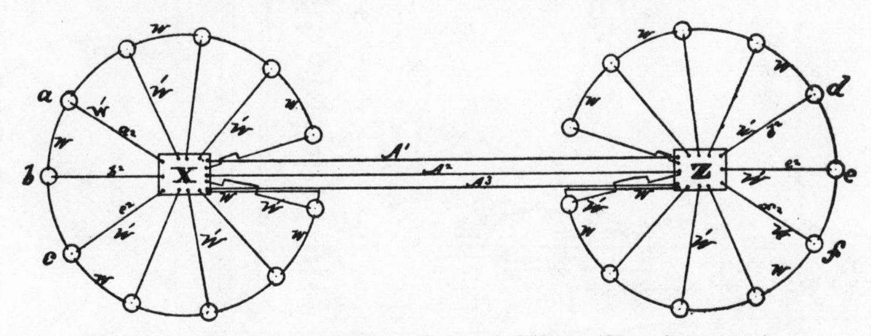

Fig. 110.—Law System Two Central Offices (Shaw's patent).

> The disconnection is accomplished either by the subscriber at station *a* notifying the operator at the central station X, and he notifying the operator at the other central station Z by the same wires as before, or by the subscriber at the station *d* notifying the operator at Z, and he notifying the operator at X.[1]

The Law system was sometimes called the 'Call Wire System,' the 'Two Wire System,' and in Great Britain the 'Mann System.'

[1] U.S. specification, No. 220,874, October 21, 1879.

The latter name was given to it under the impression that Mr. Mann was the first inventor of an improvement in the running of the

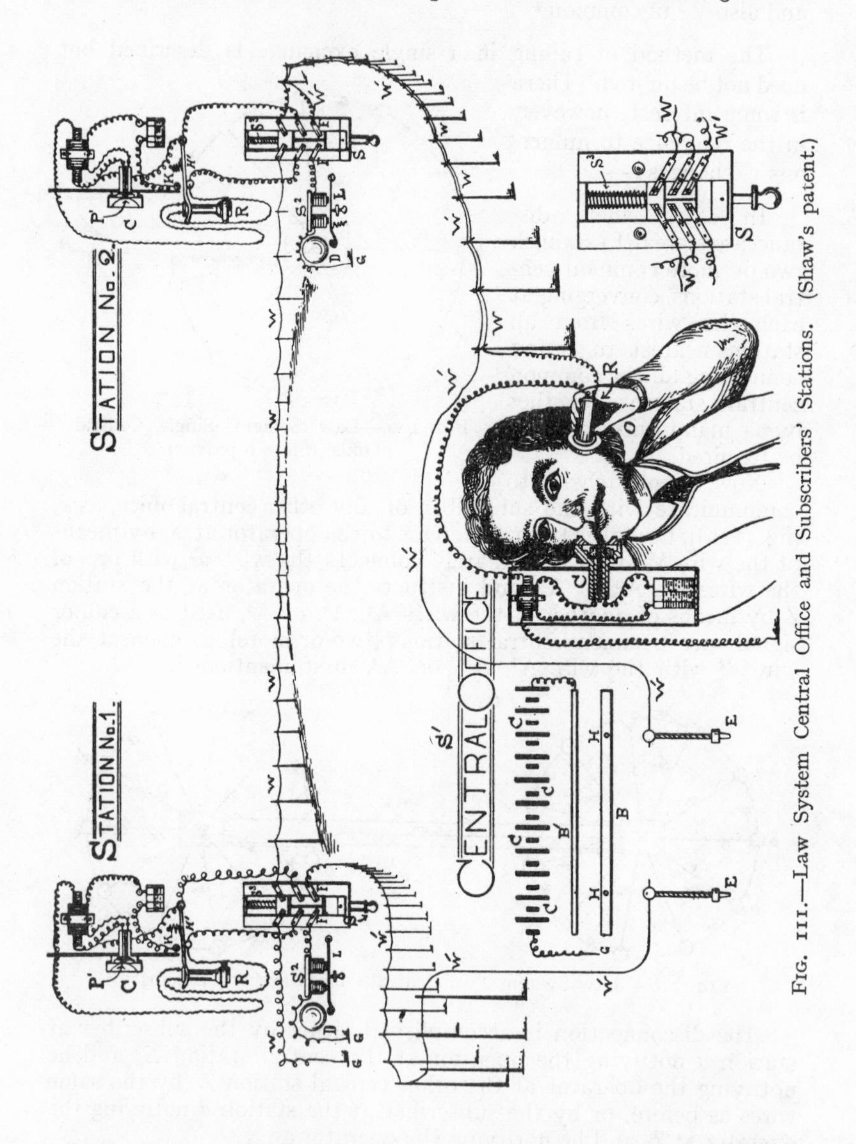

Fig. 111.—Law System Central Office and Subscribers' Stations. (Shaw's patent.)

circuit wire by taking spurs into the subscribers' offices instead of looping in and out again. The connection by derivation reduced the inconvenience from the failure of the circuit wire, but, as was shown in the course of a controversy on the subject in the London

Electrical Review of 1891 (vol. 29), the derivation method was common in the Law exchanges of the United States.

In 1887 the exchange service had been in operation ten years. The problems which those ten years had presented were of the most varied character. In the conduct of this service, which gave to the public an entirely new means of communication, the providers had to contend with conditions in which electrical and mechanical problems were intermingled. Difficult as some of these problems were, they were at least capable of settlement in the light of scientific knowledge. But the human element required more study, and the human element was at the beginning, at the end, and at the middle of the line. The last mentioned was the operator at the exchange, a servant of the company available for instruction and subservient to discipline. But the subscriber who initiated the call, and the other subscriber who received it, were very important contributors to the satisfactory completion of a connection. Upon these contributors the company relied for its revenue. Whilst discipline was desirable it could not be enforced. The subscriber had to be studied and, so far as possible, to be educated.

The systems in operation afforded material for comparison, but the absence of uniformity was a drawback. The time had come to weigh experience and to select the best under all the circumstances.

CHAPTER XXIV

TELEPHONE ENGINEERING ON A SCIENTIFIC BASIS

(The Switchboard Conference of 1887)

Telephone exchanges had grown and flourished, using different methods and materials whilst aiming at the same result. In Europe the companies or administrations carrying on the service had no relationship to one another, and there was little disposition to exchange information or suggestions. In the United States there were many exchange companies, but all had a common interest in their relationship to the parent company under whose licenses they worked. In the United States people with a common interest are accustomed to indulge in periodical conventions. The annual meetings of the National Telephone Exchange Association were called 'conventions,' and from 1880 onwards these meetings had afforded valuable opportunities for comparing experiences and obtaining information. These opportunities were taken advantage of by the progressive managers of telephone companies, and much information was freely exchanged. In some respects results were so clear as to determine the line of further work. But there was no provision for reaching definite conclusions by evidence arguments and votes, and no effective authority for enforcing any such general conclusions as might be reached.

Variety in methods and material was a natural consequence of different minds being responsible for the inauguration and carrying on of the exchanges. Whilst the parent company in the United States, for instance, was ready to advise or to help, it gave no definite instructions or even general recommendations to its licensees as to the systems which they should adopt. There is no evidence that this attitude was the result of any preconceived plan, but there can be no doubt of its advantage. The exercise of any dogmatic authority in a new enterprise is a mistake. There was no experience of telephone exchange service upon which reasoned decisions could be based. The wisdom of leaving a free choice in methods has been demonstrated in one instance by the fact that the multiple principle, a feature in switching mechanisms destined to survive, was the product of a manufacturing company not connected with the Bell Company. It was the obvious advantages of the multiple switchboard (though not at that time fully developed) and the clear evidence of the mastery which its designers had obtained over exchange problems, as well as the patents they held, that induced the American Bell Company to acquire the Western Union Company's interest in the Western Electric Manufacturing Company in 1881.

As the business progressed it became evident that there was a limit to the advantages which would result from having each exchange run on the plan which might appeal to its own manager. It seems to have been recognised that excellence could not be determined merely by individual opinion ; that there must be some one system amongst the many in operation which was better than the others ; or that some combination might be obtained which would be superior to all. Opportunity had been given for obtaining information regarding the effectiveness, the cost, the durability, and the popularity of different methods and material. The time had been sufficient to demonstrate also that experiments on the large scale in the live exchange were very costly in cash outlay and very disturbing to the service. Moreover, the advantages of uniformity in operating methods might be expected to appeal to the American Bell Telephone Company, for from the first the controlling spirits of that company, had contemplated a national system, and in pursuance of that object had retained an interest in all trunk lines as well as a substantial interest in local exchanges. By these means they were able to exert influence upon the engineering and operating methods of the various exchanges which enabled them to advise with some authority what systems should be used.

It was with the object of obtaining accurate information and determining the most suitable policy for the immediate future that it was decided to hold a conference of representative men to discuss central office methods. The credit for calling the conference was ascribed by Mr. W. D. Sargent during its sittings to Mr. Theo. N. Vail and Mr. E. J. Hall, but the latter stated that this particular conference was held at the earnest request of Mr. Hudson, then the Vice-President and General Manager of the American Bell Telephone Company, who was anxious that everything should be done to bring out all the information available for the development of the business. The proceedings of the conference are contained in a printed volume of 255 pages which has never been published. These proceedings, in combination with those of the cable conference held in the same year,[1] may be considered as placing the arts of telephone exchange engineering and management—arts, perhaps, of greater complexity

[1] Chapter xxii.

and requiring more careful thought and more freedom from prejudice than any other practical applications of science so far attained—for the first time on a scientific basis.

The conference was an early example of co-operative effort or 'team work' such as has become by this time a commonplace of telephone practice. Though representative men, not all the members grasped the principles underlying the details of apparatus or service; but the members exchanged views without restraint, and the imperfect knowledge of some only served to afford opportunities for definite demonstration by others, so that all points were brought out with clearness and precision.

Individual opinion was still of value, but henceforth it must be based on demonstrable experience or sound reasoning which should appeal to other minds initiated in the mysteries but open to conviction. Thus, in the United States at any rate, telephone engineering was emancipated from the rule of thumb or the personal predilections of individuals and placed upon a more scientific basis so far as the experience at that date permitted.

A synopsis of eight pages, prepared by Mr. Lockwood, summarises the conclusions of the conference. The reasons for the summary are thus stated:—

The full report of the proceedings, while of great value, is so long, that much time, careful research, and close attention would often be necessary to ascertain the opinion of the conference expressed or implied on any question. It has, therefore, been thought well to prepare and append a synopsis formulating the absolute conclusions reached, and also those points of agreement which the drift of the discussions indicated as representing substantially the sense of the conference.

'The art of blotting' is said by Goldsmith, on the authority of 'an eminent critic,' to be 'the most difficult of all arts.' He was himself engaged upon abridgments, which he said 'are generally more tedious than the works from which they pretend to relieve us, and they have effectually embarrassed that road which they laboured to shorten.'

This was, however, only because the art of blotting had been 'usually practised by those who found themselves unable to write.' The synopsis above referred to is an exception to Goldsmith's generalisation, having been compiled by one not only 'able to write' but also familiar with every feature of the work covered. It was probably prepared in order to 'shorten the labours' of Mr. Hudson and the executive officers of the company in acquiring a knowledge of the practical conclusions of the conference, and to such practical conclusions it is severely limited.

Since space prevents the publication of the full report, and abridgment becomes essential, I must acknowledge a temptation to incorporate the synopsis. But 'to attain the greatest number of advantages with the fewest inconveniences,' which, to quote Goldsmith again, 'is all that can be attained in an abridgment the very name of which implies imperfection,' I have concluded that it will be better to draw upon both the full report and the synopsis, making such extracts therefrom as will serve to record the conclusions reached and the reasons for them, as well as to give some insight into the conditions of the time.

The peculiarity of the Telephone Exchange Service, compared with other public services or methods of communication, lies in the combination of individuals and apparatus. In transport and in prior methods of transferring intelligence, the public perform a merely passive part. The telephone required of the public an active participation.

The individuals engaged are of two kinds—the subscribers and the operators. The operators, and the apparatus which they manipulate, are for the service of the subscribers, and, therefore, in abridging the proceedings of the conference, it will be convenient to select such points as deal with them in the following order:—

1. General.
2. Subscribers.
3. Operators and other servants of the exchange authority.
4. Apparatus.

1. *General.*—The conference [1] was held in New York on December 19, 20, and 21, 1887.

In the course of his introductory address Mr. Hall said:—

For the past ten years the telephone service has presented an unending series of problems, and to-day the complicated conditions which surround our large exchanges present difficulties still to be surmounted which will tax to the utmost our united energies. The very satisfactory results of our recent conference on telephone cables lead us to hope that this meeting may accomplish something of equal importance for the switchboard. It is only by attacking these difficulties in detail that we can cope with them successfully; and, while there are many other subjects of equal interest, let us confine the work of this meeting as closely as possible to the central office and its apparatus.

From the standpoint of a general manager the first problem connected with central office apparatus is to find a place to put it. The stuffy attics of early days, with a little cupola reached by a

[1] The members were: E. J. Hall (in the chair), E. M. Barton, T. D. Lockwood, A. S. Hibbard, W. D. Sargent, J. A. Seely, C. E. Scribner, Henry Metzger, F. G. Beach, B. E. Sunny, C. H. Wilson, Wiley W. Smith, J. P. Davis, W. J. Denver, and I. H. Farnham.

ladder through a tangled mass of wires, can no longer be the lodging-place of the modern telephone exchange. At best, it was but a boarding house, with all its discomforts and uncertainties. Now we must have homes which we can build or arrange to meet our varied and complicated needs.

Every large exchange should be located in a building owned by the company, and that building should be located and arranged after a careful study of the territory to be served and a most liberal estimate of the future needs of the business for which we are planning.

Ample space, light and air must be provided for the operating department, and convenient toilet and lunch rooms for the operators.

Overhead and underground wires must be brought in in such a manner as to be fully accessible ; suitable testing apparatus must be provided, and competent electricians and office managers employed to supervise every detail and keep the machinery up to its highest efficiency.

As to the desirability of these things there can be no question, but their practical necessity should be impressed upon the directors of our companies so that they may not make the costly mistake of equipping any large exchange office in other than permanent fireproof quarters.

In these quarters we can put no flimsy makeshifts in the way of apparatus. We must require of the manufacturers strong and durable materials, put together substantially and accurately. What those materials shall be—in what way they shall be combined and how the various parts shall be arranged to equip the ideal central office—are questions for us now to consider and decide.

As a basis for discussion, Mr. Lockwood presented a memorandum summarising previous experience and formulating certain questions for decision. The thoroughness with which the problem was attacked may be gathered from the introductory definition of the purpose for which the switchboard was required.

Switchboards.—By this term, for the purposes of this conference, we are to understand the entire apparatus of the central station, whereby any two subscribers are enabled to converse with one another.

To bring the subject intelligently before us, the first thing to do is to consider and formulate the purpose for which the central station switchboard is required.

A hasty reply to the question, 'What is the function of the switchboard ? ' would be, and often has been : 'To connect the ends of any two subscribers' lines together so that they can communicate directly with one another just as if they were united by the same wire.'

This looks like a simple operation, but we know that, while

this is a true statement of the final result, it is not the whole truth. There are a variety of sub-operations, all tending to this end. The several steps, as I understand them, are as follows :—

To receive the call signal from the subscriber.

To respond by answering, and then to receive the order.

To ascertain whether the subscriber's line wanted is in use.

If so, to notify the calling subscriber of the fact, and so for the present to end the operation.

If not, to send the call signal over the line of the subscriber wanted.

To ascertain if he answers his bell.

If not, to notify calling subscriber that No. 2 does not answer, and thus conclude the operation for the present.

If so, to connect the two lines, removing the ground terminal of both, and bring subscribers together, if necessary.

To receive disconnecting signal.

To disconnect.

To perform all these functions we want:

Call-receiving appliances.

Call-sending appliances.

Telephones to receive and forward orders.

A switchboard proper to connect wires together.

An appliance to connect and disconnect the telephones with and from any desired line.

Some appliance, mechanical or otherwise, to find out whether lines are in use.

Appliances to receive disconnecting signals.

These operations and appliances to perform them are not common to all central offices, but the five operations of call receiving, call sending, finding whether line wanted be in use, and connecting and disconnecting in some form, are inseparable from the work of the central office.

We have now, I think, to consider the principal switchboards which have been in use, especially the multiple switchboard in which all others seem to have culminated.

First.—With respect to efficiency.

Second.—With respect to economy.

Third.—With respect to permanency.

On the lines of this introductory definition Mr. Lockwood formulated twenty-two questions for the consideration of the conference. Each question had a definite and practical bearing and was considered by the members in detail, it being recognised that—

The problem of the switchboard and of central office work is not one which can be worked out by electricians alone, however capable. It is more mechanical than electrical, and it is one in which the experimental knowledge of men who have practically operated large exchanges also will count materially. Electrical and mechanical skill and experience must therefore together consider the subject.

Attention was given not only to definite practical questions, but also to the avoidance of possible dangers of a more general character. One of the questions, for example, related to the practicability of

setting aside all apparatus which as a whole is now used, and by availing ourselves of the universal stock of knowledge to devise and construct something which shall be more efficient, more economical, or more endurable, and whether in such an enterprise it would not be judicious to associate with known switchboard experts one or more persons who, while electricians or electro mechanics of acknowledged skill, have had no special views upon switchboards or switching organisations, for the purpose of obtaining absolutely new ideas unfettered by previous practice or conditions.

In introducing this question, Mr. Lockwood said that it was based upon views which Mr. Theodore N. Vail had frequently expressed to him, that nearly all switchboard experts and telephone men of that era were working in ruts,

that we have got accustomed to certain things, and we cannot see our way or feel our way or see any other possible way out of it than to go to a certain extent as our predecessors have gone, and he (Mr. Vail) thought it would be a good thing if we could find somebody somewhere who was a first-class man in every other respect but who had not devoted his attention to switchboards.

A sub-committee had been formed, consisting of Mr. Scribner, Mr. Hibbard, and Mr. Wilson, to investigate a specific question ; but, as the sub-committee was to report on the following morning, it was not practicable to add the suggested new member, but the chairman thought it was well to have Mr. Lockwood's question on record to consider. This conference and its successors, composed as they were of practical men in various branches of the service, seem to have been so constituted as to avoid the danger of the ' one idea ' or the getting into a groove.

2. *Subscribers.*—The education of the subscriber in the use of the telephone was agreed upon as a very desirable measure. Mr. Sargent estimated that fully two-thirds of the troubles experienced were directly traceable to the users. Mr. Hall explained that in Buffalo the ' chief canvasser or contract agent ' undertook the work of educating the users.

When he goes to people on this business he does not talk anything except the good of the subscriber, the defects of the service, and expressing the desire of the company to be sure that its subscribers are receiving the service for which they are paying. . . . Of course, to send around a man who did not possess a good deal of tact and judgment would simply be to cause trouble rather than to remove it.

We have found another incidental advantage in Buffalo in that missionary work, that a great many complaints, more or less well grounded, have been brought to the attention of the company that would not have been received in any other way, and a great many mistaken ideas of subscribers have been corrected; and the whole effect of the thing has been, I think, to impress them with the belief that the company are sincerely interested in giving them good service and took pains to ascertain whether the service was satisfactory to them, and if not, what they would suggest in the way of remedying it; and then, hearing their suggestions, the company has an opportunity to point out to them where the fault lies with them, or if the suggestions are good ones, and some of them have been, to follow them out and correct abuses in the service.

One of the objections to the Law system was the difficulty of educating the subscriber in its use; and another respect in which systems and subscribers were considered, was to put the proposition as to whether it would be easier to change from the Law to the more general magneto system or *vice versa*. It was recognised that subscribers objected to any change. This was alluded to by Mr. Smith, who illustrated his views with a story then current:—

Suppose I should find the Law system most desirable. As Mr. Denver intimated a moment ago, it would create a great disturbance in Boston, I know it would in my city, to attempt to change. No matter how desirable I might think it to be, or my directors might think it to be, I should hesitate to attempt to change; I should expect to find myself in the position of the boy that I read about not long ago, who was sent upstairs to bed, during the prevalence of a thunderstorm, against his vigorous protests. He was finally persuaded to go by telling him that God would be there with him, and it would be all right. He did go; his mother kissed him good-night, and he went upstairs. Then came an exceedingly loud clap of thunder, and the next thing there was a little boy heard at the top of the stairs, saying, 'Mamma, mamma, you come upstairs and stay with God awhile, I am coming down.' I think that is about the situation.

Systematic education of the subscriber had been advocated before the 1887 conference. The remarks of Mr. Fay at the 1884 meeting of the National Telephone Exchange Association give one instance. He was referring to the records of incompleted connections:—

I think that probably ninety out of every hundred of uncompleted connections result from busy wires alone, and that the uncompleted connections are particularly owing to the fact that some wires are very busy. For instance the heavy railroad wires can generally be charged with half of the uncompleted connections. In other words, they are busy all the time, and it is impossible for a man to call for a wire and get it, except by taking his turn. The drag of the subscriber on the exchange, however, is because he is slow in answering, and careless about doing his part of the work, ignorant of the way in which the machine works, hasty in making his calls, indistinct in his speech, or for some other similar reason, and he becomes an unmitigated nuisance in every way except as a payer of rentals. We have lately undertaken the missionary work of converting him and of making him more amenable to reason, of making him feel that he ought to know something about his machine, how to work it and how to get the best results out of it. We have found that the heaviest users we have make the smallest number of complaints in reference to the service of the telephone. They know how to use the telephone, and if they get into an exceptional condition, where the telephone will not work satisfactorily, they know how to get out of it. As I say, we have gone into the missionary service in that respect, and for that purpose we have four men going around the city of Chicago all the time teaching the subscribers how to operate the telephone. That has certainly been of great assistance to us.[1]

A question of great importance at the 1887 conference was: How much should the subscribers be called upon to do? Mr. Scribner pointed out that this determined the number of connections that should be required of an operator in a given time and the cost of the apparatus by which the work was done.

The least that an operator can be required to do is to get the number of the line wanted and to connect the two lines together. The operator may do more than this. She may call the wanted subscriber with a ring of the bell. To ascertain the number, to connect the lines together, to ring the wanted bell, and to wait in the line until the bell is answered and until the two subscribers get to talking, is the extreme of the ordinary requirements of an operator. But sometimes the line wanted will be found busy. In this case the calling subscriber must be informed of the fact. But in that case it may be left for the calling subscriber to call again at his convenience, or the operator may be required to remember the number of the subscriber calling, and to watch for the busy line, and when it is disengaged to get the two subscribers to talking. This represents about the extreme of requirements for connections other than toll-line connections.

Mr. Hall expressed as his judgment 'that the function of the

[1] *National Telephone Exchange Association Report*, 1884, p. 66.

telephone exchange is to receive calls from its customers, call up whoever is wanted, inform them who wants them, see that the lines are connected together and the parties properly talking before they leave it. That is the extreme case that you have stated, and I believe that anything short of that is falling that much short of the full duty of the exchange.'

The first 'conclusion indicated by drift of discussion' recorded in the synopsis is that—

It is the duty and obligation of a central station when a call is received, to follow the call through and to bring the subscribers actually into communication with one another, not merely to receive the call and ring the subscriber desired.

It is, however, recognised that there are many diverse conditions in the several exchanges, and that methods must differ to suit the varying conditions which should be studied. While in one exchange the proper procedure, where a number of calls closely follow one another, is to answer each, ring the subscriber desired, and connect the lines, returning to each a moment later to make sure that the two subscribers are in communication; it is also desirable, where time will permit, that a connection shall be followed closely by the operators until the subscribers have received the service.

It will be seen that systems which threw the work in part upon the subscribers, such as the Law system and the ring-through system, were ruled out not only on account of such particular defects as were observable in practice,[1] but also on the general ground that the service which they afforded did not attain the standard considered to be necessary.

Some economy might be obtainable by dividing the work between subscriber and operator, but that economy would be dearly purchased at the price of an imperfect or unsatisfactory service. The conference placed the standard of service high, and indicated that it was the duty of the exchange to undertake any work which might be required to attain that standard. It was an important pronouncement, for good service dominates all other considerations telephonically.

The relations of the exchange authorities to the subscribers and the manner of communicating with them were carefully considered, even to such points as the difference between 'won't answer' and 'don't answer.' Mr. Smith expressed the objection to the former. The fact was that the subscriber did not answer. 'I have spent a great deal of time to get my operators to use the term "He does not answer." That conveys an entirely different meaning.'

[1] Chapter xxiii.

3. *Operators and other servants of the company.*—The need of strict discipline in order to get an efficient service was insisted upon. Mr. Smith said :—

I have always instructed my operators as to the exact manner in which they should convey the desired information to subscribers. There are but few things they have to thus convey to them : first, whether the line is, or is not in use ; whether the line is out of order, and one or two other things ; and these are given in the same language by all of our operators, and we hold them to it strictly. We have two assistant chief operators in our Kansas City exchange, while we have from thirty-five to forty operators on duty all the while. Those two assistants are really floor walkers, who are up and down the room all the while, watching operators in their conduct of the service. As to the matter of ringing also, they are instructed to give only short rings. We believe that a short ring excites the curiosity of the subscriber, and that he will answer it quicker than he would a long one, which irritates him ; observation has proven that to be the result.

. . . . . .

We answer by asking : 'What number ?' The number being given, we repeat the number so that the subscriber may know that his order has been received and correct it if an error has been made in understanding it. If the line is in use, we say, 'They are talking.' If the line is crossed or otherwise not available by reason of trouble, we say, 'The line is out of order.' If the line is ready for use, she simply calls and passes to the next drop that is down, and so on, until she has answered all the calls before her, and then comes back over them to see if they are talking. If any of them have been connected long enough, as she may think, to be through, she tests by asking, 'Did you get them ?' provided she hears no talking.

Mr. Lockwood : She listens first ?

Mr. Smith : She listens first, and, hearing no conversation, she asks, 'Did you get them ?' We used to ask 'Are you through ?' following the example given us, but we found that irritated the subscriber, as the operator would frequently meet with a party who was just connected and it would make him very angry ; it sounded as if she wanted to hurry him up or check him off. The other question, 'Did you get them ?' seems like a desire to aid him and help him, and he answers pleasantly. When a ring-off comes, we believe that operators, in nine cases out of ten, will know positively whether it is a ring-off or a desire for a second connection, so that we take the risk of having them disconnect promptly without asking any further questions, and allowing the second connection wanted to come in the regular way on the regular subscriber's drop. Our experience is that we have made but very few mistakes in cutting off parties before they were through, and the advantage gained by promptly disconnecting is in having to answer less frequently that they are talking.

It was agreed that the exchange manager should be a man, and that he should have some degree of electrical skill. Mr. Sunny said their method in Chicago was to promote a boy from the switchboard, after he had been at it long enough to deserve promotion, and put him in the inspection and repair department.

He is educated in that department in the work of repairs, and we also make him find out how to climb a pole, so as to take out line trouble. After he has been in that department two or three years and finished the course in the exchange, if he has the requisite requirements as to force of character and so on, we make him manager . . . We have followed that system for seven or eight years, and find that we get the best results.

Mr. Seely explained that they followed a somewhat similar system in New York except as regards pole-climbing. Two of their prospective exchange managers tried to climb a pole, 'and we have them now as pensioners.' There were eleven exchanges and a superintendent of exchanges. 'The superintendent has absolute charge of all the managers in each and every exchange, hires all the operators, and discharges them, upon recommendation of the managers.'

The operators' school so familiar a feature now was less general in those days, but it existed in Chicago and is thus described by Mr. Sunny :—

We have a pretty hard time to get girls of the right character who are willing to work Sundays, and we are not getting the class of applications that we did in the early days of the telephone service. This is not because we have always tried to keep the standard of intelligence and morality high, but it comes from some other cause, of which I know nothing. We probably get one young lady out of ten that is entirely satisfactory. We have a school of instruction where new-comers are taught the business as far as possible. The report of this school runs something like this: For two years we have had about three hundred and fifty applicants, and we have only found out of the three hundred and fifty about eighty-five who were competent to do the acutal work.

Mr. HALL : That is a very good percentage—about one in four.

Mr. METZGER : What kind of a kindergarten have you there ?

Mr. SUNNY : We have a teacher. The school will accommodate about ten operators—ten applicants for positions—and we have in charge of that school a first-class operator of five or six years' experience, who was cut out for a school ma'am. She can size up an applicant in short order, and in four or five days can tell with almost absolute certainty whether that young lady is going to make an operator or not.

Mr. METZGER : Does she put her down before the regular board ?

Mr. Sunny: She sets her down before a little switchboard that we have made up for the purpose. We have three sections of switchboard, and then telephones scattered about the room, and the students talk to each other, and on the three sections of switchboard they make connections and do everything they would be called upon to do at the big board, except to make the actual connections. We find that it educates the students in the matter of hearing and talking, and handling the cords and handling the cam levers, so that when they sit down to actual work they have nothing to overcome except the momentary nervousness. In the old system we used to take a new-comer and put her on a section to answer fifty subscribers, and we used to depend upon the subscribers to educate the operator and make her competent to fill that position.

The preference for women operators was already well defined by 1887, but difficulties were experienced in the engagement of women as night operators. Mr. Smith had overcome the difficulty in part, and the need for it he explained as follows :—

I would like to say, with reference to the night operator question, that in five of our exchanges we have, within the past year, substituted women for men or boys, and with excellent results. In all of those places we are called upon to do fire alarm service, and we found it necessary to do something to make it more certain, and we have accomplished it by having women ; they are wider awake, more attentive than young men or boys. We are now considering the propriety of substituting women for night service in our Kansas City exchange, where we keep thirteen operators for the first half of the night. The only difficulty seems to be in getting the young women to their homes at that hour of the night.

It was recognised that wherever the telephone was used at night the occasion was one of urgency and as high a class of operators as could be found should be selected for the service.

In operating methods considerable attention was given to the relative advantages of having one operator carry out in its entirety a required connection or having the work divided, the conclusion reached being strongly in favour of having each operator perform all the various operations necessary to bring the subscribers of the lines assigned to her care into actual communication with the subscribers which they may call for. This in contradistinction to the mode of central station operation known as the 'division of labour' plan, in which the co-operation of two or more operators was necessary ; experience having demonstrated that by the centralised or single operator method the service is rendered more speedy ; the chance of error diminished ; in case of error, imperfect service or fault, the responsibility is easily fixed ; and the operator becoming familiarised with her subscribers is enabled to more readily understand and meet their wishes.

Operators standing at their work was advocated, one member estimating that she could work a third faster standing than sitting. High chairs permitting standing or sitting at will, or as occasion required, were advocated.

4. *Apparatus.*—The multiple switchboard had been rapidly developed and largely used. There were still, even in the conference, advocates of other plans, but the advantages of the multiple were so manifest that there was no room for doubt, and in the 'Conclusions reached by the Conference,' as narrated in Mr. Lockwood's synopsis, the multiple switchboard is given first place as follows :—

No. 1. The multiple switchboard for large central stations is a material improvement upon, and presents decided advantages in, the matter of efficiency and economical operation over the grouping or sectional switchboard and system, in which a number of lines were assigned to a single operator at a given section, and in which communications were effected between the several sections by means of trunk or transfer lines extending from one switchboard to another throughout the system. For such central stations it is decidedly to be preferred, and no circumstances can be conceived which would render a return to the grouping or trunk line boards desirable.

The position of the Trunk Board at the beginning of the line of sections has the next place.

No. 2. In central stations where the multiple switchboard is employed, that section which is devoted to extra-territorial lines, or other out-of-town service, should invariably be the first one reached by incoming subscribers' lines; for when any subscriber's line is connected with a distant point through a long line, the connection being thus made through the first spring jack, the compound line is relieved of any electrical and mechanical difficulties such as defective jacks, is rendered less liable to interference, and is made to work easier than if the long line connections were indiscriminately applied at any point of the course of the subscriber's wire through the central office switchboard.

And the imminence of metallic circuits is recognised in

No. 3. That the Western Electric Company should hereafter counsel its present or prospective customers when considering the purchase and installation of large switchboards, or switching organisations, to have such switchboards, irrespective of character, so arranged as to be readily adapted to metallic circuits.

The growth in the sizes of switchboards and the large number of spring jacks required to be employed were causing some apprehension on the score of expense, and the attention of the conference was given to the consideration of possible means of economy. To place the switchboard horizontally, instead of vertically, was regarded even by Mr. Lockwood as a 'more hopeful line for future experimentation and consideration than any other plan.' Mr. Scribner demonstrated that the operator had not the same reach on the horizontal as on the vertical board. The conference arrived at no definite conclusion on the subject, but it is to be remarked that the exchanges in the United States refrained from using flat boards. Some years later the same effort at economy was tried in Great Britain and it fell to the writer to criticise the proposal,[1] without being aware of the 1887 conference arguments. The results of the operation of the flat boards in Great Britain confirm the criticisms that were made against them, and illustrate the advantage of the conference method of reaching conclusions.

One of the principal practical features in the multiple switchboards dealt with by the conference was the cross-talk resulting from the test lines. The boards were growing in size beyond the expectations at the time of the original design, where the plug connected the subscriber's line with a test wire, which test wire remained connected during conversation, making practically 'two long open branches either in one cable with other lines or in two cables with other lines; so there is an electro-static effect between those two open branches, and the lines that are talking.' The trouble was a serious one, and the conclusion of the conference is thus recorded in the synopsis:—

No. 9. The testing arrangement which is now in general use in connection with multiple switchboards, whereby an operator at any board can determine whether a line wanted is already in use at some other switchboard, has been found to be a considerable factor in producing cross-talk, because when any two lines are united at any switchboard, the test wire of each line is temporarily united also to its own line and forms an open branch of the said line. Thus there are with each pair of connected lines which are in operation two open branches in inductive proximity (that is in the same cable) to similar branches of other lines, some of which are at any given moment probably also in operation. Any two connected lines are thus inductively united by a condenser with such other lines as happen to have their test wires in the same cables and which are at the same time being used.

The result is, that the subscribers of these lines are much disturbed by overhearing the conversation going on over other lines.

[1] *Journal of the Institution of Electrical Engineers*, April 16, 1896, xxv. 353.

This disturbance increases with the size of the switchboard, the length of test wire being correspondingly increased.

To remedy this, a plan has been devised whereby the test wire, instead of being grounded through its own main line, can be automatically and independently grounded, and thus need have no electrical connection with its own main line.

This plan is being experimentally tested, and will be adopted and recommended by the Western Electric Company if satisfactory.

The improvement was outlined to the conference by Mr. Scribner as follows :—

> It consists in providing a double plug instead of a single plug, or with a double cord or sleeve on the plug, which, when the plug is inserted, makes contact with the test ring, the sleeve being connected with one of the double cords and directly to ground ; so that when you connect together two lines, you simply connect those two lines one to the other, and ground the test wire directly, which would certainly improve the service.

Although not appearing in the conference proceedings, it may be stated here that the new plan under trial was that of Mr. Carty, described in his U.S. patent, No. 415,765.[1] In giving evidence on November 17, 1898, in patent suit against the Detroit Telephone Company, Mr. Scribner stated that Mr. Carty explained his invention to him some few months before June 1, 1885, when the application for patent was filed ; that the invention was purchased by the Western Electric Company upon his (Mr. Scribner's) recommendation. Mr. Scribner in further evidence in the same suit said :—

> The test circuit of my patent was a marked step in advance over the try circuit of Haskins and Wilson, but it remained for John J. Carty to invent the highly developed system of his patent No. 415,765, which in detail, as stated in each of the claims, has gone into use in practically all the modern multiple switchboards made by complainant.

Admitting the superiority of the multiple system, the conference were constrained to inquire what were its limits and what arrangements would need to be made when those limits were reached. Mr. Scribner reported that it was then proposed to equip the Cortland Street switchboard with over ten thousand spring jacks to each section. That number, however, had not been in operation, and Mr. Sargent assumed the then existing limit as five thousand.

Two important features in exchange planning were brought out in Mr. Scribner's paper entitled 'The Multiple Switchboard

[1] Dated November 26, 1889 (application filed June 1, 1885).

System, its Requirements and its Limitations.' Hitherto switchboards were installed without definite data as to the requirements to be met. Mr. Scribner pointed out that the cost of central office switching apparatus depended both upon the number of lines and upon

the aggregate number of connections per day or per hour. The aggregate number of connections per day determines the number of lines for which one operator can do the work, and hence the number of operators, and hence the number of multiple switchboards. Another consideration affecting the number of connections possible to be made by an operator is the efficiency of the operator. There is a great variety in this respect, and there is the possibility of increasing the average work by selection and by training of the operators.

The other important new feature was the suggestion to distribute busy lines.

The number of connections an operator may make in an exchange may be increased by equalising them among the operators. This has never been done thoroughly and systematically. A considerable aggregate gain can be made by distributing the lines among the operators so as to equalise the work among them.

The value of systematic distribution of the lines in accordance with the number of calls received over them was further emphasised during the discussion and bore immediate fruit. A number of distributing boards and intermediate distributing boards were subsequently designed and gradually some form became universally used. It is to be feared that outside the influence of the conference a considerable time elapsed before the advantages of distribution in its effect on the economy in first cost of apparatus and in operating expenses were fully realised. The prior practice is well illustrated by the remarks of Mr. Sunny :—

I think where a popular mistake is made in determining how many wires we ought to put before an operator is in the fact that the statistics from the other exchanges are looked up, and we find that some exchange in the South is making eighteen hundred connections a day by one operator, and that another exchange in the North is making eight hundred connections a day by one operator, and we immediately split that and conclude that we may give to an operator a thousand connections a day, and then we will be sure to have a pretty fair service. I think the way to determine that would be to find out how the business runs in the particular exchange. All over the country it seems to be the busy hour, or the busy hours, between nine and eleven. In Chicago the busy time is from half-past eight until half-past twelve. Now,

I would give an operator about as much as she could comfortably handle between half-past eight and half-past twelve, and she could sit with her arms folded the balance of the day. But when we average it up and give an operator eight hundred calls to answer, the chances are that you will give her too much for three hours a day, and just enough to keep her comfortably busy for the balance of the day.

The capacity of the operator of the period with the best available material as well as the principle of distribution is shown in subsequent remarks as follows :—

Mr. SCRIBNER : Is it not a question of the number of connections she can make ? It would seem to me that the whole question is, How many connections are you going to have on each line of your board during the day, and how many connections can an operator make in a day ? or take it for an hour, and the busy hours rather than for the whole day, because the connections do not come evenly during the day, but you are busier at one hour than at another. Take the business for a day, and then take the business for a busy hour, and how many connection can one operator make in an hour ? Then give her as many lines as will provide those connections. Now, we have positive knowledge on the subject of how many connections an operator can make in a busy hour. When she is rushed we have positive knowledge that she can make two hundred and fifty connections.

Mr. DENVER : Do you mean an average operator ? Would not that require quite an expert ?

Mr. SCRIBNER : It requires a good multiple switchboard operator. I do not think it would require an expert. I have myself sat with a head telephone upon my head connected in circuit with the head telephone of a multiple switchboard operator, and have checked off the number of connections she made for an hour ; and I have had the thing done by others and counts made at different exchanges, and made without the operator's knowledge, so as to prevent any special work on the part of the operator ; and these figures vary from one hundred and seventy-five to two hundred and twenty-five, and in some cases two hundred and fifty connections in an hour by an operator. Now, the only thing necessary is to give that operator a number of lines which will produce that number of connections, and if you give her a set of lines which will produce more than that during any given hour, then you break her down. If you accumulate in front of her, as is frequently the case, a large number of very busy wires, when they become very busy at the busy hours she is broken down, and other operators along the line are comparatively idle.

Mr. DENVER : The only way to give satisfactory service is to give prompt answer to the greatest number of calls in the shortest possible time ?

Mr. SCRIBNER : That is the whole secret. By distributing your busy wires in front of the different operators, giving each operator only what she can attend to easily, two hundred connections in the busy hours, which in an eight or ten hour day would amount to something like fifteen or eighteen hundred connections.

Mr. METZGER: Mr. Scribner, do I understand you to say that an operator can make two hundred connections per hour and answer the subscriber and get the party called for?

Mr. SCRIBNER: Get the party called for and get them together.

Mr. METZGER: Two hundred per hour?

Mr. SCRIBNER: Yes, sir; over two hundred.

Mr. METZGER: Where is that?

Mr. SCRIBNER: I have counted them in New Orleans, and I have counted them in Nashville, and I have counted them in Boston, which is a slower switchboard than either of these others. I have counted them in Boston, where there were from one hundred and twenty to one hundred and twenty-five per hour. I have had them counted at Omaha in the past week, where a single operator made two hundred and twenty-two connections in an hour, and did it nicely; she was not rushed; it was on a quicker board which they have in use there, which system we shall probably have a chance to discuss during the week. But the idea is that an operator can take care of, without being rushed, from two hundred to two hundred and twenty-five connections in an hour; and if she can do that, she can take care of one hundred of the busiest lines—that is, one hundred of the average lines of the busiest exchange in the country.

Mr. METZGER: Then, according to that calculation, an operator could take care of two hundred wires where the average number of connections did not exceed seven per day to a subscriber.

Mr. SCRIBNER: If they were distributed evenly over the board.

The facility which the answering jack offered for distributing without changing the subscriber's number was not universally realised. The following extract from the proceedings will serve to illustrate the prior conditions and the need of demonstrating even to a selected body the newer practical features:—

Mr. DAVIS: There is one thing that Mr. Scribner has dwelt on that has not been answered; perhaps it would be well to answer it now; and that is, the possibility of distributing your busy subscribers around through the board in such a way as to make the work more equal. I do not think in the New York exchange that can be done, because each subscriber has his telephone number printed upon his paper and on his letter heads, and he objects to changing it. I think that alone is sufficient to prevent it being done.

Mr. METZGER: I would like to state how we do it. We distribute among our busy boards the residences, so as to try and relieve the business on that board, and supposing it to accumulate during the year, when we issue a new list we make the changes in that new book, but do not actually make the changes until the book is issued.

Mr. DAVIS : I understand ; but here in New York, the practice is for a man to print on his letter heads his telephone number.

Mr. METZGER : We discourage putting the number on.

Mr. SUNNY : We have discouraged it for the past year also. We ask them to put on their cards the word 'Telephone.'

Mr. DENVER : We found trouble with it in Boston.

Mr. SUNNY : We were putting in a section of switchboard in the La Salle Street office, and the numbers ran from 2000 up. We wanted to abandon those numbers in six or eight months, and we told those people not to print those numbers on their cards, that we might have to change them. They all fell into line on that. I have seen a great deal of stationery with the word 'telephone' and the number omitted. In practice a man pays no attention to the number of the telephone printed on a card ; he does not hunt up the card to see what the number is, but goes directly to the book.

Mr. SCRIBNER : Mr. Davis assumes that it is necessary to change the numbers in order to make this redistribution ; but I do not think it is necessary to renumber the subscribers. I think the equalisation of the distribution of busy lines among quiet lines may be made without renumbering the subscribers. Where you have answering jacks and drops, the drops fall and call attention to a given answering jack. It does not matter what the number is on that drop ; in fact, the number might be left off entirely from the drop and from the answering jack, simply putting in place of the subscriber's number a number to indicate which answering jack to go into. In that case the operator would not know the number of the subscriber she was answering, would not know his exchange number ; she would simply know his number upon the board. In that way by putting them in their proper position on the face of the board for the multiple jacks, and distributing them at will through the drops to the answering jacks you would accomplish all that you wished to accomplish, and without this general tearing up of the exchange book and the renumbering of exchange subscribers.

Mr. DAVIS : I do not see how that is going to operate. Take the reverse thing ; when you call for that person, where is his number on the board ?

Mr. SCRIBNER : On the board you place his number just as it is ; on the board he will appear always in his old position ; on the multiple jack he would appear in the same position that he now holds ; but after it has passed through a series of multiple jacks, as subscriber No. 1 for example, then carry him to drop 23, but do not number him there. It does not make any difference to the operator what his number is on the multiple jack so long as she knows where to go to answer. A drop falls ; she goes into the answering jack which corresponds to that drop, and she has the subscriber calling ; she does not know who he is to be sure, but she has him and can connect him to whomever he wants. She always knows him as subscriber No. 23 on her board. That is, you take the drop, and answering spring jack and number them from 1 to 150, and every subscriber then would have really two numbers—No. 1 in the exchange, and No. 23, board 3, in the office; and, connecting with him, it would always be done just as it is now.

Mr. DAVIS: The operator would have to remember.

Mr. SCRIBNER: Remember nothing.

Mr. METZGER: Does not that compel the operator to remember the changes?

Mr. SCRIBNER: Not at all. The operator plugs into an answering spring jack which corresponds to an answering drop.

Mr. SUNNY: That is all she has to know.

Mr. SCRIBNER: All she has to do when the drop falls is to know where to find the spring jack that goes with that. As they are placed near together, and have a definite position with relation to each other, they can be numbered the same with relation to each other and not with relation to the board. She can go to the answering spring jack and find who is wanted.

Mr. DAVIS: Do you put answering spring jacks on your boards now?

Mr. SCRIBNER: That is one of Mr. Seely's inventions, and it is on either the John Street board or the Thirty-ninth Street; and it takes out that necessity of numbering the drops.

Mr. SUNNY: They may be lettered A, B, C?

Mr. SCRIBNER: They may be lettered A, B, C.

Mr. SEELY: Upstairs it is so arranged. It is immaterial where the call comes from: all the operator does is simply to answer that annunciator.

In respect to details of construction there was much discussion on spring jacks and indicators. The deleterious effect produced on talking through the ordinary indicator was beginning to be pronounced. The absence of any indicator was one of the advantages urged for the Law system, while the special form of clearing-out indicator (the 'galvanometer drop') used in the Omaha system was recognised as offering less retardation in the talking circuit. This was referred to by Mr. Sunny in the following remark:—

Mr. SUNNY: I have an idea that that kind of a disconnecting drop would not introduce so much retardation in the telephone circuit as the ordinary kind. I don't know what the resistance is, or how many convolutions of wire; but I think that a helix of wire with a loose needle in it is not so likely to retard the current as one in which the core of iron is close to the wires.

But this idea was not pursued by the conference.

In the discussion on spring jacks importance was attached to the use of German silver instead of phosphor bronze, when it was elicited from Mr. Barton that German silver was first used, but that everybody thought phosphor bronze was better. Mr. Barton said :—

German silver antedated phosphor bronze as a spring metal in electrical instruments, and the first spring jacks had German silver springs. Afterwards phosphor bronze came into use, and in the large size spring jacks they have been used right along until now, and have never given any trouble that I am aware of ; and I don't know that any change has been made or has been called for in those springs. The people who have room for spring jacks of that size have used them right along and preferred them, because there is such an abundance of metal there that it makes a contact which is very firm indeed. Now, the small spring jacks such as are mounted on rubber strips, and such as are used in New York and Boston and elsewhere, have a spring of very much less surface, and it is a matter of great deal of pains to make them all uniform. During the later period of construction of the smaller spring jacks the method of testing has been improved, so that now the practice is to try every strip of spring jacks by putting a pin into the plug hole with a weight at the other end, making a lever of the pin, and trying with a telephone at the ear of the man who is making the test to see if the contact is lifted. If the contact is impaired by a certain weight working against any spring, that spring is marked and rejected and another one is substituted which will not be lifted from its contact by that weight. In that way, by taking a great deal of pains, the narrow springs of German silver are thought to be now quite efficient, and I am in hopes that they will answer as well as the wider and stiffer springs of the old type.

Mr. Lockwood thought all would agree

that phosphor bronze is not now, and never was for that matter, adapted for the purposes of the spring, but such experience as I have had with the matter leads me to the belief that this is not because it is more easily oxidised than German silver, and not because it more easily acquires a dirty surface than German silver, but because it did not hold its resiliency. The contact between the spring and point is practically permanent ; it is permanent except at the moment it is lifted up by the plug when inserted. It hardly looks to me possible, if we have a German silver spring and a brass or a German silver point, that it will oxidise much, because the two surfaces are close together, except at the moment when the plug connection is made, and so I think if we have a first class German silver spring and any good, clean, unoxidisable metal or alloy for pin contact points, we are all right there. Now, the question comes in, What shall we do with the plug in the spring contact ? It is rather a curious thing—it was to me, at least, when I first noticed it—that I found that the main trouble in the plug contact and spring contact was not that it did not make a good connection, but that by and by, after the phosphor-bronze spring had lost its resiliency, the insertion of the plug failed to lift the spring up off the ground contact, left the ground contact there at the same time that it made a new contact with another line through the plug.

Suggestions to use platinum in spring jacks and other contacts were considered, but Mr. Scribner argued successfully against it. His argument was, briefly, that the idea of using platinum is to prevent corrosion caused by sparking when a current is broken; there was very little apparatus in telephone work where the current was broken or closed, or where sparking and consequent burning takes place. It was necessary to consider the possibility of platinum contacts in a spring jack causing trouble.

If we put platinum in a spring jack at all it should be put into the contact points and into the spring. It is rather difficult to construct a spring jack cheaply in that way. It is rather a difficult matter to get an adjustment of parts made as these parts are, by machinery, so that the two points would always meet, unless very large platinum points be used. If we put the platinum contact on one point alone, we will say on the contact point, then the spring will be likely to make as much trouble as practically the two points make now. Now, in view of the fact that the trouble occasioned is not a trouble of corrosion, but a trouble of dust or some foreign material collecting between the spring and its contact, isn't it possible that there would be just as much trouble in the spring jacks as there is now, and is it not possible that it would be more difficult to remove that trouble when it appears than it is now? The present plan is to insert from the front of the board, in case a spring jack is found open, a small piece of thin steel—not a file, but simply a strip of steel, which removes the trouble, which removes the foreign material, whatever it may be. It would be next to impossible to insert such a thing in case there were platinum points on both spring and contact, because you could not strike between the two points.

Mr. Scribner suggested putting platinum points on the anvil, so called, on which the spring rests, but no platinum on the German silver spring, thinking that German silver is free enough from oxidation to prevent trouble. The improved springs, however, were of such a character that platinum was not necessary even on the anvil contact. This was fortunate, for subscribers multiplied and multiple switchboards grew to dimensions unexpected even by the conference. Their cost was considerable, and had platinum been used, a very large amount would have been added to such cost.

The conference devoted attention to apparatus other than switchboards. Mr. Sunny presented a paper of which a part relating to the magneto bell has been quoted in Chapter xii. p. 149.

He continued :—

We have on record [in Chicago] 5300 cases of transmitter trouble for 1886, and 6300 cases ir 1887 (December estimated). This includes the transmitter, primary circuit, and battery. Five per cent. of these troubles are bad connections in the hinge, which develop after the instrument has been in service six months or a year. Of course, the majority of the trouble originates in the battery. The Leclanche form of porous cup battery is generally used, and the current that it furnishes is at its greatest strength when the circuit is first closed, and constantly loses every minute it is kept in service, so that at the end of a fifteen minute conversation a very perceptible difference in the transmission is noted. A transmitter that will automatically adjust itself to the changing condition of a battery, or a battery with more constancy is needed sorely. We can be good-natured through all the switch and battery trouble, but it is hard work to be bland and smiling when the telephone itself shows defects. Five hundred cases of telephone trouble in 1886 and six hundred in 1887 are made up very largely of shavings of rubber getting in between the diaphragm and the magnet. This trouble began about the time the use of wax was discontinued, and was found to exist in a batch fresh from the factory a month ago. The aggravating feature about this trouble is that it seems such a simple matter to remedy. The total number of cases of trouble and inspection for 1886 was 48,406 for 4460 telephones, of which thirty-five per cent. was line trouble and sixty-five per cent. instrument trouble and inspection. For the current year the total number will be 49,279, approximately, for 5100 telephones: twenty-nine per cent. line troubles and seventy-one per cent. instrument trouble and inspections. The cost will average twenty-seven cents for each, or about $9500. No man can figure the actual cost to the telephone at large of those cases of trouble. They lose us a friend here and there; they frequently determine a would-be subscriber that he does not want a telephone; and they add their no inconsiderable mite to what makes up the sum total of the cry of 'Monopoly.'

In the discussion which followed, Mr. Lockwood struck a prevalent source of trouble, remarking :—

I notice with some interest that a great number of complaints arose in practice in reference to the various kinds of telephone apparatus from the defective connection of the spring in the hinges. It appears to me that there is a point to which both the suppliers of the telephones and the manufacturers of the bells might give earnest heed. I do not think, if I may express my own opinion, that we are doing justice to the apparatus in sticking to the old spring, fastened by a screw at one end, and projecting out at the other end. I do not think that the connection is a trustworthy one; and it is not for want of knowing better that we stick to it.

I know at least three persons who are actively engaged in practical telephony, who have suggested that a better arrangement might be made by allowing the centre portion of the pintle of the hinges to be stripped of any connection with the hinges, and then have a spiral of stiff wire, brass or perhaps steel, or whatever might be found best, which should really be a portion of the circuit, and which should be extended out on both sides and soldered to the bell or transmitter connections, as it might be. I have a very strong impression that that would be a much better plan, and I should like very much to see it adopted both by the telephone manufacturers and by bell manufacturers.

This simple remedy was promptly applied to magneto bells, transmitters, and other apparatus, thereby reducing a prevalent source of trouble, and tending materially to increase the reliability of the service.

Less effective were the criticisms of the design of apparatus. Mr. Sunny opened the attack in his paper by remarking that

The field for improvement in the construction of subscribers' apparatus is a particularly broad one. The entire outfit is crude and defective, and it represents a smaller growth towards perfection than anything else that we have in the service.

And Mr. Lockwood continued it by saying :—

I think we have also got a valuable idea, at least, in the suggestion that we do not make our telephone apparatus as sightly and as attractive to the eye as it might be made.

There were practical objections to changes in design. Mr. Seely alluded to the difficulty found in New York in introducing a new kind of bell. 'If we put in a new bell for one subscriber we will have five or six hundred to put in next week of the same kind of bells, and we find it a very expensive luxury.' Other members recorded a similar experience, and there is little doubt that the continuance of the criticised type, so far as appearances are concerned, is largely to be accounted for by the disturbing effect upon existing subscribers arising from the introduction of a more attractive type to new subscribers. But so far as working parts were concerned, improvements were continually introduced as they were called for by experience or developed by invention.

Party lines existed in the earliest days of the telephone, and the apparatus in connection with them was the subject of numerous inventions mainly with a view to obtaining a system of selective signalling which should prevent all the subscribers on a line being called when only one was required. By 1887 there was sufficient experience of the original form and of its successors to show that none was satisfactory. Still the telephone manager was faced with the fact that a line was unused for a much greater period than it was in use. This seemed an economical heresy, and it was felt that the use of the telephone might be extended and the cost reduced if some means could be found for the common use of a line or lines. Like most other points, the subject was introduced to the consideration of the conference on broad grounds. Question 15 of Mr. Lockwood's paper reads :—

Whether, considering that the lines of subscribers are in many exchanges only used in the aggregate a small fraction of each day, it is feasible to relieve the central office work, and at the same time reduce the cost of construction by some system of rearranging main circuits outside ; so that a given number of circuits may suffice for a greater number of subscribers, while at the same time each subscriber may be enabled to use some wire every time he desires ?

Reference was made to a system suggested by Mr. Berthon of the Paris exchange and described by him at the telephone convention at Philadelphia in 1884.[1] It was a duplex or multiplex system on the principle of a Wheatstone Bridge. It was an idea rather than a practical application, but another method of selective signalling for party lines described by Mr. Berthon at the same meeting was an early application of a principle that in later years was much used. Mr. Berthon described the system as follows :—

One arrangement is for four subscribers on metallic circuit. We have four keys at the central office, and when we wish to call one subscriber we call him by a positive or negative current sent on one or other of the wires, the wires being grounded at subscriber's office. That current goes through the ground, positive or negative on the wire, and reaches the ground at the other end, so that with polarised relays at subscribers we can call the two subscribers branched on the first wire and the two branched on the second wire, which makes four. This is one of the best combinations we have for calling four persons on one circuit.[2]

As remarked, however, the subject was brought before the 1887 conference not merely as a question of party lines, but with a view to the possible suggestion of a 'nucleus or germ of something we have not reached yet.' The economic side impressed itself upon Mr. Hall, who remarked :—

It has always seemed to me that a system which gave, for instance,

[1] *National Telephone Exchange Association Report*, 1884, p. 133.
[2] *Ibid.* p. 136.

to a residence a private wire and all the equipment of a subscriber's and central office stations, all of that apparatus and plant being idle for twenty-three hours and fifty minutes out of the twenty-four, involved an unreasonable waste.

This argument was met by Mr. Sargent, who said :—

I have given that idea some thought, and it seems to me that it is a question of relative economy. You start out in the first place with the proposition that there is a terrible waste of wire and expense; you propose to remedy this by mechanical devices that, up to the present time, have been very imperfect, and to make them perfect, I think, would require an amount of expense that would bring you back to the first ground, and that it is a question as to whether you can produce the mechanical effects for reaching individual subscribers over one wire or whether you can better afford to multiply the wires than to pay for expense and intricate machinery.

Mr. Barton added some personal experience and, as was the case with most of his remarks, struck a definite note of a practical kind. Inventors of good ability and the best experience had devoted a great deal of attention to, and much money had been expended upon, this subject. None of the schemes heretofore proposed and tried had been, he said, on the whole satisfactory :—

I had a telephone line built to my house some years ago with the expectation and understanding that my next-door neighbour, who is an intimate friend of the family, should share in the use of the line, and an excellent type of apparatus was provided, which it was supposed would reduce the objections to that kind of partnership to a minimum; I am free to say that I would not now, under any consideration, be willing to allow any partnership in the use of my telephone line. It is not used more than ordinary residence lines are used, but when I want that line from either end I want it; I want it then; I don't want to wait until my neighbour is through; and I presume in Chicago and in other cities of the largest class there can be found subscribers enough who are willing to pay the price of telephone service including a fair price for individual lines in metallic circuits. The question of partnership lines would then include not only the items of economy in the apparatus and lines, and the maintenance of the apparatus and lines, and the maintenance of the service, but also the element of what the subscribers are willing to pay on the condition that they have their service without any drawback which can possibly be avoided. Waiting for some one else to get through is a drawback that can be avoided; and inside of the limits of one hundred dollars a year the exchanges can afford to furnish lines without that drawback.

Mr. Wilson remarked that if a great saving in investment and expense can be made by placing two or more instruments on the same line, it stood to reason that the telephone companies could make a lower rate for that service and that subscribers should be given the option of obtaining either direct or party line service, according to what they were prepared to pay.

Mr. Sunny estimated that, after deducting the cost of the special apparatus required, the extra cost of operating would wipe out the saving in ten years; and Mr. Sargent added the further element of cost of inspection, for from his own experience 'any individual bell system is going to be more than ordinary inspection.'

In the synopsis this point is included in the category of 'conclusions indicated by drift of discussion,' the result being given as follows :—

That in view of the obvious fact that in most exchanges the subscribers' lines are only used for a small fraction of the day, and that during the remainder of the time they are idle and represent unremunerative capital; it is desirable that some arrangement, or system, be devised whereby, for residence lines at least, a given number of circuits may be made to serve efficiently a greater number of subscribers.

The last speakers on the subject were Mr. Hall, Mr. Barton, and Mr. Lockwood, whose remarks are given below *in extenso* :—

Mr. Hall: Mr. Wilson brought out one very important point, that with the introduction of the metallic circuits—and that is inevitably coming—the margin of profit to the exchange on any possible rate is going to be reduced. It seems to me, however, that we may properly and rightly hope and expect that invention in the telephonic field will keep pace with the demands of the service and the exigencies of the situation, so that we can continue to do profitable business and adapt our business to the conditions as they from time to time change; and that is what I had in mind in getting up this discussion, that we should get the problem clearly before us, and not reject the idea because the methods which had been used in the past had not been successful; and that we should look forward to working out this problem in some way in the future to meet conditions which we know are going to exist.

Mr. Barton: Meanwhile the more imminent problem is to adequately and economically serve the subscribers on single lines, and then after that has been tolerably successfully worked out, the inventors, I hope, can be got to work on the problem which is now not so imminent, but which is coming along in the course of time, of giving a second grade or third grade of telephone service. The first grade, however, is lacking a good deal in many respects of efficiency and satisfaction.

Mr. Lockwood: The argument of Mr. Barton is good, but I can recollect a time when we had some switchboards which could be worked, but which had a great many troubles in them. While one set of inventors and exchange managers were trying to get the trouble out of the old switchboard, another set went to work and worked up a multiple switchboard ; but if they had waited until all the troubles were got out of the old one, it is possible that there might not have been such a large field for the new one. I think the moral we have to learn from that is, that while it is well to accomplish one branch of the business and do what we have to do now and do it as well as we can, we must not wait for the necessities of the future, or rather we must not wait for apparatus that will meet the necessities of the future, until those necessities are right upon us.

Mr. HALL : In other words, while we do not want to cross the bridge until we come to it, it is advisable now to send the engineers ahead to prepare a bridge, if there is not one there.

These remarks admirably illustrate the tone of the conference held within ten years of the inception of a new industry. There is conflict upon the pressing questions of the moment, the possible conditions of the future, but unanimity of aim to attain methods and systems which should bring the benefits of telephone service to the largest number of patrons and at rates adapted to their means or requirements. The introductory remarks of Mr. Hall on the provision, housing, and care of plant, and the paper by Mr. Scribner on the bearings of the traffic on the construction and cost of switchboards are especially noticeable at so early a date.

CHAPTER XXV

THE 'BRANCHING' SYSTEM

In its earliest days the difficulties of the telephone system were those incident to the service itself. The development of new devices to accomplish a new service, the application of new methods to accommodate the rapid increase in subscribers, kept the pioneers busy. But, very soon after the telephone, other electrical services came into public use. The civilised world with one accord awoke to the possibilities of electricity. Jablochoff, Werdermann, Brush, Rapieff, and Lontin developed their arc lamps; Gramme, Siemens, Wallace and Farmer their dynamos, all anticipating the use of such apparatus for what has since their time come to be known as 'isolated plants.' Edison and Swan followed with their incandescent lamps and the bold intention of a general supply. The telephone is probably in part responsible for the early development of the central station system of electric lighting, for it was the commercial success attending his quadruplex telegraph and his carbon telephone transmitter that encouraged financiers to supply the means for those experiments on a large scale which enabled Edison to demonstrate the possibility of central station electric supply. It was this working out and demonstration by Edison of the complete system, even more than the details of which the system was composed, that entitle him to the recognition which he then and has since received.

The applications of electricity are roughly divided into two parts—the strong current and the weak current. The telephone is included in the latter category. Supersensitive to every disturbing influence, the telephone was at the mercy of its stronger current rivals. Induction from one telephone circuit to another was simply a family trouble partly amenable to discipline, but the disturbances arising from extraneous currents were more serious. Light was followed by power; and one of the first applications of electric power was for locomotion. The earth was common to both strong and weak, and the claim that the weak came first and were entitled to protection did not avail. Lightning was a known evil which had to be guarded against from the first, but 'strong current protectors' were a later development occasioned by the application of electricity to the purposes of light and power.

In Great Britain the telegraphs were in the hands of the State, and some legal powers were obtained to protect the lines from the invading currents of stronger rivals. In the United States the legislatures held aloof and left telegraph, telephone, and electric light interests to settle their affairs amongst themselves with such assistance as the existing laws might afford. While the Federal authority in these early days did not interfere, the legislatures of a number of the States and several of the municipalities did; their interference mainly taking the direction of insisting upon the speedy removal of overhead telephone and other electric wires in large cities, and of requiring that all such street wires should be placed underground. Policies differed, the municipal authorities of New York, for instance, backed by a State law, being insistent upon immediate subterranean construction and the abolition of pole and housetop lines before the art had sufficiently developed, and without considering whether this with existent knowledge could or could not effectively and satisfactorily be done; whilst the County Council of London, on the other hand, refused underground facilities to the telephone company, even when metallic circuit apparatus and cabling systems were thoroughly developed.

The provision of return circuits for the lines was necessary for the extension and maintenance of the service as well as for the removal of the aerial wires. It was a necessity for the subscribers, the operating companies, and the amenities of a city.

The single wire, or earth return, circuit was general for local service, but metallic circuits were adopted for trunk or toll lines. The eventual use of metallic circuits for all telephone lines was long foreseen, but a period of transition had to be provided for. There were single lines and double lines, and intercommunication must be possible between them. A test also must be available to indicate the engagement of a line. A 'mixed circuit' switchboard had to be devised, and it was necessary to meet this requirement with as little change as possible in existing apparatus. British patent specification No. 9125, dated July 19, 1890,[1] describes the method adopted.

The first claim of this patent is:—

In a system of testing to determine whether a subscriber's line is in use at another section of switchboard or not, the use of a

[1] Equivalent U.S. specification, No. 442,145, dated December 1890.

retardation coil, connected from the subscriber's lines in a derived circuit, either direct or through a battery to ground.[1]

In this system the spring jacks previously in use were unaltered. Metallic circuit lines were connected one limb to the spring, and the other to the bush, of the jack. The plug was already double as required by the improved testing system of Carty,[2] and the connection with the lines was consequently by spring pressure at the tip of the plug and by butting contact at the sleeve. Whilst this was the first use of such a screen as a retardation coil permanently attached between a metallic circuit telephone line and disturbing elements, the coil itself was not new, as may be gathered from the limiting effect of the introductory words of the claim. In prior multiple test systems employing the line as part of the test circuit, the earth was there as the necessary condition of its use as a telephone line. In the case of metallic circuits in which the line was also to serve as a part of the test circuit the earth had to be attached, but in some manner not inimical to telephonic communication. The retardation coil connected to the cord circuit permitted this to be done. The coil as then made was probably more effective for the purpose than any of those preceding it, but such a coil and its effects were sufficiently well known. For example, Dr. Silvanus Thompson described for a different purpose in his British patent specification No. 3564 of March 19, 1885,[3] an 'induction plug,' and so called it 'because of the well-known property of such an electro-magnet of obstructing, by its great electro-magnetic inertia, the passage of rapidly fluctuating currents.'

A modification of the same principle was adopted by T. D. Lockwood in his U.S. patent No. 393,165, November 20, 1888 (application filed August 10, 1888), in which he describes more in detail the electrical characteristics of such coils and uses the expressive phrase 'it follows that an ordinary electro-magnet is in no inconsiderable degree opaque to telephonic or voice currents.'

The induction plug of Thompson and the 'electro-magnetic shunt' of Lockwood were connected to the line either as shunts or in derived circuit (or bridge), whilst the retardation coil of the mixed circuit test system was an attachment to a complete metallic circuit for the express purpose of adding an earth in the most harmless (or least harmful) way. That an earth might be attached to a telephone line through a high resistance relay without impeding conversation was known in the earliest days. It is shown in Scribner's 1879 British patent No. 4903,[4] but the rapid alternations

[1] British specification, No. 9125, 1890. [2] Chapter xxiv. p. 321.
[3] U.S. equivalent, No. 327,837, October 6, 1885 (application filed April 20, 1885. [4] Chapter xiv. p. 168.

of the speaking current and the choking effect of the magnetic impedance thereon were not generally understood. Had they been so, the clearing-out drop would not have been placed directly in the circuit as it was in the standard switchboard [1] and its successor the multiple. It was not until 1889 that the clearing-out drop was made of high resistance, high impedance, and 'teed on' to the line in the same way as the universal switch illustrated in Scribner's 1879 specification before referred to.

The multiple switchboard had been developed as a means of effecting the required interconnection of the lines with certainty, speed, and economy. It underwent numerous changes as a result of experience. The overhearing, due to the condenser effect in the test line, was not contemplated on its first introduction and only became apparent as exchanges grew in size. The difficulty was met by the modification of Carty affecting plugs and wiring. Other changes of a detail character were numerous, but the principle of construction remained the same as when originally introduced—the line wire was brought to the spring of a jack which made contact with a stud, and the circuit was continued from that stud to the spring of the next jack.[2]

The very facilities which the multiple switchboard offered for interconnecting the lines tended to the concentration of as many lines as possible in one exchange, and this, with the great increase in subscribers, resulted in large exchanges. As will have been seen in the last chapter, arrangements were in progress in 1887 for a switchboard of 10,000 lines, whilst there were actually in work numerous exchanges of 5000 lines. Two hundred lines to a section were very general, and on this basis a 5000-line exchange would require twenty-five sections. In every line there would thus be twenty-five points of cleavage, twenty-five spring- or hammer-contacts, and twenty-five stud- or anvil- contacts. The answering (or local section) jack [3] added one more, making twenty-six in each line, without considering the trunk line sections. The trunk lines being metallic circuit, the jacks required two springs and two studs. Assuming such an exchange of 5000 subscribers to require three sections of trunk switchboard, there would be six contacts for each line, which, added to the twenty-six previously mentioned, would make thirty-two vulnerable points within the switchboard for each subscriber's line.

Dirt has been defined as 'matter out of place.' Dust, the 'fluff' or 'lint' from cords or wearing apparel, and other material adulterants of the atmosphere are distinctly out of place between the hammer and anvil contacts of a spring jack, for, upon

[1] Chapter xiv. p. 173.
[2] Fig. 93. Chapter xix. p. 232.
[3] Chapter xix. p. 238.

the integrity of the circuit, telephonic speech depends. The multiple board, while facilitating enormously the operation of the intercommunicating service, increased also this liability to derangement. It was in a large office that the operating facilities of the multiple were most apparent, and it was in the large office that 'matter out of place' had the greatest opportunity for causing defects. Each jack was in this respect a weak link, and the larger the office the greater the number of weak links.

Spring-jack cleaners—small strips of spring steel passed to and fro between spring and contact—bellows, blowers, and other methods of removing dust particles were tried. The earliest use of the now prevalent vacuum cleaner was probably as a switchboard jack cleaner.

Mr. Wilson related to the October 1891 switchboard committee that the Chicago Company had for some time past been blowing out the dirt by means of a steam blower, but they found the relief only temporary, the displaced dust settled on the cables and found its way into the spring jacks again.

We have consulted the Sturtevant Blower people, and have them at work now making an experimental machine which it is intended will suck up the dust. The intention is to have a funnel-shaped arrangement that we can place right over the face of the spring jacks, and the suction of air will be sufficient to draw the dust out of the spring jacks. It is also expected that by adapting to it different forms of nozzles, we will be able to take up the dust between the cables on the back of the board, and, instead of discharging the dust in the room, it will be discharged either into a reservoir or receptacle for that purpose or out of the windows of the office. Just how well the machine will operate remains to be seen.

The sub-committee on the care of apparatus reported that out of a number of schemes which had been proposed for systematic cleaning of spring jacks by a draught of air, it seemed better to them to adopt that which sucked the air from the face of the spring jacks and retained the dust so collected in a receiving chamber or disposed of it out of doors, than any scheme which would simply blow the dust out of the spring jacks and into other portions of the apparatus or into the room.

The small strip of steel referred to in Mr. Scribner's remarks at the 1887 conference, quoted on page 328, was useful, but no real remedy existed to overcome the faults due to 'dirty jacks.'

The defects thus attributable were of various degrees of importance, from the complete break—which could be discovered—to the increased resistance which only slightly reduced the power of the transmitted speech. The faults also were intermittent. One might exist at one moment and at the next be removed by a current from a magneto generator burning out the obstruction.

The defects in the service caused by imperfect spring-jack contacts was one of the most important subjects considered by the next switchboard committee held on July 21, 22, and 23, 1891.[1] The proceedings of this committee were not printed, but are recorded in type-written volumes, a 'synopsis or abridgment' preceding the full report as in the previous conference.

From the synopsis I extract the following :—

Consideration of difficulties introduced with the multiple switchboard.

The present difficulty with the busy test is mainly a concomitant of the transition stage of a mixed system of metallic and earth completed circuits, and is the result of an attempt to minimise switchboard wiring. In such mixed circuits, with the present system of testing, false test signals often occur, due to the extensive employment of heavy currents in adjacent lines for many purposes, and, therefore, the trouble is, or shortly will be, a general one.

In this connection the sub-committee assigned to this matter reported that the existing device is unsatisfactory and why, and that a new form of test apparatus said to be capable of working in connection with mixed circuits had been submitted, but the committee was not yet prepared to report upon its merits.

The difficulty of maintaining clean spring jacks. The sub-committee on the care of apparatus presented its views in a preliminary way, vigorously stating the necessity of cleanliness in person and dress on the part of individuals, and also in all classes of apparatus, whether of the exchange station, the central office, or the conductor, and emphatically urging neatness in construction, thoroughness in repairs, and constant watchfulness. In pursuance of this report, which met with general concurrence, it was arranged that all members having suggestions regarding the general subject should present them to the chairman of the sub-committee.

In the more specific consideration of the especial difficulty of maintaining the efficiency of spring jacks, while the matter was referred back to the sub-committee for further consideration, the opinion of the general committee was unmistakable, that the only radical solution of the difficulty consists in the employment of a switchboard which does not involve the use of spring contacts in the main circuit.

In view of the very serious difficulties which occur in practice, by reason of unclean spring-jack contacts, this subject is regarded as being of the highest consequence, and, as has been stated, a sub-committee to take charge of it was appointed.

[1] The members were : E. J. Hall (chairman), T. D. Lockwood, C. E. Scribner, J. J. Carty, I. H. Farnham, and A. S. Hibbard. C. H. Wilson had also been nominated a member but was unable to be present.

Prior difficulties had been ingeniously overcome by minor alterations, but on this matter the committee were not prepared to consider palliatives. They looked for a 'radical solution of the difficulty,' and that involved a new design altogether—there must be 'no spring contacts in the main circuit.'

It will be recalled that the break in the jack was required in order that the indicator should be inoperative whilst the lines were engaged. If the indicators of engaged lines should be operated, there would obviously be a condition approaching chaos. The cutting out of the subscriber's indicator introduced in the standard board was a valuable feature carried on to the multiple board, where it was of even greater value in view of the possible distances which separated the indicators of the calling and the called subscriber. To effect it, however, a make and break in the spring jack was required. The possible defects from this make-and-break contact were of little consequence in the single jack of the standard switchboard or the few jacks of the earlier multiple boards. It was later, when the larger multiple boards had been for some time in operation, when the standard of service also was becoming higher and the demands of subscribers more exacting, that the jack troubles forced such attention upon them as to prompt the committee to express an opinion (which was practically a decision) that there should be no make-and-break spring contacts in the jacks.

Multiple switchboards had been used on the Law system, simple socket contacts being employed. The socket sufficed because, on the Law system, the attention of the operator was obtained through the telephone over a call wire, and since there were no indicators there was no need of a device to cut them out of circuit.

In the July 1891 conference the board on the new type was referred to as the socket system. In the attention given to it the committee organised rather than initiated the demand. This may be seen from the following extract of a letter from Mr. Barton, read at the opening of the proceedings :—

> Should an experimental switchboard be made on the socket system? If yes, then which of the, say, half a dozen available socket systems should be employed? The socket system leaves the ground on the line while subscribers are talking and holds the subscriber's drop from falling. After a selection of the most promising of the socket systems for further development and study, then it will be necessary to take some one central office and work out further details of its application to the actual conditions. This will take time and undoubtedly require further experimentation and involve discussions at subsequent meetings of the committee.

A plan proposed by Mr. Pickernell was also referred to, but no selection was made at this meeting in July. The committee resumed in October of the same year (1891), when a sub-committee submitted a skeleton specification for a 'branch terminal' switchboard. The name was thus returning to the descriptive form which was first used in the 1887 conference. The seventh point in Mr. Lockwood's paper at that conference was as follows :—

It has been frequently suggested that in a multiple switchboard single branches instead of loops may run to the several sections. The advantage of the branch compared with the loop is that spring contacts in the line circuit could be dispensed with. Is there sufficient promise in the idea to make it worth considering again?

And in the discussion Mr. Lockwood said :—

The multiple board operated in the way that the Law system is operated, or in the way that the Philadelphia and St. Louis companies are operated, is simply the multiple board operated as described in this seventh proposition, where, instead of looping to each section of the switchboard, the line is branched.

The multiple with contacts in the jacks was sometimes called the 'series' system, and by comparison the system without contacts might have been called the 'parallel.' Mr. Lockwood himself suggested that name in the course of the discussion in October 1891, when he said that in contradistinction to the ordinary or series form, that which had been variously denominated a socket board, or a parenthesis board, might better be called a parallel switchboard. He said :—

From the very beginning of the multiple switchboard use, and long before its use, the branching idea has been before the world to some extent. In fact, the very first patent ever granted broadly covering the multiple switchboard, that to Mr. Firman, was of that type. But the difficulty before us now did not come in there, and it was possible to leave all the branches normally open or discontinuous because the call came in on a separate line ; the call came in by the American district system. Later on C. C. Haskins and Mr. Wilson (here with us now) invented a kind of try signal for just such a switchboard ; and sometime subsequent Mr. Sabin of San Francisco got up a plan in which the circuit was to be closed, although a parallel switchboard with normally open branches coming into sockets was then contemplated ; and the difficulty at once appeared (which we have) that there was a line drop in there, that the line drop would come down when the second subscriber was called, and that it was a short circuit through the normal circuit of the line when the connection was on—that is, when two lines were connected together, if it was a grounded circuit, as all were then, the ground would still be left on unless some means were taken to get it off. Mr. Sabin adopted the means of closing a local circuit, which would trip the line drop and which in tripping it would take off the original ground. Of course that could be applied, if it could be applied at all, to metallic circuits. There are a great many electrical and mechanical difficulties in the case. It was not, I think, until a comparatively recent date that the present series of devices have been thought of, at least thought of systematically, with means to leave the normal circuit closed in some way, either through a magnetic resistance or through some other resistance, so as to keep the line drop from coming down and so as to ensure at the same time that the clearing-out drop shall come down whenever it is wanted. That, therefore, is the problem which is at present before us, and which is solved, or which appears to be solved, in one or two ways by the diagram and description before us. There are other ways which will come up later, but I wished to say so much to show the committee what the history of the idea has been.

The term adopted by the 1891 committee was, however, the 'branch terminal,' and this has continued to be used in the United States, but in Europe the system has generally been known under the name of 'branching,' which naturally and briefly describes the system in which the lines are simply branched into the jacks from continuous stems in contradistinction to one in which they are looped into and out of the jacks on the several sections.

The plan proposed in October 1891 included the main features of the system which was finally developed, but it included also a proposal which was speedily altered. The members of the switchboard committee were impressed with the growing cost of multiple boards due to the number of jacks, and were anxious to have a practical example of a horizontal (or 'flat') switchboard. It was realised that the absence of contacts in the jacks materially reduced the liability to troubles from dust, and it was thought that a practical test might therefore be given to the flat form. The policy of testing new plans in a single exchange under sympathetic conditions and careful observation was followed in this case. The exchange selected for the branching board was the Tremont Street office in Boston in the territory of Mr. Farnham, a member of the committee. In October 1891 the committee had recommended the adoption of the flat form. At its next meeting, in March 1892, Mr. Farnham reported that they had made up a model under working conditions, had submitted it to practical tests by operators, and given it such consideration as satisfied him that he did not wish to have the flat board in Tremont Street. The committee discussed the matter exhaustively and decided that it was not expedient to install the new board in a horizontal form, though it was advisable to continue experimental work in that direction. The first installation of the branching system was therefore of the vertical type, and the subsequent experiments with the horizontal form only demonstrated the more clearly that any advantages it might appear to have in first cost were more than negatived by the inconvenience and expense in working.

The branching system, as finally adopted, was gradually evolved. Ideas were submitted to, and discussed by, the committee. Some were rejected altogether. Others were regarded as offering promise of satisfactory development, and in these improvements were suggested.

There were two prominent exceptions to the general development of telephone systems on the earth circuit plan—the Telephone Company in Paris and the Post Office systems in Great Britain.

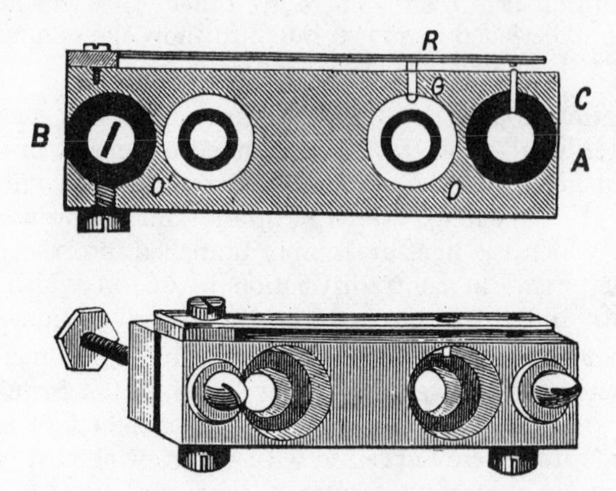

FIG. 112.—Double Jack-knife Switch, as used at Paris in 1882.

The latter were small, the principal exchange being at Newcastle-on-Tyne. The Paris system was larger. The use of metallic circuits was forced upon the Paris Telephone Company by the city authorities, whose extensive system of sewers provided an already existing spacious subway in which to lay cables as an alternative to the suspension of aerial wires. The cables first used had rubber insulated conductors encased in lead and were thus not available for very long distances, but the telephone was not then considered a long-distance instrument. Conversation was possible within the Paris limits.

A description of the Paris system appears in 'Bell's Electric Speaking Telephone' by Prescott (1884), this description being taken from the *Journal Télégraphique*, of which the date is not mentioned. It was published in the *Journal Télégraphique* of January 25, 1882 (vi. 26). Illustrations are given of the 'double jack-knife switch' (fig. 112) and of the 'peg' used therewith (fig. 113).

In general form the double jack-knife switch followed the Scribner design.[1] It was made of two metallic plates separated from each other by a thin piece of ebonite. The anterior plate is perforated by two orifices, which extend across the ebonite into the posterior plate where they are of a little smaller size.

The two plates each 'communicate with one of the wires of the double line.'

The pegs, which must be introduced in the orifices O and O′, are likewise intended for double wires. The part *a* is very much larger than the part *b*, and these two parts of the peg are carefully insulated from one another by means of the ebonite handle *c*.

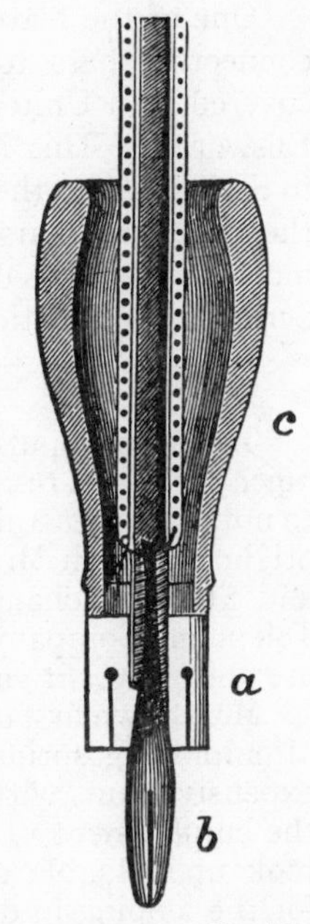

FIG. 113.—Double 'Peg,' as used at Paris in 1882.

The concentric form of double plug was very general with electrical appliances, but with wear it was quite possible to have one contact firm and good and the other imperfect. For telephone switchboard work it was early abandoned, the development being in the direction of contacts made between jack and plug at different positions in the length of the plug so that the loose fitting inevitable with wear was no impairment of any contact between plug and jack.

The indicator of the Paris system was connected permanently to one limb of the line and was in contact with the other limb through the spring. The insertion of the plug into the orifice on the right hand side of the jack raised the spring and thereby disconnected the indicator. It was customary to disconnect the indicator of the calling subscriber and to leave in circuit that of the called subscriber by inserting the second plug in the left hand orifice of the called subscriber's jack. The description says:—

It is proper to remark that the arrangement of the jack-knife switch number 15 leaves the indicator of the subscriber corresponding in the derived circuit of the two wires over which the communication is being held. [In other words, the indicator was 'in bridge.'] This arrangement does not weaken the sounds of the telephone, and we shall see, further on, that the circuit will even admit of two of these derivations without affecting the conversation.

The indicator was of the then general type and had a resistance of 200 ohms.

[1] Chapter xix. p. 226.

The earliest publication of the derived or bridged indicator is that in fig. 1 of Scribner's 1879 British patent, illustrated in fig. 62 and referred to in Chapters xiv. and xix. This fig. 62 is an outline illustration of the Universal switch. Scribner's patent is 'partly a communication from abroad' by George D. Clark, Milo G. Kellogg, and George B. Scott. The last mentioned was superintendent of the Gold and Stock Telegraph Company, and Mr. T. G. Ellsworth was the manager of that company's New York Central Office.

One of the features of the Scribner patent is the turning of the connecting bars to indicate their being in use. This feature is covered by a United States patent granted in 1880 to T. Gardner Ellsworth.[1] This is probably the portion communicated by Scott to Scribner, but the United States specification is strictly limited to the revolving bars, no mention being made of the clearing-out indicator connected in derivation which is fully described in the Scribner specification.

Prescott says:—

This idea of putting the subscribers' line to ground in the central office through a resistance, for the purpose of enabling the subscribers to notify the central office when they had finished their conversation, originated with Mr. R. G. Brown, then chief operator of the Gold and Stock Exchange, and now electrical engineer of the General Telephone Company, in Paris, where this ingenious device is still in use and proves a valuable auxiliary to the service.

Mr. Brown first conceived this idea in November 1878. The plan of employing spring jacks and other looping devices was not only expensive, but, with the kind of switchboards in use then, required the employment of two connecting bars—one for each line—which took up valuable space; and, looking about for some substitute for the looping-in devices, Mr. Brown ascertained that a telephone, or high-resistance electro-magnet, could be attached between a telephone line and the ground without perceptibly interfering with the transmission of speech, and hence adopted this plan with very excellent results.[2]

Haskins and Wilson, in their paper read before the American Electrical Society in 1879, say:—

When it is desirable to use but one line for both signalling and conversation . . . a polarised relay [is] placed in a derived circuit between line and ground. . . . This relay is wound to a resistance of 300 ohms, and remains permanently connected to the line, thus forming a slight escape when connection is made between two subscribers.

[1] U.S. specification, No. 226,991, dated April 27, 1880, (application filed December 30, 1879).

[2] *Bell's Electric Speaking Telephone*, 1884, p. 233.

> By reason of the extra currents set up by the telephonic impulses, the detriment to conversation attributable to this escape is not so great as might at first be supposed. Each electrical impulse from the induction coil of the telephone transmitter, on reaching the magnets of the polarised relay, meets an opposing momentary current, originating in the helices, and, as the telephone current is nearly instantaneous, this opposition is virtually a continuous one. Experiment at the centre of a telephone wire, three miles in length, showed that when the relays employed were wound to a resistance of 300 ohms, no difference was discernible whether its coils were inserted in the line or between line and ground in a derived circuit. With a higher resistance the comparison was in favour of the derived circuit, while with a relay of lower resistance the reverse was true.[1]

This use of the relay, as described by Haskins and Wilson, was the same as adopted in the Universal switch and described by Scribner in his 1879 British patent.

It seems probable that the indicator employed by Mr. Brown in Paris must have had less resistance and impedance and was therefore less effective for the purpose than the relay employed on the Universal switch to operate a separate indicator.

The indicator used in the British Post Office system differed from that of any other telephone administration. The indicators of both lines were directly in the circuit, with results which are thus described by Preece and Maier:—

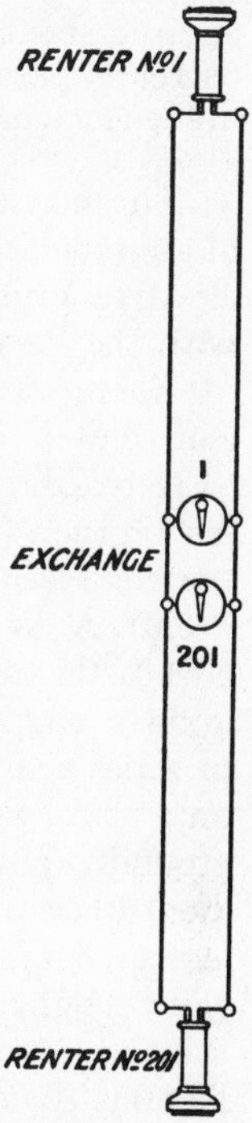

> It is found that on lines of any considerable length the disturbing influence caused by the self-induction of the indicators seriously impairs the efficiency of the circuit, and the plan of working with the indicators in 'bridge' has therefore been resorted to at several exchanges. The two indicators are by this means got out of the direct line altogether and placed across as shown in fig. [114]. The indicator for this system is similar to that already described except that the electro-magnet is lengthened and wound to a high resistance (1000 ohms). The coils also have an iron casing surrounding each, to increase their electro-magnetic inertia.[2]

Fig. 114.—British Post Office Bridged Indicators.

The bridge method of connection was also described by Preece

[1] *Journal of the American Electrical Society*, 1880, pp. 51–2.
[2] *The Telephone*, 1889, p. 223.

in a paper on 'Long Distance Telephony' at the Society of Telegraph Engineers (I.E.E.) in London on May 13, 1886. He remarked that—

while these telephones in bridge are perfectly susceptible to action from a steady current, when rapid reversals (which the telephone currents are) are sent, then the indicators in these bridges, being specially constructed of a large mass of iron—high resistance with an iron sheath, in fact—contain so much self-induction that the currents passing through from Newcastle to Stockton have no influence whatever on the telephones at the intermediate places.[1]

The increase in resistance from 200 to 1000 ohms, and the addition of an iron casing to increase the retardation, made a very much more effective bridging indicator than the one described in connection with the Paris exchange.

There is some reason to believe that Oliver Heaviside contributed to this important practical development. In his 'Electro-Magnetic Theory,' 1893,[2] Mr. Oliver Heaviside claimed for his brother (Mr. A. W. Heaviside) the first use of the bridging or parallel method.

Mr. A. W. Heaviside was in charge of the Post Office system at Newcastle-on-Tyne, where there were a number of instruments upon a single circuit. The deleterious effect of the instruments in series soon set a limit to the admissible number of instruments on a circuit and to the length of the circuit, especially when underground wires were included. Mr. A. W. Heaviside found by experiment that an immense improvement was made by putting the instruments across the line as shunts or bridges. This method was brought to Oliver Heaviside's knowledge, and he was asked for the theoretical explanation. At what period he does not say, but he adds that it was found 'the bridges caused a weakening of the intensity of the speech received when many bridges were passed,' and 'to prevent this weakening becoming inconveniently great the intermediate call instruments in bridge were purposely made to have considerable resistance and inductance. . . . This tended to prevent the currents passing along the line from entering the shunts, and especially so as regards the currents of high frequency, and allowed them to be transmitted in greater magnitude.'

These were the features which distinguished the Post Office bridge indicators described by Preece and Maier. Newcastle was the principal telephonic centre of the Post Office. Mr. A. W. Heaviside discussed these problems with his brother Oliver, and it is perhaps but a reasonable inference that for this very practical

[1] *Journal of the Society of Telegraph Engineers*, xv. 289.
[2] Vol. i. p. 434.

advance we are indebted to an independent worker in abstract science.

When writing on the subject, it is probable that Oliver Heaviside was not familiar with what had been done previously in the way of bridge connections. As we have seen from the Universal switch, from Scribner's 1879 patent, and from Prescott's claim for R. G. Brown of Paris, apparatus had been connected in derived circuit or bridge probably before there were any telephones at Newcastle-on-Tyne.

The clearing-out indicators at Manchester were connected in bridge in 1880, and so continued for several years.[1] That electro-magnets impeded the passage of telephonic currents was well understood in 1879, as may be seen from the remarks on page 6 (line 50) of the complete British specification of Scribner (No. 4903 of 1879): 'The earth connection of the coupling bar being through the coils of a magnet in no way interferes with telephonic conversation.'[2] The electro-magnet was 'teed on' (or in bridge) to earth from a single wire. Diagrammatically, to one not familiar with telephonic currents, it might appear to be a short circuit, and it is perhaps for this reason that the patentee of an early invention in a new art explained that such a derived connection to earth does not interfere with telephonic conversation because it is through the coils of an electro-magnet, clearly indicating the knowledge that the electro-magnet offers so great impedance to telephonic currents that they will not be diverted by the bridge but will continue in the main circuit comparatively unimpaired.

Whilst this bridging method of connection was old for the clearing-out indicator and the principle of its use would render it equally available for the connection of numerous instruments on a line, there is no evidence that it was so used in the United States. The series method was found to be unsatisfactory, and remedies were suggested, such as connecting the coils of the electro-magnet in parallel,[3] or shunting the coils with a condenser.[4] Both these methods were referred to by Mr. Lockwood in 1887,[5] but there is no reference to the bridging method until the paper entitled 'The New Era in Telephony' by Hibbard, Carty, and Pickernell, read before the National Telephone Exchange Association in September 1889,[6]

[1] *The Telephone*, Preece and Maier, p. 350; *Telephone Handbook*, Poole, first edition, p. 169.

[2] See also Prescott's reference to effect of magnets in telephones with multiple diaphragms, and the application of condensers when many sets of apparatus are placed in one circuit (*The Speaking Telephone*, by George B. Prescott, 1878, pp. 23 and 31).

[3] Sargent, *National Telephone Exchange Association Report*, 1883, p. 52.

[4] Ross, *ibid.*

[5] *National Telephone Exchange Association Report*, 1887, p. 57.

[6] *Ibid.* 1889, p. 39.

where the method is recommended for the connection of switchboard indicators and for the connection of two or more subscribers on the same metallic circuit. Preece, in the paper quoted, in 1886, and 'The Telephone' by Preece and Maier, published in 1889, describe the Post Office bridging indicator, and Preece, in 1886, as well as Oliver Heaviside in his 'Electro-Magnetic Theory' (1893), said that the bridging method was adopted in the Newcastle district for connecting a number of instruments on one line. In the Newcastle area underground work was general, and the limitations of number or distance would therefore be more quickly reached. The Newcastle district was consequently a likely locality for the practical application of the bridging method, and on the theoretical side the benefit to be derived from the association of the brothers Heaviside seems obvious. The description of the Post Office bridging indicator explains for the first time the theoretical advantages of the increased electro-magnetic inertia and resistance.

The iron shield around the indicator coils had also the effect of reducing the induction between neighbouring indicators when not connected in bridge. It was for the latter purpose that they were added to a strip of clearing-out drops by the Northern District Telephone Company competing with the Post Office in the Newcastle and Sunderland district. A supply of such indicators was ordered by that company from the Western Electric Company, whose further investigations and experiments demonstrated that the best effects were obtained by a single coil completely surrounded by an iron casing connected to the core. The result was the production of the indicator known in the United States as the 'Warner tubular drop' in contradistinction to the 'Warner drop' as first applied to the standard switchboard.[1]

The accompanying illustrations of the tubular indicator (figs. 115, 116, 117, and 118) are reproduced from British patent specification No. 9571 of 1889.[2]

The specification relates that the electro-magnet is formed 'of a central core of soft iron, a helix of wire, and a tube also of soft iron. The external tube and the central core are magnetically connected and together attract the armature, to which is attached the pivot arm or catch releasing the indicator shutter. This

[1] These drops were the design of Mr. J. C. Warner, who was born in London in 1822 and was employed in making electrical apparatus in the workshop in which instruments were made for Cooke and Wheatstone. Subsequently he made some apparatus for experimental use by Morse on the line between Baltimore and Washington. Mr. Warner entered the service of the Western Electric Company before the telephonic era. He installed the multiple switchboard in Melbourne in 1885, visiting London on his return.

[2] Date of application, June 8, 1889; U.S. equivalent, No. 477,616, dated June 21, 1892 (application filed, June 17, 1889).

arrangement increases the sensitiveness of the indicator and reduces the electrical induction between neighbouring indicators.' [1]

No mention is made of its special suitability for connection in derivation, but it was immediately adopted as a clearing-out drop,

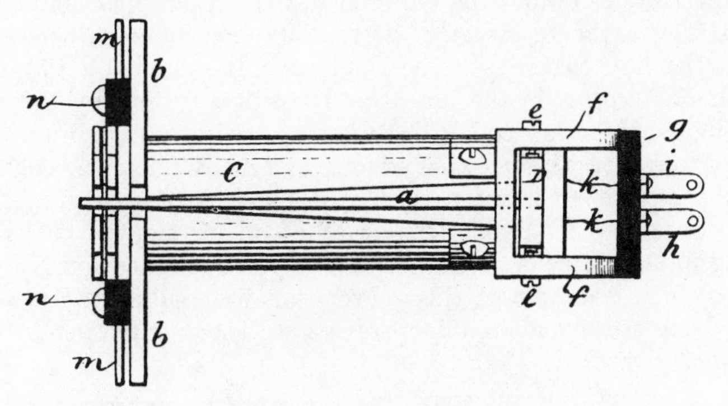

FIG. 115.—Tubular Indicator (plan).

being connected between line and earth, this feature of the Universal switch of 1878 or 1879 being thus restored in 1890.

When, therefore, the switchboard committee assembled in 1891 with the selection of a branching system as one of its principal

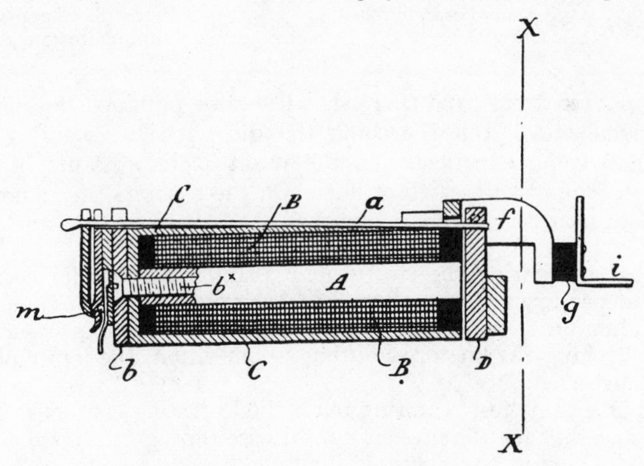

FIG. 116.—Tubular Indicator (section).

objects, a type of indicator especially suitable for bridging purposes was already in practical use. There was no question of the suitability of the Warner tubular drop for permanent connection in

[1] Date of application, June 8, 1889; U.S. equivalent, No. 477,616, dated June 21, 1892 (application filed June 17, 1889).

bridge across the two limbs of the line. But for multiple switchboard use two principal problems remained :—

(1) How to prevent the drops falling when two lines were connected ; and

(2) How to obtain the required test of an engaged line.

Hitherto the test system derived its efficacy from connection with the line circuit ; in the single circuit system by the direct earth of the line, in the mixed or transition system by the earth through retardation coil. The masses of wires through a switchboard were already somewhat serious, and many telephone engineers were reluctant to increase them ; so much so that a proposition was considered to simplify the switchboard for metallic circuits by using induction coils or translators—British specification No. 7465 (1890) is an example of this. Proposals were submitted in which the drops were rendered inoperative in a two-wire system by a

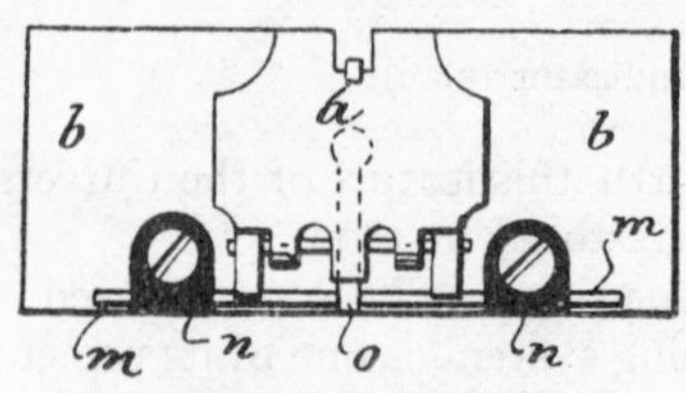

FIG. 117.—Tubular Indicator (front).

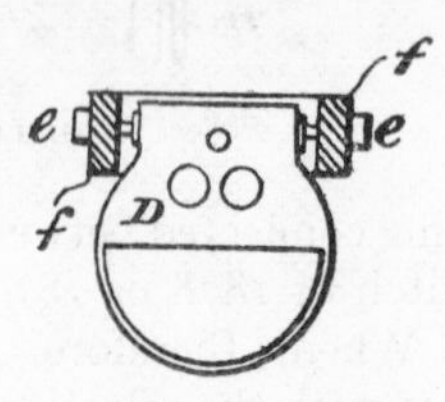

FIG. 118.—Tubular Indicator (back).

margin of current, and the busy test also obtained without an additional wire. But the committee did not like 'margins,' and had had some experience of the unsatisfactory nature of busy tests in contact with earth or line. On these points the committee (I quote from the summary) reached these conclusions :—

That comparing the two general plans suggested for securing the independence of the 'line' and 'clearing-out' drops, i.e. that depending on a margin of current, and that depending upon positively locking the line drop during a connection, the latter plan is to be preferred.

That a busy test requiring but a single testing contact is desirable, and that for the present a third wire through the switchboard for each circuit to secure a thoroughly reliable busy test is indispensable.

A double cord connection switchboard is recommended.

The views of the committee concerning 'Branch Terminal Switchboards' having thus been expressed, the subject was referred to a sub-committee consisting of Messrs. Scribner, Hibbard, and Lockwood, which sub-committee, after examining a number of plans of securing the independence of the 'clearing out' and 'call signal' annunciators, and also of busy test apparatus, decided to recommend a plan presented by the Western Electric Company, which combined a local and third wire busy test with a positive lock device for the line drop, and which also promised to produce an additional feature, i.e. the automatic restoration of the call drop.

The promise was amply fulfilled, and the branching system passed by the committee at their meeting in March 1892 was intro-

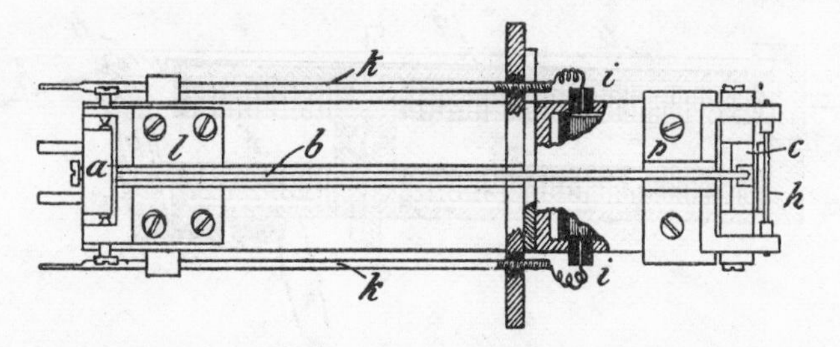

FIG. 119.—Branching Indicator (plan view from above).

duced to practical service in a form which subsequently required very little modification in essential points. The jack as submitted by the manufacturers had a stud for the test. This was objected to by members of the committee, who generally preferred a circular orifice such as had been used with the series-jack test—a wise decision, in that it offered a more suitable testing surface and

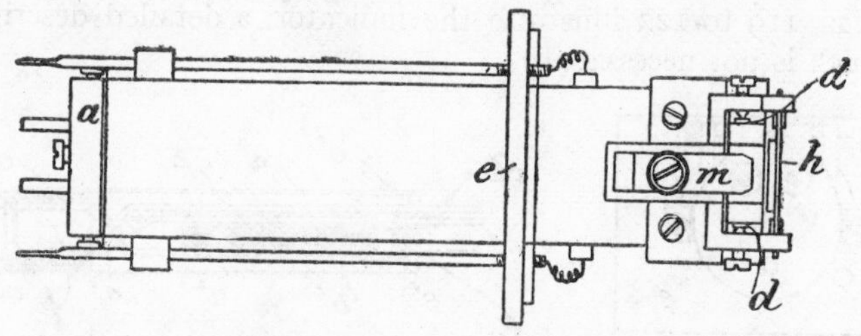

FIG. 120.—Branching Indicator (plan view from below).

involved no change to the operators in the method of testing for an engaged line.

The principle of the system is explained in British patent specification No. 4428, March 5, 1892.[1]

Quoting from the British specification the objects of the invention are :—

first, to provide circuits and mechanism whereby the individual

[1] Equivalent U.S. specification, No. 563,250.

annunciator of a line shall be automatically reset or replaced by the operation of making a connection to the line, and whereby it shall be rendered unresponsive to signalling currents during said connection; second, to provide an annunciator to respond to the signal for disconnection, and means for automatically resetting the same; third, to provide suitable means for testing at any board to determine whether a line is already in use or not; and, fourth, to

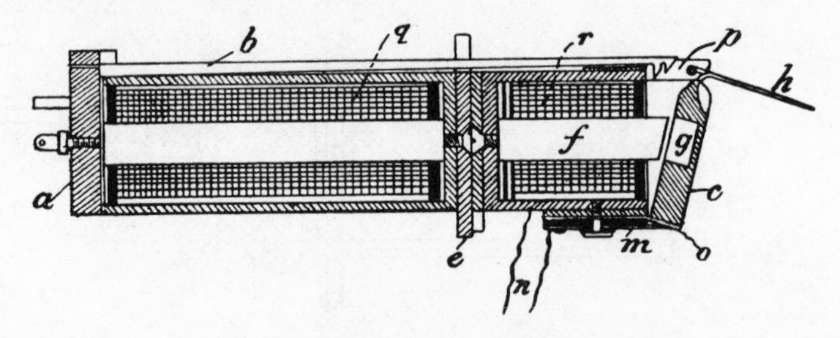

FIG. 121.—Branching Indicator (section).

avoid all connections and branches common to the different lines of the exchange.

But this specification did not include the final form of jack and indicator which were put into commercial use. These will be found in British specification No. 17,160, September 26, 1892.[1]

Figs. 119 to 122 illustrate the indicator, a detailed description of which is not necessary.

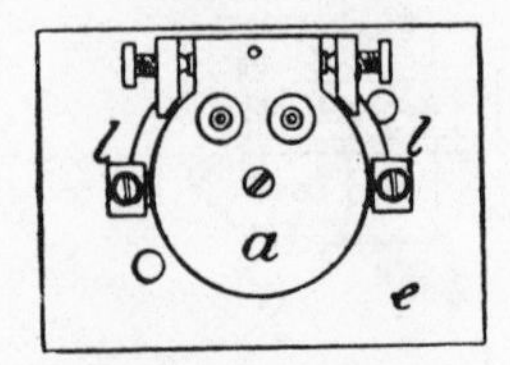

FIG. 122.—Branching Indicator (rear view).

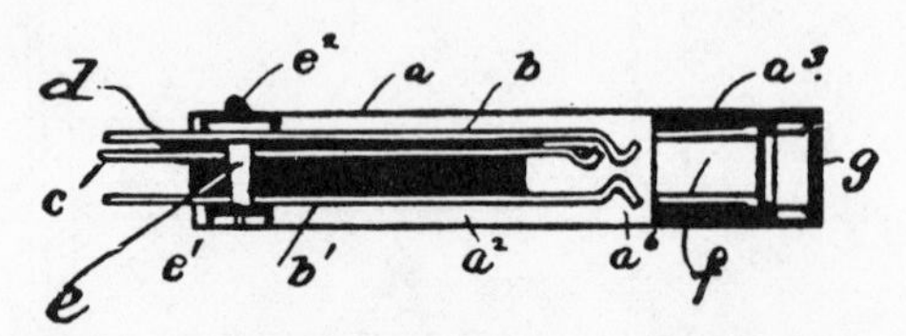

FIG. 123.—Branching Spring Jack (section).

Fig. 123 is a sectional view of the spring jack in which the short spring is connected with one limb of the circuit, the bush with the other limb. The two springs of equal length are the test and restoring circuit, normally broken but completed on the insertion of a plug.

Fig. 124 illustrates the plug and its engagement with corresponding parts of the jack.

[1] Equivalent U.S. specification, No. 533,148.

Fig. 125 shows the circuits in detail when two subscribers are connected.

The first switchboard on the branching system was installed, as already stated, in the Tremont Street office in Boston. The first in Great Britain was at Hull, the next at the Avenue Exchange, London. The first on the European continent is believed to have been at Christiania, but in that case with a smaller jack of skeleton construction.

The introduction of the branching system involved no change in the method of operation by the subscriber, who continued to turn a magneto handle for the purpose of calling the exchange and for notifying the completion of conversation. It involved no change in the methods of the operators except that it rendered unnecessary a part of their previous work, because the calling and clearing indicators were automatically restored.

Apart from the saving of actual work, the operation of making a connection was much facilitated by the certainty that the correct jack had been selected. On the older system the relationship of the jack to the indicator had to be carefully studied, while with the branching system the fact that the indicator was restored was a proof that the correct jack had been plugged into. The automatic restoration feature was in some quarters considered of such importance as to give the system the name of the 'self-restoring system'; but, as has been shown already, the automatic restoration was a development not originally contemplated. It was a great improvement in the nature of a bonus. The main feature was the jack without break—hence the 'branching system,' the introduction of which was advocated at the meeting of the switchboard committee in July 1891. The spirit animating the members and the method governing their deliberations have been referred to in the preceding chapter, so that only the practical result to which their attention was mainly directed has been recorded in this chapter. But the opening remarks of the chairman, addressed as they were to colleagues equally well informed as to progress and tendencies, expressed moreover without any regard to publication or any wider audience, afford valuable contemporary evidence of the progress between 1887 and 1891. Mr. Hall's remarks apply, of course, to the United States companies, but with minor differences they will illustrate the progress throughout the world. Mr. Hall said:—

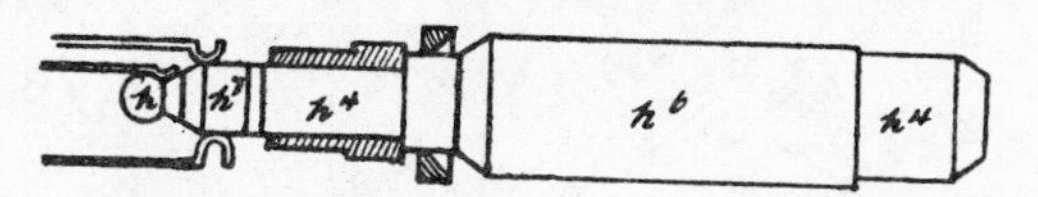

FIG. 124.—Branching Plug and Corresponding Jack-Springs.

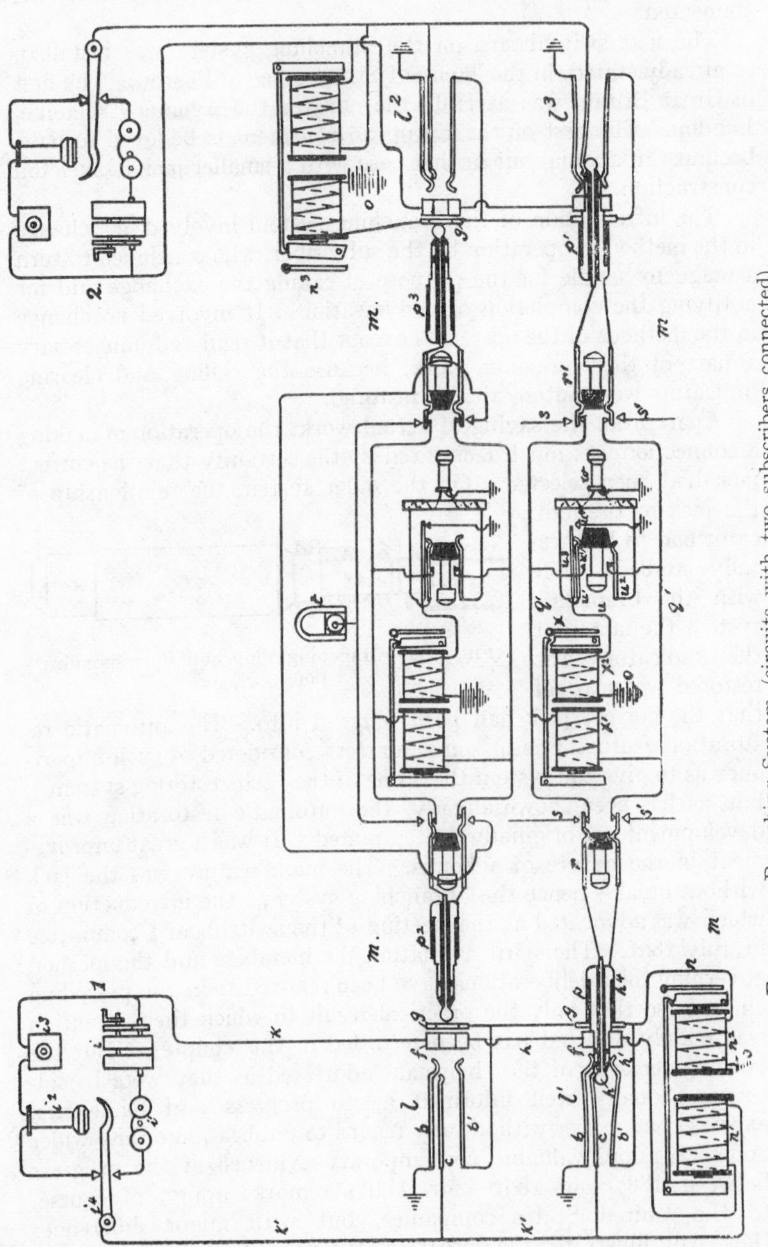

Fig. 125.—Branching System (circuits with two subscribers connected).

Since the switchboard conference held in 1887 there has been no organised work in the direction of improving and standardising our exchange apparatus, although there have been several informal conferences held and much good work has been done by companies and individuals. The service generally is on a much higher plane than it was at that time. New difficulties, however, have since presented themselves, and the very impetus which the work of that conference gave has resulted in the development of much that needs study, revision, improvement and, perhaps, repression.

In nothing has progress been more marked than in the housing of our central offices. Very many of the large offices are now in commodious and fire-proof buildings owned by the company, and a number of new buildings are being constructed. Metallic circuit service is no longer confined to the Long Distance Company, the underground problem has lost many of its terrors, and multiple switchboards are found in all the offices large enough to use them, and in some that are small enough not to.

To bring the service generally up to the high standard made possible by our present knowledge calls for the expenditure of vast sums of money, and this committee will undoubtedly do much to point out the ways in which progress and economy may go hand in hand.

While our special work is the switchboard, its functions are so combined with those of all of the other appliances that it cannot well be considered alone, and we must necessarily take account of all the elements that go to make up exchange apparatus.

There has been, I think, a tendency to glorify the switchboard beyond its deserts. It is the central and most striking feature of the exchange, but it is neither the exchange nor the most important part of it. It is the largest single piece of our exchange machinery, but the aggregate of the subscriber's apparatus, cable terminals, distributing and protecting devices, exceeds the switchboard in bulk, importance, and cost—in all but the largest exchanges.

From the engineering standpoint the problem of the switchboard for a large exchange is undoubtedly the most important that has to be solved; but I am talking now about the country at large. I have in mind some exchanges where most of the money and work have gone into the central office, and the service suffering from neglect of other matters; while other exchanges which to-day with their old board could give good service are failing to do it because the managers have asked for new switchboards and defer all improvements until they can get them.

In my judgment the one great need of our exchanges to-day, and one that could be supplied almost without cost, is *cleanliness*. I use the word in a broad sense—clean offices, clean lines, clean switchboards, clean batteries, clean terminals, and clean operating.

It seems to me that this committee might add materially to the good work which I know it will do in selecting and designing apparatus if it could devise some way of impressing on every officer and employee of our various companies the fact that no piece of electrical machinery ever has been or ever can be devised that will work well unless it is kept clean and handled properly.

The general object for which this committee was appointed is to improve our exchange service by selecting what is best from the many methods of operation and forms of apparatus in use or suggested for use. To do this—in a way to make the results available—our plans must be laid out thoroughly and systematically and our conclusions embodied in specific recommendations. It seems to me that our work naturally divides as follows :

1. To formulate the problems connected with the switchboard and other telephonic apparatus relating to exchange service.

2. To determine how far these problems are satisfactorily solved by existing apparatus and methods.

3. To pass upon the merits of new devices and methods and determine to what extent, if any, they should supplement or supersede existing ones.

4. To prepare standard specifications for all telephonic apparatus.

5. To recommend such action as may be necessary to secure the general use of standard apparatus and uniformity in methods.

6. To suggest such lines of experiment as it may seem advisable to follow up.

The subject is a large one, its subdivisions are numerous, and many of them involve technical details that cannot be considered to advantage in a general meeting. We can, however, arrange, if it be thought best, for sub-committees, either standing or special, to which matters requiring prolonged and minute study can be referred, and the reports of these sub-committees be passed on at our general meetings.

The introduction of the branching system carried with it the employment of the bridging magneto, to which Mr. Carty had devoted considerable attention ; and it also involved an improvement in distributing board apparatus, thus outlined by Mr. Scribner :—

The switchboard wires are led from a general distributing board in the usual manner. Two sets of terminals are, however, provided at the inner end of this distributing board instead of one, the wires leading to the answering jacks and drops being connected to the second set. The two sets are placed adjacent to each other so that when a line is to be operated at its regular section the connection from its terminals to the answering jack terminals may be readily made. When desired, however, the line terminals may at this board be connected to any answering jack terminals desired, and the work of intermediate distribution be thus accomplished, making the use of a second, or intermediate distributing board, unnecessary.

The branching system switchboard was in itself an important production, but it was only one example of the trend of thought of the time. When the telephone service was a novelty and something of a mystery, some latitude might be allowed in the certainty of its operation. But as the use extended and the luxury became a necessity, doubts as to whether the service was usable or would be free from interruption became unbearable. So far as human forethought and care could provide, the service must be reliable. The branching system established that reliability within the switch board. Careful construction, thorough inspection, and daily tests of the outside lines as well as the inside apparatus contributed at this period to produce that higher standard of service which recognised that every subscriber was entitled to talk clearly with any other disengaged subscriber without delay and without interruption.

The power required for locking and restoring the indicators was obtained by the use of accumulators, and thus the telephone engineer became familiar with stronger electrical currents in his own work. In the October 1891 meeting of the switchboard committee it was related that accumulators, charged direct from the Edison Company mains, were used in Chicago to supply between sixty and seventy operators' transmitters.

These were stepping-stones to further advances which were shortly to follow. Telephone engineering was growing with the growth in the numbers of stations and also in appreciation of the further advantages which might be obtained by reliance on a great single source of energy instead of a multitude of scattered batteries and generators.

CHAPTER XXVI

THE COMMON BATTERY SYSTEM

It is in the report of the meeting of the switchboard committee in May 1892 that we first find reference of a practical kind to 'Common Battery Exchange Systems.' The term 'central battery' has very generally been applied to the system, but the original term is the more correct, for a battery might be central without being common to all the subscribers. Mr. Lockwood referred to one such system in his remarks to the committee. This was the plan adopted by the London and Globe Telephone Company, who were using the original Hunnings transmitter. Mr. Lockwood said that when the Hunnings transmitter was first brought over to the United States the inventor specified that it should be used directly in the circuit with one cell of Daniell battery for every mile of line. Experiments were made between Boston and Providence. The distance is forty-four miles, and forty-four cells were accordingly inserted. These were gradually reduced to one when the transmission was found to be very good. Mr. Lockwood continued :—

In England the Globe Telephone Company worked with nothing else than the Hunnings transmitter with no induction coil, worked on a direct circuit also, and they made a great use also in connection with it—and they found that was one of the principal advantages—of the visual signal in the main line to indicate the condition of the line. They always had a galvanometer signal in their line when the line was working.

Mr. George Lee Anders was the electrician of the London and Globe Telephone Company, a reminder of which company's existence is still to be seen in the words 'Telephone Building' over the doorway of No. 31 Queen Victoria Street, London, where their offices were situated.

The use of the Hunnings transmitter without an induction coil was to them not so much a matter of choice as of compulsion. The induction coil, in combination with a transmitter, was protected by the Edison patent owned by the United Telephone Company, and the Globe Company was consequently compelled to carry on its service with direct battery working. The method in use for indicating the close of conversation has not been, so far as I can discover, a matter of record, but I think the galvanometer plan was not in operation on the 'duplicate' switchboard—an early name for the multiple—supplied to the London and Globe Company. But that it was some form of automatic clearing signal is evident from the instructions to the subscribers in the first issue of their 'Directory,' dated December 1883, in which it is stated 'Hanging up the telephones indicates that conversation is finished; therefore do not hang up your telephones until you are ready to be disconnected.'

To explain the use of the plural number, it should be said that the Hunnings transmitter supplied was of a portable form. The instructions state that 'The transmitter is the instrument with the red band.' The hand form was adopted as a means of preventing the packing of the carbon. Every time the instrument was used a movement of the carbon particles would result. In the Anders patent subsequently referred to the patentee says 'the transmitter which I employ is one of the type of what is known as Hunnings' transmitter—that is to say, one, the handling of which assures its adjustment.'

Poole refers[1] to a patent of Anders of 1882 to illustrate the difference between 'central' and 'common' battery. Anders had a separate talking battery between each pair of cords, and this, as Poole says, was a *central* battery, but not a *common* battery.

Poole also says that Anders' patent was the earliest suggestion of common battery or central energy working; but, in fact, an earlier is to be found. The United States patent No. 243,165, dated June 21, 1881, was entitled 'Centralising individual batteries of a telephone exchange,' and was granted to C. E. Scribner. Fig. 126 is the first figure of the patent. In the course of his specification he says:—

Heretofore each subscriber has been provided with one or two elements of battery in the local circuit of his transmitter and the primary of his induction coil. By the use of my invention many or all of the subscribers of a system may use the same battery for their primary currents.[2]

He relates the difficulties which hitherto had prevented the use of the transmitters of several subscribers in the circuit of a single battery, and proceeds:—

I overcome these practical difficulties by means of a Wheat-

[1] *The Practical Telephone Handbook*, fifth edition, 1912, p. 208.
[2] U.S. specification, No. 243,165, June 21, 1881 (application filed, April 4, 1881).

stone bridge placed at each subscriber's station in the circuit of the common battery. In one of the four arms of the bridge I place a battery transmitter, which may be of the form known as the 'Edison transmitter' or a microphone. The other arms of the bridge are balanced to the resistance of the transmitter by inserting resistance coils. In the cross wire of the Wheatstone bridge is placed the primary of the induction coil, and, in consequence of the bridge being balanced, there will be no current passing through the cross wire of the bridge and primary of induction coil when the system is not in use. In addition to the circuit wire to the subscribers, I run to each subscriber an individual wire for talking.

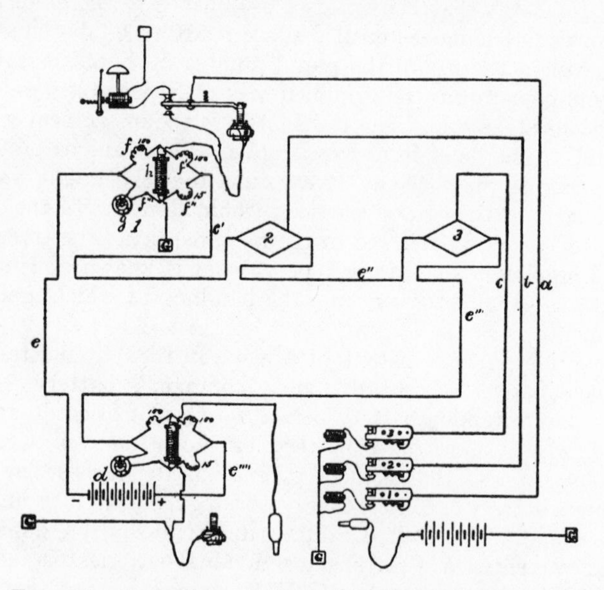

FIG. 126.—Centralising Individual Batteries (Scribner, 1881).

This individual wire passes through a switch to the call bell and to ground when the telephone receiver is on the hook, or through the receiving telephone and secondary of the induction coil when the receiving telephone is off from the hook. Speaking into the transmitter varies the resistance of that arm of the bridge in which the transmitter is inserted. This variation of the resistance of the one arm of the bridge causes the main battery current to flow through the cross wire of the bridge and through the primary of induction coil in vibrations corresponding to the diaphragm of the transmitter, and by induction these vibrations are communicated to the individual wire, and thence to the distant station to which the individual wire may be connected.[1]

[1] U.S. specification, No. 243,165, June 21, 1881 (application filed April 4, 1881).

Anders's British patent, No. 749 of 1882, includes a modification of the Hunnings transmitter, a novel battery, and an exchange system. There are two separate systems of battery. One is common to the whole of the subscribers for signalling purposes, operating the individual indicators to call and to disconnect, the simultaneous falling of the individual indicators of two connected lines being the clearing signal. The other, or speaking batteries, are central but not common. There is a battery in each pair of cords with which connections are made. The specification thus describes a complete method of centralised battery working. Taking the receiver off the hook sends a call; putting it on again sends a clearing signal. The method by which this is accomplished is that known as a fleeting contact,' the receiver switch in its passage from its lower to its higher position of rest or *vice versa* coming in contact with an earthed spring. Fleeting contacts have been demonstrated to be unsatisfactory in telephonic operations though there was no considerable experience at that date, and Anders's patent is of interest as an example of centralised energy together with simplicity of sub-station apparatus. These attempts of Scribner in 1881 and Anders in 1882 to remove the battery from the subscribers' station to the central office, whilst of historic interest, cannot be regarded as having any direct bearing on the development of common battery working as now practised, since the combination of the talking current with an automatic signal system could only result from extensive and long-continued experience of the requirements of telephone service. The circuit wire of Scribner may be regarded as economically impracticable, and the system of Anders lacked any supervisory signals.

One of the principal advantages of the common battery system as finally developed was the provision of automatic signals at the central office controlled by the respective receivers at the sub-stations and varied according to whether the receiver of either station was on or off its suspension hook. Mention should therefore be made here of the system adopted by the British Post Office, in which automatic calling signals and a single clearing signal (through the simultaneous operation of the two line indicators) were obtained, though not with either central or common battery. The Gower-Bell instrument was originally used by the Post Office. This consisted of a pencil microphone and a large Bell receiver fixed in the same case. The sounds were conducted from the receiver to the ears of the listener through two flexible tubes. When out of use the tubes rested in automatic switches—one on each side of the case. By removing the tube on the user's left hand a signal was operated at the central office. The speaking battery was brought into operation by removing the tube on the right.

When the tubes were again placed upon their respective switches the clearing signal was given automatically. The method of operation by the subscriber was therefore very similar to that subsequently developed in the common battery system, but, since local batteries were used both for calling and speaking, it was so uneconomical in construction and maintenance as not to have been adopted by any other administration. Poole says that

> although a very perfect system for a small exchange, its advantages are dearly bought, as the frequent attention necessary to the Daniell batteries and the constant waste of material constitute very serious drawbacks.[1]

Like the branching system in its operating features, the common battery system, with all the operating advantages which it afforded, was a growth. Its origin was the effort to obtain an economical and reliable source of current for working transmitters. The Leclanché quite early established itself as the most suitable all-round battery for telephone service. Compared with other forms it was cleanly and required a minimum of attention. Telephonic conversations, in the main, were short and not so frequent but that in general the intervals between conversations were of sufficient duration to allow for battery recuperation. But with the development of long-distance service the Leclanché was found less suitable. In a prolonged conversation the commencement might be satisfactory and the conclusion inaudible.

Blake transmitters were in general use for local service. The Blake transmitter was excellent with one cell of battery, but was liable to 'boil' with more. One cell did not suffice for some long-distance work. It was consequently found necessary in the United States to supply special instruments with carbon-granule transmitters for use on long-distance lines. With these sets batteries of the Fuller type were generally furnished. The maintenance and inspection of such batteries were costly to the companies as well as to some extent annoying to subscribers. The advantages which would result from getting rid of local batteries altogether were emphasised at the July 1891 meeting of the switchboard committee, particularly by Messrs. Hibbard and Carty.

Accumulators were considered and dynamos were experimented with, as appears from the reports of the committee, October 1891.

It was at the meeting of this committee held in March 1892 that the final approval was given to the details of the branching

[1] *The Practical Telephone Handbook*, second edition, 1895, p. 167.

system. Two months later, at the meeting on May 20, 1892,[1] the Chairman (Mr. Hall) said :—

We have for consideration this morning the invention of Mr. Hayes, the purpose of which is to concentrate all battery at the central office. It is not necessary perhaps to say how important such an invention would be for the business, because we all realise what an enormous part of our operating expense is due to the maintenance of batteries at subscribers' stations. If we can get rid of that, we get rid practically of 90 per cent. of our inspection and a very large proportion of course of all our difficulties, and if that enables us also to reduce largely our investment in plant, why, it is a saving still further in the right direction.

Mr. Hayes's paper was entitled 'Common Battery Exchange System,' and its purpose is indicated in the opening paragraph :—

This system contemplates the use of a large central battery placed at the telephone exchange, which furnishes all the current necessary for the operation of all the transmitters whose circuits terminate at that exchange. In general design it is similar to the common battery system which is now quite generally in use in large exchanges for furnishing the current for operators' transmitters at the switchboard.[2] In that system, however, the induction coil is placed near the transmitter, and the telephone receiver is in the secondary circuit of the induction coil, as is the common custom. In the proposed system, which we have called the 'Common Battery Exchange System,' the induction coil, instead of being near the transmitter, i.e. at the subscriber's station, is placed near the battery, at the exchange. The telephone receiver is placed in the primary circuit with the transmitter, and is operated by the current induced by the secondary into the primary circuit. In other words, the 'Common Battery Exchange System' uses a long primary circuit extending from the subscriber's station, the secondary circuit being to line, with a receiver in the primary circuit.

The battery proposed to be used consists of eight cells of storage battery arranged in series, and as many in multiple as may be found necessary to prevent cross-talk, probably five, or in all a battery of forty cells, having an electro-motive force of sixteen volts. . . . The service given to the several subscribers in an exchange will not be absolutely the same, but will vary slightly with the distance of the subscriber's station from the exchange. . . .

[1] The members of the Committee were E. J. Hall, T. D. Lockwood, C. E. Scribner, I. H. Farnham, J. J. Carty, A. S. Hibbard, F. A. Pickernell, H. V. Hayes (present) and C. H. Wilson (absent).

[2] The system referred to was devised by Mr. Carty and is the subject of U.S. patent No. 518,392, April 17, 1894, application filed April 14, 1891. The transmitters in the Chicago exchange were also operated through storage batteries. See Mr. Wilson's remarks in *National Telephone Exchange Association Report*, 1890, p. 88.

The subscriber's apparatus consists of a transmitter of the solid back type, a receiver having a moderately low resistance, the usual ringer, a gravity switch, and a condenser. . . . When the telephone is taken from the hook the circuit is closed and the condenser is cut out, thereby causing the drop at the central office to fall. . . .

The common battery system seems particularly well suited to the long distance work, as by it the expense of battery maintenance and inspection can be greatly reduced. . . .

The clearing out is accomplished by means of a galvanometer signal which is positive and sure in its operation and is sufficiently compact. . . . The advantages of this system are the decrease in cost of maintenance due to the concentration of the batteries, the freedom from the necessity of frequent inspection, cheapening of subscriber's outfit, uniform excellence of service, greater rapidity of operator's service, instantaneous detection of defective local circuit, and the fact that a poor local circuit will not unbalance a long line circuit.

After analysing the respective costs of local and centralised batteries, Mr. Hayes again referred to the economy in inspection, having regard to the fact that the apparatus would be 'reduced to the simplest instruments ever put out for telephone work.'

It would seem that an annual inspection would be all that would be required with a consequent reduction of one-fifth to one-tenth the expense.

In the course of this paper it was stated that the common battery system was developed first for 'speaking tube' purposes, an application of the telephone more generally known in Europe as 'Domestic' or 'Industrial' services. It indicates an installation suited to a house, factory or institution, by means of which different departments may intercommunicate, but does not provide for connection with a public exchange. For the purpose of explaining the genesis of the system it will be necessary to interrupt at this point the consideration of the Committee's Report, and to indicate from other sources the circumstances which led to attention being given to this application of common battery for domestic telephones. This information is derived from evidence given by Mr. Hayes[1] in a patent suit.

In 1888 and 1889 Mr. Hayes and his assistants had been giving much attention to the development of 'so-called speaking tube systems.' On August 1, 1889, Mr. A. C. White, one of the assistants (whose name is better known in connection with the development of the Solid Back transmitter), made a record of a discussion between

[1] Herzog *v.* New York Telephone Co., Circuit Court of the United States, Southern District of New York, *Defendants Record*, p. 415.

Mr. Hayes and himself, and noted suggestions which he felt would offer a satisfactory means of accomplishing the results desired. The memorandum was accompanied by a sketch (fig. 127), which is probably the first diagram of common battery system in the direct line of development.

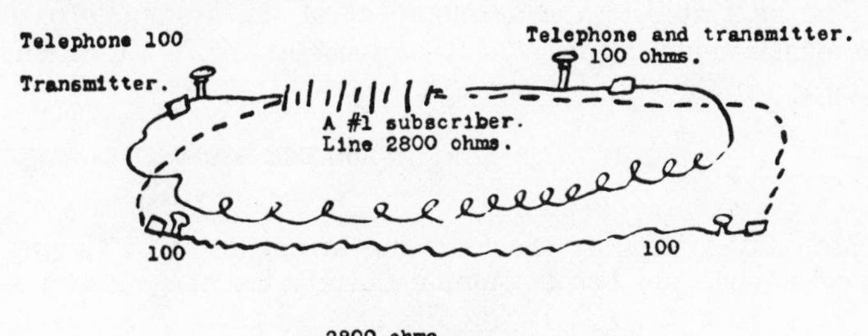

FIG. 127.—First Sketch of Common Battery System.

In the explanation of the sketch Mr. White says :—

I do not think the discharge from A's line would affect B's, as B's line is shunted by ·5 ohms the battery resistance. It seems to me that such an exchange would be highly practicable.[1]

Mr. Hayes considered the suggestion of great importance, and efforts were immediately directed towards a demonstration of its feasibility. On September 6, 1889, Mr. Hayes reported to Mr. Hudson, the President and General Manager, that Mr. White had been experimenting with transmitters used upon the line without the aid of an induction coil. Where the circuits were metallic it was possible to connect a large number of subscribers to the same battery without interference or appreciable loss of volume. The advantages to be derived from the simplicity of the subscribers' apparatus and the centralised battery were set out and further attention recommended.

The interest of the experimental department in the common battery system probably received some renewed impetus in 1891 through the consideration given to a proposal for a telephone installation in the Boston Public Library. This institution was intended to surpass any other library in its appointments, and some of the trustees discussed with Mr. Hudson the suitability of telephones for obtaining from the various departments the books requisitioned by the readers. Mr. Hudson referred the matter to Mr. Hayes, who recommended a common battery 'speaking tube'

[1] Herzog *v.* New York Telephone Co., Circuit Court of the United States, Southern District of New York, *Defendants Record*, p. 416.

system in consequence of the small dimensions of the sub-station apparatus. Telephones were not adopted by the library authorities for the purpose suggested, but the time spent over the consideration of the matter was not wasted if, as seems probable, the common battery system received renewed attention therefrom.

The next step in the development of the system is recorded in a memorandum made by another assistant, Mr. W. L. Richards, which is as follows :—

CIRCUITS FOR DIRECT TRANSMISSION

October 15, 1891.

Mr. Hayes suggests the following arrangement of circuits for Long Distance and Local Common Battery Exchange System :—

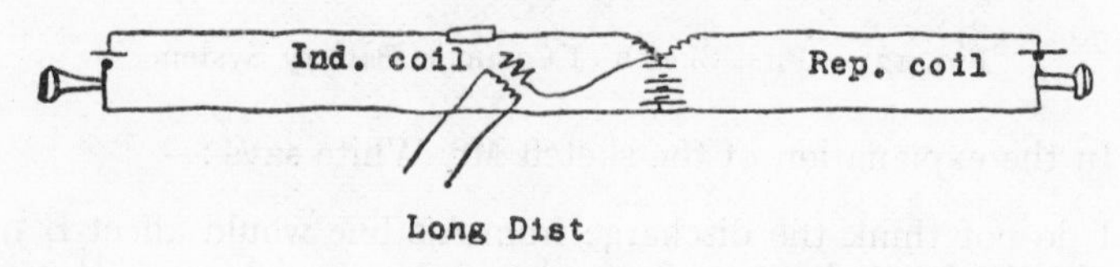

FIG. 128.

Also indicator arranged in primary circuit to indicate condition of lines :—[1]

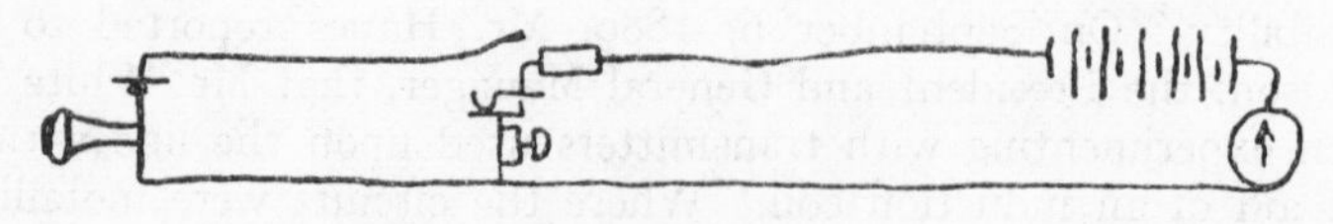

FIG. 129.

The results of the work on these lines were reported by Mr. Hayes on December 7, 1891, with the sketch fig. 130.

It is this sketch which forms fig. 1 of Hayes' United States patent (fig. 131), No. 474, 323, which was applied for about a month later (January 13, 1892), and granted May 3, 1892. The dotted lines in fig. 131 indicate alternative connections for single, instead of metallic, circuit lines.

Returning now to Mr. Hayes' paper read before the switchboard committee of March 1892, we learn that

the Local Common Battery Exchange System has not been developed to any extent as the system seemed to be more easily experimentally developed in connection with other systems. . . . We have worked

[1] Herzog *v.* New York Telephone Co., Circuit Court of the United States, Southern District of New York, *Defendants Record*, p. 421.

out two or three schemes for multiple boards which are more or less feasible, but they have not been experimented upon and have not been developed in any way. The point that I should like to

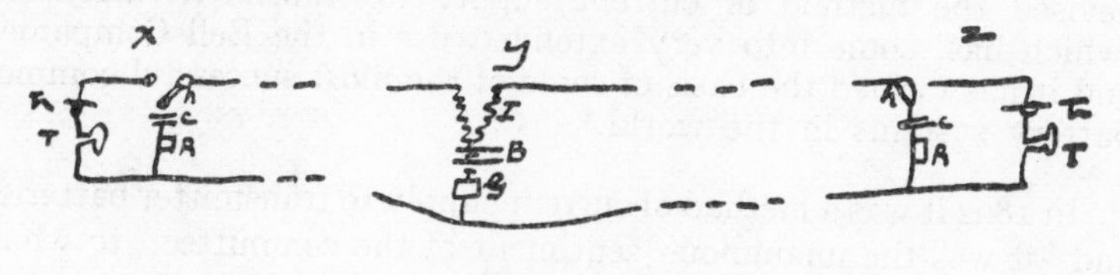

FIG. 130.

Tr—Transmitter.
T—Receiver.
R—Subscriber's Bell.
C—Condenser.
h—Gravity Switch.
B—Storage Battery.
I—Induction Coil.
G—Galvanometer Drop.
X & Z—Subscribers' stations.
Y—Central Office.

bring out would be the possibilities of improving our service and cheapening it by using first the long distance common battery exchange system. I think by that, in connection with our long lines, we could give a service equal to the service which is given to-day and decrease the expense of battery maintenance enormously.

On such lines the system had been tried on two circuits: to 127 Purchase Street, Boston, for more than eight months, and on Mr. John E. Hudson's circuit at 125 Milk Street, Boston, for about

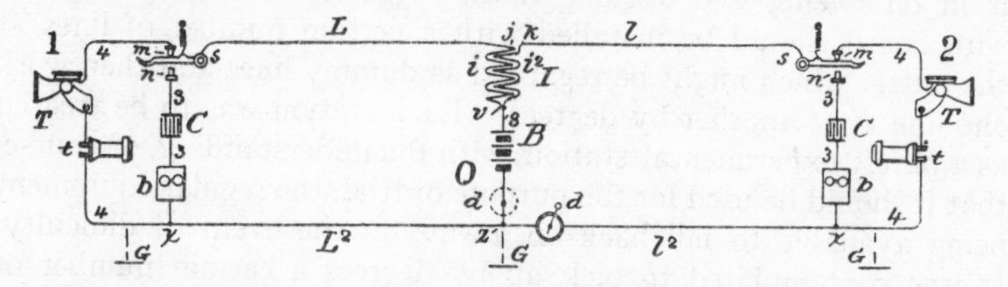

FIG. 131.—Hayes' Common Battery Patent.

three months. As a result of the practical tests which were made, although of course they were limited in number, it was considered that the system was entirely practical and commerically operative.

It will be observed that Mr. Hayes, whilst not limiting himself regarding the eventual possibilities of the system, adopted a thoroughly scientific attitude in its presentation, and suggested its use first in connection with long-distance subscribers, the maintenance of whose local batteries was troublesome and expensive. Much work yet remained to be done before the application of the system to local and general service. The plans had not been formulated, much less worked out. Mr. Hayes in 1892, when his patent was taken out[1] and his paper read before the switchboard committee, had, in the words of Kempster Miller,

> devised the method of current supply to transmitter batteries, [which has] come into very extended use in the Bell Companies, and it has formed the basis of some of the most successful common battery systems in the world.[2]

In 1892 it was a method of current supply to transmitter batteries, and 'it was the unanimous sentiment of the committee' to whom it was submitted that the scheme shown

> possesses merit which justified making active efforts to develop it, and that those efforts should now take the direction of putting it in on working lines under the conditions that would be met with in practice.

The method to be adopted was the subject of much discussion. To conduct experiments on Long Distance lines was to submit the most expensive outfit to unknown chances, and the Long Distance Company would simply be furnishing facilities for trying an experiment in the interest of the exchanges since the expense of battery maintenance was of relatively small importance compared with the rental derived from the line.

Mr. Hall, inferring that the committee did not seem to be very strongly of the opinion that they ought to take the chance of putting it in on twenty-five working lines, suggested that a twenty-five wire board should be installed with a certain number of lines at the outset which might be regarded as dummy lines and then work one line after another by degrees. Each station was to be treated as a purely experimental station, with the understanding, of course, that it should be used for the purpose of trial, the regular equipment being available to fall back on always in the event of difficulty. It was contemplated to pick up by degrees a certain number of lines, making a special study of the condition of the wire, resistance, and other factors. In that way varying conditions would be created and studied.

It was arranged to establish this special service in Boston under the immediate supervision of Mr. Hayes and his staff. Thus in May 1892 a common battery exchange was authorised and shortly after installed.

In former developments it has been easy to observe the stimulus of compulsion. Conditions of the service either as regards the growth or the discovered defects prompted remedies. There were thoughtful students and some careful studies, but the work of the

[1] U.S. specification, No. 474,323, dated May 3, 1892 (application filed January 13, 1892). [2] *American Telephone Practice*, 4th edition, p. 270.

day was too urgent to permit an undue indulgence in the luxury of forecast and the provision of improvements in advance of urgent needs. Extraneous sources could produce but little in the complex field of exchange service for the very sufficient reason of lack of experience. Thus inventive ingenuity in the operating field was largely directed to overcoming the defects and meeting the requirements of the then present or the immediate future. The telephone was a remarkable invention in more ways than one. Its production was timely, but so far from its filling a want which had been keenly felt, as has been shown in earlier chapters, the public were sceptical of its utility and by no means ready to take early advantage of the facilities it offered. The same may be said of the first exchange systems and of accessory apparatus. But once the advantages were apparent and the service fairly launched, it was as true in the nineteenth century as Farquhar expressed it in the seventeenth century that necessity was the mother of invention. It was necessary to develop methods to provide for ever-increasing demands, and it was necessary to overcome the defects which the very growth produced. Skilled men were limited in number, and they were kept very busy in supplying the needs of the moment.

There were keen students of individual problems, such as Van Rysselberg in Belgium, and men who were adopting a scientific attitude towards the exchange system as a whole, like Wietlisbach in Switzerland; but only in the United States were there organised efforts at advances and improvements in the Exchange methods.

By the year 1892, when the common battery system was presented, the companies had ceased to be surprised at increasing demands for service. They had acquired experience, and the staffs of capable experts had so far grown in numbers as to permit the leading managers and engineers to devote some time to the consideration of probable future requirements, to look to the possibly desirable, and not to limit their attention to the absolutely necessary.

One direction to which, about this period, considerable attention was given was the reconsideration of the multiple switchboard. Some practical and strictly economic minds were still exercised at the multiplication of connecting points, some of which might never be used, and advocates were found for a divided board system. It was the province of the committee to adjudicate on this question; to determine whether some of the money spent upon a multiple might not be saved by the division of the work between two or more operators. The inquiry took two forms. One was an actual analysis of the work on a multiple board, the other the preparation of definite proposals for a divided board.

It was well recognised that for celerity and certainty the one-operator method had a great advantage, but with the growth in the number of exchanges there were already a large number of calls which of necessity could not be completed by one operator. It was considered that whilst a divided board might be slower, the effect of its use would be to obtain a consistent average of service.

Fig. 132 is a facsimile of an outline diagram to illustrate Mr. Hall's paper submitted to the committee held in May 1892.

EE represents a number of sections of connecting board, and FF a row of terminal boards.

A and B are subscribers' stations. A's line goes through a spring jack at G on the connecting board, extending from there to the spring jack H at the terminal board R, and then to the

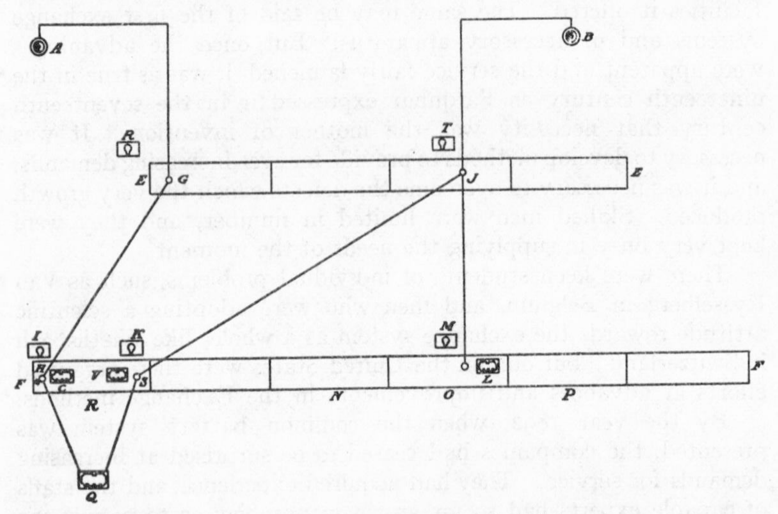

FIG. 132.—E. J. Hall's Divided Switchboard.

terminal drop C. B's line goes through a spring jack at J on the connecting board to the spring jack and terminal drop at O. A's call is received by the operator at the terminal board R, who inserts answering plug in spring jack H. This engages A's line, and since there is also a means of access to A's line at the connecting board, it becomes necessary to advise the connecting board operator of such engagement. Thus, when a plug is inserted in A's line at H, a visual signal R is automatically displayed at the connecting board.

The call of A for line B having been received by the terminal operator at R she communicates with J over an order wire. The operator at J connects line B to an office trunk extending to R. The insertion of the office trunk plug into B's jack at J disconnects the end of the line extending from J to O and actuates the visual signal M, indicating that station B is busy. In the event of B being busy at the time of A's call, the operator at J would take the plug representing the end of the office trunk and insert it in a special spring jack which automatically causes the visual signal K to flutter, and provides also an audible indication of engagement. It is not intended to describe in detail the operation of this proposed board. Many complex diagrams and elaborate descriptions were submitted to the committee and formed the subject of analysis, comparison, and debate ; but the above simple diagram and brief description comprise the essential principle of the divided board discussion to the line of progress.

In addition to the divided board above described, a system with a very similar title but on an entirely different principle was developed by M. G. Kellogg. This was called the 'Divided Multiple System,' and contemplated the division of subscribers into various multiple groups with a selective system of calling by the subscriber. Numerous patents were taken out on this system, but for a description of, and discussion on, the principal features reference may be made to the *Journal of the Institution of Electrical Engineers*, xxxii. 795. This system did not meet with the approval of telephone engineers generally, but was installed on a large scale by an 'Independent' Company at St. Louis, and though in operation for a considerable period was eventually discarded.

Other attempts to avoid the use of multiples were the express systems of Sabin and Hampton in San Francisco and Hibbard in Chicago. In the Sabin system 'small sections of switchboard were placed contiguously, each section being arranged for three operators. The two end operators were answering operators, and an intermediate operator served as a trunk operator. The answering operators answered calls from subscribers whose lines were located at the section at which they were seated and the trunk operators completed calls from other sections to all subscribers at this section.'[1]

Order wires and transfer circuits available to the various operators provided for the interconnection of the various lines. Thus an answering operator requiring a connection at the board of a trunk operator, by depressing the key leading to that operator, would give the order for the subscriber wanted, and the trunk operator would designate the transfer line to be used for the purpose of connection. Such methods were obviously cumbrous compared with the multiple, but the delay incident to connecting and disconnecting was sought to be minimised by the use of signals automatic in their operation and conveying information to the respective operators employed in a connection.

In the Hibbard express system there were some modifications

[1] Scribner, *Herzog Case*, p. 548.

of the Sabin system, but in its purpose and in the use of signals it was very similar. Whilst adopted for purposes of economy, it appeared to one who inspected it in 1895

> to be a very expensive system, seeing that when I was in the exchange nine operators were being employed to attend to rather less than 300 subscribers. . . . The board is, however, only an experimental one, and no doubt will provide valuable data for future use. I do not understand the term 'express' being applied to it, as it must, I am sure, be somewhat slower than the multiple.[1]

The chief interest in these express systems is to be found in the adoption of the method of calling that has been more particularly identified with the common battery system. The subscriber's set consisted of transmitter, telephone, ringer, and condenser.

In Chicago 'The subscriber calls by lifting his telephone from the hook, so causing a lamp to light in front of his operator.' [1]

In San Francisco the operation was the same, but the signal was electro-magnetic.

The divided board and the express systems had for their object the attainment of economy by the elimination of the multiple.

The multiple board was not deposed, but the feature that was relied upon to reduce the drawbacks attendant upon the employment of more than one operator for a call was the use of a system of visual signals which should give the respective operators information without the need of speech. Whilst it was found that these signals did not make the divided board a practical or economical device in one building where a multiple could be installed, there would be obvious advantage to their use in the transfer of calls from one exchange to another, where the employment of two operators to one call was not a question of choice but one of necessity.

These visual signals were introduced for trunk line working between exchanges. Mr. C. J. Phillips in the report quoted above described the system as in use in New York, and said that

> owing to the use of automatic signals much less talking is done on their call wires than on ours, consequently operators do not have to wait to get their calls in, and the listening operator is able to attend to a larger number of junctions (30 as against 24).[1]

In the original application of these automatic signals is to be seen an effort to provide contrivances which should make practicable a switching mechanism which was comparatively imperfect. But the evidence which demonstrated the superiority of the multiple also demonstrated the great utility of automatic signals, and telephone engineers who were anxious to obtain the most perfect

[1] C. J. Phillips, *Report to National Telephone Company in London*, 1895.

apparatus with which the public could be served were not slow to combine the best in all respects. In his patent of 1892 Mr. Hayes had provided automatic signals for calling and clearing; of the latter one only was shown in the diagram (fig. 131), but the provision of two, and their automatic operation, were contemplated in the text.

At the meeting in January 1893, Mr. Durant described to the switchboard committee[1] improvements which had been introduced in the Law system as carried out at St. Louis. These concerned the transfer of calls from one exchange to another, and since, in the Law system, the operators who attend to receiving calls cannot for a moment leave the work of receiving orders, the signals were required to give varied information. The signals proposed were incandescent lamps, and the code contemplated their use in three conditions, dark, bright, and an intermediate stage which may be called dim. The system as presented did not meet with the approval of the committee, but (quoting from the summary)

> incidentally the question was raised whether a system of glow lamp signals, like that submitted, is a perfectly safe adjunct to a central station, and the committee stated its opinion that it is.

The committee had incidentally reached an important conclusion without, perhaps, realising how near was the time when it might be put in operation. In the diagram of the divided board (fig. 132) the visual signals are apparently intended to be lamps, but in the discussion the particular form of signal was not defined.

In the summary of the January 1893 meeting it is recorded that

> at the request of the chairman Mr. Hayes reported progress in the centralised transmitter battery exchange system, and stated his conviction, that in a short time as good service could be given by it on the longest lines as could be given by the plans and apparatus now in use. . . . It was thought that by the next meeting, results could be reported in such shape as to afford a basis for profitable and instructive discussion.

It was probably not expected that a period exceeding two years would elapse before the next meeting of the committee, and in one particular it is to be regretted, since we are the less able to trace from carefully recorded sources the various stages of progress during an important period of experiment and development.

It was in May 1895, at the office of the American Bell Telephone Company, 125 Milk Street, Boston, that the committee next met.[2]

[1] Members: E. J. Hall, T. D. Lockwood, C. E. Scribner, I. H. Farnham, J. J. Carty, A. S. Hibbard, and F. E. Pickernell.

[2] The members present were: J. F. Davis (chairman), E. J. Hall, T. D. Lockwood, C. E. Scribner, I. H. Farnham, J. J. Carty, A. S. Hibbard, F. E. Pickernell, H. V. Hayes, and W. S. Ford.

The advance which had been made is indicated in the notice convening the meeting, which stated that—

Since the last meeting of the switchboard committee held in January 1893, there have been proposed, and to a considerable extent put into service, new methods or new combinations of old methods and new apparatus for operating a telephone exchange which have attracted very general attention, and have been now so far developed and tested that it is thought desirable that a meeting of the committee should be held at an early day to consider that which has been done and to mark out, as definitely as practicable, the best path to follow in further development ; also to decide what system of service and style of apparatus it is most judicious for telephone companies in immediate need of new switchboards to order.

The drift of thought is further indicated in the following :—

You are also requested to collect and bring with you, or, if unable to be at the meeting, to send any data within your reach bearing upon the design and use of automatic signals, or other improvements in equipment of central offices or subscribers' stations, and upon methods of operating. It appears to me desirable that the committee should give an opinion upon the following matters, together with such others as the committee may select.

1st. Is it recommended that on switchboards hereafter purchased provision shall be made for automatic signals in line and connecting circuits, which shall inform the operator of whether or not the subscriber's telephone is hung on the hook ?

2nd. If yes, what system of automatic signals, in the opinion of the committee, is most promising ?

3rd. Is it recommended that on switchboards hereafter purchased, provision be made for supplying current to sub-station transmitters in some other manner than by means of primary batteries at the sub-stations ?

4th. If yes, what method of supplying current to transmitters is, in the opinion of the committee, most promising ?

7th. The best method of counting calls in central offices, for statistical purposes, and manner of tabulating and analysing such counts.

The committee was unanimously of opinion that in future switchboards provision should be made for automatic signals in line and connecting circuits which would inform the operator whether or not the subscriber's telephone is hung on the hook. It was further agreed that incandescing or glow lamps should be the preferred form of signal device, and that these should be operated direct, if possible, but with relays to such an extent as might be necessary.

After discussing the subject fully the committee voted an emphatic recommendation that provision should be made for supplying current to sub-station transmitters in some other manner than by means of primary batteries at the sub-stations.

It is, however, necessary at some time to come from the general to the particular, and this the committee forthwith proceeded to do. Three plans were before them, thus related in the synopsis :—

The ordinary centralised and common transmitter supply plan was described by Mr. Hayes ; a common battery system proposed by Mr. Dean of St. Louis, by Mr. Durant ; plans in which the common source at the central station is supplemented by a Planté Cell at the sub-station by Messrs. Hayes and Scribner ; and one in which a storage battery at the sub-station is charged from the central station source while the line is at rest, and discharged to furnish transmitter current in operation, by Mr. Hibbard.

For convenience these plans were, during discussion, referred to respectively as the 'Hayes Common Battery System,' and the 'Dean,' 'Planté,' and 'Local Storage' Systems, plans or arrangements.

For further convenience we may reduce these to the first and the last.

The Hayes common battery system, as previously related, had been under test upon a few long distance subscribers' lines in Boston. It was reported to the 1895 committee by Mr. Spencer that it was first put in operation in Lexington in December 1893, and had continued in operation with satisfactory results. It was next put in at Wellesley on much longer lines and under much more trying conditions ; and about sixty stations in Philadelphia had been in operation for about one week.

The local storage system was a method which had been used in a section of the Chicago exchange. Instead of a primary battery there was a small accumulator at the sub-station. The accumulator was charged during the idle time of the line and provided current for the transmitter when the line was in use. About 100 cells had been in use for three months with satisfactory results.

The competition between common battery and local storage was decided in favour of the latter as recorded in the synopsis as follows :—

The opinion of the majority of the committee (Messrs. Carty, Hayes, and Ford dissenting and preferring the common battery system) was that the 'local storage' system was the most promising plan which had been discussed and that it should be provided in connection with switchboards for the immediate future.

In view of subsequent developments, it is of interest to refer more in detail to the opinions expressed by Mr. Carty, who (of the minority) occupied an entirely independent position so far as the development of the system was concerned. He said :—

The general result of our experience with that system leads me to believe that, providing the question of transmission is settled satisfactorily, there is nothing that we can demand of the system in the way of adapting itself to switchboards that we cannot get from it. I think that the general features of the common battery system, that we are now discussing, lend themselves beautifully to what seems to be the improved trunking and connecting systems.

The saving clause on the question of transmission was due to the fact that the tests in New York were not 'up to the mark,' and this was believed to result from the use of an unsuitable type of repeating coil, which could be remedied. Mr. Carty continued :—

Now I feel that if we were satisfied that the transmission was all right, we would go ahead in the Harlem central office, on which we are delaying work, waiting to find out the developments of this case. We would go ahead and instal, very likely, a switchboard, employing the common battery system of the Hayes type, unless something in this meeting should transpire to show that there was something better. . . . . . .

The plant in New York was intended to be the best expression which we could give of the Hayes common battery system, and we followed as closely as it was possible the latest ideas in that system up to the time of its installation. The results were somewhat uncertain as to the talking, and the plant was investigated in the ordinary way, by competent people, but those who had not special experience in the common battery system ; and, in order to be sure whether we had what Mr. Hayes would call a correct expression of his ideas, I asked one of his assistants while in New York, with Mr. Hayes's consent, to make an examination. I am awaiting his report, and I will try to see him while here.

On the following day the local storage system was under discussion and the members were asked individually to express their opinions. I give Mr. Carty's remarks *in extenso* :—

Mr. CARTY : My objections to what is known as the Hayes common battery system have been removed, with the exception of the objection that I raised to the transmission. It has not been clearly established that the transmission is all that was claimed for it by Mr. Hayes, but my experience with it, and the reports that we have had at this meeting, have simply served to cast doubt upon the curves which were shown to us. With the single exception, then, of the doubt about the character of the transmission, I would favour the Hayes common battery system, and favour it because of its beautiful simplicity, by the fact that the subscriber's station apparatus is simple, especially when you consider such apparatus as being extended to a subscriber's desk.

The absence of the battery and the wire that is needed for it is a very good point, when you consider it from the standpoint of a subscriber's station. Now, I am informed by Mr. Hibbard that the switchboard arrangement which would be suitable for the local storage system such as he advocates, would also be suitable for the Hayes common battery system. If that is the case and I had to put in a switchboard to-morrow, with no further information than that which is now before the meeting, I should arrange for a switchboard with the Hayes common battery system, and I believe that if it is not exactly right now it can be made right; but, if I was mistaken in that latter supposition—that it could be made right ultimately—why, we could then put in our local storage system at the subscriber's station. That is all.

The decision of the majority of the committee was reached with due regard to the paramount importance of transmission and on the basis of the evidence of practical results then available. The common battery system, as then submitted, was but a sketch compared to the finished picture it later became. The more credit, therefore, seems due to the minority and particularly to Mr. Carty for his perception of what it had accomplished and might be made to accomplish.

The 'beautiful simplicity' of the system, the fact that 'there is nothing we can demand of the system in the way of adapting itself to switchboards that we cannot get from it,' were observations on established facts—facts important enough not to give a verdict on the basis of other established facts—that the transmission was not in all cases satisfactory. There were reasons for thinking that the unsatisfactory features were not incident to the system but could be removed, and the faith found expression in the statement 'I believe that if it is not exactly right now it can be made right.'

The belief was justified, but how remains to be told.

The diagram of cord circuit in fig. 133 was presented to the switchboard committee of 1895:—

Here, it will be seen, the repeating coil assumes the form finally adopted, the battery being connected to the middle of the coil instead of one end as in the preceding diagrams (figs. 130, 131). The description is as follows:—

The cord circuit consists of two two-conductor cords connected to a split induction coil; at the centre of the induction coil is a battery; a clearing-out signal is placed in one wire of each cord. . . . The visual signal in the cord circuit operates so long as the subscriber is connected to the circuit. As soon as his telephone is placed upon the hook, the signal ceases to operate, indicating to his operator that disconnection is needed.

Fig. 134 is the drawing which illustrates the first of the modern type common battery patents taken out in Great Britain,[1] and may probably be regarded as describing an operative system rather than as indicating the instrumentalities to be employed.

The form of subscribers' station apparatus shown is that which was suggested in Hayes's 1891 sketch and 1892 patent—a call bell and condenser—though a high-resistance bell without condenser is mentioned as an alternative. Lamps and relays were doubtless at this time under consideration but had probably not reached a

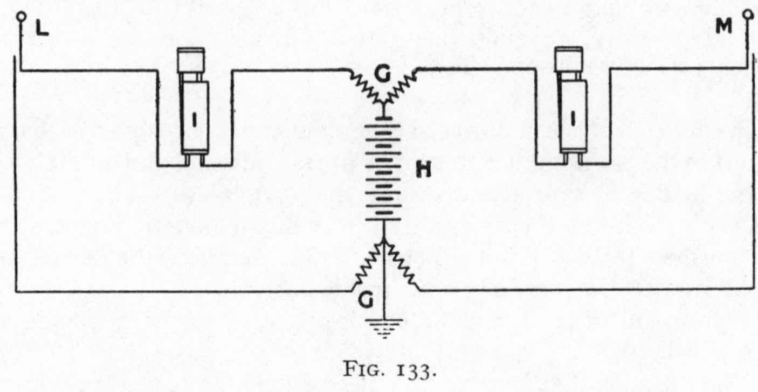

FIG. 133.

G—No. 15 Induction Coil.
H—Common talking battery.
I—Clearing-out signals.
L—Answering plug.
M—Clearing plug.

stage that could be regarded as reliable. The common battery system is therefore illustrated with devices which were known to be operative but had only been used on a small or experimental scale. The line indicator is of the branching type, the clearing-out indicators or 'visual signals V V′' are not described, but indicators of a galvanometer type had been used by Mr. Hayes for the purpose of giving the supervisory signals. This diagram probably represents the stage which the common battery system had reached early in 1895. There were two features—'a common battery or centralised source of electric current' by means of which call signals could be sent; and, secondly 'a common battery or centralised source of electric current by means of which telephonic transmitters at the subscribers' stations are energised and by means of which various automatic signals or clearing-out drops are operated in the central station at which the subscribers' lines terminate.' At this time some importance was attached to the further statement :

[1] British specification, No. 8222, April 25, 1895.

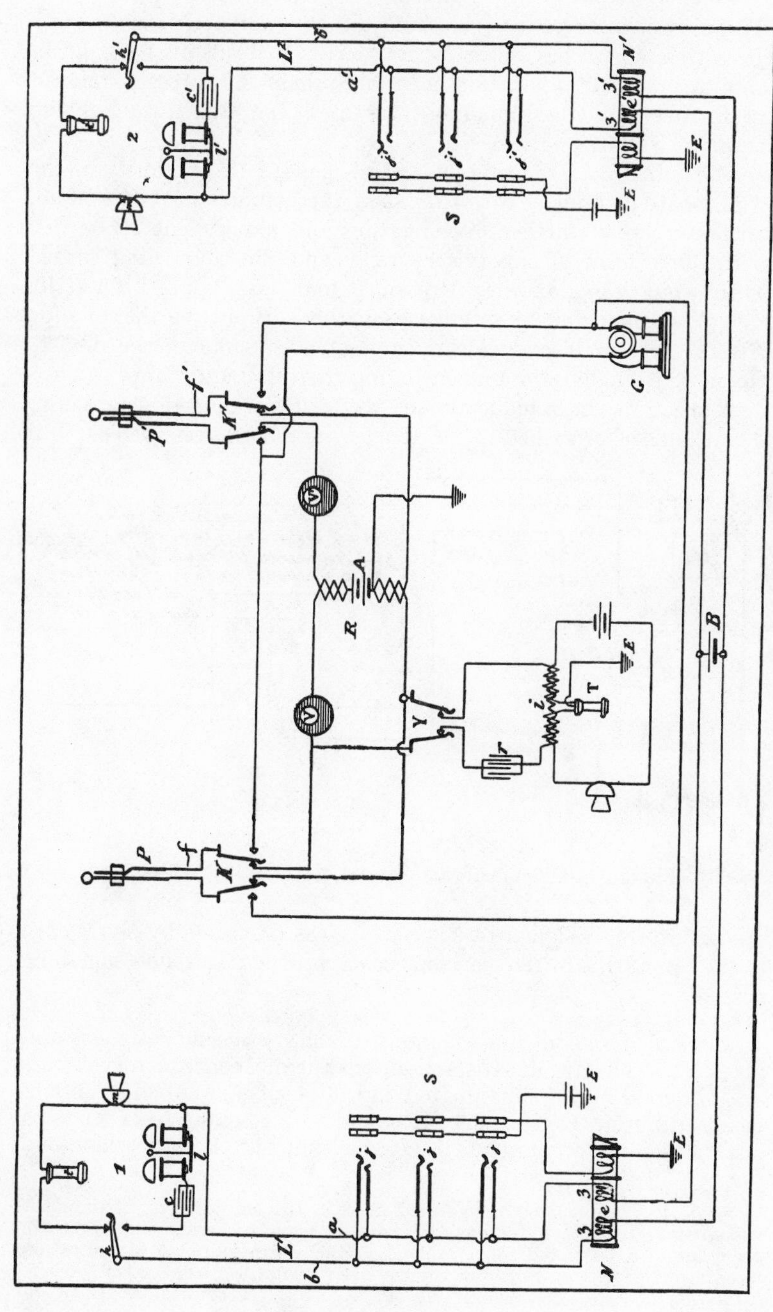

FIG. 134.—Common Battery System. First British Patent.

'These two features can be used either together or separately.' There was, as we have seen, some fear as to transmission, and in the first practical application on a large scale at Worcester (Mass.) one feature only (the signalling) was used, the transmitters being energised by local batteries.

The imperfect transmission, which caused the majority of the switchboard committee to vote against the common battery system, was found upon further investigation and experiment to be due to the first form of supervisory relay and the connecting cords.[1]

In June 1895 another British patent was applied for (No. 11,549), which includes a number of methods of putting the common battery system into operation and includes some of the devices which were finally adopted, including the relay and lamp.

Allusion has been made already to the use of incandescent lamps at St. Louis as described to the 1893 switchboard committee. But

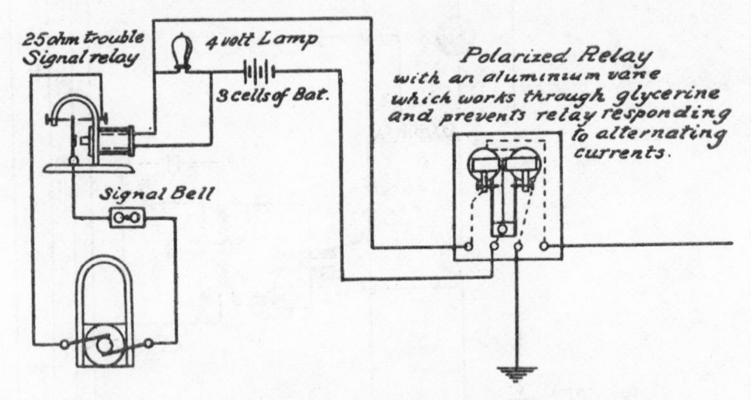

FIG. 135.—O'Connell Lamp Signal described by Wilson, 1891.

an earlier use, designed by O'Connell, was described by Mr. Wilson to the committee in 1891 in connection with a trunk line signalling system.

Several forms of target indicators have been designed and given a good degree of satisfaction, but the most satisfactory results are obtained by an illuminated indicator of simple construction, designed by Mr. O'Connell.[2] It consists of a hollow metal tube of one inch diameter, closed at the outer end, but through which end

[1] Carty, *Herzog Case*.

[2] 'J. J. O'Connell . . . suggested in 1888 the employment of an incandescent lamp upon burglar-alarm circuits in order to permit the legitimate occupant of a protected room to send to the alarm office an identification signal whereby his entrance to the premises could be made known' ('The Evolution of the Line Signal,' by A. V. Abbott, *Transactions American Institute of Electrical Engineers*, June 29, 1898, xv. 430).

is cut, stencil fashion, the number of the indicator. Within the tube, which is about two inches in length, is placed a miniature incandescent electric lamp operated in the local circuit of a relay.

Fig. 135 represents the lamp portion of the diagram accompanying Mr. Wilson's paper.

The lamps used at St. Louis in 1893 were also for trunk line signals, and were probably, like those of O'Connell, of comparatively large dimensions.

It is not possible to relate in detail the various steps in the progress, but it will readily be understood that very considerable work and expenditure were involved in the experimental development and manufacture of the lamps which were finally decided upon as satisfactory for use. A new condition had to be met. Life in lamps hitherto had been measured in hours of incandescence without regard to frequency of illumination. For switchboard use, life had rather to be measured by the number of times that the circuit could be made and broken without the utility of the filament being impaired.

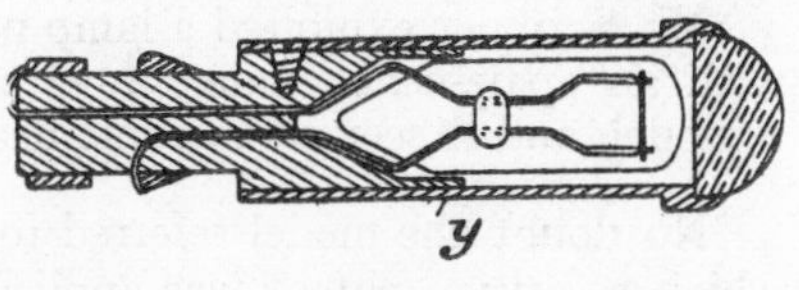

FIG. 136.

The drawings figs. 136–139 from British specification No. 11,549 of 1895 illustrate the lamp as first put into practical use in the common battery switchboard. Fig. 136 is a section drawing of the lamp,

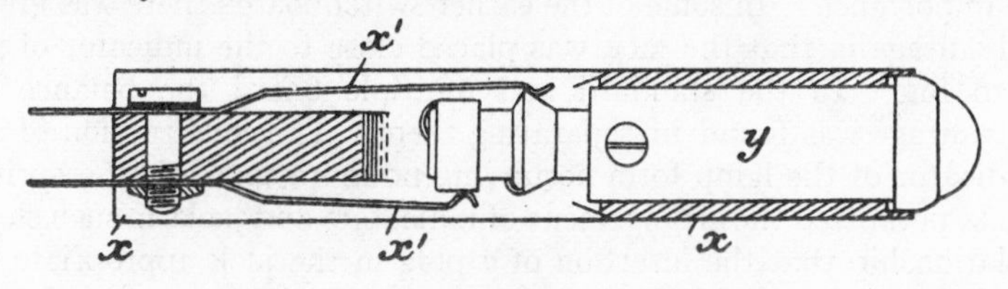

FIG. 137.

fig. 137 shows the lamp inserted into the switchboard and connected to the contact springs, and fig. 138 a part of the ebonite strip in which the lamps were mounted. An important detail in this lamp is the plano-convex lens of opal glass (fig. 139).

This lens serves two purposes: one is to disperse the light from the lamp so that the signal can be seen from the side as well as from the front; the other is to prevent the red glow of the filament, when both lamps are in parallel, from being seen and mistaken for a signal.[1]

[1] British specification, No. 11,549, 1895, p. 9, line 31.

The second purpose applied only to one of the various ways of connecting the supervisory signals and was not adopted in practice, but the illumination to the side was of great importance and has been used continuously. At night or slack periods it is necessary that an operator should be able to see the indication of a call at some distance to right or left of the indicator. The plano-convex lens provided for this condition in a very simple manner without detracting in any way from the utility of the signal to an operator immediately in front.

In the committee of May 1895 Mr. Carty said:—

Mr. Scribner exhibited a lamp not long ago which was provided with an opalescent shade. I witnessed the operation of that in a model, and it seemed to come near the ideal.

No doubt the model referred to was the plano-convex lens, for which a British patent was applied for in the month following.

The relationship of the indicator to the jack was always a matter

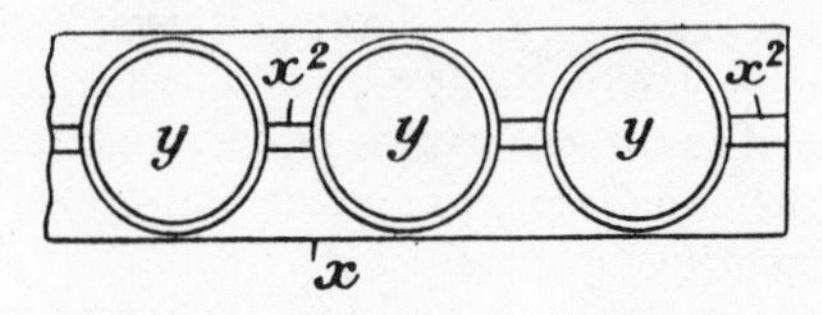

FIG. 138.

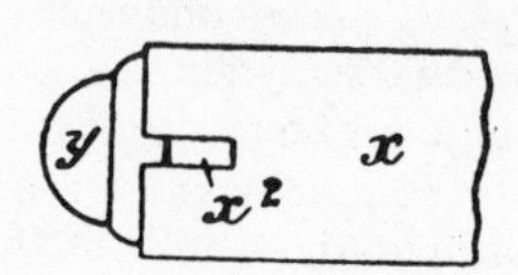

FIG. 139.

of importance. In some of the earlier switchboards there was great advantage in that the jack was placed close to the indicator of its own line. In the standard and multiple board the balance of advantage was found in separating them. The construction of an indicator of the lamp form occupying no more space than a spring jack, permitted the arrangement of indicators and jacks in such close relationship that the insertion of a plug in the jack appropriate to a glowing lamp could be carried out with a minimum of conscious effort on the part of the operator.

The lamps were mounted ten on a strip and the answering jacks also, it being preferred, as the specification states (p. 9, line 39),

to place each line lamp over the corresponding answering jack. The two clearing-out signals belonging to a pair of plugs are placed in the key-shelf in front of and in line with the plugs which they represent.

An arrangement which has not needed to be altered, except that the lamps are under instead of over the spring jacks.

The tongs shown in Fig. 140 are to enable a lamp to be 'quickly removed and replaced by another in case of failure of the lamp.'

The use of a lamp as the indicator of a telephone call was a very great advance.[1] In its primary purpose of attracting attention it was far superior to the shutter forms previously in use. Whilst the shutter indicator needed watching, the lamp seemed to speak for itself, and the announcement it was impossible to ignore. A secondary advantage lay in being able to remove the operative mechanism from the face of the switchboard. In the indicator it was not possible to separate the shutter from its electro-magnet, but when the lamp was substituted for the shutter, the relay which was the equivalent of the electro-magnet could be placed at a distance, thus leaving valuable space to be occupied by apparatus which the operator needed to handle.

In a preference for lamps as signals the members of the 1895 committee were unanimous. There were differences of opinion as to the method by which the lamps should be operated—with or without relays. The signal lamps used in Chicago were controlled by relays, the lamps being in the local circuits. Regarding them Mr. Scribner said :—

If the hook is moved too rapidly, the relay fails to open and close the lamp circuit. You can hear the vibration of the relay at the board, but the lamp does not respond.

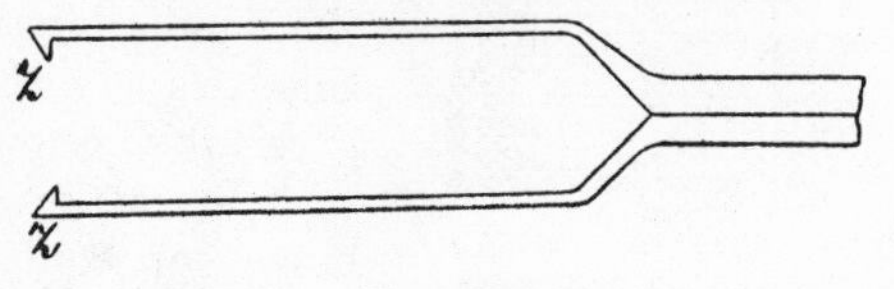

FIG. 140.

As illustrating a prevailing sentiment at the time, I continue the quotation from Mr. Carty's remarks given on p. 384.

Now if we could have such a lamp as that operated directly without any relay, I would like it. If it could not be done I think a lamp like that operated by an electro-magnet would be satisfactory. I would prefer a lamp to a ' magnetic signal.'

A lamp operated directly without any relay was thus regarded as promising for the supervisory signals. There was less objection to the relay for the line because the action would be direct and the responsiveness to quick changes would not be required. A method including this combination is illustrated in British specification, No. 11,549, June 1895, already referred to.

[1] ' The prime advantages of lamp signals are that they are extremely compact, they have no working parts, and therefore may be placed in any position, vertical, horizontal, or at an angle ; they are automatic in action or self-effacing, since the signal disappears immediately the current is cut off ; and finally they give a much more positive and assertive signal than any form of indicator.'—H. L. Webb, *Journal of the Institution of Electrical Engineers*, May 11, 1905, xxxv. 291.

It was necessary that the supervisory signals should be 'positive in their action'; that is, 'that they should by their condition indicate unequivocally to the operator either that the subscribers' telephones were on the hooks or were off the hooks.'[1] The supervisory signals were needed to be responsive to that fluttering of the hook by which subscribers were able to attract an operator's attention to a connected line, and so far relays were not responsive enough for this purpose. A new design was required. The design which in the main principle met this requirement is also included in No. 11,549 of 1895. I am not familiar with all the varieties of all the relays which have been invented for telegraphic use. Their number must be very large, but I am probably safe in saying that the armature which was attracted by the magnet had a hinge of some sort. Now the relay which was devised expressly for the common battery system had an important share in making that system practicable, and requires therefore some attention.

The illustration fig. 141 is the relay included in fig. 2 of No. 11,549 of 1895. It is extremely unlikely that Bell's experiments of twenty years before were in the mind of its designer, but in comparing this relay with its predecessors, one is irresistibly reminded of Bell's experiments with armatures in 1875.[2] Armatures with hinges or equivalent methods of support were discarded, and the disc was attached directly to the diaphragm. Similarly in the design of a relay which should be readily responsive it was found advisable to discard any support fixed to the armature. The description in the patent is very brief :—

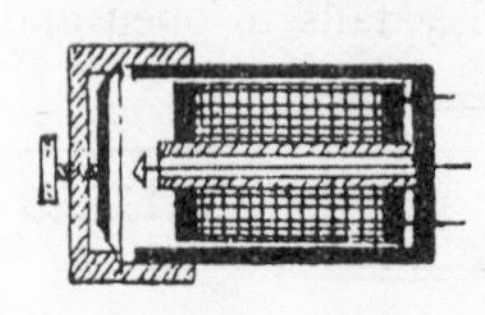

FIG. 141.—Common Battery Relay.

The relay shown in fig. [141] is a tubular magnet with a disc armature which is brought to an edge at its periphery, the disc resting on this edge in an annular groove formed between the end of the tubular magnet and a shoulder in the cap. When the armature is attracted it closes against an insulated contact stud which projects from the magnet core. When not attracted it drops away by gravity from this contact.[3]

The tubular form of this relay also recalls the iron box receiver exhibited at the Centennial,[4] though the method of operation was different. Magnetic attraction on the one hand, gravity on the other, were not seriously impeded by friction. The armature rested upon the tube, and the surface at the point of support was reduced to the smallest possible dimensions by being 'brought to an edge at its periphery.'

[1] Carty, *Herzog Case*, p. 33.
[2] Chapter vi. p. 48.
[3] P. 9, line 49 of specification.
[4] Chapter vi. p. 51.

The movement of the armature away from the electro-magnet was restricted by the 'annular groove.' It is clear that such restriction could be obtained by other means. It is also clear that the circular or disc formation of the armature was only useful with a 'tubular magnet.' In subsequent modifications the tubular form of magnet and the circular armature were abandoned, but the base of the latter was still 'brought to an edge' and became known as a 'knife edge.' It was the 'knife edge' which gave its name to this form of relay.

The inventor of the knife-edge relay was Mark A. Edson of Chicago. His United States patent specification[1] contemplated the combination of a relay and incandescent lamp in 'a single appliance or instrumentality.' The illustration fig. 142 is fig. 1 of his specification :—

The original description of the operation and statement of the advantages of his form of armature are even clearer and more

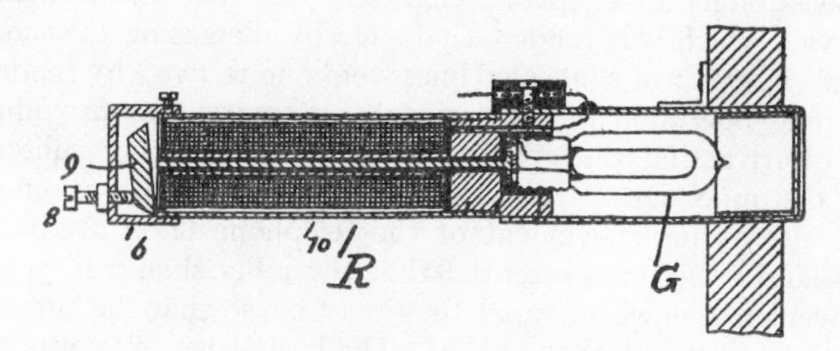

FIG. 142.—Edson Knife-edge Relay.

definite than those used in respect to the apparatus which was developed from it, and quoted above from the British specification No. 11,549 of 1895. The original inventor says :—

> 9 is a movable tilting soft iron disc armature shaped as a truncated cone, which rests upon the floor of the cap 6 by the edge of its base, its lower inner surface bearing upon the end of the metal tube 10, and being held in place and made adjustable by the adjusting screw 8. A great advantage of this construction is that the armature resting on its sharp edge on the internal periphery of a circle considerably larger than itself is almost frictionless.[2]

It is probable that the confidence inspired in the improved relay led to the modification of the circuits and the adoption of relays for controlling all the lamps. Thus the common battery line circuit within the central office approximated to that of the

[1] No. 550,260, dated November 26, 1895 (application filed May 20, 1895).
[2] U.S. specification, No. 550,260, p. 1, line 88.

branching system, where an indicator consisted of two coils known respectively as the line coil and restoring coil. A magneto-current sent over the line dropped a shutter, and the insertion of a plug in the jack of the line by an operator restored the shutter. By analogy the line relay and the cut-off relay of the common battery were the equivalents of the line and restoring coils of the branching board. A current coming over the line energises the line relay which lights the line lamp. The insertion of a plug in the appropriate jack energises the cut-off relay which breaks the circuit of the line relay and extinguishes the line lamp. The analogy of the line and cut-off relays with the coils of the branching indicator is so far carried into the construction that the line relay and the cut-off relay are mounted together, though not part of the same structure, as were the line and restoring coils of the branching indicator.

The feature of the common battery system which commended itself so strongly to telephone engineers was the improvement in the service which was rendered possible by increasing the number of signals in a pair of connected lines from one to two,[1] by rendering the normal operation of those signals automatic, yet providing a means of attracting the operator's attention by either subscriber before disconnection. The last-mentioned operation is effected by the alternating movement of the telephone hook at the sub-station, and it becomes essential that the relay should respond to the movements, however rapid they may be, so that the lamp may reflect the movement of the hook. The knife-edge relay was equal to the demand upon it, and the supervisory lamps were accordingly controlled by relays as well as the line lamps. The development of suitable relays was an important feature of the common battery system; in fact, the technical term descriptive of a full common battery equipment was the 'No. 1 Relay Board.'

Following the experimental plants previously recorded, the first large commercial installation of the common battery system was at Worcester, Mass., in June 1896. It was designed for employment in connection with common battery transmission, but when first installed the common battery was used for signalling only, local batteries being used for talking. A relay was employed for the line signal, this line relay served also for the supervisory signals.

[1] 'In my opinion the real dividing line between the periods mentioned is the introduction of double supervisory signals. . . . The weakness of the ring off indicator was not due to the fact that it failed to work when required, . . . but . . . that it [was] worked by the subscriber who had to perform a deliberate act that had no necessary connection with the work in hand ; . . . it did not give continuous supervision . . . and . . . independent knowledge of each side of the circuit.'—W. W. Cook, *Journal of the Institution of Electrical Engineers*, May 11, 1905, xxxv. 304.

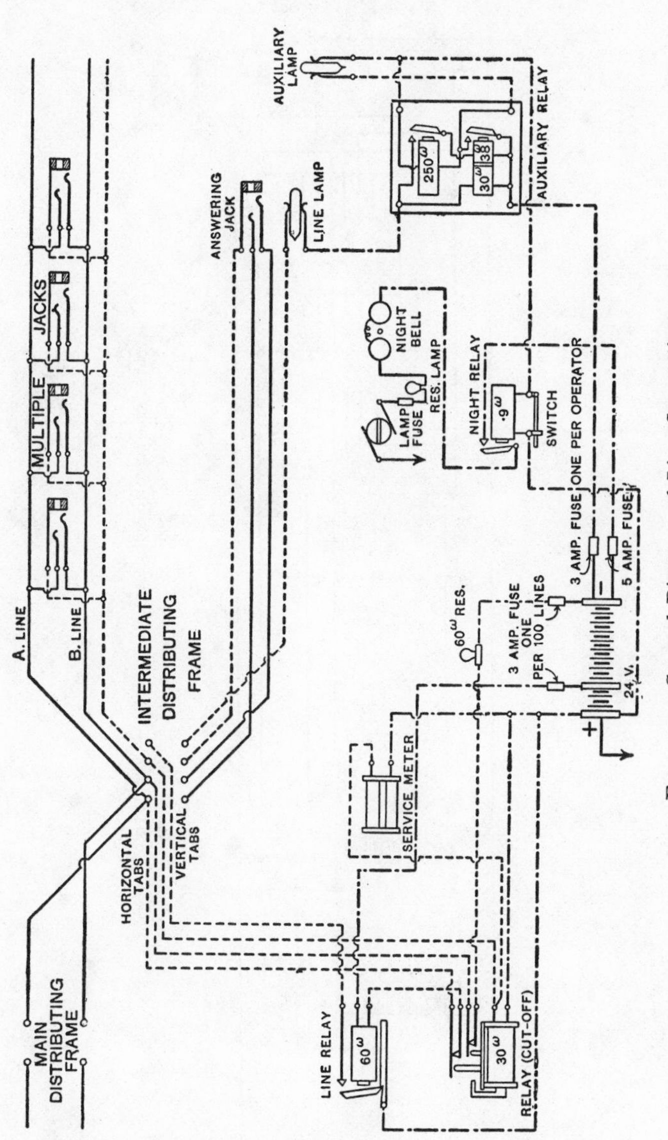

FIG. 143.—General Diagram of Line Connections.

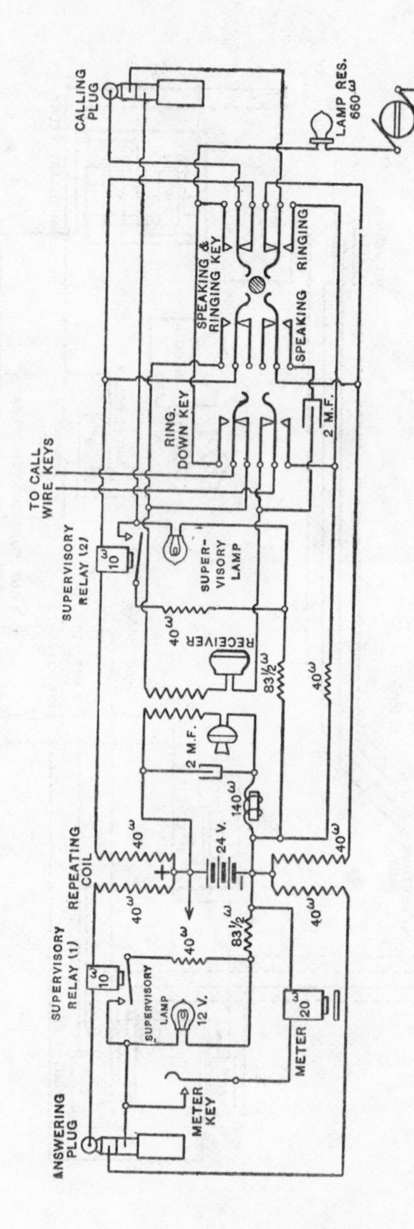

FIG. 144.—Cord Connections at Subscribers' Board.

A circuit of the supervisory lamp, which was completed when a plug was inserted in the spring jack, served to efface the line signal by virtue of a shunt circuit established around the line lamp by the supervisory lamp. Hanging the telephone upon the switch interrupted the current of the line relay which de-energized and released its armature, thus opening the contact previously made and interrupted the circuit through the line lamp. This served to increase the current through the supervisory lamp to an extent sufficient to light it.[1]

The Worcester installation was the public experimental début of the common battery system. Following the method by this time well established, a new system was devised in the laboratory tested on a sufficiently practical scale in a private exchange, and, proving successful under those tests, was put into practical use in a public exchange. The Worcester essay was successful, but showed that modifications might be effected. The modifications were tried in private and were included in the next public exchange at Louisville, opened in September 1897. In the Louisville outfit were to be found the cut-off relay and supervisory relays to control the supervisory lamps.

In these switchboards a common battery was employed for operating the line relays and the supervisory relays, and for supplying current to the subscribers' transmitters over their lines. A second common battery of a less voltage was provided for supplying the supervisory lamp with current; and still another common battery of still lower voltage for supplying the operators' transmitters and the line lamps with current, and the system has been termed the three voltage relay system.[2]

A few other central offices installed early in 1898 were of the three voltage type. Those installed later in the same year were of the single voltage type in which all of the apparatus of the equipment was operated from a single common battery.[2]

The Harlem (N.Y.) board, which was under discussion in the committee of 1895, was brought into service November 12, 1898. The plans for this board were made with due regard to the requirements of the whole New York service, and the results of its operation were such as to determine the complete conversion of the New York system from the magneto branching to the common battery system. Four central offices were completed in 1899, three in 1900, four in 1901, three in 1902, and two in 1903. These included equipment for 73,400 lines. Descriptive articles appeared in the American electrical press from time to time during this period. Other cities in the United States were also being fitted at the same time.

[1] Scribner, *Herzog Case*, p. 555. [2] *Ibid.* p. 557.

The first installation in England was at Bristol, completed in 1900; the next at the London Wall exchange in London. When Mr. Gill became engineer-in-chief of the National Telephone Company, the common battery system was adopted generally in the principal central offices of that company. Sir John Gavey decided on the adoption of the common battery system for the Post Office service in London. The first central office in Carter Lane, E.C., was opened in April 1902.

From a description of the Post Office system published in *The*

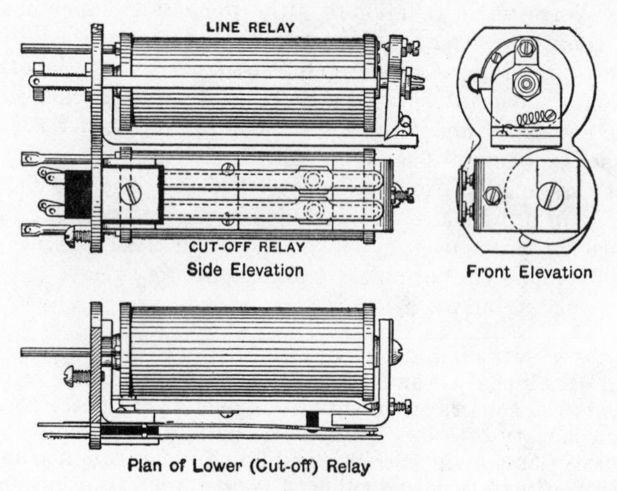

FIG. 145.—Line and Cut-off Relay.

Electrician [1] figs. 143, 144, and 145 are taken. They clearly illustrate the circuits.

The illustration (fig. 145) may be compared with fig. 141 to show the development from the disc armature.

When first introduced the jacks were larger than is now general, and the lamps were mounted ten on a strip. Later it was found possible to manufacture lamps sufficiently small to be mounted twenty on a strip of the same length as a jack strip with a similar number of spring jacks.

[1] March 14, 1902, xlviii. 808–10.

CHAPTER XXVII

AUTOMATIC AND SEMI-AUTOMATIC SWITCHBOARDS

THE switchboards so far described are elaborate machines, conveying certain information by what may be called automatic means but depending upon the service of an attendant for the connecting together of any two subscribers' lines.

We have now to consider even more elaborate machines whereby the position is to some extent reversed, the subscriber performing certain definite and conscious operations which, by means of suitable machinery, effect the connections desired. The machines have become conventionally known as automatic switchboards, and though the accuracy of the term is questioned, its adoption has become so general that it would be inconvenient at the present stage to use any other.

Telegraphic instruments sending signals without an operator have been called automatic telegraphs, as for example 'Wheatstone's Automatic,' though the paper slip used therein is previously punched by hand. Such systems have of late years been frequently called machine telegraphy. Perhaps the automatic telephone system, if it survives, will undergo a similar change. 'Automatic' is sufficiently accurate to describe the switchboard, which is really an automaton. It becomes inaccurate and somewhat misleading when applied to the system as a whole, for the so-called automatic system requires conscious effort on the part of the subscriber for every stage of operation, whilst much of the so-called manual system is carried out automatically without any conscious effort on the part of the subscriber.

One of the earliest examples of the machine or automatic type of switchboard, and the earliest in patent date, is that referred to in the following letter :—

Law Offices of Connolly Brothers,
Patents and Patent Causes,
Philadelphia.
September 17, 1884.

DEAR SIR,—We tender you and the members of the National Telephone Association a cordial invitation to inspect, criticise, and improve upon our *Automatic Telephone Switch*, which will be set up for exhibition to-day at the Electrical Exposition.

As this invention has been productive of much controversy among telephone people, many of whom asserted it was impossible to effect what it proposed, and as it *does its work*, enabling subscribers to effect connection of their lines interchangeably, without personal or manual service at the Central Office, we feel confident that an inspection of it will be interesting to yourself and the members of the Association.

The location of the exhibit is in Section U 15, near B. & O. Telegraph Office, S.E. corner of Exposition Building.

Very respectfully,

CONNOLLY BROTHERS.

Morris F. Tyler, Esq., President,
National Telegraph Exchange.[1]

The Connolly Automatic Patent is U.S. No. 222,458, dated December 9, 1879.[2] It is the joint invention of Messrs. M. D. and T. A. Connolly of Philadelphia and Thomas J. McTighe of Pittsburg, and is entitled 'Improvement in Automatic Telephone Exchanges,' suggesting that automatic exchanges of some kind were already in existence. The patentees say :—

The object of our invention is to provide what may be termed an 'automatic telephonic exchange' in which each station of the exchange is in communication with a main or principal station, through which connections are established between any two of the individual stations.[3]

Conforming to the usual practice in the preamble of patent specifications, the patentees relate the prior method and indicate their own view of the unsatisfactory results.

Under the present system in use in the principal cities having telephonic facilities, the lines from the several stations converge to a central office and terminate in a switchboard. When any individual member of the exchange desires to communicate with any other member he signals the central office, states his desires, and an attendant thereupon makes the desired connection. The operation of making these connections is now altogether a manual work, and requires not only constant attention but much dexterity in order that there shall be as little delay as possible ; but in exchanges comprising many members, the work of the central office is very great, requiring many employés to meet the wants of the community. Even then there are incessant delays, much confusion,

[1] *National Telephone Exchange Association Report*, 1884, p. 76.

[2] Application filed September 10, 1879.

[3] U.S. specification, No. 222,458, December 9, 1879 (application filed September 10, 1879).

and consequently many mistakes and annoyances which it is highly important should be obviated.[1]

The means to be adopted to obviate all these mistakes and annoyances very naturally follow :—

Our present invention contemplates the employment, in lieu of manual labour and the necessary skill and intelligence to apply it, of the capabilities of electricity and electro-magnetism whereby all the difficulties now met with are entirely overcome and the operation of the central office rendered completely automatic, rapid, and reliable. Each or any member of the exchange may, by means of local contrivances having electrical communication with the central office, place himself in communication with any other member whose line happens to be unoccupied. At the same

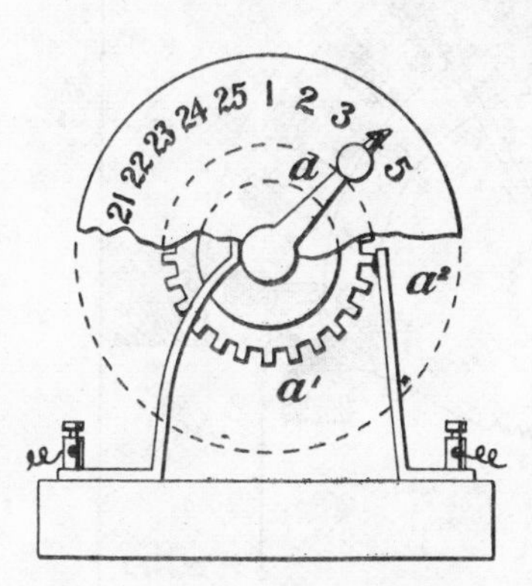

FIG. 146.—Connolly Automatic (Dial).

FIG. 147.—Froment Telegraph Transmitting Dial.

time he is enabled to entirely isolate his own and the line he desires to communicate with from all others of the exchange so that no interferences or interruptions can possibly occur. He is also enabled to signal the member to be communicated with, and in fact to place his own and the other line into the most desirable and convenient relation to each other and to the balance of the exchange as the most urgent demands of the telephonic-exchange system require.[1]

Considering the date of the invention, there are features of much interest in Connolly and McTighe's specification, but space will not permit extended reference to its mechanical details. In automatic apparatus it is natural to turn first to their limitations. 'Upon the face of the dial are indicated the numbers or letters of the different

[1] U.S. specification, No. 222,458, December 9, 1879 (application filed September 10, 1879).

stations in the system.' The relation of the pointer to the teeth on the periphery is described.

These teeth correspond in number and position with the number or letters on the dial, so that when the pointer coincides with, say, 100 on the dial, it indicates that 100 teeth have passed the spring and a like number of breaks have been made.[1]

While 100 is mentioned in the letterpress description, the dials shown in the drawings have no numbers exceeding 25, being in

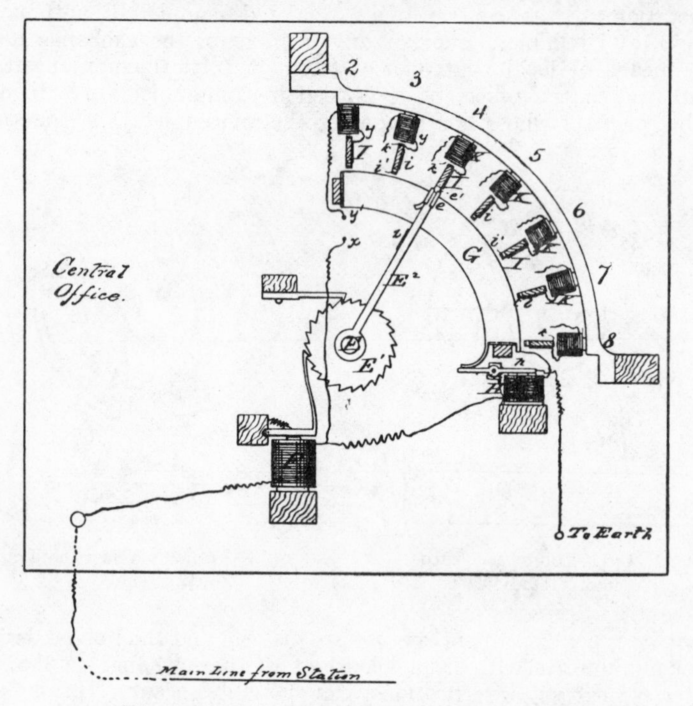

FIG. 148.—Connolly Automatic (Central Office).

this respect no advance on pulsation-transmitting devices previously used in dial telegraphs. Pouillet's 'Éléments de Physique Expérimentale,' seventh edition (1856),[2] describes the ordinary dial telegraph and particular improvements effected therein by certain specified inventors, including those of Froment presented to the Academy of Sciences in 1851.

Comparison of fig. 146 enlarged from Connolly's specification with fig. 147 from Pouillet will show that the selecting apparatus

[1] U.S. specification, No. 222,458, December 9, 1879 (application filed September 10, 1879).

[2] Vol. i. § 340.

of Connolly was practically the well-known dial telegraph which had been in existence for about thirty years.

The similarity of operation at the receiving station will also be observed from the comparison of Connolly's illustration indicating the central office (fig. 148) and that of Pouillet illustrating the receiving apparatus of an ordinary dial telegraph (fig. 149). Froment's was described in detail by Shaffner [1] and the manipulator and receiver illustrated.

The twenty-five detents in Connolly and McTighe's illustrations are probably due to the fact that they took as a model an instrument intended to indicate all the letters of the alphabet less one—the W was omitted by reason of the French origin of this invention. They describe its operation and say :—

This is to be understood as merely a suggestion as to the means of breaking the circuit and indicating the intermissions. Any of the well-known forms of dial instruments adapted to the use required may be employed, and hence our invention is not limited to any particular one.[2]

Fig. 149.—Froment Telegraph Receiving Apparatus.

The novelty of the invention begins with the multiplication of the receiving ratchet wheels E, there being one for each station and for each ratchet a pointer travelling with it, and a fixed segment G. Beyond the segment there is for each line a bar I long enough to reach from end to end of the series of segments G and arranged 'parallel with the common axis' (i.e. shown in cross-section in fig. 148). From each bar project hooks i^1 of metal,

there being as many hooks on each bar as there are ratchets and pointers—i.e. one hook for each circuit that enters the central office —and these hooks are made to stand normally, in such position that the hook e^1 of the slide e on pointer E^2 will pass through, and in electrical contact with, hook i^1 when the pointer is rotated.[2]

The contact between the pointer and the hook of bar I is made

[1] *The Telegraph Manual*, 1859, p. 374.

[2] U.S. specification, No. 222,458, December 9, 1879 (application filed September 10, 1879).

by a sleeve *e* with a longitudinal movement on the pointer arm. The sleeve *e* has a projection with a right angle bend e^1 which engages with another right angle bend i^1 on bar I. The reversing key is an important feature of the apparatus. In the words of the patentees it 'effects several remarkable results.' One of these is that the magnet K has its poles changed, attracts the bar I 'which is pivoted at i^2 (or otherwise arranged so as to be moved radially from the centre) and pulls the latter out.' The effect cannot be more graphically or succinctly described than in the language of the patentees :—

> The result is that the hooks i^1 of bar I are now out of the path of the hooks of all other pointers, and consequently the circuit thus established between the two local stations is completely isolated, and the pointer of the central office ratchet of no other local station can obtain a contact with said circuit. Hence absolute immunity against interruption, and all the annoyances of cutting in and cutting out, cross-talking, etc., is afforded, thus ensuring the utmost privacy for the stations wishing to converse.[1]

The work of the subscribers is thus described :—

> To sum up, then, the operations required are as follows: At the station desiring to call another station, operator [i.e. subscriber] whirls round his index till it arrives opposite the number of the station desired, reverses his key and then sets his switch on the call-bell. As soon as the answering signal is received, he sets his switch to the third point, which shunts the local battery into the primary of the telephone and places the secondary in the main line. After he is done talking, he switches on the battery to the main line and places his reversing key back to normal, which restores the central office to normal, and then he whirls his index around to zero and switches his bell into main.
>
> At the station called, as soon as the bell rings the operator switches his battery into the line with its direction conforming to that required to preserve the armatures at central office in the positions set by station calling. This will at once ring the bell of the station calling, after which he switches his telephone secondary to the line and battery to the telephone. After the connection is finished he switches back to the bell simply.[1]

The opinion of the inventors in regard to the capabilities of the average telephone user was evidently high, for they say :—

> These are simple operations, requiring no skill, and may be all performed successively within the period of a moment or two.[1]

[1] U.S. specification, No. 222,458, December 9, 1879 (application filed September 10, 1879).

In modern automatic phraseology their invention may be said to combine the features of a ' connector ' and a ' selector,' a point of much interest being the method by which a connected ' connector ' was removed from the path of other ' selectors.'

The Connolly system was exhibited at the Electrical Exhibition in Paris in 1881, and was described in the *Journal Télégraphique* of December 25, 1881, by M. Rothen.[1] The installation was for eight stations. The mechanism, says M. Rothen, was certainly very ingenious and worked well, but would it do the same in working on real lines ? He could not disguise from himself that success was very doubtful. It was perhaps M. Rothen's doubts of 1881 that Messrs. Connolly referred to in their letter of 1884 at the beginning of this chapter, and that prompted them to call the attention of the members of the telephone convention to the fact that ' it *does its work*.'

About a month later than the Connolly and McTighe patent date, George Westinghouse (junior) of Pittsburg, Pennsylvania, applied for patents on automatic exchange apparatus.[2] In these applications, Westinghouse did not contemplate the adoption of automatic working in a large exchange, but provided only for its use as an auxiliary in outlying areas. After describing the ordinary use of the telephone he says :—

It has also been found that in suburban or outlying villages, boroughs, etc., a few miles distant from the central exchange, a few persons frequently reside who desire to be in telephonic communication at their residences with one or more of the users of telephones in the city, but the number of such suburban residents in any one locality is frequently so small that it does not pay commercially to maintain a separate local exchange for them, and the distance is so great that the cost of a separate wire for each, leading to the main city exchange, prevents them from enjoying the desired advantages and conveniences of a home telephone. . . .

By my present invention I enable each such suburban user to call and open telephonic communication through the auxiliary exchange with the main or city exchange, and through it with any desired city user without the necessary intervention of an operator at the auxiliary exchange. In doing this he automatically locks the opening or closing connections of the other suburban users having connections with the same auxiliary exchange, so that they cannot call or interfere with his use of the line until he is through.

The invention also includes provision by which, when such user is through [i.e. completed his conversation], the operator at the main

[1] Quoted by Du Moncel, *Le Téléphone*, fourth (French) edition, p. 365.

[2] U.S. specifications, Nos. 223,201, 223,202, dated December 30, 1879 (applications filed respectively October 11 and 13, 1879).

or city exchange can restore the apparatus at the auxiliary exchange to its normal condition, so that any other suburban user can call and hold telephonic communication in like manner with a city user, and also in so doing lock out his suburban co-users as before.[1]

Westinghouse contemplated an auxiliary automatic exchange, and the system suffered from the same limitations as did the party line—the willingness of the subscribers to participate in a service in which they were liable to be locked out by their co-users.

Inventors in the field of automatic switchboards in the main limited themselves to the provision of apparatus serving the purpose so clearly described by Westinghouse in 1879.[2] Mr. D. Sinclair installed at Glasgow some experimental apparatus in 1883.[3]

The *Electrical Review* of London, November 17, 1883,[4] describes a system the English rights of which had been acquired by Mr. A. S. Paul. This, according to my recollection, was the production of Ericsson of Stockholm, and was, like Westinghouse's, intended only for auxiliary exchanges. It worked like Connolly's, by a step-by-step arrangement. There were numerous other inventions of the same class—none being adapted for use by large numbers of subscribers. Mr. Kempster Miller[5] considers that the first steps towards the realisation of the ideas advanced by Connolly and McTighe were the inventions made by Strowger, commencing with his U.S. patent No. 447, 918, dated March 10, 1891. This application was filed on March 12, 1889, approximately ten years after that of Connolly and McTighe. While an advance upon its predecessor in the more elaborate operation of the central office connecting device, it must be considered as retrograde in that, instead of the single wire of Connolly, it required five wires from the sub-station to the central station.

Fig. 150 is a reproduction of fig. 1 of Strowger's specification, No. 447,918. The method of operation is thus described in the specification :—

The person wishing to place his transmitter and ear-phone in connection with those of another, he will do so by successively pressing or depressing the keys, which cause the circuit closer C C′ [a pointer or connecting arm revolving within the cylinder A and thus not shown in the diagram] to move. For example, if telephone 288 wishes to place himself in connection with telephone 315, he will do so by pressing the key marked G′ three times, then the key marked H′ once, and then the key marked I′ five times. His circuit closer C C′ is then in contact with wire terminal

[1] U.S. specification, No. 223,201.
[2] Other U.S. patents granted to Westinghouse on the same subject were No. 224,565, February 17, 1880, and No. 237,222, February 1, 1881.
[3] Aitken's *Manual of the Telephone*, 1911, p. 576. [4] Vol. xiii. p. 382.
[5] *American Telephone Practice*, fourth ed., 1905, p. 692.

No. 315. . . . The person at telephone No. 315 will take down his ear-phone. The two are then able to converse with each other. When conversation is ended, the person calling up hangs up his ear-phone, depresses key marked P′, which causes the magnets K′ to be energised, attracting the armatures, thereby withdrawing the several pawls from their engagement with the ratchet teeth, and allow the circuit closer C C′ to fall and return to its initial point.

If a person has called up the wrong number he will push the key marked P′ and start over again.[1]

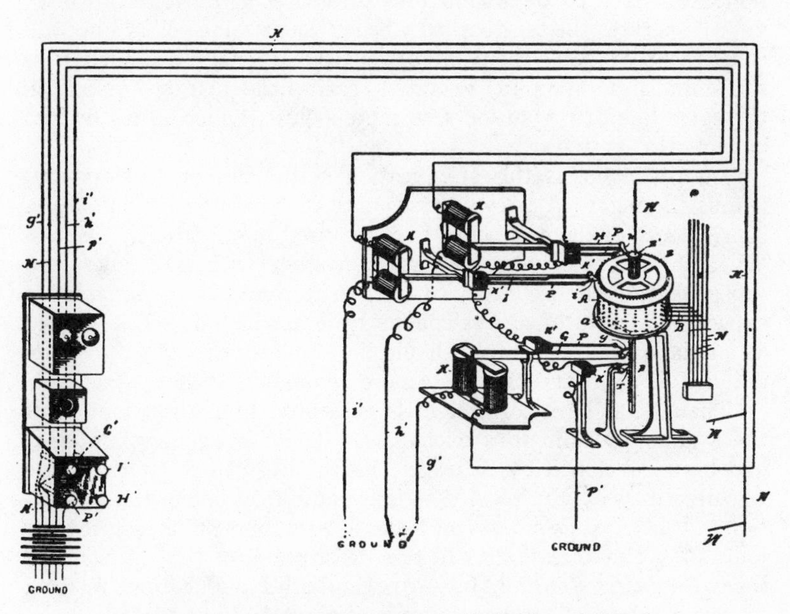

FIG. 150.—Strowger's Automatic Switchboard (1891).

Each of the keys mentioned above is connected by a separate wire to the central office. There are thus four wires to operate the exchange mechanism and one for the speaking circuit.

The connecting points on the cylinder are in hundreds around the circumference, each hundred commencing a new row. The 'circuit closing needle (C′),' which may be regarded as the equivalent of the pointer in the receiving dial telegraph, is mounted on a shaft which has two movements—vertical and circumferential. As it moves it makes contact with 'connectives "or legs,"' applicable to each of the lines on the inner circumference of the cylinder. As these 'connectives' are arranged in rows of one hundred, it follows

[1] U.S. specification, No. 447,918, p. 2, line 81.

that if the shaft to which the pointer is attached is moved upward to the extent of one row, it would have a different hundred at command. That is what happens when key G′ is operated. A current is sent which energises a magnet; the magnet operates a pawl which lifts the shaft to the extent of one row of contacts. The same happens with each repetition of the key movement, so that with three pushes of key G′ the shaft has been raised so that the needle or pointer has command of the third hundred subscribers. To the shaft are attached two cam wheels, one with ten teeth, the other with one hundred. Key H′ by a similar set of devices as those recorded for key G′ operates the ten-toothed wheel. Depressing key H′ once will consequently revolve the pointer over ten contacts. Key I′ controls the units and, becoming operated, carries the pointer forward in the same line of revolution five more points, which is its desired haven—No. 315.

Interesting as is this specification in the line of development, it must be regarded as even less practical than that of Connolly in consequence of the number of lines required to operate it. But as Wheatstone began his telegraphic system with five lines and subsequently reduced them to one, so it would be reasonable to expect that an automatic telephone system, which may be regarded as an elaborated telegraph, should be similarly brought to a more practicable form and assume a more commercial basis.

That it has been so brought is evidenced from the adoption of the system by numerous exchanges. The Strowger system, as it had been developed by Strowger, Keith, and others, at that date was described by Mr. W. Aitken in a paper read before the Institution of Electrical Engineers on May 18, 1911, and published in their journal.[1] The accessibility of this description, of that in the *Post Office Engineers Journal*, July 1912, and other publications relating to the apparatus, renders it unnecessary to give details here. Illustrating the direction of simplification by reduction of the number of lines, I extract the following from Mr. Aitken's paper:—

The greater number of exchanges are on what is known as the 3-wire system, the signalling being over the two wires alternately with earth return; but in San Francisco and other later exchanges, all signalling is effected by the intermittent opening of the circuit at the subscriber's instrument. This has greatly simplified the subscriber's instrument.[2]

One of the most important modifications is that effected in the dial sending apparatus, which forms the subject of British

[1] Vol. xlvii. p. 651.
[2] *Journal of the Institution of Electrical Engineers*, xlvii. 658.

specification No. 809 of 1898.[1] Fig. 151 ' is a front elevation showing the parts which are ordinarily visible.' If it is desired to communicate with telephone number 456 the subscriber ' inserts a finger in contact with finger hold number 4 which is pulled down around in the direction indicated by arrow 7 until the finger contacts the shoulder of the case S at S^2, when the finger is withdrawn, which releases the finger hold disc C.' A spring then returns the parts to the normal position. These operations are repeated successively for the numbers 5 and 6.

The method will recall that of the American district call box (Chapter ix), variety being obtained by repetition of movements.

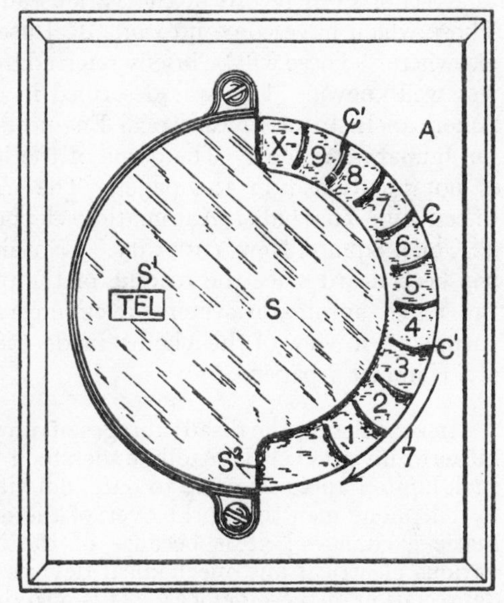

Fig. 151.—Strowger Dial-sending Apparatus (1898).

There is here an improvement both mechanically and from the subscriber's view point. The sequence of the operations follows more naturally to the subscriber than the selection of different keys and depressing them a required number of times, as indicated in the method described in the United States 1891 patent.

It will be understood that the illustration fig. 151 is from the original patent covering this feature, and that, in the meantime, considerable improvements have been effected both in the arrangement of the apparatus for the convenience of the subscriber and in the performance of the electrical operations, details of which have been published [2] as above mentioned.

Another means of sending the call adopted in other systems is that of setting up the required numbers first and then starting the train of operations. By this method a subscriber can satisfy himself before sending the call that the number is correct.

[1] U.S. equivalent, specification No. 597,062, January 11, 1898.

[2] The latest publication at the time of writing is *Automatic Telephony*, by Arthur Bessey Smith and Wilson Lee Campbell (New York: McGraw Hill Book Company), 1914.

The method first described permits the series of pulsations to operate the machinery in detachments, so that they are completed as the last number is sent. In the second method the pulsations for the whole group of numbers follow each other without any break. The detached method has been more generally used so far, but there has perhaps not yet been sufficient experience to determine definitely the balance of advantages in respect to the reliability of service and economy in time or the costs of installation and upkeep in the respective plans.

I regret that the limits of this chapter prevent any complete analysis or even record of the various automatic systems devised. Those which have come into practical use are described in detail elsewhere. These will be briefly referred to as well as another system less well known. This was described in a paper presented to the American Institute of Electrical Engineers by Mr. W. J. Hammer on January 23, 1903. The name of the inventor of the apparatus is not mentioned in the paper. The machine was being introduced by 'The Faller Automatic Telephone Exchange Company, 22 Union Square, New York City.'[1] Nothing is heard now and little has been heard since the reading of the paper above mentioned of the Faller automatic system, a circumstance which is somewhat surprising, in view of the claims made for it in the conclusion of Mr. Hammer's paper.

In considering the disadvantages of automatic exchange systems hitherto invented, it is readily understood why they have met with such limited application up to date, notwithstanding the realisation by telephone men the world over of the many advantages of automatic exchanges; it is because of the inherent delicacy of the various electrical and mechanical devices employed and the rapid depreciation of these devices. It is also due to their extreme complexity, especially in the multiplication in the numbers of wires, contacts, relays, magnets, etc. which increase out of all proportion with the increase in the number of subscribers. It is quite common to employ from five to seven or more magnets and relays in each switching device, and employing one to three of these switching devices in establishing the necessary connections, and utilising these magnets and relays for performing the actual mechanical work of lifting, rotating, and switching the mechanism employed with all the disadvantages which this entails; and, finally, it is by reason of the very high class of engineering talent essential in the designing and constructing of the complicated apparatus which has been employed.

It is self-evident to all electrical engineers who have had experience with electro-mechanical devices that it is far better to

[1] U.S. patent No. 686,892, granted to E. A. Faller, dated November 19, 1901 (application filed August 22, 1901).

employ a well-designed mechanism performing a definite cycle of operations and driven by some source of power.

This mechanical operator has been designed and constructed upon the following lines : i.e. an absolutely reliable, prompt, rapid, and uniform service in telephone exchanges of any size ; an embodiment of all essential features demanded by the exigencies of modern telephone practice ; an elimination of the unreliable, flimsy, and complex features inherent in systems heretofore presented to the public ; and, finally, the effecting of large economies in construction and maintenance charges.[1]

The power driving—making the pulsations over the line directive rather than operative, and permitting the use of more substantial contrivances—is a feature which modern systems utilise to the full.

The Lorimer[2] is one of these systems, and that developed by the Western Electric Company is another. Whilst in these systems there are many valuable original features, it is to be recalled that the use of electrical pulsations as a controlling agent to determine the movements of mechanism put into operation by other power is in no sense novel. An early example is that of the House Printing Telegraph, first designed in 1846 and protected in 1852 by a patent (U.S.), of which the first claim is—

The employment of electro-magnetic force in combination with the force of a current of air or other fluid, so that the action of the former governs or controls the action of the latter, for the purpose described.[3]

In the Faller system the called number was set up complete before any of the pulsations were sent in.

This feature also was combined in the Lorimer system invented by the Brothers Lorimer of Brantford, Ontario, the Canadian home of Alexander Graham Bell's father, and the place which divides with Boston the honour of the invention of the telephone.

Another prolific patentee of automatic systems is Clement, and a more recently developed system that of a Swedish engineer, Betulander. Commencing with a selective device after the manner of, but differing from, Strowger in its mechanical details, Betulander developed a system employing stationary relays in place of moving mechanisms.

The basic principle which permits of all the operations being carried out entirely by relays is a series of ' link lines ' comparable with a number of double-cord circuits of a manual exchange, which, at the first stage of a call, effect a connection between the calling

[1] *Transactions of the American Institute of Electrical Engineers*, xxi. 53.
[2] *Post Office Electrical Engineers Journal*, July 1913, vi. 97.
[3] Shaffner, *The Telegraph Manual*, p. 401.

line and an idle outgoing junction leading to the ten thousands, thousands or hundreds group, depending upon the size of the exchange. The subsequent stages down to the tens, and the individual unit required, are reached in exactly the same way, and when the connection is set up the various groups of common relays which searched for the idle junctions, and responded to the impulses from the subscriber's instrument from stage to stage of the call, immediately free themselves from the connection and take charge of other calls. When a connection is completed, the two subscribers are connected by a transmission circuit on the 'Stone' system equal to that employed in the best central-battery manual practice.[1]

Messrs. Rorty and Bullard, of the American Telephone and Telegraph Company, designed apparatus mainly intended for auxiliary exchanges which was put into operation in several places, one being St. Louis, in or before 1905. Their work included the application of common battery current supply to automatic service and formed the basis of a special department instituted by the Western Electric Company for the study and development of automatic apparatus. The Western Electric 'Machine Switching' Telephone System is described in *The Electrician*, December 25, 1914, and subsequent issues. The writer of the articles (G. H. Green, B.Sc.) summarises their scope as follows :—

The article[s] describe the automatic and semi-automatic telephone systems of the Western Electric Company, notable features of the system being the construction of selectors used, the method of controlling them by so-called 'sequence switches' and the use of selectors serving a larger number of lines than usual.'

It is stated that a full automatic system installed at Darlington was put into public use in October 1914, and a semi-automatic system at Anger, France. Mr. F. R. McBerty is mentioned as the chief inventor of the system.[2]

In the exhibition at Chicago in 1893, commonly known as the World's Fair, there were two exhibits of automatic switchboards, one by the American Bell Telephone Company and the other by the Strowger Company. The official record of the exhibition says of the American Bell Company's exhibit :—

A number of small towns have been equipped with automatic exchanges of this kind, as many as forty-five subscribers being connected in some cases. For purposes of exhibition only three subscribers' stations were installed, the principle of operation being the same for any number.[3]

[1] *Electrical Review*, London, July 10, 1914, lxxv. 44.
[2] *The Electrician*, lxxiv. 390, 420, 453, 494.
[3] *Electricity at the World's Columbian Exhibition*, Barrett, p. 331.

The methods of construction and operation are not described. Of the Strowger exhibit it is said:—

In this exhibit it was intended to represent an automatically working central station of a telephone system, and connected therewith were nine telephone stations outside of said area, and one telephone within, making ten in all. They were so connected up that communication could be held between any two of the ten 'phones. The space was fitted up as a central telephone office. Upon a framework of shelves were one hundred switches, but only ten were connected up, each one with its own particular telephone as regarded manipulation, but together for telephonic service. The object of these switches was to accomplish by mechanical means the work done by the operators at the central office and, in other words, to dispense with the services of the telephone girl.[1]

A description of the apparatus in general terms follows (the operation being apparently in accord with the 1891 patent above quoted), and in conclusion it is said:—

The Strowger system is applicable to trunk line service, with corresponding cheapness. Only one operator is required for a number of towns, which operator is needed to keep an account of the toll.[1]

The American Bell Telephone Company had thus apparently a number of automatic systems in operation in 1893.

It was, however, Strowger and his associates who first sought to develop an automatic system with a view to its use on a large scale. The operator was to be dispensed with. Having developed such a system up to a certain point, means had to be found for putting it into operation under real working conditions.

Immediately upon the expiry of the Bell patent in the United States, numerous companies were formed to give telephone service in competition. In general these companies followed the same operating methods as the Bell Companies, though necessarily with less experience and with some limitations due to patent control in other apparatus than the speaking instruments.

These competing companies were known as 'Independents,' meaning that they were independent of Bell influence or control. This independence was an important factor in the development of competition, the idea being that a profitable enterprise might be conducted with local influence and finance at lower rates than the Bell Companies charged. The Independent movement flourished to a considerable extent, and it took several years to demonstrate to others what was known to those who had been longer in the

[1] *Electricity at the World's Columbian Exhibition*, Barrett, p. 340.

business, that there were two important factors not present to the minds of promoters and subscribers. In the first place, competition in telephone service is wasteful and of public disadvantage, since it is not competition in any true sense, but duplication. In the second place, financial results cannot be judged without a substantial allowance for depreciation.

The public inconvenience from competing services was apparent in the early competition in the United States, to which reference was made in Chapter xvi. Three companies were at first authorised to carry on the service in Paris, but shortly after the opening of the communications the municipality and the companies themselves found that the existence of several companies in the same town was an impossibility. Fusion consequently became a necessity, and it came about under the auspices of the State. Hospitalier[1] says that the existence of separate companies was much to the detriment of the subscribers and the development of the system, but since the fusion matters had considerably improved. In other places on the Continent where competition was experienced the same result—combination—was attained. Space does not permit extended reference to the principles involved. They formed the subject of a paper which I read before the British Association in Belfast in 1902.[2] It suffices to repeat that it is now generally recognised that competition in telephone exchange service in one locality is an economic waste and a public disadvantage. Nevertheless, the post patent competition in the United States was coincident with considerable development.

From natural causes that competition has been largely reduced, and it seems probable would be still more largely reduced but for the influence of political tendencies, which retard the operation of natural laws. Widespread sentiment usually has the public welfare as a basis, however mistaken it may be, and for some years past the sentiment of a large section of the inhabitants of the United States has been against combinations in industrial enterprises. The underlying objection is the belief that such combinations are in restraint of trade and enterprise. The application of the same sentiment to competing telephone exchanges overlooks the essential difference between telephone service and the supply of other commodities, for unification upon fair and reasonable terms to both parties is in the public interest, and not—as is so generally though erroneously thought—opposed to it.

The adoption of automatic switchboards gave the independent movement an additional claim to popular support. The promoters

[1] *The Modern Applications of Electricity*, 1882, p. 390.

[2] 'The Future of the Telephone in the United Kingdom,' *Electrical Review*, September 19, 1902, li. 517; *Electrician*, September 26, 1902, xlix. 893, 907.

opposed the Bell 'monopoly'; they offered employment to local capital and influence, as other independent companies had done. In addition, they promised the abolition of 'the telephone girl.'

The operator is, in the telephone system, the representative of the organisation. Though not a visible sign, she is the audible personal connection between a subscriber and a complex system, and whether trouble be occasioned by the subscriber himself, by some sin of omission or commission on the part of his desired correspondent, by a failure on the part of other departments of the service, or a catastrophe beyond the control of all the parties, the operator, as the active agent, has to bear the blame. Telephone subscribers collectively are not discriminating. The many helpful acts of the telephone girl are taken as matters of course, and are forgotten. The deficiencies may be few, but they leave a very strong impression, and the possibility of doing without such intermediaries as operators may to certain members of the telephone public seem attractive.

Whether this be so or not, it is undoubted that automatic exchanges made greater headway, grew to larger numbers, and have continued in operation for a longer period than some competent observers contemplated. As an example of earlier opinion and changed conviction may be quoted Mr. Kempster Miller, an able and independent telephone engineer, who wrote in his 'American Telephone Practice':—

> In an early edition of this work this statement is made concerning the subject of automatic telephone switchboards. 'It is with the idea of following briefly, though not completely, the growth of an interesting phase of telephone work, rather than of attempting to chronicle any really practical developments, or of giving any hopes of its future practicability that this chapter is written.' During the last few years, and particularly within the years 1903 and 1904, conditions have greatly changed, and the automatic switchboard, or more particularly the so-called automatic exchange, has come into such prominence through the efforts of its developers and promoters as to make it appear a decidedly important factor at the present date, and one that will be of increasing importance in the future.[1]

A more extended consideration of the subject was given by the same authority at the International Electrical Congress held at St. Louis in the previous year (1904). In this paper he gave an account of certain tests and experiences, and in general recorded the impression above quoted.

The interest in automatic telephony widened and spread to

[1] *American Telephone Practice*, seventh edition, 1905, p. 691.

Europe. It was adopted in some central offices in Austria, and several practical demonstrations were made in Germany, where the Strowger inventions were acquired, and in many respects modified, by Messrs. Siemens & Halske. It formed one of the most important subjects of discussion at the International Conference in Paris in 1910, when Mr. J. J. Carty delivered an address in which he referred to the many automatic devices of the so-called manual system, and the manual operation by the subscriber of the so-called automatic system. Mr. Carty's address was a very important contribution to the automatic controversy. It was published in the technical[1] papers of the period, and is, therefore, available for reference.

The conclusion reached by Mr. Carty after very careful study was that the automatic system was not adapted to the requirements of telephone service in America. The reservation of locality was apparently made in order to restrict his remarks within the lines of his own experience as well as, perhaps, to recognise that the European experts whom he was addressing had problems with which he was unacquainted, which problems might lead them to different conclusions.

Local conditions have to be studied, and Mr. Carty very properly recognised that fact. But local conditions do not affect systems. They only affect details. Telephone service is required for the same purpose the world over. It is true that the American service is larger and more widespread. It is more general, more frequently used, and used over longer distances. If other countries have fewer numbers, less use and narrower boundaries, they have to look forward to growth in numbers and increase in use, whilst national boundaries are merely names. Telephonically, Great Britain is not separated from Europe by the Channel. Submarine cables as at present constructed limit the distance of speech, but suppose that a dry core cable suitable for submergence be evolved or that the Channel Tunnel be completed, the talking distance will be extended, the cost reduced, and the number of communications increased.

Local conditions as they are should not weigh unduly when experience on a larger scale is under consideration. Local conditions as they might become should also be brought within the field of vision.

The method which is best suited to the accomplishment of telephonic intercommunication can only be determined by reasoned argument and practical demonstration. Practical demonstrations of the automatic system are now in progress in numerous localities

[1] *Electrical Review*, London, November 18 and 25, 1910, lxvii. 844, 883; *Electrical Review and Western Electrician*, Chicago, lvii. 794.

and under varied conditions, so that in the course of time its proper rôle will certainly be discovered.

Meantime, it may be well to remember that the various switching systems which have by force of demonstrated facts retired into the background, were recommended by their advocates with much zeal and confirmed confidence. Some of the systems were, moreover, strongly supported by the statements that the subscribers preferred them to other available systems, and in the main they required the subscriber himself to undertake some of the necessary steps in the completion of a telephonic connection. It will be convenient to quote here again the standard of service established in 1887 :—

It is the duty and obligation of a central station, when a call is received, to follow the call through and bring the subscribers actually into communication with one another.[1]

The system which has so far survived is that which has enabled this standard to be attained in the most effective manner. It may be urged that the supervision has only been necessary as a check on operators' errors ; that it was adopted when telephone service was comparatively new ; and that, in the meantime, telephone subscribers have become more proficient. But whilst adopted quite early in the service, the utility of the operation of this supervision has been continuously demonstrated to the present time. The telephone subscriber of the future cannot be expected to inherit the knowledge his forbears had acquired, and simplicity in use must always be an advantage.

The supervision of a telephonic connection has been regarded since 1887 as the essential duty of the operating staff at the central office. In the automatic system supervision finds no place. The connecting mechanism is an automaton, and the 'working of the figure' is dependent upon 'the pulling of the wires.' Can the subscriber be depended upon always to pull the wires correctly ? The automatic advocate will say that if he does not the resultant defect is his own fault. In most cases this may be true, but in some cases the automaton will be at fault, and in every case the subscriber will assume so.

Caution is needed when subscribers' preferences are urged in favour of a particular system. The favourable opinions of subscribers have been advanced as a reason for the installation of systems theoretically and practically defective. What the public need, and have a right to expect, is good telephone service. The means by which it should be attained are too complex for them to be able to give an intelligent opinion upon.

[1] Chapter xxiv. p. 315.

The reason why the subscribers—in so far as their opinions have been asked—have gone wrong in the past, is simply that they have taken a purely personal view of the matter. The Law system gave a preference to the insistent caller, it gave him control of two connected lines; the ring-through system gave a facility for the continued connection of lines for which other callers were waiting.[1]

Subscribers express a preference for a particular form of transmitter, regardless of the fact that its use may offer some difficulties to the reception of the conversation by the subscriber to whom he is speaking. Telephone service is the service of a community, not of an individual; and that which is best for the community is best for the individual, in telephony as in other respects.

Another point of public preference liable to mislead is the comparison of a system well worked with some other system badly worked. The automatic system must inevitably appear promising in comparison with the most perfect manual machinery where the manual part of it is imperfectly worked.

In 1908 the New York Telephone Company published in the daily newspapers a series of what were called 'Educational Bulletins.' The following is an extract from one of these:—

> Telephone operating. Efficient telephone service is critically dependent upon good telephone operating.
>
> A telephone plant may be properly designed to meet every traffic requirement. It may be well constructed and maintained. But unless it is operated by a competent force well organised and well trained, the service will be inefficient.

A system which promises to do away with all operating may be accorded very readily a preference over a system which, however good as a machine, is badly operated; but such a comparison is misleading in its conclusions unless it be admitted that the administration responsible is unable to organise its forces so as to effect the best results which the machine is capable of giving. And in the event of that admission, it follows that the automatic preference may arise from some defect within the organisation and not a defect in the manual system itself. It is only by a very careful balance of the possibilities that it can be determined whether it were better to endeavour to improve the organisation of the expert personnel or to seek the aid of machinery operated by the unexpert subscriber. To adopt automatic is to flatter the subscriber, who may be expected of course to be appreciative, at least at the beginning, but who will certainly disclaim responsibility if failure follows.

Exclusively automatic telephone service is not at the present

[1] Chapter xxiii. p. 299.

time suggested by any automatic advocate. In all systems so far proposed, an index is provided which will connect the subscriber with a trunk or toll operator, to whom he will communicate through the telephone the trunk or toll number required. The automatic feature, then, is limited to local calls. How local has not yet been determined. The Educational Bulletin previously quoted says :—

In New York and vicinity there are 185 central offices connected by nearly 20,000 trunk [junction] lines, and serving over 400,000 telephones.

Those 185 central offices are known by the names of their locality. The local designation of an office and the numerical definition of the subscribers connected thereto are convenient methods, but numerical definition might be adopted over a wider area. In that case a limit has to be assumed for any locality. Such a limit was assumed by the Edison Telephone Company of London in 1880. In the list of subscribers dated April 28, 1880, five central offices are mentioned as being in operation, but numbers only appear against subscribers' names. There is no definition of central offices. Examination of the list by localities shows that 200 were reserved for Queen Victoria Street central office, 200 for Cornhill, and 200 for Chancery Lane. The West End exchange in Windmill Street was not then in operation. In the light of later developments, such figures show that it is well not to adopt a method which requires limitations of this kind.

So far as the machinery is concerned, it is but reasonable to expect some simplification and an extension of capabilities, but on the whole the limitations of automatic telephones are much the same as any other automatic supply. A limited number of commodities may be obtained from a limited number of shoots controlled by a limited number of slots adapted to the use of particular coins. Commodities outside that range must be asked for and supplied over the counter. It follows that the calls required outside a restricted area must be transmitted through the telephone, which fact reduces the force of the argument sometimes used, that automatic operation is called for because of the imperfect transmission of numbers through the telephone. To suggest the advisability of automatic operation because of the alleged imperfection of an instrument which is supplied for the express purpose of conversation, may seem to be a doubtful recommendation from a telephone authority, but it is to be remembered that in ordinary conversation there are aids to audition which are lacking in the recital of simple numbers. The percentages of error in transmission, verbally and by machinery, are the subjects of careful tests, so that known factors are available. The percentage of error in inception is less readily obtained, but would be the subject of fair estimate.

The obvious limitations, however, of the automatic system show that the operators are not to be extinguished. They can only be reduced—which raises the further question : Shall they be reduced by compelling the subscriber to manipulate complicated apparatus for local calls and use the simple apparatus of his telephone for extra-local calls, or shall they be reduced by providing operators with mechanism which will facilitate their work whilst leaving the subscriber to obtain all calls in the (to him) simplest possible manner through the telephone ?

The latter method is known as 'semi-automatic,' a term also lacking in accuracy but in such general circulation as to compel adoption. All makes of automatic connecting and selecting contrivances can be used for semi-automatic purposes, though some makers have given more special attention to the necessary variants. A machine which can be worked by a subscriber for the selection of a number can obviously be worked by an operator ; and, in view of the fact that the operator is a trained expert, the machine may be made more complex to cover a greater variety of connections.

In the completely manual system a connection in one central office group may require to be passed by the operator of that central office to the operator in another central office. The semi-automatic system gives the first operator control of a connecting mechanism at the second office whereby the services of the second operator are dispensed with. The existence of such machinery is unknown to the subscriber who sends his call to the operator through the telephone in the usual way. The most recent installation of the semi-automatic system is that at Newark, New Jersey. This is on the Western Electric system, and is so arranged as to provide for that supervision of a connection and that assistance by an operator which was the standard of 1887.

Technical descriptions of these systems would involve more space than can be afforded. It has been preferred to give in greater detail particulars of the earlier and less-known work and to discuss the general principles involved in the later developments, the details of which are available elsewhere.

It has been possible to record the results in those previous controversies which have divided expert telephonic opinion. The controversy regarding Automatic *v.* Manual is probably keener than any previous discussion, but it is still (1915) in progress. Results cannot be recorded for some time yet, and any definite forecast is to be deprecated. Switchboard experts were early warned of the possible danger from getting into a groove[1] and the stress

[1] Chapter xxiv. p. 312.

of the service compels the telephone engineer to embrace readily any available method of improving it. An open mind to the possibilities of new methods is essential. But a mind which is open to those possibilities and closed to the experience of the past is a danger.

The 1887 switchboard committee, from whose records I have already freely quoted, discussed the following question :—

> Some systems or organisations of central office apparatus, as we know, are extremely simple in the office and throw the complication outside on to the lines and subscribers' apparatus.
>
> Is this advantageous, or should we aim as far as possible to keep such complications in apparatus and operation as seems unavoidable under the control of the central station? [and unanimously agreed that] it was decidedly advantageous to keep all the complication necessary at a telephone exchange as close to headquarters as possible.

This question had no direct relation to automatic apparatus. The comparative merits of varied manual systems only were being considered, but the importance of the decision as a matter of general policy cannot be overlooked.

Since that time the simplification of the subscribers' apparatus has been carried greatly beyond what was then contemplated, and the service has been proportionately improved and extended. It cannot be said that automatic apparatus is simple at the central office, but any complex machinery that is reliable in working and whose use is desirable on economic grounds may be permitted there. The work which the subscriber has to do in the automatic system requires comparatively complex apparatus at his station, but it does not necessarily follow that the complexity is such as to debar its use. That is to be determined by experience. If, over the whole range of telephone service, it could be shown that with the use of complex apparatus better results were attained, it would be useless to vaunt the merits of simplicity. Such evidence on any reliable basis is lacking because the service so far given by automatic exchanges in large areas has been a competitive, and therefore a supplementary, service. Where it has stood alone the area of communication has been small with no great variety in the conditions of service. Upon simplicity in apparatus at the subscriber's station depends continuity of service and economy in upkeep. Of even greater importance is simplicity and uniformity in operation on the part of the subscriber. These may be regarded as important aids to the extension of telephonic communication. Machinery must come in telephony, as it has come in other enterprises, as a labour and time-saving appliance, but whether the machinery shall be used by experts or by the subscribers remains to be determined.

All that can be done now is to record the progress made, to anticipate further developments and inventions, and, in the light of existing apparatus and conditions, to deprecate extravagant expectations of improvement in the service by entrusting the operating to the subscribers.

But the experience of the past shows that there is something good in every system, though the good feature may be ill adapted to its originally intended use. The call wire of the Law system had to be withdrawn from the public, but formed one of the most valuable features for communication between operators as an order wire ; the efforts at the production of a divided board failed because the multiple system offered greater facilities, but the signalling system developed in connection with divided boards was a valuable branch when grafted on to the multiple stock. In a similar way it seems probable that the efforts to obtain automatic switching by mechanical means, even though it may not be adapted to subscribers' use, may be a valuable aid to the service in the hands of operators.

CHAPTER XXVIII

LONG DISTANCE SERVICE

TWENTY miles was the modest limit for telephonic conversations referred to in the first Bell circular of May 1877.[1] In his lecture before the Society of Telegraph Engineers (I.E.E.) in London on October 31, 1877, Bell referred to the natural question as to the length of line over which the telephone could be used, and said that the maximum amount of resistance through which the undulatory current would pass and yet retain sufficient force to produce an audible sound at the distant end had yet to be determined; in laboratory experiments no difficulty had been experienced in conversing through 60,000 ohms, the maximum at his disposal. Bell, however, evidently realised even then that resistance was not the only enemy the telephonic current had to encounter. The longest line through which he had attempted to converse had been about 250 miles. This was between New York and Boston. Sunday was chosen for the experiment because on that day it was probable other circuits would be unemployed. No difficulty was experienced so long as parallel lines were not carrying telegraph messages, but when business opened upon the other wires the sounds, though still audible, were much diminished. 'It seemed indeed like talking through a storm; conversation, though possible, could be carried on with difficulty, owing to the distracting nature of the interfering currents.'

At the same lecture Bell stated that Mr. Preece had informed him that conversation had been successfully carried on through a submarine cable, sixty miles in length, from Dartmouth to the Island of Guernsey by means of hand, i.e. Bell magneto, telephones.

The introduction of the battery transmitter, valuable as it was for local service, was of still more importance for long distance communication. Prescott says[2] that Edison's carbon telephone had already permitted conversation to be carried on over 500 miles of actual telegraph line, but the statement is considered to be open to some doubt. One of the earliest experiments with

[1] Chapter viii. p. 67.

[2] *Speaking Telephone, etc.*, 1878, p. 40.

this instrument outside the United States was made between London and Norwich over the private telegraph line of Messrs. J. & J. Colman on November 11, 1878. The evening was stormy and in the line there were probably numerous bad joints. What Bell said of his New York-Boston experiment would apply equally to the London-Norwich. 'It seemed indeed like talking through a storm'; but, being a participant in that experiment, I retain a very distinct recollection of the curious effect of the voice apparently forcing its way through the barrier[1] of what was then called 'induction.'

One of the noticeable facts recorded in the report of this experiment (*Morning Advertiser*, November 13, 1878) was that conversation was best heard when carried on even more quietly than the ordinary tone of voice. This fact, so early demonstrated, is to this day very difficult of belief by talkers over long lines, who are generally so impressed with the intervening distance that they almost invariably shout, with a loss of clearness, instead of adopting a conversational tone and paying special attention to distinctness of articulation. The experiment on the Norwich line and others reported about the same time brought to a head the English patent question. Though the transmitters were the Edison carbon, the receivers used were of a magneto type (I think the Phelps pattern) and the Bell patentees claimed that they were infringements. They certainly were. In consequence, commercial operations were delayed until the electro-motograph (or motophone) receiver of Edison was developed a little later. The subsequent combination of the Edison and Bell interests in both the United States and Great Britain enabled the long distance systems to be developed to the best advantage so far as the instruments were concerned. The limitations of the Government licenses under which the British companies worked prevented them from devoting much attention to long distance communications. In the United States there were no governmental restrictions. The only limits were those of science and commerce, and both of these limits had to be found by experiment.

The faith shown by the Bell interests in the utility of the telephone for intercommunication within exchange areas was equally pronounced for communication over longer distances. The contracts referred to in Chapter xv show that so early as July 1878 provision was made for the transaction of business between towns and cities within an agreed area to places outside that area; whilst a contract

[1] A very similar experience is noted in the early experiment between New York and Chicago on the Postal Telegraph Company's wire. At the 1883 meeting of the National Telephone Exchange Association, Mr. Fay said (p. 57): 'Nothing could be more noisy than that postal wire between Chicago and New York. You could hold up the receiver at some distance from your ear and hear the sputtering and frying of that Morse induction very distinctly. It sounded like hail rattling on a roof, but still you could talk past it.'

dated August 9, 1879, and signed by Mr. Vail, exclusively reserved to the Bell Company the extra-territorial or long distance rights.

By the end of 1880 the company was able to report that the business of connecting up cities and towns had been taken up with vigour, and only experience was needed to reach satisfactory results.[1] At first this work was carried out by the American Bell Company itself. Later, the American Telephone and Telegraph Company was formed for the purpose of undertaking extra-territorial work and became known as the Long Distance Company. It was in 1899 that the American Telephone and Telegraph Company, by reason of its wider statutory powers, absorbed the American Bell Telephone Company.

It was probably easier to determine the scientific limits of distant speech than to ascertain the commercial possibilities. Yet lines had to be built before either of these points could be satisfactorily gauged, for experiments over telegraph lines were fruitless; and until a suitable line had been built between two places it was impossible to tell what were its commerical possibilities. Confidence thus played a considerable part in the early investments in long distance plant.

As was pointed out in Chapter xx, iron wire was generally used for land telegraphs and was thus naturally adopted for telephones in their early commercial use. There were good reasons for the use of iron for telegraphs. The cost was low, the breaking strain high, and the conductivity sufficient for all practical purposes. If the distance were such as to trench somewhat on the margin of safety, it was very much cheaper to increase the battery power than to reduce the resistance of the line. The signals were received as transmitted, and that was all that was required. These signals were makes and breaks of the circuit not more frequent than could be conveniently produced by the manual movement of a key at the sending end or translated by the ear when drummed out by a sounder and recorded by hand at the receiving end.

The highest speed of the most expert operator would not put a severe strain on the capabilities of the circuit to respond to the required pulsations. But with the rapid alternations and infinite variations of telephonic currents the conditions were very materially altered. The waves became quickly blurred or attenuated, so that the limiting distance of speech was soon reached. Early long distance experiments were necessarily carried out upon iron wires, for none other were then erected upon telegraph lines. These, moreover, were earthed circuits, so that the distances over which commercial conversation could be carried on at all times of the day were very restricted.

[1] Chapter xv. p. 184.

That the inductive trouble could be cured by return circuits was well known, but that those return circuits needed to be arranged in a particular manner was less well known. An early instance of the practical recognition of the need of return circuits on long lines is given by Mr. H. L. Webb in the course of an article on 'Long Distance Lines' in the *Electrical Engineer* (N.Y.), May 4, 1892. In November 1881 General C. H. Barney, general manager of the Inter-State Telephone Company, issued a circular calling attention to the advantages of a projected ex-territorial exchange between Boston and Providence. The great feature of this service was to be 'a complete metallic circuit, ensuring a perfectly quiet line.'[1]

In the report of a committee of the National Telephone Exchange Association in 1884 it was recognised 'that only by the use of the parallel metallic circuits can long lines be at all times satisfactorily operated';[2] but long lines are defined as those over one hundred miles in length. It was believed that lines one hundred miles or less in length could be operated readily with a single wire, provided that first class construction were effected.

It was also reported that trials of translators or repeating coils had been made in 1879 with unsatisfactory results. Mr. Lockwood informs me that these experiments were carried out by Mr. Watson and himself 'at the Boston and Lowell terminals of the 2-wire line uniting those cities.' He adds:—

> Watson thought that it was he who had devised the translators, and I thought it was I. However, the patent was issued to him as the inventor. It is 232,788, dated Sep. 28 1880. . . . But the thing had previously been done. See Savage, *Journal Society of Telegraph Engineers*, 1878, vii. 268; *Electrician*, i. 22–23; also Sacher, *Philosophical Magazine*, fifth series, 1878, v. 158; and *Nature*, Feb. 14, 1878, 310.

These trials on the Boston-Lowell line were not published, and were probably unknown to others than the experimenters. The translator device was independently invented by Mr. A. R. Bennett,[3] and described in the London *Electrical Review*, January 13, 1883, p. 21. The early translators did not permit ringing through them, and improvements were made by the United Company in London and by the Lancashire and Cheshire Company at Liverpool, the latter city giving its name to the translator produced under the direction of Mr. Hope-Jones, in which the wires composing the core were lengthened

[1] Mr. Carty states (*Herzog Case*, p. 360) that he proposed and carried out the metallic circuiting of this line, with the result that the disturbances which had previously existed were removed. The circular was first reprinted at his suggestion in the New York *Electrical Review*, January 10, 1891, p. 239.

[2] *National Telephone Exchange Association Report*, 1884, p. 87.

[3] British specification, No. 4428, October 1881.

and brought back over the coils producing a complete magnetic circuit. The Coleman and Jackson combined translator and indicator, patented in 1889, was shielded with an iron tube. Translators were used for the purpose of connecting the earthed subscribers' lines to the metallic trunk lines. They were adopted on the New York-Philadelphia line about 1886 or 1887. The New York-Boston line was designed for use with metallic circuit subscribers' lines. It was anticipated in 1884 that this plan would eventually be necessary,[1] and while introduced originally for the purpose of talking over the long distance lines the metallic circuits were gradually extended for local work also.

The next important consideration was to determine of what metal the conductor should consist. Scientific and practical men were reaching in their respective ways the conclusion that iron must give place to copper.

Copper was invariably used for internal conductors, but this was annealed wire lacking the strength needed for suspension. In order to be available for long line use a new method of drawing had to be developed. Mr. Doolittle, a pioneer in other lines of telephonic development also, was apparently the first to undertake experiments in this direction. He says[2] that in November 1877 he 'conducted some experiments at the wire mill of the Ansonia Brass and Copper Company with the object of making available for telephone purposes the best known conductor of electricity. First, on account of its superiority as a conductor; and second, its lasting qualities.'

Copper, though known to be the best conductor of electricity and to have lasting qualities superior to those of iron, was not produced in a form suitable for line use. Mr. Doolittle, in conjunction with some manufacturers (who are far more ready than is generally supposed to facilitate research and to discover improved processes), set to work to produce a copper wire which should have the necessary tensile strength and conductivity. Tests were successful, and other manufacturers in the United States and Europe developed processes for the production of hard-drawn copper. In Europe much attention was given to the preparation of a bronze wire offering greater tensile strength at the expense of conductivity. The improvements in the process of wire drawing resulted in such a tensile strength being given to copper alone, that such alloys as phosphor bronze or silicon bronze have ceased to be generally used, except for cases where strength is of primary importance and conductivity a secondary consideration.

It was in 1883 that the American Bell Telephone Company

[1] *National Telephone Exchange Association Report*, 1884, p. 87.
[2] *Ibid.* p. 121.

decided to put up a metallic circuit copper line between New York and Boston. This was apparently regarded as a bold project, for in the following year Mr. Doolittle said: 'Our copper wires from New York to Boston which Mr. Vail had the pluck to order constructed are a perfect success. All who have used them for communication pronounce the result to be superior to the average local connection.'[1]

In 1885 it was reported that No. 14 copper wire was used between Chicago and Elgin (62 miles); and No. 12 copper from Detroit to Saginaw (100 miles). From the paper by Mr. A. S. Hibbard, in which this information is conveyed, the following conclusions, evidently reached by practice and comparison, are taken:—

It has been proven beyond dispute, that the best wire for long lines is that of the smallest diameter and of the lowest resistance. Soon after this conclusion was reached, the hard-drawn copper wire was perfected and introduced largely. The best results which have been obtained in long distance telephoning have been over this hard-drawn copper wire.

It is found advantageous to keep the wire clear of annunciator coils, ring-off drops, or electro-magnets of any kind. Such coils have a retarding effect, which is disastrous.

The introduction of resistance into the circuit has been found beneficial in suppressing inductive disturbances, but has, in a corresponding degree, weakened the telephonic effect. Leaky lines have been found less noisy than those well insulated.[2]

'The Relative Merits of Iron and Copper Wire for Telegraph Lines' was the title of a paper by Preece read before the British Association at Aberdeen in September 1885. It was fully reported in the London *Electrical Review* of September 26, 1885.[3] In this paper Mr. Preece said that copper was gradually superseding iron for aerial telegraphs, owing to its greater durability in the atmosphere. Hitherto only short lengths had been erected in smoky towns and through districts where chemical industries filled the air with gases destructive to iron wire, but the Post Office Telegraph Department had recently erected a No. 14 copper wire, ·080 inch in diameter, weighing 100 lb. per mile, between London and Newcastle, 278·08 miles in length, and it became desirable to measure very accurately its electrical elements. The paper gives full details of these measurements, which indicated great superiority over iron. The experiments were considered to prove

that the superiority of copper is not simply due to its smaller electro-static capacity and resistance, but that it is more susceptible to rapid changes of electric currents than iron. . . . Possibly the

[1] *National Telephone Exchange Association Report*, 1884, p. 121.
[2] *Ibid.* 1885, pp. 64, 65.
[3] Vol. xvii. p. 270.

magnetic susceptibility of the iron is the cause of this. The magnetisation of the iron acts as a kind of a drag on the currents. It is well known that telephones always work better on copper than on iron wires, doubtless for the same reason.

It was the requirements for high speed telegraphy rather than long distance telephony that prompted this copper line experiment in England, where, however, copper in the form of bronze was early used by the telephone companies for aerial wires in towns. The adoption of copper by the Post Office [1] stimulated manufacturers to produce qualities of wire which continually improved and rapidly attained the highest standards in tensile strength and conductivity. With the practical demonstration afforded by the New York-Boston line in 1884 the use of copper wire became general.

Whilst experience was being gained in the technical features of the long line problem, the commercial features were also being considered. It was in the course of a discussion on the hard-drawn copper question that one member of the National Telephone Exchange Association (Mr. Fay) asked: 'I would like to have somebody tell me how to make these lines pay. We can construct them and any number of them without trouble . . . . But how are we going to make them pay after they are constructed?' The receipts compared with telegraph line receipts were low; telephone use was six or seven hours per day compared with the telegraph's twenty-four; and how were they to compete with the telegraph in rates? To this question the president Mr. Tyler promptly responded. He said Mr. Fay had suggested a problem which the Bell Company was at work upon. They had not by any means solved it, but he would tell the Association the manner they were working at it.

We began by recognising the fact that a long telephone line is available for earning purposes only about six hours out of the twenty-four, and as a consequence, we did not, and have not, attempted to furnish that line on prices which are in any sense competitive with telegraph prices, for the simple reason that we knew and claim that we are going to furnish something which cannot be put in competition with the telegraph, or rather that the telegraph cannot be put in competition with us, because of the superior character of the service which we expect to furnish. We give something immediate and complete, while the telegraph gives something completed only after a very considerable lapse of time. We are working, or trying to work, this copper line between New York and Boston in a business shape.[2]

[1] It was stated in the last report that the use of copper wire had been resorted to and was likely to prove advantageous; but the further use of such wire has been prevented for the present by the abnormal rise in the price of copper.—*Thirty-fourth Report of Postmaster-General*, 1888, p. 9.

[2] *National Telephone Exchange Association Report*, 1884, pp. 123–5.

Mr. Tyler indicated that the circumstances at the terminals and at intermediate points on the line had been very carefully considered. The line would be run to metallic switchboards and the subscribers at either end would have metallic circuit lines. He added :—

We expect to get from the American Bell system in connection with the Hunnings transmitter, results which will be so satisfactory to the owners of establishments which require service that they will gladly accept it at remunerative rates.[1]

The long lines were embarked upon with the same confidence as the local service and for the same reason. The telephone was so superior to the telegraph that there was no question of competition. Let the service be good, and the public would be content to pay a remunerative rate. It was a sound policy, but, without that confidence in advance, long distance telephony would have been much deferred. The utility of the telephone having been demonstrated for local work, it was obvious that its value must increase in proportion to the distance, and thenceforward long lines grew in number and in length.

The local companies connected up distant points in their own territories, and the Long Distance Company built the lines and handled the traffic between points in different territories. In Mr. Hibbard's paper previously quoted it is said that :—

The exchanges which show the greatest amount of average profit in toll business are those ranging from 100 to 500 subscribers. Cities of from 10,000 to 50,000 population appear to hold that peculiar business relationship with the surrounding country which is most essential to toll telephone business. The question of earnings depends so much upon local influences that it is quite impossible to produce any general average representative of the results in all parts of the country.

In the practical operation of long lines in connection with telephone exchanges, one of the most important features is the attention and co-operation of an intelligent operator. This is particularly necessary in making prompt connections and discon nections, and also in what is called ' getting the parties together.'

The network of lines and wires doubles and trebles in extent each year. Cities, towns, villages, and hamlets are awakening to the fact that inter-communication is indispensable. No such pressure was ever before brought to bear on any one department of business. For the solution of this problem the public looks to the American Bell Telephone Company and its licensees in the United States. It is with much confidence that I venture the opinion that the expectation will be satisfactorily and speedily realised.[2]

[1] *National Telephone Exchange Association Report*, 1884, pp. 123–5.
[2] *Ibid.* 1885, pp. 60–6.

The long distance lines of the licensee companies had been mostly single lines. The Boston and New York line had demonstrated, according to the annual report of the American Bell Telephone Company for the year 1884, that there could

no longer be a question of the success of long distance lines between all points within two hundred and three hundred miles of each other, and even more where the demand for such service is of a character to warrant the building of metallic circuit systems. Such systems will be very costly, and our present licensed companies are neither provided with capital nor organised for the purpose of undertaking a work of the proportions which commercial success in this business will involve. It, therefore, becomes necessary for our company to take the lead in building lines of connection between the large cities, and a beginning will be made this summer (1885) between New York and Philadelphia.

But it had been realised that more capital would be required, and application had been made to the Massachusetts legislature for the necessary authority. The report for the year 1885 notified that this authority was refused and a new company had been organised under the laws of New York called the American Telephone and Telegraph Company.

The following figures [1] indicate the length of extra-territorial and toll lines at the respective dates :—

January 1,	1883	13,653 miles of wire.
,,	1884	29,359 ,,
,,	1885	35,631 ,,
,,	1890	86,728 ,,
,,	1895	180,557 ,,
,,	1900	501,832 ,,
,,	1905	1,121,228 ,,
,,	1910	1,804,552 ,,
,,	1915	2,437,697 ,,

Mention of the long distance service is made in the various reports of the Company as follows :—

That for the year 1886 records delay in getting terminal facilities in Philadelphia and the decision to proceed with the Boston-New York line without waiting for the commercial results of the Philadelphia line.

1887. The long line service had been extended from New York to Albany and to Boston. The success electrically as well as commercially was beyond the company's expectations. The intentions for the year were to complete New York and Boston, extend from

[1] Taken from the annual reports of the American Bell Telephone Company and its successor, the American Telephone and Telegraph Company.

Albany towards Buffalo, and to build a line from Chicago to Milwaukee. The estimated cost being one million dollars.

1888. The service on the long lines had been excellent and the business was increasing, but the best results could not be expected until the principal exchanges should have adopted the metallic circuit system. The success of the long lines had convinced the directors that the system should be extended. The advantage to the public of the proposed development was so plain that they hoped the authority to increase capital would be generally approved.

1889. Philadelphia to Baltimore and Washington ; a second line, N.Y. to Boston ; lines connecting Philadelphia with Reading, Easton, and Scranton ; and connected with its eastern lines Newport, R.I., and Fall River and New Bedford, Mass.

1890. From Scranton, Penn., to Syracuse, N.Y. From Buffalo via Erie, Penn., south to New Castle, Penn., taking in the towns of Western New York and Western Pennsylvania.

In the report for the year 1892 it is recorded that the gap between Cleveland, Ohio, and Hammond, Ill., had been filled, thus giving through service between Boston, New York, and Chicago.

> To determine the feasibility of transmitting speech over the theretofore unapproached distance of 1000 miles—the limit of the successful transmission of speech had not before exceeded 500 miles—a special experimental circuit, consisting of two number 8 hard-drawn copper wires, was constructed. . . . The success was so complete that the lines were at once extended to Boston, and two more circuits ordered strung from New York to Chicago. . . . The line from New York to Chicago was formally opened to the public on October 18 last (1892), and the [extension of the Chicago] line to Boston on February 7, 1893.[1]

The photograph of the inventor of the telephone opening the New York-Chicago line, reproduced in fig. 152, was widely published in the periodical literature of the time, but may fittingly be incorporated here.

Whilst the companies in the United States were erecting lines, studying the business relationships of localities, and developing the long distance service in every possible way between 1878 and 1884, the companies in Great Britain were unable to make any move in the same direction. This was due to the limitations of the first licenses granted by the Post Office in 1881, which restricted the area of telephone exchange communication to a radius of four or five miles from a defined centre. The Department reserved to itself the right of connecting the exchanges by trunk lines, but

[1] *American Bell Telephone Company Report*, 1892, p. 10.

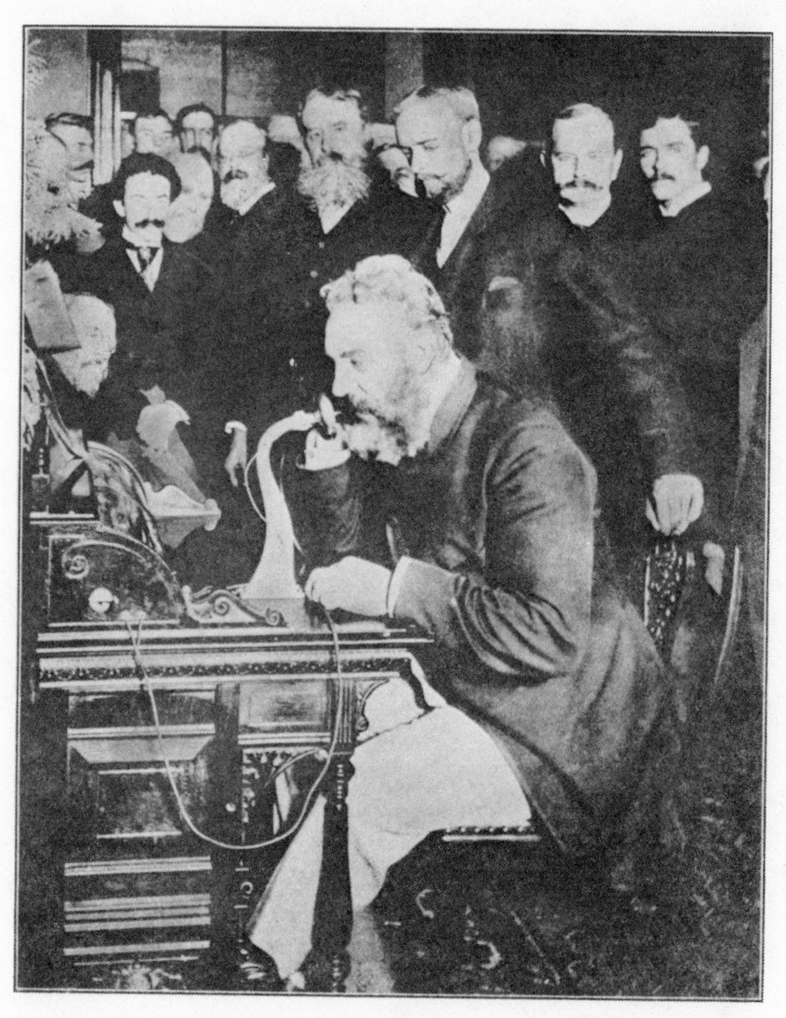

FIG. 152.—Professor Bell inaugurating the New York-Chicago Line, October 18, 1892.

upon terms which did not permit development. On such trunk lines as were rented in the National Company's district there was a loss of £684 in 1883, and the loss on the Edinburgh-Glasgow line was £238 in 1884. An example of the efforts of the companies to promote through communications and of the Post Office to retard them may be seen in the action of the Lancashire and Cheshire Company.

Under date of April 18, 1883, the secretary of the Post Office intimated to the Lancashire and Cheshire Telephonic Exchange Company revised terms proposed for providing trunk wires, whereby the company should pay £10 per mile of double wire, should obtain as many subscribers as they could for a minimum period of one month, at such annual rate as the company liked to charge, the surplus of such charge over the £10 rate being divided between the company and the Government. The agreement having been practically concluded, the company notified the Department that they intended to join the Manchester and Oldham exchanges, to throw open communication between the two places exactly as if they formed parts of the same town, making no extra charge to subscribers for the use of the trunk wire, the whole cost being borne by the company. Such a proposal was an enlightened one on the part of the company and was admitted to be within the terms of the agreement, but it was a use not contemplated by the Department and was not allowed. An amendment was proposed (May 1884) providing that a rental for trunk wire use should not be less than 10s. per mile per subscriber. The company complained that the restrictions under which they had been labouring from the first had most effectually prevented the adoption and use of the telephone by the merchants and manufacturers of the district.

> The trunk wires form the connections between the exchange centres of the towns in our district, and it is of great commercial importance and would be of inestimable value to the commercial community if . . . we could put the whole of the towns of Lancashire in communication with each other, and it would have a very important bearing upon the conduct of business generally and give facilities to business men which under the present condition of telegraphic charges and telephonic restrictions are not available.[1]

Another illustration, in the National Company's district, is given in the *Electrical Review* of January 13, 1883:—

> Although the Government ultimately agreed to grant licenses to the various telephone companies to open exchanges in most of

[1] Memorandum by the chairman, Lancashire and Cheshire Company, May 1884.

the important towns of the United Kingdom, and to connect therewith subscribers within distances varying from three to five miles of their exchanges, the companies were positively forbidden to join their systems in different towns, however desirable it might be in the interests of the commercial community that such a connection should be established.

As an illustration may be cited the exchanges opened by the National Telephone Company in Glasgow, Paisley, Greenock, Dumbarton, Coatbridge, and Hamilton. The business relations between these towns are of the most intimate character, it being not uncommon for one firm to possess branches, works, or offices in two or more of them, and daily customers in all; and yet the company was not allowed to provide anything but local exchange communication.

After much negotiation, the Post Office agreed to erect trunk wires on their own poles, stipulating that a separate trunk wire should be provided for every eight subscribers desiring to avail themselves of the privilege, and that for every wire the company should pay the Government an annual rental of £320 between Glasgow and Greenock, £104 between Glasgow and Paisley, £128 between Glasgow and Hamilton, and £120 between Glasgow and Coatbridge.

Subsequently a trunk wire between Glasgow and Edinburgh was projected, and the Government tax was fixed at £560 per annum. Similar arrangements were made for Bradford, Leeds, Huddersfield, Birmingham, Wolverhampton, and other important towns; and notwithstanding the heavy rates that the Government tax rendered necessary, amounting in the case of Edinburgh to no less than £110 per annum for each subscriber, the most prominent merchants quickly took advantage of the opportunity of obtaining free communication with their distant customers.[1]

The memorandum of the Lancashire and Cheshire Company before referred to, is included in a statement entitled 'The Post Office and the Telephone Companies,' of which the opening paragraph is as follows :—

The Post Office having for some time past shown so fixed a determination to impede and to obstruct the development of telephonic facilities to the public, and private remonstrances having proved so completely unavailing, no resource seems open but an appeal to the fair-play and common sense of Parliament, the Press, and the Public.

The Postmaster-General at this time was Mr. Fawcett, whose name is more familiar as a political economist than as the political head of a revenue-earning department. Though labouring under

[1] *Electrical Review*, London, January 13, 1883, xii. 21.

the physical disability of total blindness, Mr. Fawcett was an able administrator, and for its telephonic development the United Kingdom was indebted to the far-sighted policy which he adopted in removing the restrictions as to areas and permitting the licensees to develop the business in any way they liked, subject to the royalty of 10 per cent. on their gross receipts. New licenses were issued in November 1884 on the lines laid down by Mr. Fawcett, but these licenses were signed by Mr. Shaw Lefevre, who had succeeded to the office of Postmaster-General. The new license permitted the connecting up of any areas that justified the commercial investment, and, though the time lost could not be regained, the service was energetically promoted. By 1887 lines were constructed from Dundee to Ayr. Edinburgh and Glasgow; Paisley and Perth; Leeds, Bradford, Manchester, Wigan, and Liverpool were not only in communication with each other but were the centres to which were connected neighbouring towns and villages. In May 1886 there were forty-two wires between Manchester and Liverpool. Bristol, Cardiff, Newport, and Swansea were connected.

Much as was done in connecting up local centres, London remained without the means of communicating with the principal provincial towns. The first trunk line from London was that to Brighton erected by the Post Office under the terms of the original license and opened to the public on December 17, 1884.

In 1892 an agreement was arrived at between the Tel phone Company and the Government whereby the latter resumed its position as the trunk line authority and the former restricted its activities to local service. By the Telegraph Act of 1892 Parliament provided £1,000,000 for the acquisition of the trunk lines and the provision of a backbone system connecting England, Scotland, and Ireland together.

On Wednesday, June 26, 1895, the telephone trunk lines of the Post Office from London were inaugurated. These trunk lines (including the Channel cable) amounted to a total of 10,754 miles of wire. The company's trunk lines were transferred to the Post Office in 1896, and the wire mileage scheduled in the agreement was 27,606 miles 592 yards.

International telephonic communication was inaugurated by the line between Paris and Brussels which was opened to the public on February 24, 1887. In a communication to the International Society of Electricians on May 4, 1887, M. de la Touanne stated that the total length of wire (double line) was 628 kilometres. Silicon bronze was used in the French portion of the line and phosphor bronze in the Belgian. The diameter of the wire was 3 millimetres; the breaking strain 42 to 45 kilogrammes per square mm.; conductivity 96 to 98 per cent. of pure copper.

Within the Paris limits the line was underground. For this Fortin-Hermann cable was used—a lead-sheathed cable in which the insulation was obtained by means of beads of paraffined wood strung upon the copper conductors. The capacity was ·045 to ·04 of a microfarad per kilometre. Subsequently a line was constructed between Brussels and Amsterdam. In April 1887 the Paris and Amsterdam lines could be connected so that France could speak to Holland through Belgium. The success of the Paris-Brussels line was such that within a few months arrangements were made for the construction of a second circuit. On December 31, 1891 (Montillot states), there were three circuits.

The first long distance cable was that between Dover and Calais, a necessary link in the line between London and Paris. This cable was laid in 1891.

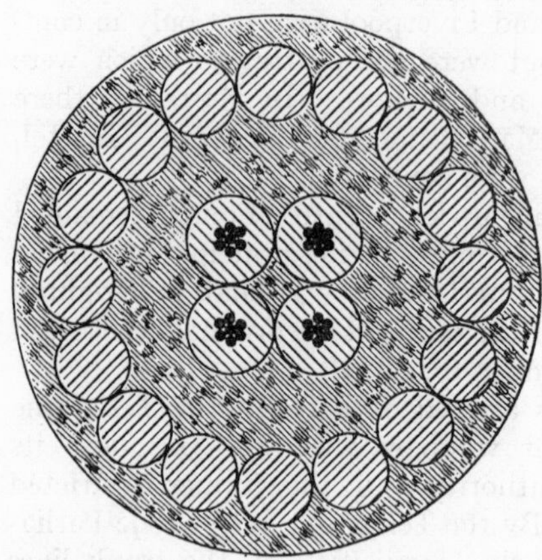

FIG. 153.—Dover-Calais Cable.

The data available in 1890 when the first Anglo-French cable . . . was designed by Mr. H. R. Kempe . . . were naturally somewhat incomplete, but the results obtained with the first telephonic circuit between London and Paris have proved eminently satisfactory from the date of its first use.[1]

Fig. 152 is a sectional illustration of the 1891 Anglo-French cable, and the following short description thereof is quoted from Major O'Meara's paper read before the Institution of Electrical Engineers in 1910.

This was a 4-core cable differing but little from those in use at the time for telegraphic purposes, except in the size of the cores. Each core consisted of a stranded copper conductor formed of seven equal wires, having a total weight of 160 lb. per knot, covered with 300 lb. of gutta-percha per knot. The extreme diameters of the copper strand (*d*) and of the gutta-percha covering (D) were 0·108 and 0·390 in. respectively, giving a ratio $\frac{D}{d} = 3{\cdot}61$.

The measured resistance of the conductor at 75° F. was 7·453

[1] Major O'Meara, *Journal of the Institution of Electrical Engineers*, December 15, 1910, xlvi. 310.

ohms, and the measured electro-static capacity 0·275 micro-farad per knot. The sheathing consisted of sixteen galvanised iron wires, each 0·280 in. in diameter, put on with a lay of 18 in., and the external diameter of the finished cable was about 2·2 in.[1]

Great care was taken in the design and construction of the London-Paris line, and good commercial speech was obtained. The line was officially inaugurated on March 18, 1891, the first message transmitted being one from the Prince of Wales (subsequently King Edward VII.) to M. Carnot, President of the French Republic. Fig. 156 is a pictorial representation of the opening ceremony from the *Illustrated London News* of March 28, 1891.[2]

This line is an example of enterprise on the part of the Governments concerned, for it cannot be said that the commercial community clamoured for its construction. When it was finished there were the usual doubts expressed by commercial men as to whether the line was wanted. But the calls were numerous and the enterprise of the Governments justified.

In respect to charges and facilities, the Government repaired the unfortunate attitude of earlier years as soon as the trunk lines were acquired by them. Reasonable rates were established. The scale of charges laid down in the Treasury minute defining the Government's telephone policy dated May 27, 1892, were: For any distance not exceeding 20 miles, 3*d.*; exceeding 20 and not exceeding 40 miles, 6*d.* Every additional 40 miles or fraction thereof, 6*d.*

The following are some of the specimen rates given in the Post Office memorandum of June 25, 1895:—

		s.	*d.*
London—	Cardiff	2	0
	Edinburgh	4	6
	Glasgow	4	6
	Liverpool	2	6
	Nottingham	1	6
Edinburgh—	Glasgow	0	6
	Liverpool	2	6

With the exception perhaps of Van Rysselberghe's arrangement for simultaneous telegraphy and telephony—to which some of the Belgian enterprise in trunk line service may be attributed—there were, thus far, no important new inventions applied to long distance lines. The best conducting material had been selected; the best arrangement of the conductors adopted; the greatest care taken in insulation and construction; all involving scientific research, engineering application, and commercial initiative, and

[1] *Journal of the Institution of Electrical Engineers*, xlvi. 334.
[2] Vol. xcviii. p. 400.

leading to extended distances and improved speech. But important developments were impending, and for their genesis it will be

FIG. 154.—Inauguration of London-Paris Line. (From the *Illustrated London News*, March 28, 1891.)

necessary to go back some years from the period of the practical accomplishments above recorded.

In January 1886 Professor Hughes selected as the subject of his inaugural address to the members of the Society of Telegraph Engineers 'The Self-Induction of an Electric Current in Relation to the Nature and Form of its Conductor.'[1] This paper gave rise to a very interesting discussion, in which the practical bearing as regards telegraphic and telephonic transmission were well brought out by the various speakers. But one critic reserved his remarks for the columns of the press. In the *Electrician* of April 23, 1886, Oliver Heaviside writes :—

We read in the page of history of a monarch who was *supra grammaticam.* All truly great men are like that monarch. They have their own grammars, syntaxes, and dictionaries. They cannot be judged by ordinary standards, but require interpretation. Fortunately the liberty of private interpretation is conserved. No man has a more peculiar grammar than Prof. Hughes. Hence he is liable, in a most unusual degree, to be misunderstood. . . . The very first step to the understanding of a writer is to find out what he means.[2]

The subject was continued more or less directly through a number of articles, until on June 3, 1887, he opens another article with the statement :—

Although there is more to be said on the subject of induction balances, I put the matter on the shelf now, on account of the pressure of a load of matter that has come back to me under rather curious circumstances.[3]

In the meantime other discussions had taken place in the same theatre.

In May 1886 Preece read a paper entitled 'Long-Distance Telephony' in which he indicated that the difficulty in speaking to a distance was not a difficulty with apparatus at all, but with the conductor and its environment.

In January 1887 Silvanus Thompson related his 'Telephonic Investigations,'[4] as a result of which he stated his conviction that the success of long range telephony depended upon the production of a transmitter permitting the use of higher battery power and of a receiver which, though not necessarily more sensitive to small currents, would have a higher electrical and mechanical efficiency.

It was in the course of the discussion on Thompson's paper that

[1] *Journal of the Society of Telegraph Engineers*, xv. 6.
[2] *Electrician*, xvi. 471. *Electrical Papers*, ii. 28.
[3] *Electrician*, xix. 79. *Electrical Papers*, ii. 119.
[4] *Journal of the Society of Telegraph Engineers*, xvi. 42.

Preece propounded the theory [1] that the electro-static capacity and the resistance of the circuit determined the limiting distance of speech.

We may now continue the quotation from the introduction of Heaviside's June 3 *Electrician* article, of which a part was quoted on p. 433. Heaviside continues :—

> In the present Section I shall take a brief survey of the question of long distance telephony and its prospects, and of signalling in general.[2]

In this article criticism of Hughes gives place to criticism of Preece and his K.R. law. The importance of the time constant was referred to by Thompson in the discussion on Hughes's inaugural address.[3] Heaviside deals with the same point. He says (criticising Preece) :—

> But all telegraph circuits are not submarine cables, for one thing ; and even if they were, they would behave very differently according to the way they were worked, and especially as regards the rapidity with which electrical waves were sent into them. . . . Now a telegraph circuit, when reduced to its simplest elements, ignoring all interferences, and some corrections due to the diffusion of current in the wires in time, still has no less than four electrical constants which may be most conveniently reckoned per unit length of circuit viz., its resistance, inductance, permittance or electro-static capacity, and leakage conductance. . . . In the case of an Atlantic cable it is only possible (at present) to get a small number of waves through per second, because, first, the attenuation is so great ; and next, it increases so fast with the frequency, thus leading to a most prodigious distortion in the shape of irregular waves as they travel along. Of course we may *send* as many waves as we please per second, but they will not be utilisable at the distant end. This distortion is a rather important matter. Mere attenuation, if not carried too far, would not do any harm. . . . Within the limits of approximately constant attenuation, the distortion is small. This is what is wanted in telephony to be good. Lowering the resistance is perhaps the most important thing of all. . . . Increasing the inductance is another way of improving things.[4]

The direction in which these investigations were to bear fruit was by the increase, and the suitable distribution along the conductors, of the proper amount of inductance. Major O'Meara says that in 1887

[1] *Journal of the Society of Telegraph Engineers*, xvi. 84, 265.
[2] *Electrical Papers*, ii. 119.
[3] *Journal of the Society of Telegraph Engineers*, xv. 79.
[4] *Electrical Papers*, ii. 120, 121.

Oliver Heaviside gave the essential parts of the theory of telephonic transmission, pointing out the importance and beneficial effects of self-induction and stating the relation which must exist between the constants of a circuit in order that electrical waves of all frequencies may be transmitted without distortion. His investigations show great power of mathematical analysis and a wonderful insight into complex electrical phenomena.[1]

But Heaviside's contribution to the art received little practical attention, whilst the K.R. law was much discussed. As we have seen (Chapter xxii) a modification of this 'law' was adopted by one of the members of the cable committee which met in New York in September 1887. Other members criticised the conclusions vigorously, and the committee generally may be said to have regarded this so-called law as unsatisfactory. But no reference was made to Heaviside's article. It is true that it had only been available for two or three months, but this practical body of 'practicians' were anxious for light on their problems. The reason for the difference in the attention given may perhaps be found in the facts that the K.R. law was announced before a scientific body, was stated in specific terms in a form that could be understood and tested, whilst Heaviside's views appeared in the pages of a technical journal, were expressed in more abstruse form, and were rather suggestions regarding the utility of inductance than a statement of a method adapted for practical application.

It would almost seem that Heaviside had determined upon a line of demarcation between theory and practice, discovery and invention. Regarding the conversion of possibilities into actualities, he says elsewhere and at a later date :—

> I am not much concerned in this part of the question. It is for practicians to find out practical ways of doing things that theory proves to be possible, or not to find them if they should be impracticable.[2]

Heaviside, though a caustic critic, would appear to have absorbed that characteristic of unselfishness which is common to the theorist or discoverer in contrast to the materialism of the inventor, or, to use his own word, the 'practician.' The declaration is definitely made in the Preface to his 'Electrical Papers' (1892). He is discussing the reasons for their republication, and says :—

> As regards the question 'Will it pay ?' little need be said. For, fifthly, however absurd it may seem, I do in all seriousness hereby declare that I am animated mainly by philanthropic motives. I desire to do good to my fellow-creatures, even to the *Cui bonos*.[3]

[1] *Journal of the Institution of Electrical Engineers*, xlvi. 311.
[2] *Electro-Magnetic Theory*, i. 441. [3] *Electrical Papers*, i. preface, p. 6.

Tyndall says [1] of Faraday that 'if he had allowed his vision to be disturbed by considerations regarding the practical use of his discoveries, those discoveries would never have been made by him.'

But such an arbitrary line of demarcation may be carried too far. The theory or the discovery may not be adapted to independent development. When Bell asked Professor Henry for advice as to whether he should publish his discovery and let others work it out, or proceed to put it in practice himself, Henry advised him to go ahead himself.[2] In Henry's judgment Bell was the one best equipped to reach the goal. In a similar way Heaviside was best equipped at this time to reach a practical result.

Bell's conception of the undulatory current as a means of speech transmission required a mental picture of what went on within the wire. Heaviside's conception was of a similar nature. He visualised the wave feature of the transmission and developed an important theory. Had he proceeded with it further he would have made a valuable invention. He had been a practician himself, and the nature of his work would enable him to appreciate the value of a practical application of his theoretical deductions. He recognised that his theory was in the nature of a paradox, but unfortunately he did not recognise that this fact might render it a less attractive field of experiment to others. There was not at this period any urgent demand for lines of a length exceeding those over which speech could be transmitted with then existing knowledge. Another aspect of the same subject—i.e. the greater economy which might be effected in short lines—had not been alluded to. On the practical side, therefore, there was no immediate pressure for the following up of the theory, and it lay practically dormant for twelve years.

On March 22, 1899, Michael Idvorsky Pupin, who was not a 'practician' at all but a professor of mathematics, read a paper before the American Institute of Electrical Engineers entitled 'Propagation of Long Electrical Waves.'

In the tribute which Pupin pays to Heaviside it is impossible not to recall the latter's comment on Hughes. Heaviside criticised Hughes's 'grammar,' giving of course a far wider meaning to that word than is ordinarily applied; and Pupin criticised Heaviside's 'arithmetic,' by which must be assumed no ordinary use of figures. Pupin says in the course of his paper :—

Mr. Oliver Heaviside has done much to introduce the living language of physics in place of the sign language of mathematical analysis. But Mr. Heaviside's English is often much clearer than his arithmetic, such at any rate seems to be the general impression, so that much yet remains to be done even after Mr. Heaviside's most

[1] *Faraday as a Discoverer*, p. 35. [2] Chapter iv. p. 40.

brilliant epoch of intense activity and radical reforms in the field of long wave propagation. That which remains to be done is not so much on the purely mathematical side of it, for that is pretty well understood now, and has been so ever since the time of Lagrange and Fourier. It is the physical side of the theory which needs cultivation.[1]

On December 14, 1899, Pupin applied for a patent[2] covering an

invention [which] consists in an improvement in the construction and installation of conductors for the transmission of electrical energy by means of electrical waves, whereby, by decreasing the current necessary to transmit the amount of energy required, the attenuation of such waves is reduced, and therefore the efficiency of transmission is increased.[3]

The first claim of the British patent is as follows :—

In a system of electrical wave transmission a non-uniform wave conductor consisting of a conductor having reactance sources distributed at points along its length in such manner that the resulting wave-conductor is equivalent, within proper limits, to its corresponding uniform conductor, but of increased effective inductance, substantially as described.[4]

Heaviside adopted the analogy of a vibrating string, the effect of its construction and the attenuation of the wave propagated along it which is produced by opposing forces. Pupin uses the same analogy, but goes further and indicates how a thin string should be weighted at intervals so as to vibrate in the same manner as would a thicker string possessing definite characteristics. He says :—

In fig. [155], A″ B″ C″ is a tuning-fork rigidly fixed at its neck C″. The full line B″ D″ represents a heavy, flexible, inextensible string which is under tension and fixed to D″. The circles, equally distributed over the string B″ D″, represent equal masses attached to the string. Let now the tuning-fork vibrate with a suitable period so as to develop in the beaded string a vibration the wave-length of which is equal to or greater than the distance B″ D″, somewhat as shown in fig. [156]. The vibration of the beaded string (represented in fig. [156]) will then, to within an accuracy of a fraction of one per cent., be the same as that of a uniform string of the same length, tension, frictional resistance, and mass which the beaded string has.[5]

The mechanical vibration in such a string is a perfect analogy to the electrical vibration in an electrical conductor represented

[1] *Transactions, American Institute Electrical Engineers*, xvi. 98.
[2] British, No. 12,733 of 1900; U.S. Nos. 652,230 and 652,231, June 19, 1900.
[3] British specification, No. 12,733 of 1900, p. 1, line 7.
[4] *Ibid.* p. 12, line 25.
[5] *Ibid.* p. 7, line 33.

in fig. [157]. In this diagram the alternator E at H is supposed to develop approximately a simple harmonic electromotive force. One pole of the alternator is grounded at G, the other pole is connected to a wire conductor. At equal distances 1, 2, . . . 10, 11, 12 are inserted in series with the line 12 equal coils. Suppose, now, that the electromotive force impressed by the alternator develops in the conductor an electrical vibration $\frac{3}{4}$ the wave-length of which covers the distance, or a greater distance. Then the

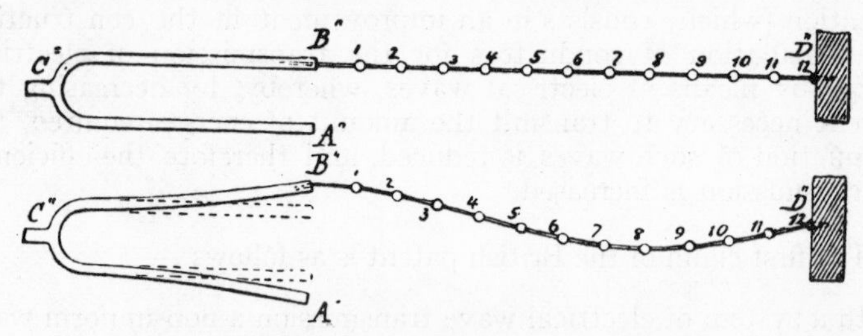

Figs. 155 and 156.

law of flow of the current in this conductor will be the same as the law of distribution of velocity in the beaded string in fig. [155] and fig. [156]. This mechanical analogy, besides being instructive, offers also an inexpensive method of studying the flow of current in long wire conductors. A few experiments with beaded strings excited by tuning-forks will convince one soon of the soundness of the physical basis on which rests the invention described in this application.[1]

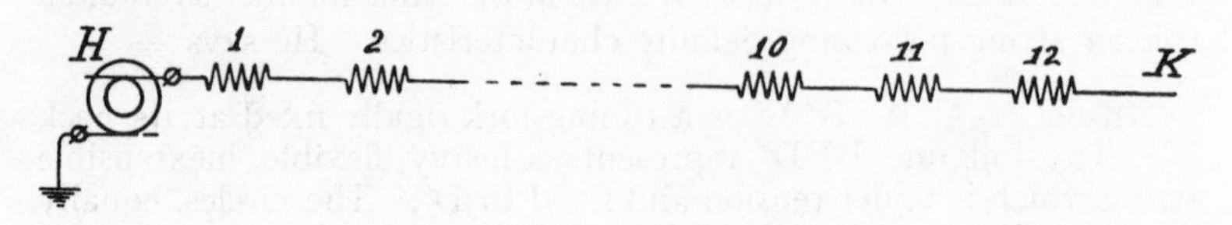

Fig. 157.

The analogy of the weighted string has been carried into the terminology of the art. The string was assumed to be actually loaded with beads. The inductance inserted in the conductor is the equivalent of the bead. Thus we get the term a 'loaded' line.

What should be the size of the beads or the inductance of the coils and at what distance should they be spaced? Answers to these questions were essential before any practical use could be made of the theory.

It was Pupin who first gave a practical rule for the spacing out of loading coils. It was quite one thing to say 'add inductance

[1] British specification, No. 12,733 of 1900, p. 7, line 42.

to cables to neutralise capacity,' but it is quite another thing to give practical engineers rules which can be followed in actual engineering for doing it.[1]

But the practical application needed more than paper work. The best form of coils and the most suitable method of applying them to the line had to be developed. Pupin's patents were acquired in the United States by the American Bell Telephone Company, whose engineers, in conjunction with those of their manufacturing branch, proceeded to put into practical form the coils mathematically defined by Pupin. This work involved much research and experiment, and, though not a matter of published record, is understood to have cost a very large sum before the development was sufficiently far advanced to be applied in service. But so satisfied were their engineers with the mathematical demonstration and the advantages which would follow its practical application, that the company had no misgivings about embarking upon any expenditure that might be required to develop suitable instrumentalities.

Particulars of the practical application of loading coils and some of the theoretical conditions were given by H. V. Hayes before the International Electrical Congress at St. Louis in 1904. In the discussion on this paper Mr. Gherardi stated that the first commercial application of loading to underground cables was made under the joint direction of Mr. Carty and himself on a cable extending between Cortlandt Street, New York, and Newark, New Jersey, the length of the cable being about ten miles. The same grade of transmission was obtained over these loaded circuits as was obtained over five miles of unloaded cable. The date of this application in practice was August 1902.

The loading of aerial lines made slow progress on account of the variable insulation and atmospheric interruptions. The comparatively constant insulation of cables, on the other hand, permitted immediate and extensive application to underground lines. In an article on 'Long Distance Telephony' published in 1892 will be found a remark which illustrates the limitations in the use of cables. After noting the improvement recently effected in the reduction of capacity, the writer continues :—

With all the improvement that can be made, however, underground cables must always constitute a serious obstacle to long distance telephony. The only places where these must be used are the large cities, such as New York, Boston, Philadelphia, and so on.[2]

This accurately represents the position at the time and for ten years following. One of the first long distance lines was that

[1] Dr. Fleming, *Journal of the Institution of Electrical Engineers*, xlvi. 357.
[2] H. L. Webb, *Electrical Engineer*, New York, May 4, 1892, xiii. 452.

between New York and Philadelphia, erected in 1885 for thirteen circuits. It was on the same route that it was determined to adopt the loading method, and instead of limiting the use of cable to the terminal cities to extend it throughout the whole distance.

The reason for this was the same as for the original use of cables for subscribers' lines, the congestion on routes :—

Some time ago the increasing circuit congestion on some of the more important toll line routes of the American Telephone and Telegraph Company made it imperative that relief be provided in a form other than by additional pole lines. In some cases it was found practically impossible to secure additional pole line routes, and in all such cases the importance of the heavy traffic was too great to hazard the continual danger of paralysis during the stormy winter months when line breaks due to sleet and snow were to be feared.[1]

Loading had been demonstrated to be commercially practicable in short lengths of cable. In the attempt to extend its utility to long distance work new problems needed to be solved, but confidence in their satisfactory solution was felt. Careful consideration, however, needed to be given to the economic side. This also was determined in favour of the construction of an underground route between New York and Philadelphia, a distance of approximately ninety miles. The work was completed in 1905. Six through ducts and additional ducts for local requirements along the route were laid. The cable first installed contained ninety-three pairs of No. 14 B. and S. and nineteen pairs of No. 16 B. and S. conductors. Toroidal inductance coils were inserted at distances of approximately 1¼ mile.

The transmission over the ninety miles of No. 14 conductors was equal to eleven miles of standard cable, and over the No. 16 conductors to fifteen miles of standard cable.

The success of the New York-Philadelphia cable prompted the construction of underground routes between New York-Washington (235 miles) and New York-Boston (240 miles). The cable used on these routes is composed of fourteen pairs of 166 lb., forty-two pairs of 83 lb., and eighteen pairs of 40 lb., conductors. The inductance coils are spaced 1·4 mile apart.

This cable was brought into service in sections. The section between Philadelphia and Washington (136 miles) was opened in June 1912 ; that from Boston to New York on February 17, 1914. On the 26th of the same month through service between Boston and Washington was obtained.

[1] *Long Distance Telephony in America*, by Dr. Jewett. International Congress of Electrical Engineers. Turin. September 1911.

The progress which had been effected in loading and in cable design may be further illustrated by a quotation from the official memorandum prepared for the inauguration of the Telephone Trunk Lines of the British Post Office in June 1895 :—

For the 'backbone' line from London through Leeds to Edinburgh, Glasgow, and Dublin, the heaviest copper wire ever erected has been used. It weighs 800 lb. per mile. Efforts have been made to reduce the underground portion as much as possible, as underground wires seriously impair the efficiency and clearness of telephone speaking.

On September 8, 1915, at the invitation of Sir William Slingo, engineer-in-chief of the Post Office, representatives of the technical press and a few others were enabled to converse over a cable which had been laid between London and Birmingham, a distance of 110 miles. Connecting different circuits in the cable, experiments were made over distances of 220, 440, 660, and 880 miles. The last-mentioned distance was beyond the limit of commercial conversation, but very satisfactory results were attained at 660 miles. The other circuits of the cable were in use at the same time, but no over-hearing (or cross talk) was observable. The immunity from disturbance was absolute and complete. The cable is composed of two pairs of 300 lb., fourteen pairs of 200 lb., twelve pairs of 150 lb., and twenty-four pairs of 100 lb., conductors. The spacing of the inductance coils is 2½ miles.

It was designed by the engineers of the Post Office, manufactured, laid, jointed, and loaded by the Western Electric Co., Limited. A similar cable extending from Birmingham to Liverpool is being manufactured and laid by the British Insulated and Helsby Cables, Limited. The tests made demonstrated that conversation between London and Aberdeen would be satisfactorily obtained over underground conductors for the entire distance. Twenty years before 'efforts had been made to reduce the underground portion' as much as possible in consequence of the serious effect which underground wires had in impairing 'the efficiency and clearness of telephone speaking.' The change is due to the loading and to the improvements in the design and manufacture of cables.

The first loaded cable laid under water was that connecting Friedrichshafen and Romanshorn on Lake Constance in 1906. This was a lead-covered cable, the core being specially protected with steel wire spirals inside the lead tube. Particulars of the cable, its construction, and laying are given in the *Electrician* of May 24, 1907.[1]

[1] Vol. lix. p. 217.

The first submarine cable of the ordinary type with gutta-percha insulation to be loaded was that between Dover and Calais, laid in May 1910. This is described in detail by Major O'Meara, C.M.G., in a paper read before the Institution of Electrical Engineers on December 15, 1910.[1] The insertion of the necessary inductance in a cable which had to be paid out of a moving ship involved some new and interesting problems fully related in the paper. The cable was manufactured by Messrs. Siemens Bros. & Co., Ltd., and has continued in successful operation.

Another Channel loaded cable was laid in 1911, between England and Belgium, 47·89 miles (of 1760 yards), with results which prompted the Post Office to avoid the then existing circuitous route from London to Dublin via Portpatrick and Larne, and to lay a direct cable nearly 74 miles long between Nevin in Carnarvonshire and Howth near Dublin. This cable was laid in December 1913.

On January 1, 1914, the continental service from London was, by reason of the improvement effected by the loaded cable, extended to Basel, Geneva, and Lausanne, at a charge for three minutes of 7*s*. 6*d*. by day and 4*s*. 6*d*. by night.

A cable thirty-five miles long joining Vancouver and Victoria, British Columbia, was laid in the Gulf of Georgia in 1913. Owing to the depth of water (1300 ft. maximum) continuous-loading was adopted for this cable. Its length is 28·3 nauts. There are two metallic circuits and it is worked with a phantom circuit. The cable was manufactured by Henley's Telegraph Works Co., and is described in the *Electrician* of August 22, 1913.

Though not strictly related to the title of this chapter, it will be convenient to refer here to the advantageous use of loading coils on short lines. Having adopted a certain standard of transmission, it is the province of the engineer to determine the economical constitution of the circuit over which it is to be obtained. Reverting to the analogy of the vibrating string :—For the propagation of a particular wave a string of large mass may be required, but the same result may be attained with a string of smaller mass if suitably loaded. For 'string' read 'conductor,' and the analogy is converted into practical terms. Suitable loading permits the use of smaller conductors.

It was for this purpose that the first commercial use of Pupin coils was made in Great Britain by the National Telephone Co. The company had no long distance lines under their control, but Mr. Gill, their engineer-in-chief, promptly recognised the economical advantage of loading for cable lines long or short, and made extensive

[1] *Journal of the Institution of Electrical Engineers*, xlvi. 309.

use of the system. The first long distance line to be loaded with Pupin coils by the Post Office was that between Liverpool and Manchester.

But the best results from loading for long distance work were in the nature of an unexpected development. To make three circuits out of two was the subject of a patent granted to F. Jacob in 1882.[1] This method of making up a larger number of circuits than the physical lines, separately considered, provide, has sometimes been called duplex or multiplex telephony, the added circuits being described as 'phantom circuits.' In 1884 M. Ducousso installed a Morse telegraph circuit between the Gobelins and the Rue Caumartin offices in Paris, superimposed upon a metallic circuit underground line about ten kilometres long.[2] The use as speaking circuits commercially was much later.

Though designated a 'phantom' this circuit is, of course, really more substantial than the 'physical' circuits out of which it is composed, and it will readily be understood that the superior transmission results are due to the superior physical conditions of the so-called 'phantom' line, each limb being composed of double the quantity of copper that constitutes the conducting medium of the separate lines, with a corresponding reduction in resistance. To combine lines in this manner is comparatively simple, but to arrange the loading so that it becomes effective for the 'phantom' and not detrimental to the 'physical' circuits is a difficult problem. Much study and experiment were required before the system could be put into practical use.

The extension of cable circuits and of long distance aerial lines was steadily and systematically proceeded with. In 1910 Mr. Carty said :—

We already have an effective long distance service through underground cables of the Pupin type from New York to Philadelphia (90 miles), and good talking with prompt connection is an every-day matter between New York and Boston (235 miles). Our long distance wires extend to Chicago and other more distant western cities, and to Washington, Baltimore, Atlanta, and other places in the far south. At the present time we are extending an underground cable of the Pupin type from New York to Washington (235 miles), and are making surveys and plans for an extension from New York to Boston. More than this, by the adoption of phantom loaded overhead circuits between New York and Chicago, and by similar extensions westward as far as Omaha and thence to the Rocky Mountains, we expect by the first of January next

[1] British specification, No. 231, dated January 17, 1882.

[2] Devaux Charbonnel, *Bulletin Société Institut Électricien*, May 1903, p. 265.

[1911] to have so greatly extended our 'Long Distance' frontier that conversation may be held between Denver, Colo., and New York City, a distance of 2200 miles.[1]

The anticipation was fulfilled, but Denver was only part way across the continent, and the goal which management and engineers had set themselves was transcontinental conversation. On January 25, 1915, the line connecting the Atlantic and Pacific coasts was opened for commercial use. Alexander Graham Bell was at New York and Thomas Augustus Watson, his co-worker in 1875, at San Francisco. Two other lines played an important part in the inaugural ceremony. One to Washington for the President of the United States, the other to Jekyl Island, Florida, where Mr. Vail, the President of the Telephone Company, was staying. Over the transcontinental line President Wilson exchanged congratulatory messages with the Mayor of San Francisco and the President of the Panama-Pacific Exhibition. He also engaged in simultaneous conversation with Bell and Watson. Washington was then switched on to Jekyl Island, when President Wilson informed Mr. Vail that he had just been speaking across the continent, extended his congratulations 'on the consummation of this remarkable work,' and expressed the hope that Mr. Vail would speedily be restored to health. The connection of the lines to San Francisco and Jekyl Island through New York also enabled Bell, Watson, and Vail to engage in conversation.

Later on the same day Boston was connected with San Francisco and Jekyl Island also brought into circuit.

The length of the line New York to San Francisco is 3400 miles. The gauge of the wire is No. 8 B.W.G., 0·165 in. diam., 870 lb. per circuit mile.

The photograph reproduced in fig. 158 was taken at New York. Reading from the observer's left to right there will be seen :—Mr. Carty, Chief Engineer; Hon. George McAneny, President Board of Aldermen, New York City; Mr. U. N. Bethell, Senior Vice-President American Telephone and Telegraph Company; Dr. Bell; Hon. John Purroy Mitchel, Mayor of New York; Mr. C. E. Yost, President Nebraska Telephone Company, Omaha; and Hon. W. A. Prendergast, Controller of the City of New York. The picture on the wall is a portrait of Mr. Vail. At Dr. Bell's right hand is a standard desk transmitter and receiver; on his left a reproduction, trom the original in the Smithsonian Institute at Washington, of the early telephone illustrated on page 43. Through this reproduction Bell spoke to Watson at San Francisco.

[1] J. J. Carty: Discussion at International Conference of European Telephone and Telegraph Administrations held at the Sorbonne, Paris, September, 1910.

FIG. 158.—Inauguration of New York-San Francisco Line, January 25, 1915.

The following represent the principal stages in the development of commercial conversation over long distance lines in the United States:—

1876. Boston to Cambridge, 2 miles.
1882. Boston to Providence, 45 miles
1884. New York to Boston, 235 miles
1892. New York to Chicago, 900 miles.
1911. New York to Denver, 2100 miles.
1913. New York to Salt Lake City, 2600 miles.
1915. New York to San Francisco, 3400 miles.

The connection of the Florida line with San Francisco made a total of 4300 miles, the longest talking distance yet achieved. Simultaneous conversation was effected between New York, San Francisco, and Jekyl Island. The three participants in this conversation were thus in the extreme east, the farthest west, and the most southerly states of the Union respectively. And they were—Bell, who thirty-nine years before had invented the telephone; Watson, his assistant in the development of that invention; and Vail, one of the early organisers in the United States of the industry which has grown out of that invention. The conjunction of men and circumstance is remarkable, as well as the mutual recognition of the fact that individual work ceased with the invention, and that the subsequent developments resulted from collective effort. In Bell's own words, addressed to those present at the inaugural ceremony from the platform (illustrated in fig. 158), 'The telephone system of to-day is the product of many minds, and not the product of one alone.'

CHAPTER XXIX

INSTRUMENTS

WHILST the scientists whose views were quoted in the early part of the preceding chapter produced the impression that one regarded improvement as attainable only in the line and the other only in the instruments, it is probable that in a less controversial atmosphere both would have admitted quite readily that there was room for improvement in all three elements—transmitter, conductor, and receiver. It must have been as obvious then as now that a transmitter which would permit a more powerful current to be sent to line would produce a more powerful effect in a receiver on a short line or afford a larger margin for attenuation on a long line. It was as obvious then as now that a poor battery resulted in poor transmission, and it would seem equally to be obvious that results depended upon not one, but all the components.

The demand for modified instruments put forward by Thompson in 1887 had been expressed by practical telephonists in 1885. Hibbard in the paper already quoted referred to the disturbances or noises on long wires:

> The most natural and general remedies which suggest themselves [he said] are a louder transmitter and a less sensitive receiver, a receiver which will be affected only by the particular current governed by the transmitter worked in conjunction with it.[1]

The latter condition was hoped for, though it is difficult to see upon what ground it could be expected. If the receiver were not sensitive, it could not operate as a reproducer of speech vibrations. Since its very excellence depended upon this sensitiveness, it could hardly be expected to receive the vibrations intended for it, and reject those which were in the nature of trespassers. The transmitter, on the other hand, afforded a far more hopeful field. There was practical evidence, in the introduction of the battery transmitter, of the superior results which would follow the use of a more powerful

[1] *National Telephone Exchange Association Report*, 1885, p. 65.

current than the magneto transmitter afforded. Several varieties of battery transmitter were tried over the New York-Chicago line in 1883, and the best results were said to have been obtained with what was known as a double-Edison. Two transmitters, each having its own battery, were combined with one mouth-piece, as indicated in the illustration fig. 159.[1]

The line was single, having been constructed for telegraphic purposes. The results obtained from it were unsatisfactory, but the possibility of transmitting speech over a line approximately one thousand miles in length was demonstrated. If reliance is to be placed on the statement that the double-Edison gave the best results, it is probably due to the increased battery power available. The grouping of a number of transmitting devices was tried in both magneto and battery telephones, but abandoned. It seems probable that synchronism was unattainable, and that consequently the vibrations were out of phase. Whatever the reason, none of these instruments survived.

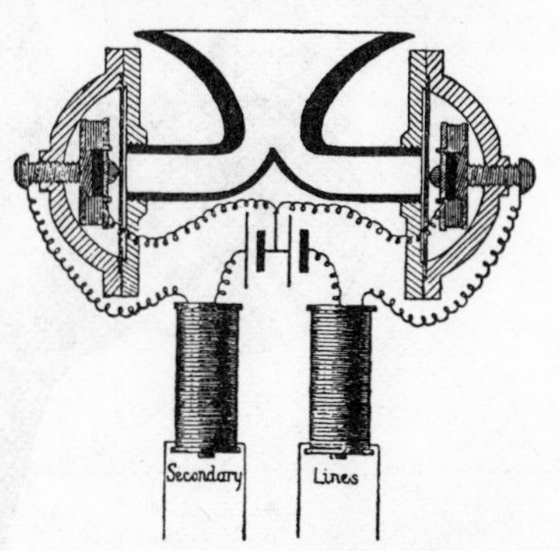

FIG. 159.—Double Edison Transmitter (1883).

The direction of development lay in modifications of existing types of transmitters. As explained in Chapter xl, these were broadly three: the Edison, the Blake, and the Hunnings.

Of these, as they left their inventors' hands, only the Blake was complete and practicable. The composition of the Edison variable resistance was unsuitable, and the construction of the Hunnings imperfect, but as a medium of varying resistance the granular carbon of the Hunnings was recognised as offering advantages hitherto missed, and inventors in Europe and America turned their attention to it. No Hunnings transmitter at that time, or since, has transmitted speech with greater clearness or fidelity than the Blake, whose merits in that respect have been expressed in earlier chapters. For short distances it was perfect, but for long distances it failed. The single cell did not send sufficiently powerful currents to line, and the instrument itself was deranged if another cell was added.

Granular carbon was then resorted to, and it was used in a great variety of ways. The packing of the carbon was an early difficulty. Attempts to avoid this by placing the carbon cell in

[1] From the London *Electrical Review*, June 16, 1883, xii. 491, reproduced from New York *Electrical World*.

a horizontal instead of vertical position were made by Blake, Lyon and Kellar, and Thornberry in the United States and by Husbands, Ericsson, Berliner and others in Europe. The substitution of a thin carbon disc for the metallic diaphragm permitted the return to the vertical form. Patterns too numerous for individual mention were made in Europe, but perhaps the best known was that of Ericsson. Another was the Deckert, an Austrian design, introduced into England by the General Electric Company,

FIG. 160.

under the name of the Hunnings-Cone. In these, and most others of the same type, including the Delville of the Antwerp Company, the front and back electrodes were separated by discs of felt or similar substances, the spaces between being filled by granules of carbon. There was a danger of lack of insulation by a separating substance such as felt absorbing moisture and there was an absence of uniform results by the nature of the construction.

The engineering department of the American Bell Telephone company sought to obtain a transmitter suitable for long distance work, capable of using greater battery power, and of such mechanical construction as would ensure uniformity and precision. The result of their investigations is comprised in the U.S. patent

No. 485,311, November 1, 1892,[1] granted to A. C. White. This instrument was known as the 'solid back' transmitter, a name which has been retained for this particular make, though it really describes any transmitter which has a fixed back electrode instead of the movable one of the Blake transmitter.

In his specification, White indicates that the invention consists in details of construction, whereby the tendency of the finely divided conducting material to pack is reduced or overcome and undue heating of essential parts of the instrument prevented.

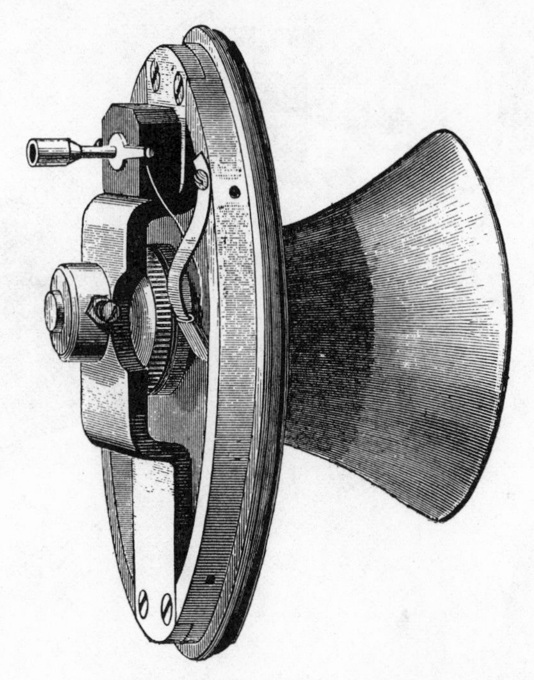

FIG. 161.

This instrument was described in the *Scientific American* of February 16, 1895. The illustrations (figs. 160 to 166) accompanying the description are in greater detail than the figures of the patent specification. At the time that this transmitter was introduced, it was customary for a long distance subscriber to be furnished with a special instrument, which was similar in appearance to a Davenport writing desk.[2] In the lower part the batteries were placed; in the upper part were the generator, ringer, coil, and switch. The swan-necked arm was mounted on the desk top, bringing the

[1] Application filed March 24, 1892.

[2] Illustrated in fig. 152.

transmitter itself to a convenient height. The granulated carbon used was described in the specification as 'anthracite carbon.' 'The secret of its manufacture,' says the *Scientific American*, 'is not known to the public, but it has been ascertained that coked Schuylkill anthracite coal will answer the same purpose.' Not only the substance but also the size of the granules was important. 'The carbon granules are screened through a wire sieve of 60 mesh; smaller and larger granules than those which pass through this mesh are rejected.'

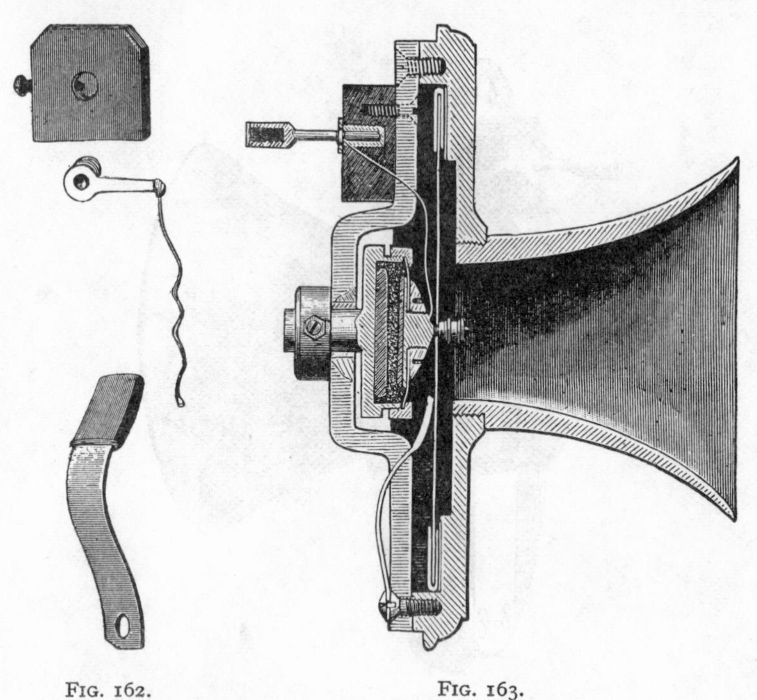

FIG. 162. FIG. 163.

Comparison of the section of this transmitter with that of the Edison[1] shows a great similarity in appearance. The 'solid back' may be said to have brought the transmitter to operate as Edison had intended. His theory contemplated a change of conductivity in the tension regulator under the influence of the diaphragm, but the composition of his tension regulator was not elastic, and the effect of the diaphragm upon it was in one direction only. Instead of a box with a drop lid, as in Edison's, White provided a cell partially lined with carbon at its back, and at its

[1] Chapter x. p. 111, fig. 32.

front a small mica diaphragm also partially lined with carbon. Instead of the plastic but inelastic mass of lamp black, White provided sufficient carbon granules nearly, but not quite, to fill

FIG. 164.

the carbon cell. The small mica diaphragm was bolted at its centre to the large diaphragm, so that every movement of the latter was conveyed to the former. Regarding the whole of the

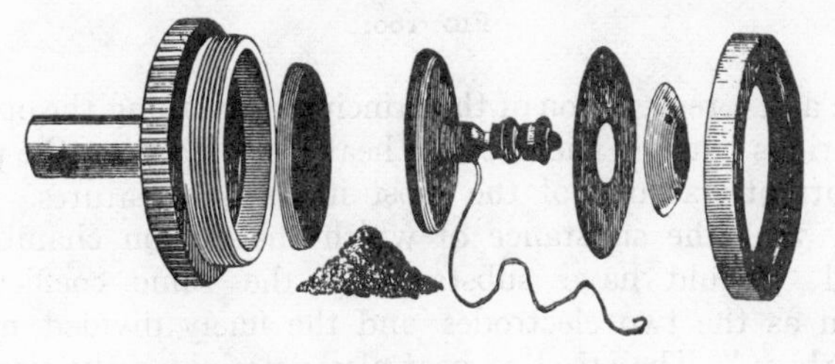

FIG. 165.

carbon cell as a 'tension regulator,' the effect is just that contemplated by Edison, though the researches of Hughes and the suggestion of Hunnings were necessary to explain the operation and produce effective results. By dint of great precision in the manufacture of the carbon cell, and granules, and in the quantity of the latter placed within the cell, as well as by reason of the mechanical construction, which was entirely free from any substance affected by moisture, or altering its condition with use, it was possible to produce these transmitters with remarkable uniformity, and certainty of results. The rubber ring surrounding the diaphragm and the damping spring applied thereto, which were so effective in the Blake, were incorporated in the 'solid back.'

As in the invention of the telephone itself, and of other definite advances in the lines or apparatus, the improvement effected by White in the transmitter was essentially due to a study of the

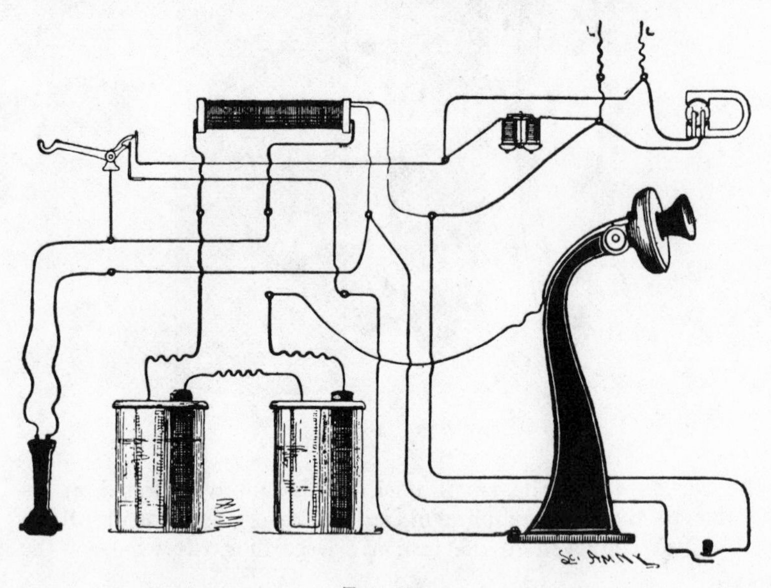

Fig. 166.

problems and a recognition of the principles governing the operation of the various parts. The effect of heat resulting from the passage of the current was one of the most important features. It was necessary that the substance of which the carbon chamber was composed 'should have substantially the same coefficient of expansion as the two electrodes and the finely divided material between them.' The effect of heat also determined the size of the front and back electrodes. These are of less diameter than the box within which they are placed. The reason is clearly stated in the specification. 'The considerable space around the periphery of the two electrodes' is 'to receive the finely divided conducting material out of the direct path of the electric current. The finely divided material in this part of the chamber not becoming so much heated in the operation of the instrument as the portion between the electrodes, the latter portion is permitted to expand into the former, and so offer less disturbance to the electrodes, due to its change in temperature.'

It is further stated that this form of construction also prevents the clogging of the carbon granules between the front electrode and the walls of the chamber. Where the electrode is substantially of the same diameter as the walls, a clogging takes places which interferes with the proper movement of the electrode. Presumably experiment determined the statement that 'the electrodes of whatever material made should be highly polished for the best results.' The importance of a high polish in the Blake carbon was previously well known, but the operation of this instrument was so different that the experience could hardly apply.

It will be understood that the illustrations given are those of the original design. The present model is smaller and more compact. A number of changes in detail have been made, all tending to increase the talking range or add in some respect to the efficacy of the instrument, but the general principle of construction remains the same. Its efficacy is mainly due to the capacity to send to line a stronger current than its predecessors. It was designed to be comparatively sluggish in its microphonic effects. The transmitters which had preceded it, like the Blake, the pencil microphones and the many varieties of the Hunnings using granular carbon, had been designed so that they could be spoken to at a distance. This required a sensitive action on the part of the 'tension regulator' (as Edison called the microphonic part of his transmitter), which was useful in one respect but detrimental in another, since it permitted the transmitter to be affected by extraneous noises which were transmitted to the line together with the speech vibrations, resulting in confusion at the receiving end. The 'solid back' was supplied with a mouthpiece to direct the sound waves to the diaphragm, but this mouthpiece was of such length and shape as to minimise as much as possible the effects of resonance. The resonance of an air chamber may effect a reinforcement in one tone or partial tone, which would result in impairment of articulation. The frequently inarticulate transmission of some of the early forms of pencil microphones was probably due to the sympathetic vibration of the air in the box within which the transmitter was mounted. Considerable trouble was taken to obtain a particular kind of wood supposed to be the most suitable for diaphragms (which was probably of small importance), and little, if any, consideration given to the elimination of resonance, which was probably very important. With the 'solid back' the speaker is required to talk quite close to the instrument, and when this condition is complied with, a quiet tone suffices for the longest lines.

The magneto telephone of Bell's 1877 patent was adapted to use either one or both poles. Commercially the double pole form was reserved for a transmitter, being mounted in a case and known as the 'box telephone.' The receiver or 'hand telephone' was a single pole instrument. Later the double pole feature was applied to the hand instrument which had settled down into the position of the receiver, and the double pole feature has now been in general use for many years. The receiver also has undergone a number of changes in detail. It has been stated that fifty-three types and styles have been introduced in the United States since 1877, and that of those in use in 1914 none were in use prior to 1902. These changes increased the efficiency, but did not modify the underlying principles defined by Bell.

The suitability of the names 'transmitter' and 'receiver' applied to those respective instruments has been questioned. Sir Frederick Bramwell, for instance, says:—

> I think it well before I go further if I state definitely that which I intend to convey when I speak of a 'transmitter' and of a 'receiver.' I should not have thought this necessary had I not found that educated men, unacquainted with telephonic matters, confounded the two terms, and moreover, were able to defend their want of discrimination. They said why should not the instrument which receives the speech of a speaker be called a 'receiver' and not a 'transmitter,' and why should not the instrument which delivers the reproduction of that speech be called a 'transmitter' and not a 'receiver.'[1]

It may be assumed that such an argument was not considered when the names were adopted. They were simply transferred to the telephone from telegraphic usage, where the instrument at the sending end of the line was the transmitting instrument or transmitter, and that at the receiving end of the line was the receiving instrument or receiver.

Magneto Call Bells

It was stated in Chapter xiii, where the early forms of magneto bells were described, that much attention was being given to their improvement and development. Primarily required to generate a current, the Siemens armature early established its superiority, and the improvements in the generator were mainly in features of construction, leading to increased power, the strength and permanence of magnets, and other details. Improvements in gear-cutting

[1] *The Practical Applications of Electricity*, Institution Civil Engineers, 1884, p. 27.

machines and minor improvements to be noted later permitted the production of toothed gear wheels effecting such smooth working that the superior mechanical results and lasting power of cut gear were enabled to overcome the less reliable friction and band gears which had been rather preferred in earlier years. The receiving part of the magneto, termed the ringer, has retained the principle of Watson described in Chapter xiii (p. 146).

But, in addition to the generator and the ringer, there had been collected within the magneto case the switch and its numerous contacts by whose means the circuits were changed automatically

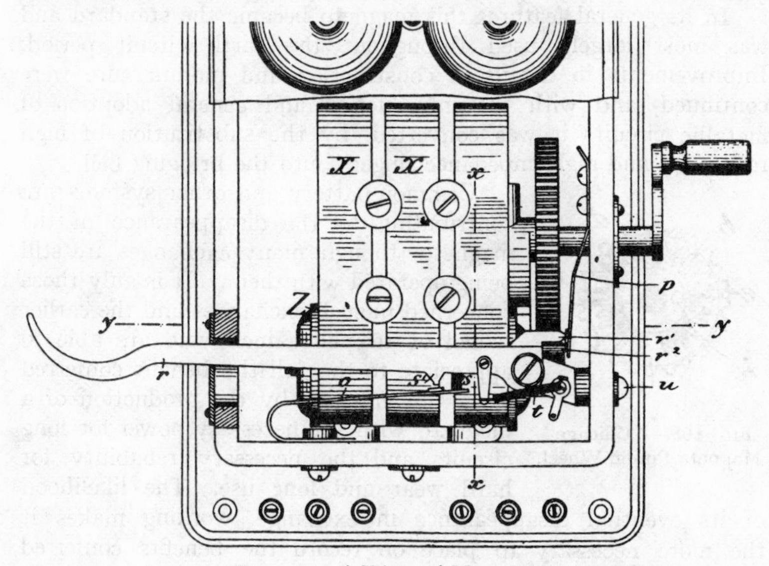

Fig. 167.—'Chicago' Magneto.

so far as the subscribers' conscious thought was concerned, from the ringing to the speaking condition. The reports quoted in Chapter xiii indicate the effects of wear and tear upon imperfect designs and unsuitable materials. Experience had shown the points of weakness, and organised experiment and ingenuity had found the remedies.

The British patent specification No. 19,045 of 1888 [1] illustrates what was commercially known as the 'Chicago' magneto. The principal new features of design were in the switch and the gear. The springs of the switch arm were arranged in such a manner that the force of one set was in opposition to that of the other. The switch arm (*r*) (fig. 167) was provided with insulating studs or

[1] U.S. equivalent, No. 425,058, April 8, 1890 (application filed February 1, 1888).

projections (r^1 r^2) opposite the corresponding springs (p). Good electrical contacts and actual electrical separation, according to the position of the switch arm, were thus ensured. The insulating projections which effected the electrical separation afforded a medium by which the mechanical force of the springs out of circuit was utilised for producing the efficient electrical contact of the springs in circuit.

The improvement in the gear consisted of a coiled spring in the pinion wheel (fig. 168) adapted to take up the starting movement of the driving wheel and thus contribute to the smoothness of working.

In its general features this magneto became the standard and was most largely used throughout the earth circuit period. Improvements in details of construction and manufacture were continued, and with the introduction and general adoption of metallic circuits it was converted, by the substitution of high resistance and high impedance ringers, into the bridging bell.

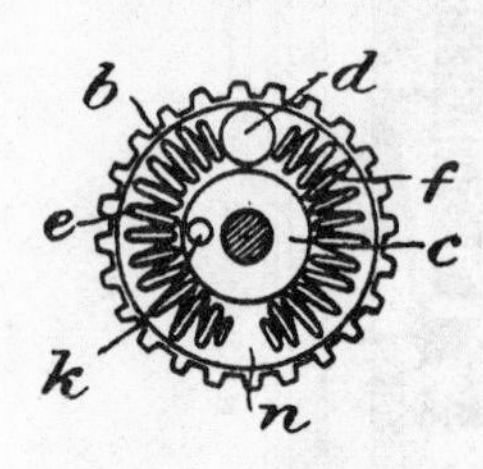

FIG. 168.—' Chicago ' Magneto Pinion Wheel.

Common battery and other systems are contributing to the disappearance of the magneto, though many exchanges are still being operated with them. It is only those concerned in such exchanges and the earlier school of telephone men who are able to appreciate to the full the benefit conferred upon the industry by the production of a magneto with the necessary power for long circuits and the necessary reliability for hard wear and long use. The likelihood of its eventual disappearance in exchange working makes it the more necessary to place on record the benefits conferred by the production of an instrument of much intrinsic merit as a result of sound design and improvements continually effected in manufacturing methods.

In the United States all the telephonic instruments were designed with a strict view to practical use. Little if any attention was given to æsthetic considerations—unless it be urged, as it well may, that the form of the greatest utility is in fact the most artistic. But even simplicity may be overdone, and in Europe more ornamental types were called for. In Germany the instruments were usually rather overloaded with ornament. In Scandinavia perhaps still more so, but of the latter there was one example—an Ericsson desk magneto set[1]—which was very popular not only in Sweden but in other parts of the world. In this magneto the

[1] Illustrations of this and other instruments used in Europe will be found in *The Telephone Systems of the Continent of Europe*, by A. R. Bennett. London: Longmans (1895).

permanent magnets formed a base for the whole construction. The combination of transmitter and receiver on one handle very generally used in Europe was known as a micro-telephone. Such a combination may be noted in the illustration of the Universal switch.[1] It was then known as the switchman's telephone. A model called the Berthon-Ader was much used in France. In fact, the convenience of the combined instruments rendered them very popular with subscribers everywhere. The discontinuance of their use in large systems is due to the greater efficiency of the separate instruments. The explanation is quite simple: the necessary adjustments for mouth and ear are easy with separate instruments and impossible when they are rigidly combined.

Condensers

The condenser, which is mounted in the wall instrument, or in the bell box of a desk instrument, in common battery subscribers' sets, has been developed especially for telephone use.

The earliest use of a condenser with a telephone was in a shunt circuit to the call bell, the object being 'to avoid the previously noticed inductive difficulties which present themselves when many sets of apparatus are placed in one circuit.'[2] The purpose here was to afford a path for the speaking currents, so that they should not be retarded by the impedance of the call bell.

Another early suggestion for the use of a condenser with a telephone is that of Black and Rosebrugh,[3] the object in this case being apparently to use a line for both telegraphic and telephonic purposes. The same invention was communicated to Mr. H. J. Haddon and a British patent applied for, but it was not proceeded with beyond the provisional stage, perhaps because of Prescott's prior publication. In the provisional specification one of the objects of the invention is said to be 'to enable telephones or other secondary current apparatus to be connected with the wire of any galvanic circuit without interfering with the action of said galvanic circuit'; and in the first claim of the American specification the converse is added, 'And so that the galvanic circuit may be broken between the points where the two ends of the derived line are attached without breaking the secondary current circuit.' The objects in both these proposals differed from that of the common battery use, but they are early examples of utilising the functions of a condenser to pass rapid alternations and to bar steady currents. It is this characteristic which enables the condenser to serve so useful a purpose in modern telephony.

[1] Chapter xiv. p. 169. [2] *The Speaking Telephone*, etc., Prescott, 1878.
[3] U.S. specification, No. 212,433, February 18, 1879 (application filed June 4, 1878).

When the common battery system was first introduced, it was an open question whether a high resistance or a condenser should be employed. With the high resistance method, there would be a small current flowing through each instrument, which in the aggregate would lead to considerable waste. The condenser method stopped the steady current completely, and there would thus be no waste: but condensers at that time were bulky and expensive. To be suitable for use in subscribers' instruments they needed to be small and cheap. The Cortlandt Street common battery installation [1] employed high resistance. Mr. Carty states [2] that the abandonment of that method was due to the improvements which were made in the manufacture of condensers.

After a great deal of experimentation, involving very careful research, a circuit was devised whereby, owing to the employment of the condenser, the transmission of speech by means of the common battery system was greatly improved. The effect of the condenser upon the transmission of speech when placed in the circuit as described is to sharpen up the tones of the voice and render them more distinct and penetrating. In the above-mentioned arrangement of circuit a very beautiful combination is represented, the condenser, while serving the function of breaking the common battery path, also served the function, owing to the peculiar arrangement in the circuit, of improving the transmission.[3]

The construction of a condenser as described by Black and Rosebrugh in 1878 would apply equally well to 1896:—'The condenser is constructed of alternate layers of tinfoil and thin plates of mica, gutta-percha, or paper saturated with paraffin, arranged like the leaves of an interleaved book.'

It was the need of a small and economical condenser without too great a refinement in its 'capacity' standard that prompted investigation and experiment by Messrs. Lee, Westcott, and Robes of the American Bell Company's engineering staff, resulting in the method of manufacture described in their United States specification, No. 575,653, January 19, 1897.[4] Following the analogy of the interleaved book, the development may be compared with the advances in printing from the hand press upon a sheet, to the rotary press upon a continuous roll, of paper. Paper from one roll and tinfoil from another roll were wound together for such a length as the 'capacity' required; were then immersed in melted paraffin wax and compressed, resulting in a condenser which filled the double requirement of small size and low cost. Paper in continuous rolls was no new feature of manufacture, but the preparation of the tinfoil involved considerable experiment and development. The

[1] Chapter xxvi. p. 378.
[2] *Herzog Case*, pp. 117–18.
[3] Carty, *Herzog Case*, p. 119.
[4] Application filed July 8, 1896.

omission of the separate sheet of tinfoil was suggested by Mansbridge, who patented the application of a metallic solution directly to the paper, such as had previously been accomplished for tea wrappers. The process and the advantages claimed for it were fully described by Mr. Mansbridge before the Institution of Electrical Engineers on May 7, 1908.[1]

Relays or Repeaters

Though relay and repeater are practically synonymous terms, some convenience attaches to the custom which has arisen of using both words to describe an instrument which relays or repeats telephonic speech. So many relays are employed in telephone switching, for example, that 'telephone relay' is not sufficiently descriptive, and 'repeater' has become to some extent identified with repeating coils so that 'telephone repeater' is also open to some misunderstanding. But the combination of the two words has served to identify that type of apparatus which, by the application of renewed power, carries the speech to a greater distance than may be possible with the appliances at the original transmitting station.

The first suggestion for the application of relays to the transmission of speech was made in anticipation of a demand rather than as a means of supplying a discovered want. No commercial lines existed which exceeded the capacity of transmitter and receiver without extraneous assistance. But inasmuch as relays had become of great utility in telegraphy, it was apparently assumed that they would offer similar advantages in telephony.

On July 8, 1879, E. T. Gilliland filed an application[2] for a patent on 'New and useful improvements in Reproducers of Undulatory Electric Waves or Telephone Repeaters.' The Gilliland repeater was practically a Bell receiver and Blake transmitter with a diaphragm in common. It was assumed that the vibration of the diaphragm by the receiver section would serve to operate the transmitter section and permit the retransmission with renewed battery power over the remaining part of the line. No special difficulties were contemplated in the performance of these functions by such apparatus, and consequently no provision was made to overcome them. Whilst the Gilliland relay itself calls for little attention, the plan which had been worked out for speaking in both directions will be of interest. Fig. 169 is a reproduction of the seventh figure of Gilliland's 1881 patent.

Edison proposed the use of a phonograph as a mechanical

[1] *Journal of the Institution of Electrical Engineers*, xli. 535.
[2] U.S. specification, No. 247,631 dated September 27, 1881.

relay. An experiment with the phonograph and also with the electro-motograph as a relay was made by Hammer in 1889, and described before the American Institute of Electrical Engineers in 1901.[1] But, so far as is known, none of these devices were commercially used, and it is very doubtful if they were capable of giving effective retransmission.

Some time before 1905 Shreeve developed a relay or repeater, also of the combined receiver-transmitter type,[2] which provided means for overcoming some of the difficulties incident to this form of construction, and which rendered efficient service under some conditions.

In 1910 S. G. Brown brought out a relay with some modifications in the receiving magnet and the transmitting electrodes which was fully described by him before the Institution of Electrical Engineers,[3] and in 1911 he patented [4] further improvements, 'the chief object being to provide a compact efficient instrument of this kind, and in particular a reliable and highly sensitive form of variable resistance device for inclusion in the local circuit.' Modifications in the receiver-microphone form of relay were made by Stone, who enclosed the working parts within a vacuum; and by Erdmann, who interposed an air-cushion which was varied in sympathy with the movements of the receiving diaphragm and enlarged the area of operation upon the transmitting diaphragm of the relay.[5]

Efforts have been made to substitute for the microphone, which requires mechanical movement, a current changer which could be brought into operation entirely by electrical influence and without mechanical aid.

Lee de Forest,[6] Reisz and Lieben,[7] and Cooper-Hewitt may be mentioned as workers in this direction.[8] The first three have obtained practical results with relays of the lamp type which are stated by Dr. Fleming to be developments of his vacuum oscillation valve.[9] Cooper-Hewitt arranged a mercury vapour tube within the field of the receiver magnet whose variations produced corresponding variations in the resistance of the mercury tube, which acted as the transmitter of the relay.

It has been stated by the President of the de Forest Radio

[1] *Transactions of the American Institute of Electrical Engineers*, xviii. 112.

[2] British specifications, Nos. 9605, 9606, of 1905.

[3] *Journal of the Institution of Electrical Engineers*, xlv. 590.

[4] British specification, No. 27,953, 1911.

[5] See *American Telephone Practice*, 4th edition, pp. 748–9.

[6] *Electrician*, February 6, 1914, lxxii, 726. [7] *Ibid.*, March 13, 1914.

[8] Edison, Richardson, Willows and others have also done valuable pioneer work in connection with these lamps or valves.

[9] *Proc. Physical Society of London*, March 23, 1906, xx. 177.

Telephone and Telegraph Company[1] that the Audion amplifier (as their type of relay is called) is used on the New York-San Francisco line, but whether in its original form or with modifications introduced by the engineers of the American Telephone and Telegraph Co. is not stated.

The relay or repeater has not yet reached the stage of an ordinary commercial instrument, but the persistent efforts to utilise any and all means to extend or improve telephonic transmission leave little room for doubt that means will be found to render such apparatus generally available. Perhaps the greater promise may be considered to lie with the non-mechanical type; but it is too early to regard that as certain. The Shreeve form has been used with some good results, and S. G. Brown states[2] that his receiver-microphone relay (presumably the 1911 type) has been used by the British Post Office to carry traffic on trunk lines.

The relay itself is not alone to be considered. It has been remarked that 'difficulty has been experienced in utilising relays on commercial circuits owing to the necessity of relaying the speech currents in both directions.'[3] This difficulty arises in balance systems from the reaction between receiver and transmitter known as 'howling' or 'humming';[4] and in switch systems from the possible operation of the switch in the direction contrary to that required for the relay. The switch illustrated in fig. 169 is of interest as an early application of a manually operated switch.

A subscriber at Chicago and another at Cincinnati are shown in communication through relays or repeaters placed at Indianapolis. The retransmitting instruments are in duplicate, G being allotted to the Chicago subscriber and *g* to the Cincinnati subscriber. F and *f* are the induction coils of the respective instruments. Z is a switch controlled by the keys B and *b* at the respective subscriber's stations. Chicago would receive through *g* and transmit through G. Cincinnati, on the contrary, would transmit through *g* and receive through G on the assumption that there was a singleness of purpose on the part of the subscribers, and that each depressed his key on talking and released it when listening. Such a plan makes no provision for the common failing of both desiring to talk at once, but it is an interesting example at so early a date of a practical provision for the operation of relays in both directions by means of a switch.

There have been suggestions to make the operation of such a

[1] *Electrician*, April 16, 1915. [2] *Ibid.*, April 23, 1915, p. 99.

[3] 'Application of Telephone Relays to Commercial Circuits,' by C. Robinson, B.A., and R. M. Chamney, A.M.I.C.E., *Electrican*, August 14, 1914.

[4] Note on a humming telephone by F. Gill read before the Dublin section of I.E.E., *Electrical Review* (London), May 31, 1901, xlviii. 951.

switch automatic. In 1912 S. G. Brown for example took out a patent[1] for appliances which enable the speaker's voice to do that which Gilliland required to be done by the speaker's hand. A

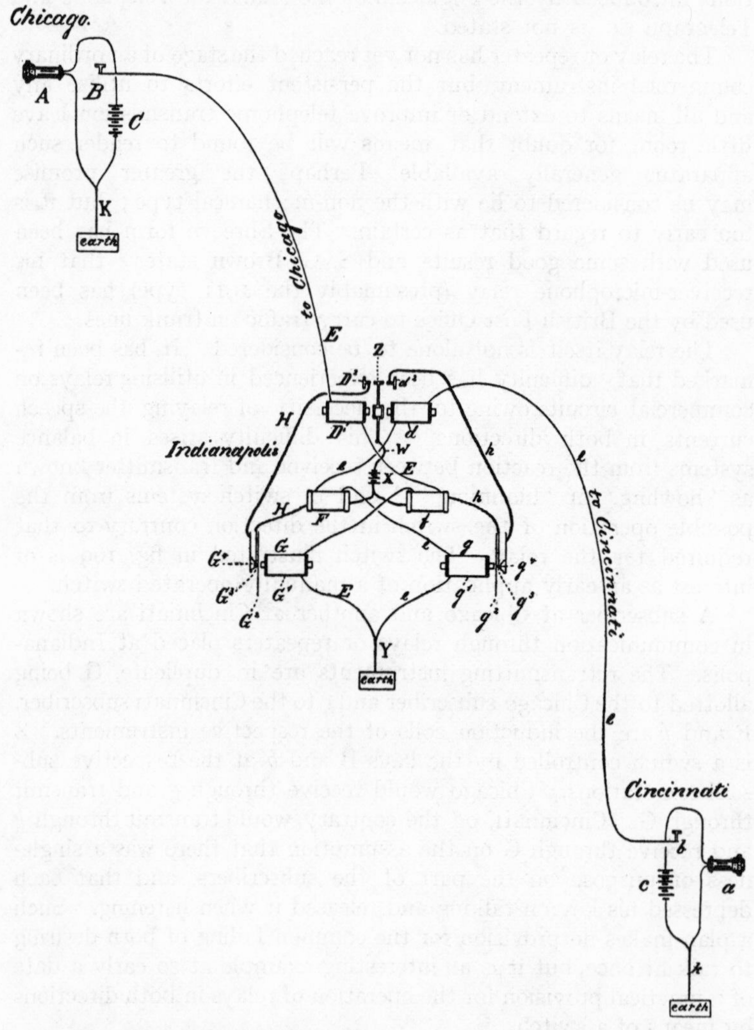

FIG. 169.—Gilliland Relay or Repeater (1879).

sensitive pilot relay operated by voice currents actuates another relay which diverts the line to the speaking relay or repeater which should be brought into operation. All prior experience would

[1] British specification, No. 9179, of 1912.

tend to show that if such diversion of circuits should be desired the switch must be rendered automatic in its operation, but this automatic operation would seem to require a greater degree of co-operative compliance between the correspondents than is necessary in ordinary conversation. Since a voice relay operates by minute current variations it would also seem to follow that inductive disturbances might actuate the relay as readily as the voice. Hitherto the necessary conditions for reciprocal working have been met by some balance system such as described by Messrs. Robinson and Chamney in the paper previously referred to.

CHAPTER XXX

RATES

THE telephone was an invention which produced a result that could not be attained before. The invention was the subject of a patent, and although the monopoly of a valuable invention can usually be obtained only as the result of keen struggle, yet, when obtained, the monopoly given by a patent is complete. Popular clamour sometimes breaks out against a patentee's charges, but Law recognises the importance of maintaining with the utmost integrity the rights of an inventor, or of his assigns, to the creation of his brain. It is assumed, and generally rightly assumed, that there is a relationship between the price the public can pay for a patented article and the rate it is offered at by the patentee which will satisfactorily adjust the conditions between the two parties to the bargain.

Bell's early associates were strongly antagonistic to his telephonic aims,[1] because they regarded them as impossible of attainment and therefore offering no profitable result. Perhaps for that very reason they could the more readily appreciate the value of the invention when it had been made. And they believed firmly in the validity and strength of their patents.

As patentees they knew their rights, and determined on a policy of exploitation which retained to themselves the ownership of the instruments and permitted to the public only the use thereof. They adopted, moreover, discriminating charges according to the purpose for which the instruments were used. The circular signed by Hubbard and issued in May 1877 is given in full in Chapter viii.[2] From this it has been seen that the terms for renting two telephones for social purposes were $20 a year, and for business purposes $40 a year, extra instruments on the same line $10 each. These prices included the keeping of the instruments in good working order.

[1] Chapter iv. p. 39. [2] P. 67.

At this time there were no exchanges in operation, nor had any been proposed, except in a very general way by Bell at his lectures during the same month of May 1877 as related in Chapter ix. The service contemplated was that which was later known as 'private lines,' the instruments being used exclusively by the parties at each end of the line or at intermediate stations, if such existed. But exchanges soon followed, and when charges had to be suggested, the private line circular had its influence. An annual rental was adopted, and the distinction between private residences and business premises was continued. The original rates were undoubtedly selected with the idea that they would return a handsome profit to the providers, whilst affording a convenience to the user which was cheap at the price. The growth in the number of subscribers was at first regarded with unqualified satisfaction by the exchange companies, and there is no doubt that in some small exchanges the subscribers provided in their advance subscriptions a considerable portion of the cost of constructing their lines. There was ignorance on both sides—the subscribers did not know the construction cost, and neither they nor the exchange companies knew what the working costs would be. On the whole it is well that there was this ignorance, for it helped to swell the subscribers' lists and to demonstrate with growing use the value of the service.

The exchange companies started in America almost invariably on the basis of a definite annual payment without regard to the extent of use. European and other exchange companies followed suit. A prior meaning of the word 'exchange' was synonymous with that of a club; a subscription was charged for the use of the exchange or club, and membership of a telephone exchange was charged for upon the club method. The terms varied considerably in the amounts charged, the length of line allowed, and some other conditions. When competition was offered it was, as usual, a competition in rates. In London the Bell Company charged £20, the Edison Company £12. At the time of their amalgamation some experience had been gained and the £20 rate was continued. The London and Globe Company subsequently started a competition on a £12 rate. This competition was closed by virtue of patent rights and the £20 rate continued. In large provincial towns in Great Britain the rates were £12.

The difference between the rates established in large and small towns seems to have been due to an intuitive appreciation of the respective costs of construction and value of service, rather than to a recognition of the additional cost of operation in a large centre. The rates had to be fixed in advance, before service was given, and it was not until the exchanges were in operation that experience could be gained. But it was not long before exchange managers found it necessary to go seriously into the question of rates.

At the first meeting of the National Telephone Exchange Association in September 1880, rates formed an important item of discussion. One member expressed the view that the inventive genius of Professor Bell in giving us the telephone had been equalled, if not excelled, by that of the gentlemen upon whom had devolved the duty of making rates. This ingenuity was shown in the multiplicity and complexity of the methods by which they had twisted ten figures and a few words to express every conceivable combination of prices and terms.[1] The statement of another member may be taken as illustrative of the method by which some rates became established. He related that they started to canvass St. Louis early in 1878, having no knowledge of any other exchange. They proposed rates of $100 a year. The Western Union Company started in competition, and subscribers signed their papers at $50 a year. The original company then reduced their rates to $50 and went into operation on that. After the exchange had about sixty subscribers connected and in working order, the rate was raised from $50 to $60, 'and the public stood it very nobly.' When the transmitter was introduced, $20 additional was charged.

> And not long ago, when the consolidation of the two companies was made, although the two exchanges are not to this day consolidated so far as regards the interchange of communication between the two, we then raised the rate to $100, and the subscribers took that, and, as far as I can see, I believe the subscribers will pay just exactly what the business is worth to them. That is our experience.[2]

Information had been obtained from various exchanges. It was conveniently summarised by another member and is abstracted as follows :—

> I have attempted to tabulate the returns that were given, but I find it rather a difficult matter. I find, however, that the rates vary from $22 to $80, and that, as a rule, they are paid quarterly in advance. To this there are some exceptions, some being monthly payments and one semi-annual payment. The lower rates are generally those of circuit [i.e. party] lines. In individual lines there is in each instance an indication of an increase in price in proportion to the increase in distance from the central office. In some of the reports I find some six or seven different rates. In some instances a rate is given where a telephone alone is used, and another rate where a transmitter is used, another rate on circuit lines, and another rate for individual lines.

[1] Hall, *National Telephone Exchange Association Report*, 1880, p. 164.
[2] Durant, *National Telephone Exchange Association Report*, 1880, p. 169.

The rates are increased pretty uniformly when the number of subscribers increase, and I believe that, as a rule, they are increased about $5 a month for every additional 100 subscribers. If the rates uniformly should be $42 on circuit lines, I think it would strike pretty near the average. Of course, in circuit lines, there would be no variation in the price in regard to distance, as a rule. As an average I should say for individual lines within the first half mile—this varies, sometimes a mile and sometimes half a mile—the rate averages about $50, and this price increases as the number of subscribers increases: If we were trying to fix our rates on an average we should find we might perhaps safely adopt: for circuit lines $42; for individual lines $50, within the first half mile, increasing at the rate of $1 a month or $12 a year for each additional half mile of distance.

Now for an increase of subscribers, as it has been universally experienced, the cost of the individual increases. As the subscribers increase you might make the basis something like this: Say for the first 300, $42 [party line] and $50 [individual line], and for the next hundred increase $5—that is, for the first half mile, $55; 500, $60; 600, $65; 700, $70; 800, $75; 900, $80; then 1000, $85; 1100, $90; 1200, $95; so that at 1300 subscribers the rate would be $100. I do not make this as a suggestion but as a conclusion we might arrive at if we adopted these figures as the basis of the experience of the various exchanges in the work they have done.[1]

The opinion of the public on telephone rates has followed a fairly consistent line through all the vicissitudes and successes of the Telephone Exchange industry. Without any data to assist their deliberations, they have invariably reached the conclusion that rates are too high. It is of interest, therefore, to read in the report of this 1880 meeting the remark of a leading member:—

The figures given by the various exchanges show that nearly all are groping in the dark for some system which shall make the compensation for the service given equitable, both for the public and our stockholders. Up to the present time, notwithstanding popular belief to the contrary, the public has had all of the profit and but little of the expense.[2]

There was something more than a ray of light sent forth to illuminate the rate question at this meeting, but it involved, in the majority of exchanges, an important change of principle. An increase in price from one fixed sum to another is sometimes more easily effected than a change in principle, however equitable it may

[1] Mr. Twiss, *National Telephone Exchange Association Report*, 1880, pp. 171–72.
[2] Hall, *National Telephone Exchange Association Report*, 1880, p. 164.

be. In the majority of exchanges in the United States—and their example, as previously remarked, was followed in Europe—the principle in force was that of an annual charge. The public had become accustomed to it, and in all probability the exchange companies placed some reliance on the certainty of income which followed a definite annual payment by each subscriber. Whatever the reason, the system continued, and we will follow the efforts of the companies to reach, on that principle, a basis which should afford them a satisfactory return for their expenditure and a suitable reward for their experience and skill.

In the report of the committee on rates in 1881 the same complaint is repeated. 'The solid fact remains, that the public is not paying a fair price for the service rendered by the great majority of exchanges.' [1]

The rates then prevalent, together with other data, are summarised by the Committee as follows [2]:—

The present report is based on returns from 271 exchanges, and the figures *very* carefully revised. For the proper consideration of the returns we have divided exchanges into four grades, and given the average figures for each grade separately ; no comparison between large and small exchanges can be of any possible value, based as their figures are on entirely different sets of conditions. Grade number 1 contains all cities having over 150,000 inhabitants ; number 2, all from 50,000 to 150,000 ; number 3, from 10,000 to 50,000 ; and number 4, all under 10,000.

Grade	1st	2nd	3rd	4th
No. of exchanges reported	14	23	140	94
Average population .	400,000	83,683	19,680	5,778
,, no. of subscribers	1,080	478	130	39
,, price per year .	$80.00	$50.00	$41.16	$38.40
,, no. of operators .	46	11	3	1-2/3
,, no. calls per month	178,000	50,000	17,800	5,125
,, calls per subscriber	165	105	137(?)	131(?)
,, ,, operator .	3,870	4,545	5,934	3,163
Subscribers to each operator	23½	43-5/11	43-1/3	24
Proportion on single wires	75%	40%(?)	67%	64%
Ratio of subscribers to population in fractions of one per cent. . .	·27	·57	·66	·67

It is remarked that

the record must be accepted with some allowance [because] many

[1] *National Telephone Exchange Association Report,* April 1881, p. 121.
[2] *Ibid.* 1881, p. 119.

exchanges sent but partial reports, and others were evidently made up by guess-work rather than from the exchange books, especially is this so in the record of calls per month, as but few exchanges are in a position to give actual figures, and all estimates are *largely* in excess of the real facts.[1]

Hence presumably the notes of interrogation in the table.

In 1882 it was reported that:

7	Exchanges	charge	$100.00	or more
3	,,	,,	90.00	,,
5	,,	,,	80.00	,,
5	,,	,,	72.00	
38	,,	,,	60.00	
37	,,	,,	50.00	
70	,,	,,	48.00	(The favourite country rate)
19	,,	,,	40.00	or more
18	,,	,,	30.00	,,
3	,,	,,	25.00	,,
205 . . .	Total number reporting.[2]			

From the detailed list the following are abstracted:—

New York,	2,875	Subscribers	$150	Business,	$150	Residence.
Chicago	2,596	,,	125	,,	100	,,
Philadelphia	1,804	,,	120	,,	100	,,
Boston	1,138	,,	120	,,	96	,,

A number of examples are given of exchanges proposing to increase rates. Increases, in fact, were necessary and were fairly general.

The development of the long distance service and the placing of the wires underground contributed to the need for metallic circuits. The effect of this condition on the provision of apparatus which would enable communication to be effected between all subscribers, whether fitted with single or double lines, has been referred to already.[3] The condition also affected rates. A subscriber had the option of taking an earth circuit line at one rate or a metallic circuit at a higher rate. If he required long distance service the latter was a necessity, though for local calls, when connected with an earth circuit subscriber, he had not the benefit of being 'off the ground.'

The flat rate system of charging for telephone service continued in general operation until 1894. By this system the companies had given a telephone service to a certain number of subscribers, and (subject to sundry unimportant variations, such as the distance of the station from the exchange) each of those subscribers had paid the same amount of money, regardless of the amount of service

[1] *National Telephone Exchange Association Report*, 1881, p. 119.
[2] *Ibid.* 1882, p. 120. [3] Chapter xxv. p. 336.

demanded and obtained. No indications are forthcoming from the reports that the companies were losing money at the rates charged. The total income set off by the total outgoings left a sufficient margin to return a profit on the capital invested. But the development was slow. In the case of New York the system increased 'from 7454 stations at the end of 1888, when the removal of the overhead lines was well under way, to 9914 at the end of 1893, an increase for five years of under 2500 stations.'[1]

The circumstances which prompted the telephone company operating in New York to re-consider the basis of charging at this time were not made public. A fixed annual sum, regardless of the number of calls, was the general rule, except that a limit was placed upon the number of originating calls which might be sent over one line in order that calls for that line might have a reasonable chance of finding it disengaged. A subscriber who exceeded the limit had his attention called to the inconvenience which he and his correspondents suffered from the line being engaged when wanted, and he was requested to subscribe for an additional line.

As has been said, this system was general, but there were exceptions. In 1881 Boston, Buffalo, San Francisco, and Rutland charged on the measured rate system. In 1882 all had changed to the flat rate except Buffalo. They so continued until 1894, except that in 1884 San Francisco, whilst retaining the flat rate for residence, adopted the measured rate for business, telephones. In 1887 Indianapolis, Rochester, Fort Wayne, La Fayette, Richmond and Terre Haute adopted the measured rate, and so continued for 1888. In 1889 Rochester returned to the flat rate, and the other cities followed suit in 1890.

In 1888 Logansport, Crawfordsville, Kokomo, Madison, Marion, Muncie, Peru, Shelbyville, and Vincennes (all in the state of Indiana) adopted the measured rate, retaining it for 1889. In 1890 all reverted to the flat rate except Shelbyville and Vincennes, which also followed suit in 1891.

A 'modified tariff system' was introduced for business telephones in Milwaukee in 1892, at Sacrameto in 1893, and in Los Angeles, Oakland, San Diego, San José, Seattle, and Tacoma in 1894.

But at Buffalo almost throughout the exchange period, and at San Francisco in part, the message had been made the basis of charge. It was to emerge from its position of being exceptional and to become more general. The ray of light which was to illuminate the rate question was really emitted at the very beginning of the industry, though many years elapsed before its illuminating effects became fruitful.

At the first meeting of the National Telephone Exchange Associa-

[1] H. L. Webb, *Electrical World and Engineer*, April 5, 1902, p. 588.

tion, the following letter, written on February 9, 1880, was read by Mr. E. J. Hall. The sentences not in inverted commas were presumably interpolations made by Mr. Hall during the reading.

'February 9, 1880.

'THEODORE N. VAIL, Esq.,
'General Manager, American Bell Telephone Company,
'Boston, Mass.

'Dear Sir,

'Yours of the 5th received. I am glad that my suggestion regarding branch offices meets your approval.'

That was with reference to opening branch offices, the way of running the business, charges, and so on.

'Still, as you say, we must ultimately come to a charge for each service. The branch office plan is a restrictive measure, bound to be unpopular and is against the real interests, i.e., the extension of the business.

'I now submit the outline of a plan which seems to me to meet all of the objections to the ticket plan of charging for each message.'

I will say that this plan was simply suggestive. The figures were not intended to be used in any case.

'1st. Charge five dollars for putting in an instrument to new customers, and in all cases when machines are taken out and replaced.

'2nd. Make no charge for rent but sell tickets according to a fixed schedule, and require customers to buy in advance of their wants, making the minimum amount enough to save the company from loss.

400	tickets at	10	cents	$40.00
500	,,	9½	,,	47.50
600	,,	9	,,	54.00
700	,,	8½	,,	59.50
800	,,	8	,,	64.00
900	,,	7½	,,	67.50
1000	,,	7	,,	70.00
1250	,,	6½	,,	81.25
1500	,,	6	,,	90.00
2000	,,	5½	,,	110.00
2500	,,	5	,,	125.00
3000	,,	4½	,,	135.00
4000	,,	4	,,	160.00

'Each customer has his choice of any number in this list, but he must buy not less than four hundred, and all tickets must be used within one year, i.e., you simply give a receipt for the number of messages paid for and charge them up as used. This meets all the objections in regard to collecting, in regard to several parties using one machine, in regard to dead beats, etc., etc.

'The interest of the company is to have machines used as much as possible; the subscriber's interest to limit the use.'

That is just the reverse of what is true now. The company's

interest is to restrict the use, but the subscriber's interest is to extend it, or practically he is indifferent about it.

'I have worked out most of the details in regard to the keeping accounts, form of tickets, leases, etc., and do not see anything impracticable, but cannot go fully into the matter in a letter. The schedule I send you was simply to illustrate.

'Culbertson and myself leave for Chicago, Cincinnati, etc., to-night. Will be back in about a week, when I shall be glad to hear from you in regard to the adoption of this plan. I feel sure that it must come to this everywhere, and the sooner the better for all concerned. We employ now nearly fifty operators, and the evil increases every day. The branch office plan is only a poor makeshift. . . . Doubtless you will see many points in this to be considered, but I believe you will quickly see how to overcome any difficulties, and I think it will be for the interest of all of us to take up this matter as soon as possible.

'Yours respectfully,
'EDW. J. HALL, Jr.,
'General Manager.' [1]

From this letter it would seem that Mr. Vail had expressed the opinion that it would be necessary to 'ultimately come to a charge for each service,' but it was evidently not considered expedient to adopt it generally forthwith. Mr. Vail ceased to be general manager in 1885, but resumed control as president of the company in 1907. In the meantime the 'ultimately' arrived, so far as New York was concerned, in 1894.

For the greater part of this period Buffalo had been charging the measured rate. They started, like other exchanges, with a flat rate in November 1878. They found the rates unremunerative and decided at the end of 1880 to introduce the measured rate.[2]

In Boston also it was reported that they had adopted a rate of $50 per annum and 5 cents per message, with discounts of 20 per cent. for calls exceeding 200 per month, 30 per cent. exceeding 300 calls per month.[3]

The effect on the use of the telephone may be observed from the following extract from the report of the committee on exchange statistics of the National Telephone Exchange Association in 1882:—

I expected to find that the larger exchanges, as would be natural, made the most connections. This is, in general true; still, there are some wide differences. . . . I have observed in the exchanges where it is a matter of notoriety that their system has settled

[1] *National Telephone Exchange Association Report*, 1880, p. 175.
[2] *Ibid.* 1881, p. 125.
[3] *Ibid.* 1880, p. 167. In 1882 Boston adopted the then prevailing flat rate system.

down, and their service has improved, an increase of service performed, and that in those exchanges where the service has reached its greatest perfection the highest number of connections were called for. There are two notable exceptions to this. They are the exchanges of Buffalo and San Francisco. Buffalo with 1047 subscribers makes for each of them but $3\frac{67}{100}$ connections per day. San Francisco with 1294 subscribers, but $1\frac{4}{100}$ connections per day for each—an astonishing result.

This is directly attributable to the presence of the tariff system (or system of charging according to number of connections made) in those cities. I call attention to it, because the question of charges is one in which we are all interested, and it seems to me that an interesting discussion might be evoked as to the true policy of telephone exchanges—whether it is better to make a tariff charge, and in that way limit the service, and, as I think, the usefulness of the telephone, or to charge an annual rental giving an unlimited service.[1]

Here was the probable explanation of the general continuance of the flat rate plan—the fear that the usefulness of the telephone would be impaired by giving the subscriber the opportunity of reflecting that its every use was a definite expense. The superiority of telephone service to telegraph service tended to the elimination of the message idea and to the encouragement of the club subscription plan, so that use should be free from any restrictive sentiment.

A similar view to that expressed by Mr. Fay in the United States in 1882 may be found in London ten years later. Commenting upon an invention which proposed to record not only the connections but also the times occupied, the London *Electrical Review* remarked that the application of such a system 'would certainly tend to restrict the freedom with which the telephone would be used.' The drawbacks attending the placing of complicated apparatus at subscribers' stations were considered, and,

whilst there is much to be said for the justice of a system by which a subscriber pays for all the service he receives and no more than he receives, something may also be said of the peculiar applicability of a rental system to telephone service, whereby both parties to a conversation are placed on equal terms. In meter systems the burden is placed on the sender, whilst the benefit of the communication is often more on the side of the receiver.

We are far from saying that the system of rentals, which is at present more general, is necessarily the best, and we think attempts which are being made in some places to alter it should be watched with interest; but telephone subscribers who feel that they are paying more than their share should console themselves

[1] Mr. Fay, *National Telephone Exchange Association Report*, 1882, pp. 44–5.

for a while with the reflection that complaints of a similar nature are very old, and are justified on the principle of a certain parable about the workers in a vineyard.[1]

To those who were already subscribers such consolation may have sufficed, but the effect of dividing the total outlay for the service by the total number of subscribers resulted in the establishment of a rate which was necessarily out of proportion to the expected use of a new subscriber.

And it was in the large city that this was more especially felt. The rate question becomes prominent in the American Bell Telephone Company's Report for the year 1892, where it is said :—

Setting aside the consideration that, in the large city containing some thousands of exchange stations, the use made by each subscriber and the value to him of the telephone facilities must inevitably be much greater than they would be in an exchange of less importance, it is also the fact that, in exchanges above the limit of a few hundred subscribers, the cost of maintaining and operating each station bears a close relation to the numerical size of the exchange and the extent to which each station is used, rising in the largest cities to an amount several times as great as the cost on the average of the small exchanges.[2]

The high cost in the large city was having its effect on development. In general it had been clearly established that the new subscriber was comparatively a small user and the older subscriber a large user. That the larger the number of subscribers who were connected to the service the greater was the advantage to all, needed no demonstration. Some system was clearly necessary which should be fair to existing users and not act as a bar to the introduction of new users.

The forecast of an impending change is to be found in the American Bell Telephone Company's Report for the year 1893 :—

The general system of the operating companies throughout the country has been to charge the subscriber a fixed sum for exchange service, graduated according to the class of equipment which he elects and leaving him free to use the service as much or as little as he chooses.

On the whole this plan appears to have worked to the satisfaction of subscribers, and in a measure distributes equitably the charge which the public pays for telephone service, since in the smaller exchanges, where the average use is the least, the charges are correspondingly small, while in the great cities, where of necessity the rates are highest, the average use is also greatest.

[1] *Electrical Review*, London, January 15, 1892, xxx. 57.
[2] *American Bell Telephone Company's Report*, 1892, p. 9.

It is by no means certain, however, that in time to come the companies and their patrons may not find a system of tariff rates—that is, the charging the subscriber according to the number of times his telephone is used—more equitable than the present system of uniform charges.

It is true that by this method the amount paid by the individual subscriber would more nearly represent the service rendered, even taking into consideration the fact that the party called by telephone may derive equal advantage from the conversation with the one who calls.

In considering this question, the wishes of the great body of the subscribers should have a large, possibly a controlling, influence.

Some experiments are in hand to see if, in the great cities, the case of those who want only a limited use of the telephone at a less rate than is fixed for an unlimited use can be met.[1]

The measured rate was introduced into New York in June 1894.

In the year 1895 the system gained 3000 stations (equal to the total gain of the previous six years), and of the 13,345 stations in service at the end of 1895, eighteen months after the adoption of the first message rate schedule, 7016 (or over 50 per cent.) of the total were at message rates.[2]

In the report for the year 1894 the matter is again referred to, it being stated that several of the companies, especially in the larger cities, had been led

to consider the adoption of a plan for measured service, of which advantage can be taken by the smaller users. . . .

The plan has already been put in operation in New York, Brooklyn, Boston, and some other exchanges. The choice is offered to subscribers of paying either the fixed yearly rate, with unlimited local service, or a graduated scale of charges dependent upon the extent of use within the year, the cost per connection lessening with the increase in the number of calls for which contract is made.[3]

Reference continued to be made to the subject in the reports for subsequent years, the growth in the number of subscribers for the year 1897 being attributed in part 'to the efforts of the operating companies to bring the service within the reach of the numerous class of those who have need for a limited use of the system.[4]

The enormous influence of the measured rate on development—

[1] *American Bell Telephone Company's Report*, 1893, pp. 13–14.
[2] H. L. Webb, *Electrical World and Engineer*, April 5, 1902, xxxix. 588.
[3] *American Bell Telephone Company's Report*, 1894, p. 11.
[4] *Ibid.* 1897, p. 8.

i.e. the growth in new subscribers—is now a matter of record. How far it was foreseen it is not possible to say, since the estimates upon which the change was based have not been made public. But it may be remarked that the evidence in the case of the one city which had almost from the first adopted the message rate system, would not lead to expectations of great increase. The National Telephone Exchange Association published annual statistics, including the ratio of subscribers to population in various cities. The numbers given for the population change in some years in a manner which suggests that the areas are not always the same. There were probably accessions of territory or other changes which prevent accurate comparisons. But comparing Buffalo with other cities of approximately the same size and, so far as a stranger can judge, of the same character, for two-year periods we obtain the following figures :—

Number of Inhabitants per Telephone

—	Year 1883	Year 1885	Year 1887	Year 1889
Buffalo	153	101	88	84
Cleveland	147	98	—	106
Detroit	38	59	93	86
Louisville	—	106	94	77
Providence	68	48	50	49

With one telephone to every eighty-four inhabitants, Buffalo, on the measured rate system, was less developed than Louisville and Providence with one in seventy-seven and forty-nine respectively. Comparisons of this kind and especially at this distant date have to be taken with much qualification. There may have been some local conditions which explain the variations. The figures would, however, seem to show that the officials of the New York Company must have been guided by estimates and the preference for a sound principle, for there was no such outstanding superiority in the number of subscribers in Buffalo as to point definitely to the desirability of adopting the plan in force there.

It is probable that in the earlier years, when the use of the service needed to be encouraged, there was some general advantage in the flat rate plan. To those who were connected to a central office there need be no hesitation in use, no bar to the acquisition of the 'telephone habit.' But by 1893 this habit had been acquired by some subscribers to a degree which largely exceeded in cost the amount they paid for the service. By that time also it was fully recognised by telephone authorities that the amount of use bore a very definite relation to the cost of the service, and it was probably realised that a continuance of the flat rate system must act as a bar to the extension of the service amongst new subscribers.

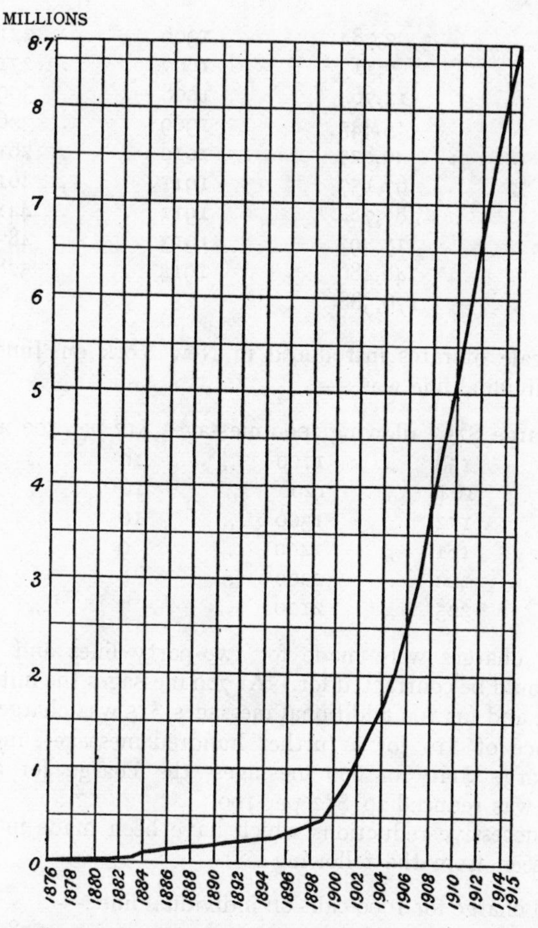

FIG. 170.—Station development in United States—Bell Companies only. From the introduction of the Telephone to January 1, 1915.

The diagram fig. 170 indicates the development in the United States for subscriber stations of the Bell Telephone Companies only from the commercial introduction of the telephone to January 1, 1915. The rapid increase which followed the introduction of the measured rate system of charging will be readily noticed.

The station development within the territory now comprising

Greater New York, for five-yearly periods from 1880 to 1900 and annually thereafter, is shown by the numbers of subscribers at the respective periods as follows :—

1880	2,784	1906	221,696
1885	7,113	1907	271,697
1890	11,764	1908	309,099
1895	15,445	1909	326,822
1900	49,827	1910	261,745
1901	67,157	1911	401,859
1902	87,784	1912	441,128
1903	117,492	1913	483,853
1904	148,486	1914	528,819
1905	176,588		

The scale of rates introduced in New York on June 1, 1894, for an individual line was :—

Initial charge	$150	allowing	1000	messages	$12	per 100 additional
	158	,,	1100	,,	10	,,
	164	,,	1200	,,	10	,,
	172	,,	1300	,,	10	,,
	181	,,	1400	,,	9	,,
	220	,,	2000	,,	8	,,
	225	,,	2100	,,	7	,,

Lower charges were made for two-party lines and a smaller number could be contracted for. At 700 messages the initial charge was $100, and for 100 additional messages $15 was charged. With an advance of $10 for a further hundred messages, making the initial charge $110 for 800 messages, the charge for additional messages was reduced to $12 per 100.

The successive reductions which have been made in the rates may be seen from the following :—

Initial charge for 1000 calls on individual line :—

1894	$150	1905	$78
1895	120	1906	66
1899	99	1914	66 (10% disc)
1901	90	1915	51

In New York the measured rate now commences with an initial charge for 800 messages of $40.

Until 1895 additional messages were charged per 100. On March 27, 1895, the schedule was altered to the single message. In the following table the earlier method has been converted to the single message rate :—

Year	Exceeding 1000	Exceeding 1500	Exceeding 2000	Exceeding 3000
1894	12 cents	9 cents	8 cents	—
1895	8 ,,	7 ,,	7 ,,	—
1897	7 ,,	6 ,,	5 ,,	—
1899	8 ,,	7 ,,	6 ,,	—
1901	7 ,,	7 ,,	6 ,,	—
May 1, 1905	6 ,,	6 ,,	6 ,,	—
June 1, 1905	5 ,,	5 ,,	5 ,,	—
1906	5 ,,	5 ,,	5 ,,	4 cents
1914	10 per cent	discount.		
July 1,1915	5 cents	4 cents	4 ,,	3 ,,

The above rates apply to individual lines, but very much of the New York service is undertaken through private branch exchanges. These are referred to later in this chapter, though the charges made may more conveniently be given here.

On June 1, 1906, the following schedule was adopted :—

Minimum equipment, consisting of a switchboard, an operator's telephone, two trunk lines to the central office, two stations located on the premises, and 3600 local messages per annum	$204.00
Additional trunk lines to central office, each	24.00
Additional stations on the premises, each	6.00
Additional messages, each	4 c.
If contracted for in advance in lots of 400, each	3 c.

On February 1, 1914, a reduction of 10% was made on these charges.

On July 1, 1915, the following schedule was adopted :—

Minimum equipment, consisting of monitor switchboard (limited to seven stations and three trunks), two stations, one trunk, and 2400 local messages per annum	$126.00
Same as above, with cord switchboard and operating set (limited to thirty station drops) in place of monitor board	$132.00
Stations in connection with above switchboard :	
First 10, each	$6.00
Second 10, each	$4.80
Above 20, each	3.60
Additional working drops in switchboards having more than thirty station drops, each station drop	1.20
Additional trunks, each	24.00
Additional local messages when contracted for in advance in blocks of 300 each	.02½
Additional local messages not constructed for in advance, each	.03

For rate purposes Greater New York is divided into zones with varying rates and conditions, which are too numerous to be set out in detail. Those included above may be regarded as characteristic rates.

In the American Bell Telephone Company's Report for 1893, quoted earlier in this chapter (p. 475), reference was made to the controlling influence in such matters of the wishes of the great body of subscribers. It is presumably due to such influences that the measured service rate is far from universal in the Bell companies of the United States. Thus, in all the cities which had a population of over 50,000, a mixed service of flat rate and measured rate was general. As at June 1, 1914, the flat rate was the only service available for business in 34 cities, and for residences in 68. Both flat rate and measured service were available for business in 76 cities, for residences in 53 cities. The measured service rate was the only service available in 11 cities for business, and not in any case for residences except in certain zones in the city of New York. Thus, out of 121 cities concerned, only 34 have 'pure' schedules, which are all for flat rate service, and 87 have 'mixed' schedules in which both flat rate and measured service appear.

In Europe the earliest interest in the message rate system of charging was shown in Switzerland. Dr. Rothen, joint director of the Swiss telegraphs, in the course of a series of articles on the telephone service under his charge, in the *Journal Télégraphique*, discussed the rate question, August 25, 1883.[1] He pointed out then, what was so often forgotten in comparisons of later years elsewhere, that figures alone do not suffice to make comparisons, because the value of money varies with localities. It was possible, for instance, that a rate of £15 or £16 sterling (375 or 400 francs) in London was not really a higher price than 150 francs in Switzerland. From another point of view, said Dr. Rothen, the present system of payment (by a flat rate) presents a substantial defect. Some large establishments, such as banking houses, may make a continuous use of the service, and for them the telephone station has a value perhaps ten times that which it offers to a private resident. The latter may make use of it once or twice a day for social purposes, whilst other subscribers transact by telephone a great part of their business. Was it equitable, he asked, that the same remuneration should be demanded for services so different? Nor did Dr. Rothen overlook the effect on development. Could one expect an extension of the telephone amongst those classes who could only make a small use of the service, if the expenses of the system were to be divided equally amongst the total number of subscribers, thus favouring

[1] Vol. vii. p. 192.

the minority who make a large use to the detriment of the majority who employ it much less frequently?

But some years elapsed before a change was made. In June 1889 a federal law was passed 'which introduced material modifications' in the methods of charging. Eight hundred conversations were allowed for 120 francs (second year 100 francs, third and subsequent years 80 francs) 'but any number in excess of that 800 was subject to a toll' of five centimes per conversation. In 1894 that law was modified by another 'which did away with the maximum number of free conversations and imposed a toll on every conversation . . . and it imposed the modified rates as follows: For the first year a subscription of 100 francs, second year 70 francs, third and subsequent years 40 francs, and in addition a toll of 10 centimes for every conversation. In 1889 the number of local calls per subscriber per annum with unlimited user was 880; in 1890, when a limit to free calls was introduced, it was 546; in 1897 the number was 544. The subscribers' stations increased from 9203 in 1889 to 32,252 in 1897.[1]

An early example of the partial application of measured service is found in the case of the Grand Duchy of Luxemburg, where cafés and similar places of public resort were allowed a limited number of calls for a regular annual subscription, excess calls being paid for according to their number.[2]

The first attempt to introduce the measured rate in Great Britain was in 1892. A study of the problem had satisfied Mr. Gaine, who had in March of that year joined the National Company as general manager, that the measured rate system was the fairest and best method of charging for service. A proposal to introduce the system in Sheffield led to an agitation, a public meeting (May 9, 1892), and a withdrawal. An incident subsequent to the meeting was related by Mr. Gaine to the House of Commons Committee of 1898. Coming down the stairs after the meeting, when he was enlarging on the beauties of the system, 'one man said he did not care, *etcetera*, it would cost him £95 a year instead of the £8 or £10 he was then paying, and he was not going to have it.'[3]

The *etcetera* probably represents a colloquial expression considered to be unsuited to parliamentary records.

Subscribers who were large users anticipated that the service would be more costly. Since the large users are generally the most influential individuals or firms in a locality, it follows that the tendency of their opposition is to perpetuate a system which they

[1] Evidence of Sir John Gavey, *Report of Select Committee, House of Commons*, 1898 (No. 383), pp. 419–22.

[2] *Histoire de la Téléphonie et Exploitation des Téléphones en France et à l'Étranger.* Brault, 1890, p. 383.

[3] *Report of Select Committee, House of Commons*, 1898, Q. 7616.

assume to be beneficial to themselves, regardless of the fact that there is a greater potential benefit in the enlargement of the circle of their communications. To the influential character of the large user is perhaps to be ascribed the attitude of Chambers of Commerce referred to in the following extract from the London *Electrician*, in 1907 :—

If a subscriber who uses his telephone to a large extent is given a rate which is unremunerative to the company, the small user is simply penalised to make up the difference. Naturally the large user objects to any movement that may tend to increase his payments. This is only what must be expected, but it is certainly strange that Chambers of Commerce, which are supposed to have the welfare of the small as well as the large trader at heart, should engage in agitation against the introduction of sound business principles.[1]

The attempt to introduce the message rate in the Commonwealth of Australia was the subject of some controversy resulting in the submission of a report (C. 8985) by Mr. John Hesketh, the chief electrical engineer, on June 30, 1910 which is a reasoned argument for, and contains a very valuable compendium of experts' opinions on, the measured rate system.

The rates at present ruling in Great Britain are as follows :—

The London Telephone area is for rate purposes divided into two sections :—

(*a*) The area within the County of London, which includes the City of London, the City of Westminster, the numerous surrounding boroughs and the inner suburbs.

(*b*) The area outside the county, comprising the outer suburbs and adjacent towns.

On the measured rate system the charges are as follows :—

—	Annual subscription	Additional per message	Minimum for messages per annum
	£		£ s. d.
Individual line (*a*) . .	5	*a* to *a* 1*d*., *a* to *b* 2*d*.	1 10 0
Individual line (*b*) . .	4	1*d*. on same exchange 2*d*. on any other exchange	1 10 0
Two-party line (*a*) . .	3	*a* to *a* 1*d*., *a* to *b* 2*d*. . .	3 0 0
Two-party line (*b*) . .	3	1*d*. on same exchange 2*d*. on any other exchange	3 0 0
Ten-party line (*b*) only .	2	1*d*. on same exchange 2*d*. on any other exchange	3 0 0

[1] *Electrician*, October 4, 1907.

In the provinces, with a few exceptions, the measured service rates are :—

—	Subscription rate per annum	Number of calls allowed	Calls not purchased in advance each
Individual line :—	£ s. d.		
Business	6 0 0	500	2*d*.
	8 5 0	1100	2*d*.
	9 0 0	1300	1*d*.
	15 12 0	4600	1*d*.
Residence	5 0 0	300	2*d*.
	6 0 0	600	2*d*.
	7 0 0	900	2*d*.
Two-party line :—			
Business	5 12 0	500	2*d*. up to 1400 and 1*d*. thereafter
Residence	5 0 0	400	
Four-party line :—			
Residence	4 0 0	300	2*d*.

Private branch exchange rates, with a few provincial exceptions are as follows :—

—	London	Provinces
	£ s. d.	£ s. d.
For each line to the public exchange per annum	4 0 0	3 10 0
For internal lines connected to the private branch exchange :—		
10 lines and under, each per annum	2 0 0	1 12 0
101 lines and over, each per annum	1 5 0	1 0 0
(Proportionate rates for intermediate numbers.)		
Exchange Calls :—		
3000 per annum (minimum)	11 0 0	8 0 0
Additional calls in blocks of 500 at per 100	0 5 4	
,, 300 ,,		0 4 0
Junction fees	1*d*. for each call from *a* to *b* or from *b* to any other exchange in the London area.	1*d*., 2*d*., or 3*d*., according to distance and exchange areas.
Extra charges are made when the length of exchange lines exceed	2 miles	1 mile
And when the length of internal lines exceed an average of	110 yards	110 yards

The minimum number of exchange lines is two, and the maximum number of calls to be made over them is 9200. When that number is exceeded the lines are to be increased as follows :—

3 lines for a maximum of		14,000 calls.	
4 ,,	,,	20,000	,,
5 ,,	,,	27,000	,,
6 ,,	,,	35,000	,,
7 ,,	,,	44,000	,,
8 ,,	,,	54,000	,,

and thereafter one additional line for each 10,000 calls.

In London unlimited service (flat rate), £17 per annum for first line, £14 for each additional line. Hotels, commercial exchanges, restaurants, public houses, etc., are not accepted for flat rate or party line service.

In the provinces unlimited service is only allowed for residences, and £8 per annum is charged.

Extra charges are payable when the line extends beyond a radius of two miles from the Exchange in London and one mile in the provinces.

Party lines are not available at the central and certain other exchanges in London.

In Appendix B particulars are given of the increases in the British telephone charges proposed to be made on or before 1st November 1915, in consequence of the financial position due to the war.

Meters

The recording of the connections was effected in the first instance by the operator, who made out a ticket showing the calling and the called numbers. These were collected every hour and stamped with the date and the time. To avoid ledger accounts a box was provided for each subscriber. Into these boxes the tickets were placed. At the end of the month the tickets were counted and the subscriber apprised the number of calls made during the month, the number which he had used prior to that month, and the balance to his credit. Twenty days were allowed to make claims for rebates or mistakes. The tickets were preserved for that period and then, failing any claims, were filed away or destroyed.[1]

Such a method was expensive, and proposals for automatic records were numerous. The first is found as an appendix to the report of the proceedings of the National Telephone Exchange Association of 1880.

Apart from the suggested mechanism, the proposal indicates an

[1] *National Telephone Exchange Association Report*, 1880, pp. 177–8.

appreciation of the importance of the rate question, and handles the subject with so much intelligence and foresight that it is desirable to transcribe it in full :—

A Telephone Service Meter—By H. L. Bailey.

The correct determination of value received from a telephone company by a subscriber is by no means a simple matter, when it has for its object the establishment of a system of rates which shall be at once equitable and universal in their application. Among the complexities which enter into such a generalisation are the following :—

1. The varied extent to which the different subscribers use their instruments and lines.

2. The differences of distance through which connections are made, whereby trunk lines are monopolised to a greater or less degree.

3. The lengths of lines from different subscribers to the district exchange, with relative costs of establishing district connection.

4. The varied character of the service rendered, owing to aborted connections, arising from imperfections of instruments or their adjustment, occupancy at the time being by other subscribers of lines called for ; and in the larger cities where the difficulties of construction are comparatively great, and where there is such an intricate maze of wires, the frequency of crossed lines is an item for serious consideration. Thus, as the length of lines between subscribers and the district exchanges, with relative cost of maintenance, is subject to great variation, and the number of connections effected by different subscribers will vary from twenty-five to twelve hundred per month, any definite universal rate per year, month or week, is manifestly inequitable, and will certainly restrict the largest profitable extension of business.

The continued and varied observations of telephone systems in New York, with a knowledge of the well-known facts which I have just presented, suggested to me the idea of automatically registering the actual service rendered by companies to each subscriber. The expediency and ease with which it may be applied to different instruments and systems I trust will be apparent.

In the first efforts I canvassed the question of a meter attachment to the call mechanism of different telephone sets, but experience in matters pertaining to telephone service almost immediately condemned its practicability. The registration of calls would not accomplish the object sought, as many calls that are made prove abortive and do not result in effecting the desired connections, a registration of which must be accomplished as a first step toward the solution of the problem.

Among subscribers finding a fixed rate charged against them for each connection, there would naturally exist a tendency to transact the maximum amount of business at each connection, which would, by monopoly of trunk wires and service of central office operators for a relatively greater length of time, increase the cost of such connections to the company. Therefore time enters as an important element and its registration as a second step.

The fact that a subscriber, seeing for himself a registry of the service he receives, would be infinitely more satisfied than by trusting to his recollection, presents as a third step the advisability of placing the meter at his telephone rather than at the central office. But then it is evident that the central office should be cognizant of the correctness of the registry, and have it so under control that any dishonest attempt on the part of the subscriber could be at once met by refusing connection or cutting him off.

Finally, the problem would be solved by an instrument simple in construction, easy of access for inspection, and of little or no liability to derangement, the function of which, in combination with a system, would fulfil the requirements in the four preceding paragraphs.

To a description of such an instrument and system I have the pleasure of calling your attention, and had hoped to exhibit the same in actual operation at this time, but delay in its completion has prevented me from doing so.

The telephone service meter (for as such I designate the instrument in question), although admitting of varied adaptation, I have for convenience attached to one of the push-magneto outfits. such as are at present in use by the Metropolitan Telephone and Telegraph Company of New York, and shall here describe it as so applied. It consists of a suitable case containing the mechanism, clamped and hinged over the face of the push-magneto box concealing the push button of the same. Its anterior face exhibits a crank with two stops and a slot through which the figures indicating the number of connections are exposed.

Extending from the right lateral surface of the meter case is the dial of a calendar clock, upon which is registered the aggregate time of all connections for any given period. The internal mechanism consists of the necessary shafts, wheels, cams, etc., for the accomplishment of the following operations for each revolution of the crank :—

1. The registry of a single count by the meter mechanism.

2. The actuation of the clock movement at the first half revolution, and the stoppage of the same at the completion.

3. The automatic winding of the clock.

4. The rendering of certain sounds, which shall notify the central office of proper registration.

5. The control of the circuits so as to accomplish the notification to central office of the above, and also through modifications to enlarge the adaptability of the meter to different systems.

6. The automatic indication to central office when the subscriber shall have finished.

As to capacity, the meter will register numerically 10,000 connections and 43,000 minutes, which would seem to meet all requirements, at least for monthly inspections.

The system consists simply in the central office operator

requesting the subscriber to register at the time of connecting him with the party called for, and should the desired connection involve the use of long lines—such as to an adjacent town—the subscriber would be told to register two, three, or four times, and in this case also the operator would be cognisant of its correct performance. By this method there would be no registration without actual connection, and *vice versa*.

A slight modification of the meter would adapt it to private lines. In such case either party can call and be called, but must register in order to talk together ; in this case both register.

The very considerable expense for stationery, and the large aggregate of time consumed by employees in the present toll systems, deserve serious considerations, and it is evident that a means of effecting a similar result automatically and in a thoroughly reliable manner, is of great value.

I have been asked how it could be managed in cases where subscribers use their telephone infrequently ; their aggregate tolls per month, based on a universal rate, would be too small to pay for putting in and maintaining their district connection. The answer is—by charging a certain amount per month on all instruments without regard to the amount of their use, in addition to the standard toll rate. In conclusion I would state that the meter itself would furnish the necessary statistics for the calculation of a just and uniform standard rate and system of charges.

New York, September 8, 1880.[1]

In spite of this early suggestion for the use of a meter system, little if any application was made of it in practice. Commenting on the Buffalo system of charging, Wietlisbach wrote in 1888 that 'to apply this system generally it is indispensable to have a very simple and very certain apparatus to register automatically the number of communications and permit the subscriber to know the number at any moment.' He states that many instruments of the kind had been constructed. Numerous patents were taken out indicating a general interest in such automatic-recording apparatus, but I have been unable to discover any practical use of them in a commercial telephone system. Wietlisbach remarks that registers of this kind are complicated and, applied generally, would introduce into the service a frequent source of trouble. It is probably to this risk that is to be ascribed the omission to use such appliances in the earlier period when the apparatus necessary for calling and speaking gave quite sufficient trouble without the company or administration seeking to increase the risks by adding machinery which could be done without. The register was not necessary, and it was not advisable to encumber the station with apparatus liable to derangement and tending to interrupt the service.

[1] *National Telephone Exchange Association Report*, 1880, p. 190.

The coin-collecting box, on the other hand, served the useful purpose of 'paying as you go,' and Wietlisbach mentions that the Pan-Electric Telephone Company of St. Louis (Missouri) employed such a contrivance. The company had a very brief existence, and it is doubtful if the coin boxes were actually in use. At the top of the box enclosing the magneto-electric generator there was an aperture of such dimensions as to afford passage to a nickel or five-cent piece. The weight of the coin served to complete the circuit normally broken between the subscriber's station and the central office. The subscriber could then call the central and speak. When the conversation was finished, the operator at the central office transmitted a current which caused the coin to fall into a receptacle under the call box. From time to time an employee of the company was to visit the subscriber and collect the money from the box.

James W. See of Hamilton, Ohio, contemplated the use of tickets instead of coins, and describes a box for the purpose.[1]

Numerous devices for the use of coins have been made and largely operated. Chicago was at one time a great centre for 'nickel in the slot' subscribers' stations. Elsewhere they have been more generally employed in public pay stations. In Great Britain they have been largely used for this purpose, and patents were taken out by the officials of various subsidiary companies—as, for example, Smith and Sinclair of the National Company at Glasgow, and Poole and McIver of the Lancashire and Cheshire Company at Manchester.

The lack of a cheap and reliable means of counting the messages was the reason given by von Stephan in 1891 for not introducing into Germany the message rate which he knew to be more equitable than the flat rate.[2]

Mr. Bailey required the operator to request the subscriber to register 'at the time of connecting him with the party called for.' As a general rule the advocate of meters aims at some automatic method of registration, regardless of the fact that what is wanted is to register connections and not applications for connection. This essential feature was overlooked in both the first and second patents for a telephone call register granted to Charles J. Bell[3] and Sumner Tainter[4] respectively in 1880. British specification No. 2289 of 1895 (a period when the message rate system was being revived) is an example of a 'toll counter' intended to be

[1] U.S. specification, No. 237,327, February 1, 1881 (application filed September 28, 1880).

[2] Holcombe, *Public Ownership of Telephones on the Continent of Europe*, p. 139.

[3] U.S. specification, No. 229,302, June 29, 1880 (application filed March 16, 1880).

[4] U.S. specification, No. 229,495, June 29, 1880 (application filed March 24, 1880).

placed at the subscriber's station. It is applied to the branching system. The line was blocked at the central office on making the connection. The block was removed by the calling subscriber on operating his counter, thus ensuring that the call was registered before conversation could be obtained. It was never put into use so far as I am aware.

The common battery system, which was regarded as so adaptable to switchboard requirements that there was nothing that could be demanded from it which could not be obtained,[1] afforded the first practicable opportunity of registering calls by meter. British specification No. 1168 of 1890 is an early example. The development of this which was applied in practice is described in the various telephone text-books.[2]

The registration is effected by the operator, but the meter cannot register unless two lines are connected. The fact that the meter is at the central office and not at the subscriber's instrument is the cause of some dissatisfaction, and it is to be feared that Sir William Preece was over-sanguine when he said that 'a month or two's experience will instil faith in the accuracy of the meter.'[3]

It is nevertheless true that the liability to error is in favour of the subscriber, since the omission to carry out an act is a more common human failing than to perform an act more frequently than necessary. It would certainly be possible to provide an indicator at the subscriber's station which should move step by step with that at the central office, but the reasons against its general adoption given by Wietlisbach in 1888 are still operative. It is not desirable to install unnecessary apparatus liable to derangement and, consequently, to impair or interrupt the service.

Private Branch Exchanges

The development of private branch exchanges on the large scale now prevalent is an indirect result of the measured rate system. Private exchanges were referred to at the 1890 meeting of the National Telephone Exchange Association as being in existence in Chicago, but the reference was incidental to a discussion on single *v.* double cord switchboards, and the introductory remarks indicate that such private exchanges were not general. Mr. Wilson said :—

In case some gentlemen here may not understand what I mean by private exchanges, I will say that we have quite a number of

[1] Chapter xxvi. p. 378.

[2] *The Practical Telephone Handbook* (1912), p. 295. Aitken's *Manual of the Telephone*, 1911, p. 230. Kempster Miller in *American Telephone Practice*, fourth edition (1905), describes (p. 482) the apparatus and method of operation more fully.

[3] *Journal of the Institution of Electrical Engineers*, xli. 122, March 12, 1908.

subscribers—like, for instance, the Illinois Steel Company, which has its general office in Chicago with about twenty telephones in use. They have three rolling mills in and around the city, and they have lines connecting them. We run all those lines into the private exchange located in their general office, and we give them three trunk lines from that switchboard into our office. We call that a private exchange.[1]

Other installations of a similar character were mentioned, and it was remarked 'we are putting in quite a good many of these outfits.'

Small switchboards on subscribers' premises were in early years very general in Great Britain. An annual charge was made for each instrument, and a switchboard provided by means of which any extension instrument could be connected with another or with the exchange line. This afforded local intercommunication, and in addition increased the number of points from which the exchange line could be obtained, which was equivalent, by increasing the facilities, to increasing the number of calls. A very common size of switchboard was that for five lines. The use of the exchange line was afforded to those five lines at exactly the same amount as for one line, though it would be reasonable to expect that the increased facilities must result in a greater number of calls. The extra fee paid for the extension instrument may have sufficed for the local intercommunication, but was inadequate for the extra exchange service. Companies which analysed their expense items were averse to the inordinate increase of extension instruments.

Under the message rate system [on the contrary] it is obviously for the interests of all concerned that the use of the telephone should in every manner be encouraged. For this reason it became feasible and desirable to install as many auxiliary instruments as possible.[2]

Comparison of this statement of 1906 with the letter and comments of Mr. Hall in 1880 on p. 471 will show how sound in principle were the proposals underlying the measured rate system, and how accurately the effect on the public was foreseen when the change was effected in New York in 1894. On this point I cannot refrain from one further quotation from the 1880 discussion. Mr. Hall, in replying to some criticisms, said:—

The statistics so far indicate the number of calls would be largely diminished, but if we can get everybody in the city to using the

[1] *National Telephone Exchange Association Report*, 1890, p. 41.
[2] Mr. Carty, *Transactions of the American Institute of Electrical Engineers*, February 23, 1906, xxv. 94.

telephone freely, generally, and commonly, they would understand it was something for their benefit; that every time they used it the company made something out of it. Our interest is first the interest of the public, to make everything open and free, if it is paid for, and to use as much as possible, to work with them, instead of against them; and if a man comes in and says he does not use a telephone very much—as you probably all have heard from different parties—he can get it and pay just what he uses it. If he doesn't use it enough to pay the minimum price, then he had better not take it at all.[1]

With every call paid for, it became the interest of the company to encourage the equipment of stations with as many instruments as the subscriber desired; every installation was a potential source of exchange revenue.

The large blocks of offices, hotels, and other buildings common in the United States, afforded a very large field for the development of the private branch exchange which was introduced in New York in 1896 and spread rapidly, so that in every business office of importance and every hotel of the higher class there is no room without an instrument that will enable the occupant to talk over the telephone system of the country.

The private branch exchange realised the economic ideal of early telephonists. All the efforts at party lines and similar contrivances were prompted by the desire to afford telephone facilities at a minimum cost; to save something in line construction for the user of an instrument. The private branch exchange attains this ideal by having perhaps hundreds of stations with only enough lines to the central office to accommodate the normal traffic. The system is largely adopted in Great Britain and is rapidly extending. The advantages of the internal or domestic service of telephones as an educative influence leading to the adoption of exchange service were referred to by Bell in his letter to the Capitalists of the Electric Telephone Co. (p. 89). In the same letter is also to be found the suggestion of a toll (or message rate) system as an alternative to a fixed annual rental.

The measured rate, as now applied, is not strictly scientific, in that it does not provide for all the factors. Mr. Webb says that under the flat rate the wrong unit was taken:—

The subscriber's station is taken as the unit of cost instead of the message, which is the true unit, and, since the cost of carrying the messages varies with the distance over which it is carried, the really scientific unit is the message-mile.[2]

[1] *National Telephone Exchange Association Report*, 1880, p. 180.
[2] *The Telephone Service*, p. 87.

As pointed out by Mr. Bailey in 1880, time also enters as an important element, so that the really scientific unit ought rather to be the message-mile-minute. Whatever modifications the future may show to be necessary, it will probably be found inadvisable to apply unduly an ultra-scientific method of charging. A system which takes into account every factor is usually either impossible of computation in practice or too complicated to be understandable by the user. The maximum demand system of charging for electric light is very scientific but very unpopular. The method of charging a small annual sum for the facility of use and a further sum according to the quantity of current used is known in electric light circles as 'The Telephone System'[1] being comparable to the measured rate system.

A certain amount of give and take between supplier and user is necessary in most services, but simplicity in methods of charges is desirable for popular appreciation and popular satisfaction. The measured rate is scientific enough to be roughly fair and simple enough to be generally understood and appreciated.

That the system might have been generally adopted earlier is evident from the early quotations that I have given. Yet it would not be wise to assume that it was unduly delayed. The subsequent growth must not be ascribed to one cause alone. The improved service afforded by the common battery system, the meter facilities permitted by its use, and the telephone habit engendered by unrestricted talks, together with the improved engineering generally, must all be considered as having contributed to the expansion which occurred from 1894 onwards. The fact that in all cases the introduction of the measured rate reduced materially the number of calls—cut out the 'frivolous and unnecessary talks,' as one expert has described it[2]—is evidence that there was some educational value in the prior flat rate. But it was an education which the companies were paying for and could not longer continue. It may have been valuable as all education should be. But it had to end as all education that is paid for must. In the application of the measured rate is to be found the great benefit of the democratic telephone, available to all for the receipt of calls at a minimum cost and for the origination of calls at a cost which is proportional to their number.

[1] Seabrook, *Journal of the Institution of Electrical Engineers*, xlviii. 391.

[2] Sir John Gavey : *Report of Select Committee, House of Commons*, 1898, Q. 8013.

CHAPTER XXXI

THE ECONOMICS OF THE TELEPHONE

So far as I am aware, the first use of the words at the head of this chapter was as the title of a letter to the press of May 10, 1879, over the signature of 'Arnold White,' who was at that time taking an active part in the promotion of the interests of the Edison Telephone Company of London; who continued his connection with the telephone until shortly after the amalgamation of the two English companies in 1880, and subsequently became active in philanthropic, political, and literary work. His letter incidentally dwelt upon the superiority of the Edison over other forms of telephone—a belief which was so impressed upon Mr. Bernard Shaw by the American artificers at a later date.[1] Continuing, Mr. White remarked that—

> hitherto the various forms of magneto electric communication have been, to the English public, little more than philosophical toys, admirably ingenious in their construction, but devoid of the interest attaching to instrumentalities closely affecting the conditions of daily life. In Mr. Edison's mechanical reproduction of the voice, we appear to have acquired an addition to the customary means of communication which it is impossible to doubt will not only largely supersede the more costly and more cumbrous machinery of the telegraph, but will obviate the necessity for much of that rapid journeying which forms so large a part of the work and the worry of common life.[2]

Explanations of the exchange system and a reference to the large number of subscribers connected in Chicago alone on the 'earlier description of telephone' adopted 'prior to Mr. Edison's solution of the question of practical telephonic communication,' together with a forecast of the results which must follow the general use of the telephone, are then given.

It seemed to Mr. White at that time that these beneficial results

[1] Chapter xvii. p. 202. [2] *The Times* (letter), May 10, 1879.

were dependent upon the use of the 'loud speaking telephone' of Mr. Edison, i.e. the electro-motograph receiver. We know differently now. But that is an unimportant detail, for we know also that telephonic intercommunication has produced economic results exceeding any forecasts that the most enthusiastic seer dared indulge in. Following a description of the exchange system and particulars of its adoption in various American cities, Mr. White considered other applications: the Fire Brigade, Railways, Lloyd's agents, House of Commons reporting, and so on. Exchanges at Chislehurst, Harrow, and Norwood were hinted at, and also the resulting benefit to a busy man who could remain away from town without neglecting his affairs. We were all enthusiasts then, but we did not contemplate the busy man in a London office casually asking his secretary to call up a correspondent in Brussels or the merchant of New York holding converse with a customer at San Francisco.

The letter prompted a leading article which commenced :—

According to a letter from Mr. White . . . which we publish this morning, the practical completion of the loud speaking telephone may possibly lead the way to those developments of telephonic intercourse which Professor Graham Bell, the first inventor, sketched out in his original scheme for the application of his discovery in this country. According to that scheme, companies were to be formed which would establish, first in London and afterwards in provincial towns, a sufficient number of telephone stations, all communicating with a central office. . . . The two persons chiefly interested might then hold a private conversation, at the close of which they would intimate to the respective district clerks that the connection might be interrupted. As soon as the system had taken sufficient root in the metropolis, the central station there was in its turn to be connected with that of every provincial town, so that ultimately every house in England, above a certain value, would be capable of being brought into direct speaking communication with every other house of the same kind.[1]

Professor Bell's original circular or prospectus to the Capitalists of the Electric Telephone Company is given in Chapter ix[2] from a copy retained by him. It was apparently never published, nor have I been able so far to find any public pronouncement of a similar kind in this country. It is clear, however, that Bell's proposals were familiar to the leader writer of *The Times* either from the circular mentioned or from some other source as yet untraced. It would appear also that the directors of the Bell Company in London did not anticipate communication between different towns

[1] *The Times* (article), May 10, 1879. [2] Page 89.

directly from subscribers' own offices or, if they did, they considered it wise to be moderate in the official expression of their expectations. In the annex to a petition to Parliament against the Telegraph Bill of 1878, they set out the aim and scope of the company. The local exchange and the call office were outlined, and it was suggested that in the same way

> different towns may be joined together such as Manchester, Liverpool, Glasgow and London. A Manchester merchant, for example, now requires to come to London to see someone on business. He consumes ten hours at least in travelling, has to spend one night in town, and is at heavy expense, apart from the loss of valuable time. On the other hand he can arrange that he and his correspondent shall be at the telephone office at a certain time, and thus settle their business by direct *viva voce* communication.[1]

The article on Mr. White's letter continues :—

> The comparatively limited 'exchanges' of which he speaks appear to be parts of a somewhat less ambitious scheme than that of Professor Bell, and some of his projected applications will probably present difficulties which it will be the province of ingenuity to overcome. . . . With the new instrument hanging over his desk, the merchant or the banker will be liable to perpetual interruptions from telephonists, who will begin with some such phrase as 'Oh, by the bye.' But there are limits to human endurance, and those who are threatened by such an evil will probably discover some means of keeping it within reasonable bounds. It would scarcely fail to happen, by way of compensation, that the general employment of the telephone would materially diminish the wear and tear of life in one important respect, by rendering it possible to avoid travelling in many cases in which it is now inevitable. It is manifest that a large proportion of the railway travelling of this country, and a still larger proportion of the cab journeys of the metropolis, are undertaken solely in order to enable two or more persons to hold verbal conference; and that if such conference could be held in spite of the obstacle now presented by intervening space, the journeys would be rendered superfluous.[2]

It has been said[3] on the authority of one who participated in these early events that the directors of the Edison Telephone Company of London could not see in the telephone much more than an auxiliary for getting out promptly in the next morning's papers the midnight debates in Parliament. The prominent mention of subsidiary services might seem to support this statement, but I cannot but think that it is founded upon imperfect

[1] *Telegraphic Journal* August 1, 1878, vi. 305.
[2] *The Times* (article), May 10, 1879.
[3] *Edison: His Life and Inventions*, p. 192.

recollection in the case at least of the most prominent director—Lord Avebury. A humorous friend once remarked that he was regarded in financial circles as a scientist and in scientific circles as a financier. In truth he was an expert in both, and to the combination the progress of the telephone exchange system in Great Britain is much indebted. His connection with the Company indicated a much wider perception of the utility of the telephone, and it ensured the provision of the capital, which was a prime necessity of the undertaking. It is the merchant adventurer spirit which in the quest for gain advances the progress of the world. To Mr. James Brand and his associates on the Bell Company side and to Lord Avebury on the Edison Company side the telephone in Great Britain owes much. Similarly, but to a larger extent because of the wider growth, the United States is indebted to Colonel W. H. Forbes and the group of Boston capitalists whose names are unknown to me, but who, having exercised judgment, took risks in the establishment of a new enterprise which has become great beyond expectations and performs important public services. What was then speculative has now acquired a stability which justifies the further considerable investment that the continued development of such enterprises will require.

From the use of the telephone 'some modifications in the ordinary conditions of life' were looked for. It is in respect to these modifications and not to questions of internal management that the phrase 'economics of the telephone' is here used. What is the position of the telephone in the conduct of the world's affairs, and what its influence on individual and national efficiency?

It is possible for each individual to form some estimate for himself, by recording particulars of the persons with whom he has been in conversation each day, the localities from which his correspondents spoke to him, the time occupied in the conversations, and their cost. Against that record let him place the time which would have been occupied in travel, and the cost of such travel to obtain such conversations if no telephone existed. This provides a comparison of the communications themselves. To this let there be added an estimate of the results attained by the promptitude of the telephonic communications over the delayed results of personal travel, the effect of a prompt dispatch of goods, a timely correction of error, or a clear understanding of some doubtful point. Let it be admitted that all telephonic conversations need not necessarily be replaced by personal converse, that the purpose of some could well be effected by telegraph or otherwise, then let the analysis be so amended. Whether it be business or social affairs, a large percentage remains in which it is the personal converse which is of value, and without the telephone this could not be obtained except at a cost in time and money which would render it impossible for an individual to do what he can now accomplish with ease.

In the Bell system of the United States there were in the year 1912 8,472,000,000 connections. These do not include those of the allied or 'connecting' companies nor of the independent organisations.[1] And on the National Company's wires in Great Britain there were made in the year 1911 nearly 1,600,000,000 connections.[2] What imposing figures might be compiled by adding the rest of the world! But they are not to be obtained so easily nor perhaps with such reliability, and ten thousand million talks in two countries for one year give room enough for thought on the revolution effected in means of communication. Every one of those ten thousand million talks saved somebody's time and somebody's money. How much time and how much money can never be computed with accuracy, but some estimate may be formed readily by anyone who will make the record of his own experience which I have suggested. To give free rein to thought and indulge in speculative estimates would be easy, but would also be contrary to the traditions which seem to have governed the telephone once it had started on its practical career. Those responsible for the development of the machinery or the conduct of the service have been content for the most part to pursue their work steadfastly but modestly, and no literary genius has been moved to sing their praises nor any statistical enthusiast given thought to the economic effects resulting from the electrical extension of human converse. Other means of communication have had their rhapsodists who have indulged in records and estimates material and moral, and who have dwelt on the revolutions such means have effected. Railways produced a full crop of such moralisings from popular writers. Even such a subject as 'Metropolitan Railway Stations' was considered justification for the following :—

The revolution which the metropolitan railways have wrought in our locomotive capabilities sinks into comparative insignificance when we contemplate the revolution they must yet work in mental and moral phenomena—blending together more and more intimately all countries and peoples, all religions, philosophies, feelings, tastes, customs, and manners, through the agency of the great social harmoniser—personal converse.[3]

'Revolution' was the favourite word to describe any change or improvement in the means of communication, but it may be

[1] *American Telephone and Telegraph Company Report, for* 1912, p. 4.
[2] Mr. Franklin's speech at the shareholders' meeting, January 9, 1912, p. 8.
[3] Knight's *Cyclopædia of London,* 1851.

applied with far greater accuracy to the telephone, for after its invention 'the great social harmoniser—personal converse' no longer required that the individual should be transferred. The telephone has effected such a revolution that it is the conversation which is transferred whilst the individuals remain immobile. It is this fact which has been so happily crystallised by somebody in the phrase 'Don't travel—telephone.'

We need not follow the example of the railway enthusiast and claim for the telephone the obliteration of national jealousies. As a means of international communication the telephone is just as available for a declaration of war as for the polite utterances which lead to an *entente cordiale*. But as an instrument of national efficiency there is no possibility of diverse opinion. The economic results, though never formulated and seldom considered, are clear. An individual will in the course of an hour hold personal converse with correspondents by means of the telephone which before its advent would have occupied perhaps a week. The cost in money may be as pence to pounds. The efficiency of the individual represented by work accomplished is raised by an enormous percentage. The freedom of social communications is also a national asset though less demonstrable on a commercial basis. All those communications, however, which allay personal anxiety are translatable into terms of efficiency for productive work.

The value of the telephone system as a means of national efficiency is proportional to the extent of its adoption, the character of the service given, and the frequency of its use.

CHAPTER XXXII

THE TELEPHONE AND GOVERNMENTS[1]

THE important place which the telephone occupies as an aid to national economy has been gained by the efforts of private enterprise without, as a general rule, the assistance or encouragement of governments. Yet the attitudes of governments have had an important, and sometimes a controlling, influence, so that the relations of governments and the telephone have been the subject of much discussion, more especially in recent times.

The telephone followed the telegraph, and in Europe the telegraph had been annexed already as a branch of the various governments' activities. The rivalries of nations on the Continent controlled the construction of strategic railways and largely determined the question of Government ownership. The telegraphs fell naturally under the same influence. In Great Britain, however, the telegraphs were acquired by the Government without regard to their utility for purposes of defence and solely with the object of consolidating and extending them as a means of commercial and social communication.

The telephone had an advantage over the telegraph in the conditions prevailing at the time of its introduction. The claims put forward by the promoters of telephonic communication related entirely to the arts of peace, and their arguments were submitted to business people. There was no need to convince a government, for financial aid was not sought and there was free trade in electrical communication in the land of the telephone's invention. The electric telegraph, on the contrary, was largely identified with warlike purposes. In 1814 Wedgwood submitted plans to the Admiralty and was informed that ' the war being at an end, and money scarce, the old system (of shutter semaphores) was sufficient for the country.'[2] Two years later (July 11, 1816) Ronalds approached the same body and was informed ' that telegraphs of any kind are

[1] This chapter was written before August 1914, and consequently without any contemplation of subsequent events in Europe. The references to war as to other points of argument are to be read only as abstract and general, without reference to any particular circumstances or conditions.

[2] *A History of Electric Telegraphy to the year* 1837, Fahie, p. 124.

now wholly unnecessary, and that no other than the one now in use will be adopted.'[1]

Even Cooke, in his 'pamphlet or sketch of 1836,' which was in the nature of a prospectus, placed in the forefront 'the affairs of the Government,' second and third places being given to 'the commercial world' and 'the private individual.' He recognised that though carried into execution by a company, such an enterprise as he advocated should always be under the control of the Government so that 'in case of dangerous riots or popular excitement the earliest intimation thereof should be conveyed to the ear of Government alone, and a check put to the circulation of unnecessary alarm.'[2] Wars could not well be left out of promoters' plans in 1814–16, nor riots in 1836; but in the United States in 1877 and Europe a year or two later, neither wars nor civil disturbances were prominent in men's minds, so that the use of the telephone was proposed simply as a superior means of commercial and social communication. It was to do wonders, but rather as a beneficent fairy's wand than a destructive ogre's weapon.

When the telephonic missionaries from the United States sought to repeat abroad the successes which they had so quickly attained at home, they were met with unexpected conditions. In Great Britain freedom was assumed and exchanges were started through the enterprise of financiers and scientific men, who foresaw their value. Freedom was assumed because of the difference supposed to exist between a telephone and a telegraph, but identity was established in law. The judgment of Mr. Justice Stephen in the case of The Attorney General *v.* The Edison Telephone Company of London, Ltd., tried before Baron Pollock and himself in 1880, shows that within the definition of the Act of 1869 'any apparatus for transmitting messages by electric signals is a telegraph whether a wire is used or not, and that any apparatus of which a wire used for telegraphic communication is an essential part is a telegraph whether the communication is made by electricity or not.' This decision settled the question of Government control of the telephone in Great Britain, for though notice of appeal was given it was not proceeded with because the Government undertook to give the company a license to work exchanges. Private lines were not included in the Government monopoly where the terminal premises were in the same ownership. These were known as A to A lines. A to B lines implied different ownership, and communication from one member of the public to another could only be

[1] *A History of Electric Telegraphy to the year* 1837, Fahie, p. 136. The original letter is in the Ronalds Library of the Institution of Electrical Engineers.

[2] *The Electric Telegraph: Was it invented by Professor Wheatstone?* Part ii. p. 250.

furnished by the Government itself or by a licensee of the Government.

The underlying policy, so far as there was a policy, was clear enough. The telephone was a new invention, or in its exchange feature a new application arising out of an improved telegraph—to adopt the legal definition. By general consent governments are not adapted to develop new ideas, and therefore the telephone was regarded, as at widely different periods East India and Rhodesia were regarded, as suitable for development by a Company of Adventurers.

The German Government alone in Europe worked the new invention itself from the start. The Balkan countries had Government exchanges, but at a much later date. Most of the other countries adopted the policy of allowing private enterprise to develop the system while assuring to themselves in the form of royalty a share in the gains. In Norway, Sweden, and Denmark, no royalties were claimed. The periods of the licenses and the terms upon which they were allowed to be exercised varied in different countries. Governments which had given concessions for short periods, as in France and Switzerland, took the earliest opportunity to regain possession and actively exploit in the belief that such a course would be for the benefit of the State. In Switzerland only the Zurich exchange had been started by a company, and in Hungary only that at Budapest.

Russia gave licenses for a certain number of years with a provision that the plant should revert to the State without compensation at the end of the period. The earlier rates in St. Petersburg and Moscow are said by Mr. Bennett[1] to 'have the distinction of being the highest in Europe—£25 per annum.' But these rates were required not only to pay for the service but also to extinguish the capital cost. With such an obligation they must be considered as far from unreasonable.[2]

Italy adopted a similar condition for licenses which run their full term, but compensation was provided for in event of earlier purchase. Confiscation has thus been avoided in the case of those properties already acquired by the Italian Government, who, on the recommendations of a Royal Commission appointed in 1910, contemplated resuming entire control of the service, but in 1912 decided to encourage private enterprise in part. Since, however, the conditions of the licenses in respect to confiscation have not been changed, there is

[1] *Telephone Systems of the Continent of Europe*, p. 316.

[2] On the expiry of the concession the St. Petersburg Exchange was taken over by the municipality; Moscow, under a new concession, was developed by a private company. The rates in both were reduced.

no security for capital, the business cannot be developed properly, and the industries of Italy, now undergoing a remarkable revival, are handicapped by lack of telephonic communication on an adequate scale, especially so in regard to long distance service.

The provision for exploitation for a period with reversion of the plant at the end of the period, was the introduction into the telephone license of a provision which was not uncommon nor entirely unreasonable in some other enterprises. A dock, for instance, or a railway between definite points is a construction once for all. It has to be maintained, but it does not grow. A telephone exchange, on the contrary, is an enterprise of constant growth, and if it is to provide for the needs of the locality, requires that new investment must be made up to the very end of the license period when repayment of capital from income is impossible. The provision was introduced by the governments and accepted by the licensees in mutual ignorance of this feature, for the service was a new one and none foresaw, or could be expected to foresee, its ultimate growth. The remarkable fact is that at this date, when the principles are so clearly understandable, any government should seek to maintain such an impossible condition.

The relations of European governments to the telephone have been the subject of volumes by one British and two American authors. A. R. Bennett gives in 'The Telephone Systems of the Continent of Europe'[1] a variety of information relating to the different countries, and more especially advocates mutual or co-operative systems. Hugo R. Meyer has limited his inquiry to 'Public Ownership and the Telephone in Great Britain.'[2] Dr. A. N. Holcombe has devoted his attention to 'Public Ownership of Telephones on the Continent of Europe,'[3] and indicates the scope of the book by adopting as a motto the words of Dunoyer, 'Je ne propose rien . . . j'expose.' Mr. Meyer's work has a more definite aim, and from the facts which he recites are drawn conclusions condemnatory of the British Government's attitude.

Whilst the conditions for carrying on the business varied, there was one respect in which the attitude of all the governments was the same. Their conception of their duty to the public was the conception which might have been expected from an individual. It was the right to conserve the profits arising from an existing business carried on by them. Thus of the English system Mr. Lee says:—

In order to protect telegraph revenue the transmission of 'written messages' by telephone was sternly prohibited, and so it came about

[1] Longmans, 1895. [2] Macmillan & Co., 1907.
[3] Houghton Mifflin Company, Boston and New York, 1911.

that the telephone was used for more or less casual conversations and the telegraph for 'written messages.' There is now no reason why such an artificial distinction should be maintained, and we are rapidly coming to an appreciation of the telephone as a telegraph instrument similar to that which was well developed in Germany and Switzerland ten years ago.[1]

The development here referred to is the equipment of telegraph offices with telephone instruments, and it existed in New Zealand in 1880 when Dr. Lemon, the superintendent of telegraphs, introduced telephones to take the place of the Morse and skilled operators in small outlying telegraph stations where receipts did not cover expenses. The statement quoted, however, shows that, because the telegraph was a Government enterprise, the use of the telephone was restricted by an artificial distinction and its utility to the public so far reduced.

The limitations of the British Government regarding long distance service, referred to in Chapter xxviii,[2] were also adopted with a view to the protection of the telegraph revenue. The State being the proprietors of the telegraph service, it was natural that the Department responsible should make every endeavour to protect the income to be derived from that service. The officials were doing their duty to the Department, and at first sight appeared to be protecting the interests of the State. When the Department had at its head a distinguished political economist, a broader view was taken. Mr. Fawcett seemed to recognise that the telephone had a 'value in use,' and that it was not to the benefit of the State to deprive itself of that value by placing artificial restrictions upon the use. A public preference for telephonic over telegraphic communication would have been in itself a proof of economic value to the State, but the users being a minority the proprietary interest of the majority in the State telegraphs was considered sufficient justification to the Government for adopting repressive measures.

General ignorance of the nature of telephone exchange service contributed to a continuance of unsatisfactory conditions. Jealous of monopolies, anxious for cheapness, and grudging the suppliers of a valuable service a reasonable reward for their enterprise, public opinion found expression in a desire for competition. This was applying to the telephone the experience of other enterprises—a very natural thing to do in its early stages. Competitors promised lower rates, the public expected increased facilities, and the Government hoped to reduce the value of the company's plant by preventing the continuance of monopoly conditions. That these hopes were delusive was known to those who had made a close study of telephone

[1] *Economics of Telegraphs and Telephones*, Pitman, 1913, p. 6.
[2] P. 426.

exchange applications. But such people were few in number and their opinions were in the main considered to be biased by their interests. When, therefore, the Duke of Marlborough in 1891–2 revived the New Telephone Company, it was assumed that the company offered an opportunity for general benefit, and there was some surprise and regret when at a later stage the National Company was associated with the Duke in its ownership and management.

The Lords of the Treasury in their minute of May 23, 1892, said: 'It is impossible to continue the present system under which the telegraph revenue is seriously suffering, while, on the other hand, the extension of telephones is checked in a manner which cannot be permanently maintained.' The loss of telegraph revenue was mainly observable in the inter-urban traffic such as that between Liverpool and Manchester. It will be noted that this loss could not have occurred under the terms of the original license, but developed with the wider powers conceded by Mr. Fawcett in 1884. There is thus brought into prominence the fact that as guardians of a particular enterprise the telegraph officials were estimating accurately and providing properly in their limited sphere, whilst Mr. Fawcett was taking a wider and wiser perception of a government's duty.

After taking into consideration all existing circumstances [said Mr. Fawcett] I have come to the conclusion that, in order that the public may enjoy the facilities with regard to telephonic communication which, in my opinion, they may fairly claim, it is desirable to give greater freedom to private enterprise by relaxing some of the conditions contained in the licenses which have been granted to private telephone companies, subject of course to the due protection for the revenue.[1]

Contemporary comment on the possible national loss from the protection of the revenue was not wanting. In a leading article on the same day (August 8, 1884) *The Times* remarked that as a thoroughgoing economist Mr. Fawcett must feel very keenly that he was not in the right place as the apologist of monopoly. Competition in telephone service was then believed in both by Mr. Fawcett and *The Times*, but the latter expressed the view that—

The Department will have, sooner or later, to get rid of the notion that the telephone must be kept under strict control, lest it should interfere with the financial working of the Government monopoly of the telegraphs and should make it result in a greater loss than heretofore. The experience of the United States has proved, at all events, that the telephone and the telegraph do not necessarily interfere with one another but that they may develop

[1] *The Times*, August 8, 1884.

vast and independent areas of new business at the same time. Be that as it may, the Post Office is bound not to stand in the way of any changes in the conduct of business which may be naturally produced and must be for the general advantage. The present Postmaster-General is the last man in the world who would wish so to control the enterprise and invention which, when left unembarrassed, can do more for civilisation, commerce, and social convenience in a single year than the most admirable of bureaucrats in a century.[1]

Subsequent impediments to the service had for their object a reduction in the price to be paid for the company's plant, when, on the expiry of the license, the service reverted to the State.

In 1898 the Duke of Norfolk was Postmaster-General, and Mr. R. W. Hanbury, the Financial Secretary of the Treasury, represented the Department in the House of Commons. Mr. Meyer[2] attributes Mr. Hanbury's telephonic activities to 'political ambition,' but it may be correct to assume that he was genuinely concerned for the finances of his Department at a future date and considered that it was necessary to take some protective measures forthwith. He had apparently reached the conclusion that there was danger in leaving the company to develop the business alone. His speech in the House of Commons on April 1, 1898, was largely devoted to *ex parte* statements tending to depreciate the value of the company's property. The price of the shares was immediately depressed. His view was that in the absence of any agreement for terms of purchase, the company, as practically the sole licensees, would be in a position to exact a sum for their plant in excess of its actual value. He concluded that this danger might be averted by competition, but as, by this time, it was evident that competition by companies was not to be relied upon, Mr. Hanbury fostered the ambitions of municipalities to conduct their own telephone services, and the committee subsequently reported on August 9, 1898,[3] in favour of granting to municipalities the powers to carry on telephone exchange service.

The National Company claimed that when they agreed to part with their trunk lines in 1892, it was understood that competition was not contemplated. The company's view of the evidence is shortly stated by the secretary as follows :—

Mr. J. S. Forbes told the committee that he relied on the declaration of policy made in the House of Commons in March 1892, that assurances were given him that the Postmaster-General would not grant licenses in respect of areas in which the company was already established if it was giving a reasonably efficient service.

[1] *The Times*, August 8, 1884.
[2] *Public Ownership and the Telephone in Great Britain*, p. 1. [3] No. 383.

. . . Mr. Forbes' evidence was confirmed by Mr. W. E. L. Gaine, the general manager, who in answer to questions said that in 1892 the Company desired to have a clear explanation of what the intentions and policy of the Government were going to be. That the Company then occupied a position that was absolutely unassailable by competition in having possession of the trunk lines and the power and the right of opening exchanges anywhere in the country. That explanations were given as to what the policy of the Government was intended to be, and on the faith of which the bargain was made. That assurances were given in his presence by the Postmaster General that it was not the intention of the Government to enter into general competition with the Company, but to co-operate with it.[1]

The documents produced to the committee showed that the Post Office had fully reserved the right to compete or license others to compete, but the evidence of permanent officials also showed that they had by this time ceased to regard competitive services as advantageous or desirable. Such a view was coming to be generally held by those whose studies and experience qualified them to reach a sound conclusion. The evidence and the circumstances indicate the probability that, whilst powers were reserved, there was no intention to use them unless special circumstances arose. In the absence, however, of a definite undertaking or of a sufficient indication of an implied understanding, the committee recommended municipal competition. It has been claimed that—

Municipal telephones have served a good purpose, the kindness with which they have been treated on their purchase by the Post Office is evidence of the fact that this is recognised. They have been factors in making possible the two agreements we have now to discuss, and this may be said with perfect frankness while contending that there never was a ghost of a chance that municipal enterprise would offer the solution of the problem as a whole.[2]

The agreements referred to were those of 1901 and 1905. The first related to the London system and its purchase terms; the second to the purchase terms for the balance of the plant of the company. Mr. Meyer says:—

So far as metropolitan London is concerned, the Telegraph Act 1899 achieved nothing that benefited the public and was at the same time fair to the National Telephone Company, that could not have been attained in 1899 by means of friendly negotiation with the National Telephone Company.[3]

[1] 'The History of the National Telephone Company,' by Albert Anns, secretary of the company—*National Telephone Journal*, December 1911.

[2] *The Economics of Telegraphs and Telephones*, Lee, p. 11.

[3] *Public Ownership and the Telephone in Great Britain*, p. 292.

It is impossible of course to adopt such a statement as a fact, since no efforts at negotiation appear to have been made, but as an expression of opinion Mr. Meyer's remarks may be endorsed and extended beyond metropolitan London. There is little doubt that the Telegraph Act of 1899 achieved nothing in any part of Great Britain that could not have been attained by negotiation. In 1884 the political head of the Department imposed on the permanent officials a broad policy. Again in 1898 the political head of the Department imposed a policy upon the permanent officials, but it was a narrow policy, unsound in principle and in tactics. It was a mistake to allow his individual judgment to take precedence over that of his responsible officers on a point which, though not ordinarily recognised as technical, required experience which a temporary political head could not expect to possess. It was a less excusable mistake to assume that any permanent good could result from a policy which was contrary to a sound sense of justice.

The extent to which the municipal competition contributed to the completion of the subsequent agreements is open to argument, but, assuming that the purchase price was thereby reduced, there yet remains to consider what was the loss to the nation by the duplication of plants, the disturbance of the systems, and the lack of development which was the necessary result of the discouragement of investment. Public opinion naturally supported the protection which was sought for the State's interests, and no criticism of departmental policy could withstand the obvious fact that it was the duty of the servants of the State to acquire at the most reasonable rate the plant with which the service was to be continued.

The course adopted by the State under Mr. Hanbury's direction was unjust both to the company and to the permanent officers of the Department, whose policy was doubtless well considered and well founded. So far as that policy can be assumed from the evidence given before the parliamentary committee, it would appear to have been based upon the recognition that competitive exchanges were not advantageous, but that the ability to compete was nevertheless an important weapon that could be used at the proper time (which had not then arrived) as an aid to negotiation. It is probable, as Mr. Meyer says, that such a method of dealing with the problem would have produced more satisfactory results, and it would certainly have promoted development to a far larger extent than the competition which never had the 'ghost of a chance' to solve 'the problem as a whole' but was initiated for the purpose of depreciation.

It is probable also that Mr. Hanbury's method of dealing with the policy of the permanent officials in 1898 prevented at a later date the development of a proposal having as its basis the welfare of the telephone service. The officials were restricted to the obvious as it appears to a politician and not to the obvious as it might appear to an expert. Consequently in 1911 and the critical period preceding it, the energies of the Department were devoted to reducing to the utmost possible extent the purchase price of the plant. For some years before the purchase the company of necessity limited its capital expenditure to work which could be relied upon to be productive and be repaid, and for eighteen months after the transfer the expert staff of the Department was engaged in the work of the arbitration. It was probably apparent to those responsible that the service must suffer by the interruption of activity, but any policy which recognised this fact and sought to prevent it could not easily be made obvious to the politician or the public, and Mr. Hanbury's treatment of officials who went beyond the immediately obvious would not be encouraging to their successors. The service under Government management was certainly given a bad start. It was the subject of criticism which, as usual, was largely uninformed and in some respects unreasonable, but where the criticisms had a basis of fact they necessarily fell upon an operating and executive staff by whom they were undeserved.

Nor was it only governments who were responsible for restrictive measures. London probably has to-day more overhead wires than any other important city of the world. Their existence is due to the interpretation which the members of the London County Council and the City Corporation placed upon their duty to their constituents. Underground facilities, if offered at all, were coupled with onerous, if not impossible, conditions having for their main object, to prevent adding value to the Telephone Company's property. The possibility of profit to a trading company was always in the forefront. To reduce the chance of that profit underground facilities were refused. The loss to their constituents from the absence of facilities, and the lack of progress, were curiously overlooked by the ratepayers' representatives, but their refusal carried with it the inability to connect subscribers or to embark upon the comprehensive plans which are essential in telephonic enterprises. It carried also an added expense in the subsequent transformation of those wires which were compulsorily erected in an admittedly unsatisfactory way.

These were minor errors, but they were in line with the major error of most if not all governments, whether national or local, in overlooking the value of the telephone in use.

In the important stages of telephonic enterprise in Great Britain, the action of Mr. Fawcett stands out as that of a statesman adjusting the policy of his Department to accord with public needs. A royalty of 10 per cent. had been demanded and approximately a four-mile limit of distance of communication had been fixed. Mr. Fawcett retained the royalty and abolished the limit. The company, on their part, never ceased to grumble at having to pay 10 per cent. They called it a tax upon the company for which the Government gave nothing. It was a tax rather upon the telephone user. Nor can it be said that the Government gave nothing for it. In fact they gave the company the right to live and earn a dividend. The company might have earned a larger dividend or given a cheaper service if no royalty had to be paid. But the Government as the supreme power and the operative body in the transmission of telegraphic signals were justified in claiming a royalty of some kind. It might have been fixed, of course, at a more moderate figure, yet it was not the royalty but the manifold restrictions and interferences that limited the use of the telephone and its development in the United Kingdom.

The development of the telephone in the United States has been far greater in proportion to population than in any other country. Freedom from Government interference in its earlier years must be regarded as a potent cause. The exchange system was started as a business for purposes of profit. The fact that the Government was not engaged in any telegraphic service rendered it unnecessary for them to adopt repressive measures. Thus the economic law of supply and demand had free play, with results in national benefit which have far exceeded those obtained elsewhere. On the expiry of the patents, competition became general, but it was the competition of private enterprise.

That the exceptional development in the United States is due to private enterprise, or would be likely to be stayed by Government management, is by no means universally recognised. Early in 1914 a report was prepared by a committee of United States postal officials under the direction of the Postmaster-General, recommending Government management of telephones.[1] The report on its publication did not receive much attention, but so far as public opinion was expressed it was generally adverse to Government management. Where the contrary opinion is held it is doubtless largely due to the success which has been attained under private enterprise. This apparent paradox may perhaps be explained by the considerations which follow.

The operations in the United States of trading companies endeavouring to monopolise sources of supply or methods of distribution have directed public attention to some developments which are detrimental to the smaller trader and, in general belief, the public as well. The merits of this question need not be discussed. It suffices to say that there is no ordinary business which depends

[1] *Sen. Doc.* 399, 63 Cong. 2nd Sess.

for its maximum utility upon unity of ownership or control. But the success of the telephone industry of the United States and the superiority of the service there rendered to the public are due to the unity of ownership or control which the American Telephone and Telegraph Company has established. It will have been seen from earlier chapters that this is not the result of accident or fortuitious circumstance, but is due to the foresight and skill of those who conceived the importance of intercommunicating, with freedom and certainty, over the widest possible areas. With divergent engineering or management methods this would not have been possible. With unity of control it has been effectually accomplished. But its very accomplishment has resulted in the development of a company employing an enormous capital and a very large staff. And it is probably due in some measure to the size of the business that any suggestions are made for its nationalisation.

The subject of Government management is necessarily one upon which there are strongly divergent views, generally due to political predilection. The individualist can see no possibility of a government doing satisfactory work in public service enterprise whilst the collectivist resents the introduction of an intermediary between the public and the service. The results in all these enterprises depend upon the Government and the character of the enterprise. Some governments succeed where others fail, and some enterprises are able to thrive in official hands where others wither.

Argument on the general question would be out of place here, and attention will only be given to the characteristic features of telephone service. Comparisons are often made, but are seldom accurate because so many considerations arise. Geographical conditions, the nature of the governments concerned, the settled or unsettled condition of the localities, and the characteristics of the peoples have to be taken into account in comparisons of telephonic development and service. Locomotion may be so difficult and communication so necessary that the use of the telephone may increase in spite of lack of enterprise. As Mr. Forbes once said to the shareholders of the National Telephone Company: English people might complain less if they were as thoroughly dragooned as their German cousins. But with all the varying results and divergent conditions, there would seem to be a consensus of opinion, confirmed by statistics and experience, that development is greater and service better under private enterprise than under Government management. On the other hand, the development which has been attained indicates that telephone service is a necessity of progress, and experience has demonstrated that the service cannot be economically furnished on the competitive conditions which serve to control ordinary private enterprise. It is the fact of monopoly which offers encouragement to the advocates of Government ownership and management.

It has been remarked that with the important exception of Germany, European governments exploited the telephone on the leasehold principle. The leases varied in duration, but with few and comparatively unimportant, exceptions they have fallen in and the freeholders are managing the properties with varying results.

In Chapter xxiv (p. 309) reference was made to a peculiarity of the telephone exchange service compared with other public services or methods of communication. It was remarked that in transport, in the post office (whether of letters or of parcels), and in telegraphic communication, the public performs a merely passive part. The telephone exchange service requires of the public an active participation. This fact has a more important bearing on Government management than is immediately apparent, though it is by no means a modern discovery. In a paper on 'Multiple Switch Boards' by C. C. Haskins and C. H. Wilson, read before the American Electrical Society in 1879,' it is said :—

A wide difference is discernible between telephonic and telegraphic communications. In the latter the correspondent writes his message, pays for it, and having required the promise that it shall be sent immediately, leaves it with the receiving clerk. A few seconds' or a few minutes' delay is of little consequence, and in any event is generally unknown to the writer. With the telephonic correspondent the case is far different. He is his own operator, and the delay, if any, he is painfully cognisant of, while the annoyance of waiting magnifies the time lost many fold.[1]

This paper was in part translated by Du Moncel and published in *La Lumière Électrique* of March 26, 1881, the last paragraph on retranslation being rendered as 'any delay occasions vexation and impatience.'[2]

The argument was used to illustrate the value of the multiple switchboard, but it is capable of far wider application.

The participation of the subscriber in a telephonic communication which his Government undertakes to obtain for him means that in the event of any difficulty the Government is the subject of criticism under circumstances which promote irritability. Criticism of his Government is the natural prerogative of every citizen, but when it applies to a method of communication, and is made by one smarting under some serious delay or difficulty which comes home to him so closely as with the telephone, that criticism is liable

[1] *Journal of the American Electrical Society*, 1880, iii. 44.
[2] *La Lumière Électrique*, March 26, 1881, iii. 225.

to take a form which is neither respectful to a Government nor productive of any good result. The critic generally has his own remedy to recommend; his complaints find expression in Parliament or Press, and under the pressure of public opinion, a Government department may be called upon to carry out some measure which experts are aware may not conduce to real improvement. In such ways the parliamentary interference, which is generally regarded as a safeguard, may be a cause of increased difficulty.

The fact that the service is undertaken without any incentive of gain leads the subscriber to infer that his interests are not being studied. Appeals to the self-interest of a private supplier are expected to be productive of some good, whilst from a public body reasonable attention to complaints is regarded as hopeless. The very general assumption that Government officials habitually disregard complaints or are inattentive to public demands is by no means correct—so far as Great Britain is concerned at any rate. Brault, writing of the German telephone system in 1890, says that the establishment of the lines had been rapid but the material was inferior and the service defective. He expressed the opinion that the autocratic methods of Government officials lacked the adaptability which was necessary to meet the requirements of subscribers.[1] The organisation of a Government department is necessarily less flexible than that of a private company. The will may be there, but the way is more difficult. The inertia is greater, the start less prompt, and the accomplishment slower and less certain. Criticism of Government service is more ample, less judicial, and far less effective than that of private enterprises.

The relationship of the employee to the service has been referred to in previous chapters. There is much reason to believe that the conditions of Government employment do not permit the exercise of the initiative, discipline, and control which are essential to obtain the best results from the working force. More freedom of selection, promotion, reprimand, and dismissal are needed than is compatible with Government employment. The efforts at the utilisation of automatic switchboards now noticeable amongst Government telephone administrations are probably to be accounted for in part by the inefficiency and cost of Government employees. This inefficiency is not of the individual, but results from the absence of the discipline, control, and incentive which are available in private enterprise.

Another feature which is more pronounced in the telephone, than

[1] 'L'établissement des lignes a été rapide mais le matériel est inférieur et le service défectueux; les habitudes autoritaires sont incompatibles avec la souplesse qui serait nécessaire pour donner satisfaction aux exigences de la clientèle.'—*Histoire de la Téléphonie, &c.*, Brault, 1890, p. 304.

in any other business, is that its capital account is never closed. The continual need of money is one which, to the popular mind, renders the telephone service especially suited to Government adoption. But this is a popular error. Governments have many demands for money, but not an unlimited supply. Their appropriations have to be determined by the budgets, and though contemplated expenditure may be of a productive nature, it is by no means certain that the funds required will be available. Being an internal matter, expenditure on the telephone must always be secondary to external demands. Thus in time of war it is certain that, with Government service, telephonic development must largely cease, however great the demand may be. This effect was to be observed in Italy during the war in Tripoli. Another example of the inability of a Government to permit supply to keep pace with demand is that of Japan. The technical qualifications of the telephone staff of the Department of Communications of Japan are of a very high order, but the capabilities of the Government to invest the necessary amount for telephone development are so far below the requirements, that a large waiting list of subscribers exists, and has so existed for many years. It is the rule to connect the lines in the order of date of application, but there is a power of delegation, and it has become common for an urgent subscriber to pay a premium to an earlier but less insistent applicant in order to obtain a transfer of the latter's right. It is probably due to the Government's difficulty in allocating the necessary funds that an important financial journal published in Tokio[1] recommended a conservative attitude regarding extensions, and by way of placating the local sentiment for progress desired the Japanese people 'to bear in mind that such a great and rich house of business as the Bank of England has not a single telephone on their premises.'

The demand for internal communications may or may not increase during a time of war, or under other circumstances when a Government is unable or unwilling to provide the resources, but there can be no doubt that the low position of some countries in telephonic development is due to the lack of the necessary appropriations by the Governments. Under private enterprise, on the other hand, the internal communications will extend if the demand exists, for the demand indicates that it will pay, and capital will flow where its remuneration is assured. Whatever the demand may be, it is certain that any large expenditure incurred on behalf of a whole nation will result in a deficiency for enterprises intended for the use only of a part of the inhabitants. Though the war may be over, 'money,' as Wedgwood was informed, 'will be scarce,' and

[1] *The Oriental Economist*, September 25, 1903.

internal development must suffer at a time when its recuperative power may be particularly required.

Whilst this applies to internal enterprises in general it is likely to have especial force in the case of the telephone, where a capital investment is required for the individual user, or subscriber as he is generally called. A telegraph line, for example, is not laid down for an individual. It serves a community, and the expenditure upon it is not incurred for a particular person. In the case of a telephone, however, the capital expenditure on the line and a substantial portion of the apparatus in the exchange is incurred on behalf of an individual subscriber. This investment is made by the Company or Government on the assumption that the receipts from the service will suffice to pay for the running cost of the service and interest upon the capital investment. In the earlier years of the telephone exchange it was not unusual for the subscriber to pay a part of the capital expenditure. In France, for example, the subscriber bought his own instrument or sub-station outfit; and in other countries attempts at 'mutual' exchanges were based on the co-operative supply of capital by subscribers. But experience has demonstrated the failure of both these expedients. Whilst a portion of the expenditure can be allocated to the individual, yet another portion is collective. Beyond the terminal of the subscriber at the central office are junction or trunk and toll lines, with their accessory apparatus. The payment by the subscriber of the capital cost of sub-station apparatus implied a choice, and individual choice is not conducive to collective efficiency, for the interlocking relationships of apparatus with each other and with the plant are so sensitive, and so important, that satisfactory results can only be obtained by a rigid compliance with the requirements defined by a competent central authority. Thus, in France the selection of apparatus and the payment of the capital cost thereof by the subscriber have been found to be unsatisfactory.

As with money so with men, the popular assumption is that a Government can at all times command the services of the men required to carry on their work. But in this, also, the popular assumption is wrong. Many examples might be given, but it will suffice to take one of early and another of more recent date to illustrate this fact.

Very shortly after the British Government had taken over the telegraphs (1870) an urgent appeal was made to them to provide telegraphic communication at some points on the Cornish coast, with a view to obtaining promptly the supply of rocket apparatus or other life-saving appliances. To this appeal Mr. Scudamore replied that the Department could not then undertake any further work in which engineers were required. The circumstances of the appeal and the Post Office reply are set forth in a guide book to the Mullion district, written by a late rector of the parish, whose melancholy duty it was to read the burial service over the numerous victims of that treacherous coast, and whose anxiety was so great as to render him unable to understand that the Government's reasons for delay were sufficient.

The second example is from the speech of Mr. Hobhouse, the Postmaster-General, in the House of Commons on April 30, 1914. In the course of his speech moving the Post Office vote, he said it was impossible to secure enough skilled workmen to maintain and extend the telephone service sufficiently quickly, but they were being trained at schools in London and elsewhere, and it was hoped to turn them out at the rate of three or four thousand a year.[1]

A sudden great demand in any branch of skilled work may overtax the available supply of labour, whether under the control of Governments or of private enterprise; but under Governments the tendency is for the supply to be rigid, whilst under private enterprise it is more flexible, and may be more readily recruited and trained. The circumstances with labour are the same as with capital. It will flow to channels affording the best prospects of employment and remuneration. The channels vary, and with a free market the supply varies with the demand.

The report of the American Telephone and Telegraph Company for the year 1913 is largely devoted to the objections which may be urged against Government ownership, and a recognition of the value of suitable regulation. A brief but bold paragraph may be quoted:

> All monopolies should be regulated. Government ownership would be an unregulated monopoly.

In theory the public control Government monopolies through Parliament, but it cannot be overlooked that in practice the control so exercised may be, and often is, either ineffective or mischievous A Postmaster-General who has to deal in Parliament with complaints of the telephone service, when he is himself the defendant, occupies a position which is not to be envied. His position is very different when he acts in a judicial capacity, and has only to see that regulations made in the public interest for the efficient conduct of the service are properly carried out by others.

In those countries in which the Governments have not already undertaken the telephone service, the value of private management is being recognised, and will probably be retained with such regulation as may seem advisable. The method of regulation will need to be carefully considered. The regulation of rates is generally

[1] *The Times* (Parliamentary Report), May 1, 1914.

assumed to be synonymous with reduction of rates. A governmental authority may seek popularity by attempting to obtain something for nothing. But this is a fatal error, telephonically as otherwise. The prosperity of the provider is essential to the success of the service. Fixity of tenure and confidence in investment are necessary to development. Regulations which take into account all the conditions, and are so made as to be fair to the supplier and the user, may be expected to be productive of good results. Regulations which unduly hamper the commercial freedom of the suppliers would only continue in another form the evils which have unfortunately been perpetrated with good but mistaken intentions.

CHAPTER XXXIII

CONCLUSION

ACCORDING to the statistics which are given in detail in Appendix A, there were on January 1, 1914, 14,888,550 stations in connection with telephone exchanges in the world, and an estimated investment of £417,906,800, converting into sterling at the rate of five dollars to the pound the total as compiled of $2,089,534,000.

And the man to whose genius we are indebted for the creation of the means happily still lives. The marvellous invention which in 1876 aroused the enthusiam of Sir William Thomson at the Philadelphia Exhibition is the basis of an enormous industry involving plant and apparatus, of which the talking instruments, though of first importance, are at once the simplest in construction and represent the smallest part of the investment.

It happens but seldom in the history of an invention that within the lifetime of the inventor the investment resulting from it is so vast or the application so wide and general, as that of Alexander Graham Bell. And it may confidently be claimed that no invention has ever effected so much economic advantage or public benefit at so small a cost to the individual participant.

To Bell we pay first homage as creator, for to him alone the discovery of the essential principle and the invention of the requisite means to apply that principle are due. Law and Science have recognised that fact so fully that a restatement of it would seem to be unnecessary. It is nevertheless the case that the limited scope of Bell's British patents, due in part to the circumstances under which they were taken out and in part to the prior publication of Bell's own work, has rendered the completeness of his mastery of the problem less generally well known here than in the United States. Numerous legal decisions there have made a much wider circle familiar with the fact of Bell's discovery of the principle as well as the invention of the means. In Great Britain, Bell, for technical reasons, was unable to obtain that control which was given him in the United States as the discoverer of a new art. Scientists recognised Bell as the first inventor of the speaking telephone, but even amongst scientists the fact that Bell's original United States patent indicated the utility of a variable resistance transmitter, such as the microphone in its telephonic usage, was by no means universally known. Whether due to imperfect knowledge or simply to unhappy expression it is not possible to say, but when in 1913 the Royal Society intended to confer an honour upon Bell by awarding him a medal, they did so in terms which are open to the interpretation that Bell's work was less complete and less individual than was really the case. It is always difficult in a brief statement to safeguard every circumstance, and the Royal Society's award may be open to explanation which accords with the facts, but since it is also open to an interpretation which is inaccurate it must be considered. The purpose of the Hughes medal and the terms of the award are given in note 2 on page 191. To award a medal to Bell for his '*share* in the invention of the telephone' and to place especial emphasis on the 'construction of the Telephone Receiver' is to minimise the great work of Bell and to magnify the smaller. To remember the magneto telephone in its secondary status as a receiver and to forget that in its primary use it demonstrated the most remarkable discovery of modern times is indeed to exalt the form and to neglect the spirit. A century hence some Beckmann of the future may be writing a history of inventions. The telephone will, without doubt, have at least a chapter to itself. Shall it then be said on the authority of the records of the Royal Society that credit may be given to Bell for the invention of an ear trumpet, but that he shared with others the invention of a speaking trumpet? The double purpose served by the magneto telephone was the subject of the earliest notice and the greatest enthusiasm amongst scientists. The Bell receiver was the Bell transmitter—but not the only Bell transmitter. The Bell ear trumpet was the Bell speaking trumpet. It was the telephone. Who unmasked the hidden mystery—disclosed the governing scientific fact? Who indicated the devices necessary for applying it? No one who contributed an essential element to such a discovery or such an invention should go unnamed. If we know that it was Bell and some others, a responsibility rests upon us to lighten the labours of posterity by saying who the others were. Thorough exploration has revealed that there were no others. It was Bell and Bell alone who made the 'original discovery' as well as the effective 'application' such as Hughes desired to be commemorated by his bequest. The sources of Bell's inspiration and the extent of the assistance received from others are related with fullness and gratitude by Bell himself. They do not constitute any share in the invention of the telephone.

The transmission of speech once effected and the means once disclosed improvements—such, for example, as that discovered by Hughes himself—are natural consequences. The successive contributors in their respective fields and their contributions have been mentioned in this book, so far as its scope permits. But it has not been possible to indicate to the full the credit which is due to those pioneers whose foresight, faith, and determination promoted the telephone exchange system as an industry, realised its widespread utility, recognised its engineering character, and organised and systematised its development and many, if not most, of its inventions. Gardiner G. Hubbard, Bell's father-in-law, who was impatient at any suggestion of a talking instrument before its accomplishment, promptly grasped the commercial possibilities when the invention was made. He began the business side with such care that few mistakes had to be corrected. In the case of an industry which has become so great a public service enterprise his letter of August 1877 (given on page 71) is remarkable, in that the first condition for the retention of exclusive exchange rights is that the agent should 'serve the public promptly and faithfully.' Foreseeing large developments and the need of organisation and control, he sought to place the management of the United States company in efficient hands. And, again to indicate the short period within which vast work has been accomplished, it may be recalled that Theodore N. Vail, who was selected by Hubbard to organise the Bell system in the United States, still directs with vigour the affairs of the American Telephone and Telegraph Company. In the earlier period of his telephonic career, when the Bell Company was small and poor, he and his firm, sagacious, far-seeing President, Col. W. H. Forbes, had to fight the Western Union Telegraph Company, which was large and rich. In the later period the Bell Company had become so large that Mr. Vail was able to acquire a controlling interest in the Western Union Company, as a comparatively small investment for the Bell Company. He gave practical expression to his belief that a combination of the telephone and telegraph services was to the public advantage; but the Government of the United States held other views, and by arrangement the Telephone Company agreed to dispose of its telegraph shares and to relinquish any interest in the management of the Western Union Company.

How much of the prompt success of the telephone exchange system in the United States is due to the readiness of the American people to adopt new methods, how much to the freedom during its early years from governmental control, or how much to the energy and ability of the promoters, it is not possible to say; but a large share of credit must be given to the Company and to those individuals who at various times controlled it.

Mr. Vail, who relinquished the management for a time, was succeeded by the late Mr. Hudson, who had acted as counsel for the company, and Mr. Hudson, after a brief interval, by Mr. Fish, also a lawyer, practising at the Bar in patent cases. Mr. Fish retired and Mr. Vail again returned to the control as president.[1] Each of these leaders had a problem of his period. Mr. Vail, in the first instance, to organise a novel business under great disadvantages; Mr. Hudson to continue an established business during a period of patent control; Mr. Fish to extend energetically the Bell business against the encroachments of competition; and Mr. Vail again to develop still further the business whose foundations he had himself helped to lay. The policies of each of these leaders differed in a way which the circumstances of their respective times required, but throughout the whole period the annual reports indicate a consistent policy on essential points: the business must be based upon intercommunication which should be not only local but national so far as the capacity of the instruments permitted or might permit—for no finality was recognised. The satisfaction of its patrons was ever a consideration, and the development of methods and apparatus to meet new conditions regarded as an essential and constant care.

As a monopolist company the American Bell Telephone Company (like every other telephone company elsewhere) was attacked by every telephone man with a grievance and every subscriber who hoped by competition to obtain cheaper service; but, surveying the whole period, it will be seen that whilst looking for a proper return upon its shareholders' investment the company recognised from the first its obligations to the public as the provider of a public utility; and its obligations, not only to the shareholders, but also to the industry, to further its material advancement in every possible way.

[1] It will be remembered that the present American Telephone and Telegraph Company is the successor of the American Bell Telephone Company, which in its turn succeeded the National Bell Telephone Company. The successive presidents of this continuous organisation have been:—

William H. Forbes, from formation to September 1, 1887.
Howard Stockton, from September 1, 1887, to April 1, 1889.
John E. Hudson, from April 1, 1889, to October 1, 1900.
Alexander Cochrane, from October 1, 1900, to April 17, 1901.
Frederick P. Fish, from April 17, 1901, to May 7, 1907.
Theodore N. Vail, from May 7, 1907, who still holds the office.

Mr. Vail was general manager of the National Bell Telephone Company in 1878 and of its successor, the American Bell Telephone Company, until May 1885. Mr. Hudson succeeded Mr. Vail as general manager in May 1885, and continued in that position until he became president in April 1889. In the photograph fig. 152 (facing page 426) may be observed two men immediately behind Bell. The one on the observer's left, who is looking out of the picture, is John E. Hudson. The one on the right, and looking down at Bell, is E. J. Hall, whose activity and interest in the industry are abundantly shown in the frequent quotations from his speeches and writings in the preceding pages.

A similar policy characterised the managment of the National Company in Great Britain by Mr. Gaine, who also was a lawyer, having relinquished the post of Town Clerk of Blackburn to join the Telephone Company. On the European Continent the name of Mr. Cedergren of Stockholm will be remembered as that of an enthusiastic exponent of the advantages of widespread development.

Though Mr. Gaine had problems that differed materially from those in the United States, he also recognised the dependence on public satisfaction and the importance of engineering details. Whilst Mr. Vail and his engineer, Mr. Carty, could consider expenditure without regard to period of redemption, Mr. Gaine and his engineer, Mr. Gill, had to consider the limited tenure of the company's license and the probabilities of repayment within that period. To plan how to get the best results for the least expenditure is the province of the telephone engineer as of any other engineer, but the problem is in many respects more complicated. It is dependent upon more unknown factors than in most engineering enterprises. There is need to plan for years ahead on the basis of some established facts and other assumed data. In the previous pages there have been described certain details of apparatus and materials, the professional control of any branch of which would confer the title of telephone engineer, but it is in a wider sense that the title is used here. All the apparatus described may be considered as only the raw material which the engineer has to select and use in types or quantities which will vary according to the conditions of each case. One type may be economical to install and expensive to operate or maintain ; another expensive to install and economical to operate or maintain. Conduits, cables, wires, will vary with the period to be provided for and the growth expected. In earlier times estimates were necessarily made, but less comprehensively. The considerations which entered into the erection of a central office building were alluded to in a paper read by Mr. F. A. Pickernell in 1890.

Before a telephone company is in a position to erect a suitable building it is necessary to make definite estimates of the probable extent of the underground work, the ultimate number of the subscribers it will be desirable to handle at the proposed exchange, the probable number of trunks to other exchanges, the probable number of extra-territorial lines, the amount of private wire business, the probable average number of calls per subscriber per day when the switchboard attains its ultimate capacity. A careful estimate of this kind determines the ultimate capacity of the switchboards, underground cable runs, etc., and consequently fixes the minimum size of the building.[1]

[1] *National Telephone Exchange Association Report*, 1890, p. 65.

In outside construction work also similar forecasts were made; poles were not merely erected for the immediate equipment of lines nor conduits laid simply for the existing number of cables. Judgment was exercised as to the spare provision for extensions, and the basis of the judgment was experience in the past and estimate for the future.

In 1902–3 the engineering department of the American Telephone and Telegraph Company adopted a method of organised studies for development which, whilst taking into account certain known particulars, applied a comprehensive system of what might be called organised estimates to those particulars which, though unknown, were necessary to be taken into account; and this was applied not merely to sections, but to the exchange outfit as a whole.

A treatise in itself would be required to describe the 'fundamental plan,' as the study came to be called, and then considerable practical experience of the work would be necessary to make use of it. All that can usefully be done here is to endeavour to indicate its purpose, which is to determine the basis upon which an exchange system should be constructed, reconstructed, or developed, so as to give the best results in the most economical way. The number of central offices in a given area, and the best position for each central office, is ascertained with precision. It may not be possible to obtain the premises in the particular position required. It then remains to discover what diversion from the perfect plan offers the least disadvantages.

To construct a conduit without any spare pipes would be to save a proportionately small immediate expenditure at the expense of a much larger expenditure in the future more or less near. To include spare pipes for the expected increase over an unduly long period would involve too heavy a burden by reason of capital charges. Engineering may thus be defined as the means of determining with precision how to expend money to the greatest advantage. To this end the co-operation of the engineers with the management is essential, in order to obtain the best results. In relating the introduction of the common battery system [1] it was remarked that the first exchange in New York to be fitted was Harlem, and that its equipment was so organised as to fit in with all the other exchanges if, as was expected, the service were satisfactory. To provide for the apparatus in all these Exchanges it was necessary to make a forecast, and it is probable that about this time (1898) the principles of the development plan were being developed, but it was not until 1903 that formal notes on the subject were prepared. Some of the diagrams accompanying

[1] Chapter xxvi, p. 391.

these notes are dated September 1902, and it is probable, as already remarked, that they were commenced much earlier.[1] It is of the greatest importance in telephone undertakings to take long views and provide for all the conditions that are known or can be estimated. The population at the moment is known, the subscribers at the moment are known. The population for the period determined upon—perhaps twenty or thirty years—must be estimated and the number of subscribers also. In the case of New York and other cities in which reconstruction work has been undertaken, exchanges already existed, so that practical data were available as a guide to build on.

An interesting example of the provision of a telephone exchange plant in a large city where no previous telephonic experience was available, is supplied in the case of Constantinople. It is recorded that

> The art of printing had now been invented for more than two hundred and fifty years, and every other state in Europe had adopted the important discovery. The Turks alone rejected it, and assigned as a reason, that it was an impious innovation.[2]

An earlier author suggests that vested interests may have had some influence in the propagation of the idea that an invention which reduced the demand for writers was one of an irreligious character. Referring to the Turks he says:—

> Printing they reject ; perhaps for feare lest the universalitie of leaning [? learning] should subvert their false grounded religion and policy ; which is better preserved by an ignorant obedience : moreover a number that live by writing would be undone, who are, for the most part, of the Priesthood.[3]

Considering that Turkey rejected the steamboat for decades and the printing press for centuries, it was not to be expected that she should make early use of the telephone, though it was not alleged, so far as I am aware, that the telephone was 'an impious innovation.' The Sultan, Abdul Hamid, continued the traditional policy of his country and was averse to the adoption of modern inventions generally—at least those of a peaceful character. After his fall, the Young Turks decided to grant a concession, which was acquired by a company whose

[1] For a more detailed exposition of the principles and their application, see a paper by J. J. Carty entitled 'Telephone Engineering' in the *Transactions of the American Institute of Electrical Engineers*, February 23, 1906, xxv. 81.

[2] *Constantinople and the Scenery of the Seven Churches of Asia Minor*, Rev. Robert Walsh (1839 ?), i. 23.

[3] *A Relation of a Journey begun An. Dom.* 1610, p. 72. By George Sandys. London, 1637.

consulting engineers, Messrs. Gill & Cook[1] with the assistance of a staff of engineers, studied the topography of the city, estimated the number of subscribers, the localities of such subscribers, and the number of calls they would send, the number of exchanges and their appropriate positions, the routes, the conduits, the cables and every requisite. The determination of the system to be used and the material of the equipment were settled by these preliminary estimates.

As a telephonic 'development study' this example was exceptional. No other large city in the world remained without a telephone. The opportunity was thus afforded of establishing a telephonic system *de novo* with all the knowledge that thirty years' experience had given, though necessarily governed by the conditions of the concession. On the other hand there was no telephonic experience whatever in the locality to form the ground-work of a development study.

Upon the accuracy of a development study and the skill and experience in utilising it depend economical construction. There may be considerable waste of money in the case of exchanges which are allowed just to grow until they grow out of their equipment and then are supplied with a new equipment in whole or in part with the same disregard of the future and what it has in store.

The 'standard of transmission' regarded as suitable for the exchange also has a considerable influence upon the material to be installed. 'It is exceedingly important on the long lines . . . but a very great deal of the money side of it is on the short lines.'[2]

The basis and application of the standard are dealt with in papers by B. S. Cohen and G. M. Shepherd ('Telephonic Transmission Measurements'[3]), and A. J. Aldridge ('Practical Application of Telephone Transmission Calculations'[4]). The specification and rules forming part of the agreement for the acquisition of the National Company's plant by the British Post Office[5] is one of the practical applications of the transmission standards. In order 'to prevent some statements passing into history wrongly' Mr. Gill, in the course of the discussion on Mr. Aldridge's paper, said:—

The standard cable and the standard instrument were . . . first used in the United States by the American Telephone and Telegraph Company's engineers, and when a standard had to be started in this country in 1904 some of us thought it was more

[1] Then engineer-in-chief and assistant engineer respectively of the National Telephone Company.

[2] F. Gill. *Journal of the Institution of Electrical Engineers*, li. 433.

[3] *Journal of the Institution of Electrical Engineers*, xxxix. 503.

[4] *Ibid.*, li. 390.

[5] Given in full by Poole (*Practical Telephone Handbook*, 1912, *Appendix*) and in abstract by Aitken (*Manual of the Telephone*, 1911, p. 35).

desirable to adopt a standard which was already in use, which had done very good work, on which a great deal of expenditure had been made, and which would form the basis of an International standard, rather than set up some standard of our own.[1]

The investigations related in the papers were made in the research laboratory of the National Telephone Company, and the results were obtained from very numerous tests and experiments. In these the ear must be the final arbiter, but by means of the oscillograph of Duddell the eye is able to render useful help, the comparison of the speech waves recorded by the oscillograph at the transmitting and receiving ends of the line giving a graphic indication of the losses and distortions suffered in transmission.

The recognition of the importance of transmission, development, and similar studies undertaken by the higher branches of the telephone engineering organisation, is by no means so general as it should be amongst telephone administrations. There are probably many cases in which a revision would give valuable results.

The telephone exchange service is not only responsible for the introduction of a new type of engineering, but has also led to the development of new methods in other branches. The engineers have at least the advantage of dealing with material things and forces. But what can the layman be expected to understand by the 'manager of the traffic branch'?—a traffic in a commodity so intangible as talks! Yet the 'talk' is the basis of the telephone business, and the traffic manager weighs and measures the talks, their number, duration, and distances. He provides much of the information upon which the engineer works and determines the personnel by whose energies the communications are controlled.

In the period covered since the telephone exchange service was established, there has been a remarkable change in the relations of employers and employed; women have undertaken work previously allotted to men, and in most large industries the welfare of the employees has received considerable attention. If they did not lead (and in some cases I am disposed to think they did) the telephone companies followed very closely the general trend in this direction. The superiority of women as exchange operators was promptly recognised,[2] though the suitability of some of the early central offices for occupation by refined and educated women may have been open to question. The requirements of the service as well as the tendency of social thought influenced the environment. The palm tree in the Boston switchroom of 1885 (fig. 103) may be regarded as a token of these tendencies.

[1] *Journal of the Institution of Electrical Engineers*, li. 432.
[2] Chapter xxiv. p. 318.

Gradually the improvement in the conditions and the establishment of more comfortable surroundings reacted upon the classes available for employment, so that the companies were able to obtain operators of education and high efficiency.

In these respects the United States telephone companies set a standard which the progressive telephone administrations throughout the world did their best to follow. There is probably no other class of women workers so well cared for or by whom such care is so well requited.

Much the same may be said for the other members of the staff. A pension arrangement was early adopted by the National Telephone Company in Great Britain for the principal members of their staff, but, though coming later, the scheme of the American Telephone and Telegraph Company was even wider in its scope, embracing all classes of employees.

The telephone industry has both contributed to, and profited by, the developments in manufacturing methods over the period covered. Machinery had been utilised in the manufacture of certain parts of telegraph instruments such as sounders, but only to a small extent. The number required did not justify the application of special machinery for the purpose. One key and one sounder served for the messages of a community. The telephone service required that each member of the public making use of it should be provided with even more complex instruments than a key and sounder. The telephonic apparatus had to be a domestic instrument. The age of domestic machinery was already well advanced. The sewing machine had led the way, and in the manufacture of this labour-saving appliance much other labour-saving machinery had been utilised. One of the early factories for telephonic apparatus (Gilliland, Indianapolis) was converted from a sewing machine factory. Gradually, but relatively rapidly, to the manufacture of telephonic apparatus was applied all the labour-saving devices which could profitably be employed in the turning out of material in quantity. Suitable machines which existed were used, and for processes in which no suitable machinery existed new machines were invented. Reference has been made to many new inventions for telephonic apparatus. Perhaps a work of similar proportions might be devoted to the descriptions of new machinery invented to produce such apparatus. It would certainly be an interesting study to follow the successive improvements in manufacturing methods which enabled the operating companies to keep pace with public requirements and to offer at a popular price the use of a highly developed scientific instrument in connection with an elaborate permutation system. The close relationship between the operating (or consuming) and the manufacturing (or producing) interests in the United States has been suggested by independent observers as one of the reasons for the rapid telephonic development there and the establishment of standards for the rest of the world. The standard of finish applied to such apparatus in America and Europe differed. Whilst sufficiently well finished in the essentials of working parts, the external finish of American apparatus was criticised by European users in early years. Later, as the influence of the operating companies extended the standard of finish of American apparatus was higher than that which some of the European organisations were content to use.

In an earlier chapter I have referred to the fact that the claims put forward by the originators of telephonic communication related entirely to the arts of peace. But all the arts of peace that can be commandeered by the forces of war are ruthlessly pressed into service, and so important a means of communication as the telephone was readily seized upon. Field sets were early designed and promptly used. The Japanese in their war with Russia used the telephone extensively, but in the war now in progress the telephone is a leading feature. The General's head-quarters are a telephone exchange whose radiating wires pulsate with information sent in and orders sent out upon which the fate of nations depends. Writing on his experiences at the front Rudyard Kipling says:—

> Pick up the chain anywhere you please, you shall find the same observation-post, table, map, observer, and telephonist.[1]

Conveying all languages with equal facility, the telephone is ready to render service to all armies impartially, but the army that fails to utilise it to the full will certainly have occasion for regret.

Yet, though of inestimable service in war, it is in the arts of peace that its triumphs are the more pronounced, its services the more widespread. Each day throughout the world tens of millions of conversations are held between millions of stations, over distances varying from a few yards to thousands of miles. Of the millions of people who hourly use the telephone, how many are familiar with the circumstances of its inception? How many can give thought to that day in June 1875 when the plucking of a reed resolved in Bell's informed mind that what he knew to be possible in theory was also effective in practice? How many of those who daily call for one subscriber after another, near or far, know the method by which the interchanges are made or consider to whom we are indebted for them? How many of those who daily grumble at the cost give thought to the convenience received, the skill and care

[1] *Daily Telegraph*, September 8, 1915.

displayed in providing the service, or comprehend the value in investment placed at their disposal?

Each day the number of users will grow, and the more familiar the service becomes, the less thought will be given to it. Each day the past further recedes and the difficulties in recalling its incidents increase. For those who may care to know, the foregoing pages have been written with a view to recording, so far as it is possible in so limited a space, the evolution of an epoch-making invention and the development of an industry of incalculable advantage to the human race.

APPENDICES

A. TELEPHONE STATISTICS OF THE WORLD

B. PROPOSED INCREASES IN TELEPHONE RATES IN GREAT BRITAIN FROM NOVEMBER 1, 1915

TELEPHONE DEVELOPMENT OF THE WORLD, BY COUNTRIES—JANUARY 1, 1914

—	Number of Telephones			Per Cent. of Total World	Telephones per 100 Population	Population per Square Mile
	Government Systems	Private Companies	Total			
NORTH AMERICA.						
United States	—	9,542,017	9,542,017	64·09	9·7	33·
Canada	106,183	393,591	499,774	3·36	6·5	2·
Central America	4,326	3,548	7,874	·05	·1	27·
Mexico	1,319	40,542	41,861	·28	·3	20·
Other N. A. Places *	20	2,318	2,338	·02	·7	·4
West Indies:						
Cuba	299	15,798	16,097	·11	·7	49·
Porto Rico	300	4,088	4,388	·03	·4	342·
Other W. I. Places *	2,018	4,581	6,599	·04	·1	114·
Total	114,465	10,006,483	10,120,948	67·98	7·5	16·
SOUTH AMERICA.						
Argentine	—	74,296	74,296	·50	·9	8·
Bolivia	—	2,500	2,500	·02	·1	4·
Brazil	1,165	38,018	39,183	·26	·2	7·
Chile (July 31, 1914)	—	19,709	19,709	·13	·6	12·
Colombia	—	3,177	3,177	·02	·1	12·
Ecuador	481	2,445	2,926	·02	·2	13·
Paraguay	129	370	499	·01	·1	4·
Peru *	—	4,000	4,000	·03	·1	7·
Uruguay	—	13,599	13,599	·09	1·0	18·
Venezuela	341	4,688	5,029	·03	·2	7·
Other Places (including Falkland Is.)	1,413	—	1,413	·01	·3	2·
Total	3,529	162,802	166,331	1·12	·3	8·
EUROPE.						
Austria	172,344	—	172,344	1·16	·6	253·
Bosnia *	1,200	—	1,200	·01	·1	101·
Belgium *	65,000	—	65,000	·44	·9	672·
Bulgaria	3,608	—	3,608	·02	·1	110·
Denmark (March 31, 1914)	1,586	127,691	129,277	·87	4·5	189·
France *	330,000	—	330,000	2·22	·8	192·
German Empire	1,420,100	—	1,420,100	9·54	2·1	323·
Great Britain (March 31, 1914)	780,512	—	780,512	5·24	1·7	381·
Greece *	3,200	—	3,200	·02	·1	104·
Hungary	84,040	—	84,040	·56	·4	170·
Italy (June 30, 1913)	61,978	29,742	91,720	·62	·3	311·
Luxemburg	4,239	—	4,239	·03	1·6	268·

* Partly estimated.

TELEPHONE DEVELOPMENT OF THE WORLD, BY COUNTRIES—JANUARY 1, 1914 (*continued*)

—	Number of Telephones			Per Cent. of Total World	Telephones per 100 Population	Population per Square Mile
	Government Systems	Private Companies	Total			
EUROPE—*continued*						
Netherlands . .	76,267	10,223	86,490	·58	1·4	471·
Norway * . . .	40,120	42,430	82,550	·55	3·4	20·
Portugal . . .	1,203	7,647	8,850	·06	·2	165·
Roumania * . .	20,000	—	20,000	·13	·3	141·
Russia (European) .	157,710	162,148	319,858	2·15	·2	73·
Finland * . .	—	40,000	40,000	·27	1·2	26·
Servia * . . .	3,700	—	3,700	·02	·1	134·
Spain * . . .	2,722	31,278	34,000	·23	·2	105·
Sweden . . .	158,171	74,837	233,008	1·56	4·1	36·
Switzerland . .	96,624	—	96,624	·65	2·5	251·
Other Places . .	1,485	904	2,389	·02	·1	57·
Total . .	3,485,809	526,900	4,012,709	26·95	·8	121·
ASIA.						
British India . .	6,504	11,193	17,697	·12	·01	227·
China * . . .	13,517	13,492	27,009	·18	·01	167·
Japan (March 31, 1914)	219,551	—	219,551	1·47	·4	356·
Russia (Asiatic) .	9,423	7,181	16,604	·11	·1	3·
Other Places * . .	22,110	3,114	25,224	·17	·01	29·
Total . .	271,105	34,980	306,085	2·05	·04	53·
AFRICA.						
Egypt	4,949	12,310	17,259	·12	·1	31·
Union of South Africa	28,889	—	28,889	·19	·5	13·
Other Places * . .	18,089	859	18,948	·13	·02	11·
Total . .	51,927	13,169	65,096	·44	·05	12·
OCEANIA.						
Australia . . .	137,485	—	137,485	·92	2·8	2·
Dutch East Indies * .	11,393	3,450	14,843	·10	·04	51·
Hawaii	—	7,284	7,284	·05	3·5	32·
New Zealand (March 31, 1914)	49,415	—	49,415	·33	4·6	10·
Philippine Islands .	1,779	4,979	6,758	·05	·1	75·
Other Places * . .	1,371	225	1,596	·01	·1	8·
Total .	201,443	15,938	217,381	1·46	·4	13·
Total World .	4,128,278	10,760,272	14,888,550	100·00	·9	33·

* Partly estimated.

TELEPHONE AND TELEGRAPH GROSS EARNINGS OF THE WORLD, BY COUNTRIES—FISCAL YEAR 1913

(Estimated where necessary)

—	Gross Earnings				Per Cent. of Total		Telephone Earnings per Telephone
	Telephone		Telegraph (See Note)	Total	Telephone	Telegraph	
NORTH AMERICA.	Service Operated by	$	$	$			$
United States	P.	305,400,000	51,300,000	356,700,000	85·6	14·4	33·00
Canada*	P. G.	16,400,000	5,568,000	21,968,000	74·7	25·3	35·30
Central America	P. G.	300,000	1,825,000	2,125,000	14·1	85·9	40·20
Mexico	P. G.	1,198,000	1,208,000	2,406,000	49·8	50·2	29·50
Other N. A. Places	P. G.	66,700	121,200	187,900	35·5	64·5	29·90
West Indies:							
Cuba	P. G.	957,000	419,000	1,376,000	69·5	30·5	68·10
Porto Rico	P. G.	219,100	49,100	268,200	81·7	18·3	58·00
Other W. I. Places	P. G.	168,400	84,600	253,000	66·6	33·4	28·20
Total		324,709,200	60,574,900	385,284,100	84·3	15·7	33·10
SOUTH AMERICA.							
Argentine	P.	3,756,000	5,473,000	9,229,000	40·7	59·3	54·40
Bolivia	P.	66,000	142,000	208,000	31·7	68·3	26·40
Brazil	P. G.	1,500,000	4,045,000	5,545,000	27·1	72·9	47·50
Chile	P.	823,000	482,000	1,305,000	63·1	36·9	39·50
Colombia	P.	90,000	464,000	554,000	16·2	83·8	36·20
Ecuador	P. G.	66,000	106,000	172,000	38·4	61·6	23·50
Paraguay	P. G.	22,000	57,000	79,000	27·8	72·2	44·50
Peru	P.	168,000	138,000	306,000	54·9	45·1	42·00
Uruguay	P.	540,000	154,000	694,000	77·8	22·2	46·80
Venezuela	P. G.	211,000	198,000	409,000	51·6	48·4	43·60
Other Places (incl. Falkland. Is.)	G.	21,500	11,000	32,500	66·2	33·8	15·50
Total		7,263,500	11,270,000	18,533,500	39·2	60·8	47·90
EUROPE.							
Austria	G.	5,766,700	3,835,800	9,602,500	60·1	39·9	36·30
Bosnia	G.	41,600	222,000	263,600	15·8	84·2	37·60
Belgium	G.	3,115,700	1,280,000	4,395,700	70·9	29·1	50·40
Bulgaria	G.	107,400	524,100	631,500	17·0	83·0	31·30
Denmark	P. G.	2,968,300	538,400	3,506,700	84·6	15·4	24·00
France	G.	12,713,200	9,500,000	22,213,200	57·2	42·8	40·80
German Empire	G.	46,249,900	9,725,100	55,975,000	82·6	17·4	34·00
Great Britain	G.	30,462,900	15,165,900	45,628,800	66·8	33·2	40·10
Greece	G.	60,900	420,000	480,900	12·7	87·3	29·00
Hungary	G.	3,315,400	2,100,000	5,415,400	61·2	38·8	41·50
Italy	P. G.	3,867,800	2,842,900	6,710,700	57·6	42·4	43·80

Note.—Telegraph service is operated by Governments, except in the United States.

P.—Private Companies. G.—Government.

P. G. Private Companies and Government. See first Table for statistics as to proportion of telephones operated by Government.

* Telegraph earnings are for year ended June 30, 1914. Earnings of Pacific Cable Board excluded.

TELEPHONE AND TELEGRAPH GROSS EARNINGS OF THE WORLD, BY COUNTRIES—FISCAL YEAR 1913—*continued*

(Estimated where necessary.)

—	Service Operated by	Gross Earnings			Per Cent. of Total		Telephone Earnings per Telephone
		Telephone	Telegraph (See Note)	Total	Tele-phone	Tele-graph	
EUROPE.—*continued*		$	$	$			$
Luxemburg	G.	86,300	22,000	108,300	79·7	20·3	21·20
Netherlands	P. G.	2,896,200	1,005,400	3,901,600	74·2	25·8	35·00
Norway	P. G.	1,584,600	815,200	2,399,800	66·0	34·0	21·60
Portugal	P. G.	317,500	725,000	1,042,500	30·5	69·5	43·80
Roumania	G.	456,500	800,000	1,256,500	36·3	63·7	23·20
Russia and Finland*	P. G.	10,248,800	17,983,200	28,232,000	36·3	63·7	30·70
Servia	G.	111,400	170,000	281,400	39·6	60·4	30·50
Spain	P. G.	1,475,000	1,850,000	3,325,000	44·4	55·6	50·00
Sweden	P. G.	5,203,000	770,700	5,973,700	87·1	12·9	23·40
Switzerland	G.	2,760,000	886,500	3,646,500	75·7	24·3	29·50
Other Places	P. G.	62,400	29,300	91,700	68·0	32·0	27·80
Total		133,871,500	71,211,500	205,083,000	65·3	34·7	35·00
ASIA.							
British India	P. G.	772,400	3,698,900	4,471,300	17·3	82·7	48·60
China	P. G.	792,000	5,274,300	6,066,300	13·1	86·9	33·00
Japan	G.	7,273,900	5,276,200	12,550,100	58·0	42·0	35·40
Other Places*	P. G.	1,776,000	1,867,200	3,643,200	48·7	51·3	48·00
Total		10,614,300	16,116,600	26,730,900	39·7	60·3	37·60
AFRICA.							
Egypt	P. G.	811,200	461,000	1,272,200	63·8	36·2	48·60
Union of South Africa	G.	1,407,600	2,067,600	3,475,200	40·5	59·5	53·00
Other Places	P. G.	561,000	2,081,200	2,642,200	21·2	78·8	33·00
Total		2,779,800	4,609,800	7,389,600	37·6	62·4	46·10
OCEANIA.							
Australia	G.	4,200,700	3,958,900	8,159,600	51·5	48·5	35·40
Dutch East Indies	P. G.	842,500	752,000	1,594,500	52·8	47·2	67·60
Hawaii	P.	333,000	—	333,000	100·0	—	50·00
New Zealand	G.	1,130,800†	1,627,700	2,758,500	41·0†	59·0	24·50†
Philippine Islands	P. G.	238,000	282,900	520,900	45·7	54·3	44·80
Other Places	P. G.	42,900	17,300	60,200	71·3	28·7	29·60
Total		6,787,900	6,638,800	13,426,700	50·6	49·4	35·80
Total World		486,026,200	170,421,600	656,447,800	74·0	26·0	34·00

Note.—Telegraph service is operated by Governments, except in the United States.

P.—Private Companies. G.—Government.

P. G.—Private Companies and Government. See first Table for statistics as to proportion of telephones operated by Government.

* Telegraph earnings of Asiatic Russia included in telegraph earnings of European Russia.

† Earnings from exchange service only. Toll and long distance earnings not reported.

TELEPHONE PLANT INVESTMENT OF THE WORLD, BY COUNTRIES—JANUARY 1, 1914

(Estimated where necessary)

—	Service Operated by	Plant Investment	Per Cent. of Total World	Invest-ment per Telephone
NORTH AMERICA.		$		$
United States	P.	1,149,900,000	55·03	121
Canada	P. G.	74,466,000	3·56	149
Central America	P. G.	913,000	·04	116
Mexico	P. G.	5,264,000	·25	126
Other North America Places	P. G.	398,000	·02	170
West Indies:				
Cuba	P. G.	3,858,000	·19	240
Porto Rico	P. G.	621,000	·03	142
Other West Indies Places	P. G.	844,000	·04	128
Total		1,236,264,000	59·16	122
SOUTH AMERICA.				
Argentine	P.	15,800,000	·76	213
Bolivia	P.	375,000	·02	150
Brazil	P. G.	11,013,800	·53	281
Chile	P.	2,153,000	·10	109
Colombia	P.	336,600	·01	106
Ecuador	P. G.	377,000	·02	129
Paraguay	P. G.	70,800		142
Peru	P.	531,600	·02	133
Uruguay	P.	1,608,000	·08	118
Venezuela	P. G.	1,092,800	·05	217
Other Places (incl. Falkland Is.)	G.	158,400	·01	112
Total		33,517,000	1·60	201
EUROPE.				
Austria	G.	39,382,000	1·88	234
Bosnia	G.	420,000	·02	350
Belgium	G.	14,495,000	·69	223
Bulgaria	G.	658,000	·03	182
Denmark	P. G.	17,060,000	·82	132
France	G.	81,840,000	3·92	248
German Empire	G.	278,340,000	13·32	196
Great Britain	G.	143,655,000	6·87	184
Greece	G.	560,000	·03	175
Hungary	G.	16,388,000	·78	195
Italy	P. G.	12,092,000	·58	132
Luxemburg	G.	694,000	·03	164

P.—Private Companies. G.—Government.
P. G.—Private Companies and Government. See first Table for statistics as to proportion of telephones operated by Government.

TELEPHONE PLANT INVESTMENT OF THE WORLD, BY COUNTRIES—JANUARY 1, 1914—*continued*

—	Service Operated by	Plant Investment	Per Cent. of Total World	Invest-ment per Telephone
EUROPE.—*continued*		$		$
Netherlands	P. G.	12,992,000	·62	150
Norway	P. G.	10,768,000	·52	140
Portugal	P. G.	1,502,000	·07	170
Roumania	G.	3,500,000	·17	175
Russia	P. G.	45,583,000	2·18	143
Finland	P.	4,279,000	·21	107
Servia	G.	925,000	·04	250
Spain	P. G.	5,100,000	·24	150
Sweden	P. G.	25,595,000	1·23	110
Switzerland	G.	18,524,000	·89	192
Other Places	P. G.	372,000	·02	156
Total		734,724,000	35·16	186
ASIA.				
British India	P. G.	2,655,000	·13	150
China	P. G.	4,456,000	·21	165
Japan	G.	23,597,000	1·13	107
Other Places	P. G.	5,856,000	·28	140
Total		36,564,000	1·75	119
AFRICA.				
Egypt	P. G.	1,948,000	·09	113
Union of South Africa	G.	8,745,000	·42	303
Other Places	P. G.	2,653,000	·13	140
Total		13,346,000	·64	205
OCEANIA.				
Australia	G.	24,458,000	1·17	189
Dutch East Indies	P. G.	2,387,000	·12	161
Hawaii	P.	980,000	·05	135
New Zealand	G.	6,047,000	·29	122*
Philippine Islands	P. G.	1,040,000	·05	154
Other Places	P. G.	207,000	·01	130
Total		35,119,000	1·69	162
Total World		2,089,534,000	100·00	140

P.—Private Companies. G.—Government.
P. G.—Private Companies and Government. See first Table for statistics as to proportion of telephones operated by Government.
* Investment in exchange plant only.

TELEPHONE DEVELOPMENT OF IMPORTANT CITIES—EUROPE, AUSTRALIA NEW ZEALAND, JAPAN, AND THE UNITED STATES—JANUARY 1, 1914

Country and City (or Exchange Area)	Population Estimated (City or Exchange Area)	Number of Telephones	Telephones per 100 Population
AUSTRALIA.			
Adelaide	201,000	8,720	4·3
Brisbane	160,000	6,671	4·2
Melbourne	651,000	27,490	4·2
Sydney	725,400	34,566	4·8
AUSTRIA.			
Lemberg	210,737	4,749	2·3
Prague	458,195	10,310	2·3
Triest	239,692	5,324	2·2
Vienna	2,092,382	64,438	3·2
BELGIUM.*			
Antwerp	486,829	8,020	1·6
Brussels	838,681	21,470	2·6
Ghent	291,656	2,938	1·0
Liège	339,937	5,060	1·5
BULGARIA.			
Sofia	103,000	1,599	1·5
DENMARK.			
Copenhagen	621,000	55,080	8·9
FRANCE.*			
Bordeaux	266,000	5,090	1·9
Lille	223,000	3,826	1·7
Lyons	547,000	7,039	1·3
Marseilles	565,000	7,735	1·4
Paris	2,940,000	95,033	3·2
GERMAN EMPIRE.			
Berlin	2,363,000	154,800	6·6
Breslau	545,000	20,573	3·8
Chemnitz	315,000	10,820	3·4
Cologne	552,000	26,422	4·8
Dresden	562,000	25,721	4·6
Dusseldorf	411,000	19,133	4·7
Essen	322,000	11,342	3·5
Frankfort	445,000	28,932	6·5
Hamburg-Altona	1,310,000	77,322	5·9
Hannover	323,000	16,194	5·0
Leipzig	622,000	31,176	5·0
Magdeburg	294,000	10,201	3·5
Munich	629,000	34,323	5·5
Nuremburg	367,000	15,354	4·2
Stuttgart	306,000	20,929	6·8
GREAT BRITAIN AND IRELAND.†			
Belfast	475,000	8,580	1·8
Birmingham	1,145,000	19,780	1·7
Blackburn	340,000	4,615	1·4
Bolton	335,000	4,171	1·2
Bradford	475,000	12,243	2·6
Bristol	440,000	9,056	2·1
Dublin	455,000	9,692	2·1
Edinburgh	515,000	15,258	3·0
Glasgow	1,190,000	40,849	3·4
Leeds	590,000	10,864	1·8
Liverpool	1,160,000	34,053	2·9
London	7,300,000	258,895	3·5
Manchester	1,255,000	31,443	2·5
Newcastle	650,000	11,561	1·8
Nottingham	470,000	8,574	1·8
Sheffield	715,000	11,354	1·6

* Statistics as of January 1, 1913. † Statistics as of March 31, 1914.

TELEPHONE DEVELOPMENT OF IMPORTANT CITIES—EUROPE, AUSTRALIA, NEW ZEALAND, JAPAN AND THE UNITED STATES—JANUARY 1, 1914—*continued*

Country and City (or Exchange Area)	Population Estimated (City or Exchange Area)	Number of Telephones	Telephones per 100 Population
GREECE.*			
Athens	167,000	854	0·5
HUNGARY.			
Budapest	880,000	27,944	3·2
Szegedin	118,000	1,500	1·3
ITALY.†			
Milan	599,000	12,709	2·1
Naples	723,000	4,774	0·7
Palermo	342,000	1,586	0·5
Rome	539,000	11,719	2·2
Turin	428,000	6,217	1·5
JAPAN.			
Kobe	440,766	5,892	1·3
Kyoto	508,068	10,447	2·1
Nagoya	447,951	5,696	1·3
Osaka	1,387,366	21,787	1·6
Tokio	2,445,048	43,681	1·8
Yokohama	424,369	4,825	1·1
NETHERLANDS.			
Amsterdam	595,000	17,212	2·9
The Hague	302,000	12,823	4·2
Rotterdam	459,000	13,630	3·0
NEW ZEALAND.‡			
Auckland	109,300	6,722	6·2
Christchurch	86,140	4,927	5·7
NORWAY.†			
Christiania	247,488	20,699	8·4
PORTUGAL.			
Lisbon	435,000	5,394	1·2
ROUMANIA.*			
Bukarest	338,000	4,983	1·5
RUSSIA.			
Kief	506,060	5,143	1·0
Lodz	415,604	4,503	1·1
Moscow	1,617,157	49,848	3·1
Odessa	620,155	7,712	1·2
Petrograd	2,018,596	54,815	2·7
Warsaw	872,478	31,952	3·7
SPAIN.*			
Barcelona	587,000	4,547	0·8
Madrid	600,000	4,365	0·7
SWEDEN.			
Goteborg (Gothenburg)	178,030	13,672	7·7
Stockholm	354,783	85,641§	24·1§
SWITZERLAND.			
Basel	140,000	7,669	5·5
Zurich	201,000	13,565	6·7
UNITED STATES.			
Total of the 12 cities with over 500,000 population	16,330,000	1,849,518	11·3
Total of the 33 cities with over 200,000 population	23,000,900	2,749,785	12·0

* Statistics as of January 1, 1913. † Statistics as of June 30, 1913.
‡ Statistics as of March 31, 1914. § 70 % of this development is secured by a private company.

TELEPHONE DEVELOPMENT—URBAN AND RURAL—EUROPE, AUSTRALIA, NEW ZEALAND, JAPAN, AND THE UNITED STATES—JANUARY 1, 1914.

Countries	Service Operated by	Number of Telephones		Telephones per 100 Population	
		In cities of over 100,000 population	Outside of cities of over 100,000 population	In cities of over 100,000 population	Outside of cities of over 100,000 population
Austria	G.	95,053	77,291	2·8	·30
Bosnia*	G.	—	1,200	—	·06
Belgium*	G.	43,600	21,400	2·0	·39
Bulgaria	G.	1,599	2,009	1·5	·04
Denmark (Mar. 31, 1914)	P. G.	55,080	74,197	8·9	3·33
France*	G.	154,000	176,000	2·6	·52
German Empire	G.	743,246	676,854	4·9	1.29
Great Britain (March 31, 1914)	G.	579,686	200,826	2·6	.83
Greece*	G.	900	2,300	0.3	·06
Hungary	G.	29,444	54,596	3·0	·27
Italy (June 30, 1913)	P. G.	53,937	37,783	1·3	·12
Luxemburg	G.	—	4,239	—	1·59
Netherlands	P. G.	46,777	39,713	3·2	·84
Norway*	P. G.	20,699	61,851	8·4	2·82
Portugal	P. G.	7,647	1,203	1·2	·02
Roumania*	G.	5,200	14,800	1·5	·20
Russia	P. G.	200,029	119,829	2·1	·08
Finland*	P.	9,957	30,043	6·5	·96
Servia*	G.	—	3,700	—	·08
Spain*	P. G.	12,850	21,150	0·6	·12
Sweden	P. G.	99,313	133,695	18·6	2·62
Switzerland	G.	30,177	66,447	6·3	1·96
Other places	P. G.	—	2,389	—	·09
Total Europe		2,189,194	1,823,515	3·0	·45
Australia	G.	83,807	53,678	4·5	1·77
New Zealand (March 31, 1914)	G.	6,722	42,693	6·2	4·40
Japan (March 31, 1914)	G.	99,645	119,906	1·5	·26
United States	P.	3,339,806	6,202,211	11·9	8·88

* Partly estimated. P.—Private Companies. G.—Government.
P. G.—Private Companies and Government. See first Table for statistics as to proportion of telephones operated by Government.

NOTE.—The foregoing particulars have been collected from authoritative sources by the statisticians of the American Telephone and Telegraph Company, who have kindly acceded to my request for permission to use them.

APPENDIX B

Proposed Increases in Telephone Rates in Great Britain from November 1, 1915.

EXCHANGE SERVICE

Unlimited (or Flat) Rates:	Present Rates.	Proposed Rates.
London	£17 per annum; £14 per annum for second and subsequent connections.	£20 per annum; £17 per annum for second and subsequent connections.
Provinces (Unlimited Service rates have not been available for new subscribers in the Provinces since 1908; and it is not proposed to make them available now.)	In most cases £10 per annum, with £8 10s. per annum for second and subsequent connections; but many old subscribers retain obsolete rates lower than these, varying down to £5 per annum.	£12 per annum; £10 per annum for second and subsequent connections. (For existing flat-rate subscribers only.)
Do. Party Lines.	Various; in most cases £6 for a two-party line, £4 for a four-party line.	Two-party lines £7 10s. Four-party lines £5. (For existing flat-rate subscribers only.)

	1915–16	1916–17
	£	£
Additional Revenue	25,000	280,000
Saving in Expenditure	—	—
Totals	£25,000	£280,000

TRUNK SERVICE

	Present s. d.	Proposed s. d.
For 25 miles or under	0 3	0 4
,, 50 ,, ,, ,,	0 6	0 8
,, 75 ,, ,, ,,	0 9	1 0
,, 100 ,, ,, ,,	1 0	1 4
Every additional 40 miles or fraction thereof.	0 6	0 8

(There are a few Trunk fees at 1*d.* and 2*d.*; these it is proposed to make 2*d.* and 3*d.* respectively.)

	1915–16	1916–17
	£	£
Additional Revenue	120,000	290,000
Saving in Expenditure	—	—
Totals	£120,000	£290,000

CALL OFFICE FEES

	Present Rates	Proposed Rates
	s. d.	s. d.
London . . .	0 2	0 3
In addition to the Trunk fee in the case of Trunk Calls.		
Provinces :—	s. d.	s. d.
Local Call (in addition to any Junction fee).	0 1	0 2
Trunk Call (in addition to the Trunk fee).	0 2	0 2

	1915–16	1916–17
	£	£
Additional Revenue . . .	60,000	140,000
Saving in Expenditure . . .	—	—
Totals . . .	£60,000	£140,000

Summary of Gains to the Exchequer

	1915–16 (5 Months)		1916–17	
	Additional Revenue	Saving in Expenditure	Additional Revenue	Saving in Expenditure
Telephones	£	£	£	£
Exchange Service (Flat Rate Subscribers) . .	25,000	—	280,000	—
Trunk Service . . .	120,000	—	290,000	—
Call Office Fees . .	60,000	—	140,000	—
Totals . .	£205,000	—	£710,000	—

NOTE

The above estimates are based on the assumption that the new charges will be operative from November 1, 1915. In some instances, especially where statutory authority is not required, it may be possible to introduce the new rates before that date, but the effect on the anticipated revenue would not be considerable.

From 'Statement showing the proposed increases in Postal, Telegraph, and Telephonic charges, and the additional revenue and saving in expenditure estimated therefrom for the years 1915–16 and 1916–17. Presented to both Houses of Parliament by Command of His Majesty, September, 1915.' [Cd. 8067.]

INDEX

PRINTED BY
SPOTTISWOODE AND CO. LTD., COLCHESTER
LONDON AND ETON, ENGLAND

TECHNOLOGY AND SOCIETY

An Arno Press Collection

Ardrey, R[obert] L. **American Agricultural Implements.** In two parts. 1894

Arnold, Horace Lucien and Fay Leone Faurote. **Ford Methods and the Ford Shops.** 1915

Baron, Stanley [Wade]. **Brewed in America:** A History of Beer and Ale in the United States. 1962

Bathe, Greville and Dorothy. **Oliver Evans:** A Chronicle of Early American Engineering. 1935

Bendure, Zelma and Gladys Pfeiffer. **America's Fabrics:** Origin and History, Manufacture, Characteristics and Uses. 1946

Bichowsky, F. Russell. **Industrial Research.** 1942

Bigelow, Jacob. **The Useful Arts:** Considered in Connexion with the Applications of Science. 1840. Two volumes in one

Birkmire, William H. **Skeleton Construction in Buildings.** 1894

Boyd, T[homas] A[lvin]. **Professional Amateur:** The Biography of Charles Franklin Kettering. 1957

Bright, Arthur A[aron], Jr. **The Electric-Lamp Industry:** Technological Change and Economic Development from 1800 to 1947. 1949

Bruce, Alfred and Harold Sandbank. **The History of Prefabrication.** 1943

Carr, Charles C[arl]. **Alcoa, An American Enterprise.** 1952

Cooley, Mortimer E. **Scientific Blacksmith.** 1947

Davis, Charles Thomas. **The Manufacture of Paper.** 1886

Deane, Samuel. **The New-England Farmer,** or Georgical Dictionary. 1822

Dyer, Henry. **The Evolution of Industry.** 1895

Epstein, Ralph C. **The Automobile Industry:** Its Economic and Commercial Development. 1928

Ericsson, Henry. **Sixty Years a Builder:** The Autobiography of Henry Ericsson. 1942

Evans, Oliver. **The Young Mill-Wright and Miller's Guide.** 1850

Ewbank, Thomas. **A Descriptive and Historical Account of Hydraulic and Other Machines for Raising Water,** Ancient and Modern. 1842

Field, Henry M. **The Story of the Atlantic Telegraph.** 1893

Fleming, A. P. M. **Industrial Research in the United States of America.** 1917

Van Gelder, Arthur Pine and Hugo Schlatter. **History of the Explosives Industry in America.** 1927

Hall, Courtney Robert. **History of American Industrial Science.** 1954

Hungerford, Edward. **The Story of Public Utilities.** 1928

Hungerford, Edward. **The Story of the Baltimore and Ohio Railroad, 1827-1927.** 1928

Husband, Joseph. **The Story of the Pullman Car.** 1917

Ingels, Margaret. **Willis Haviland Carrier, Father of Air Conditioning.** 1952

Kingsbury, J[ohn] E. **The Telephone and Telephone Exchanges:** Their Invention and Development. 1915

Labatut, Jean and Wheaton J. Lane, eds. **Highways in Our National Life:** A Symposium. 1950

Lathrop, William G[ilbert]. **The Brass Industry in the United States.** 1926

Lesley, Robert W., John B. Lober and George S. Bartlett. **History of the Portland Cement Industry in the United States.** 1924

Marcosson, Isaac F. **Wherever Men Trade:** The Romance of the Cash Register. 1945

Miles, Henry A[dolphus]. **Lowell, As It Was, and As It Is**. 1845

Morison, George S. **The New Epoch:** As Developed by the Manufacture of Power. 1903

Olmsted, Denison. **Memoir of Eli Whitney, Esq.** 1846

Passer, Harold C. **The Electrical Manufacturers, 1875-1900.** 1953

Prescott, George B[artlett] **Bell's Electric Speaking Telephone.** 1884

Prout, Henry G. **A Life of George Westinghouse.** 1921

Randall, Frank A. **History of the Development of Building Construction in Chicago.** 1949

Riley, John J. **A History of the American Soft Drink Industry:** Bottled Carbonated Beverages, 1807-1957. 1958

Salem, F[rederick] W[illiam]. **Beer, Its History and Its Economic Value as a National Beverage.** 1880

Smith, Edgar F. **Chemistry in America.** 1914

Steinman, D[avid] B[arnard]. **The Builders of the Bridge:** The Story of John Roebling and His Son. 1950

Taylor, F[rank] Sherwood. **A History of Industrial Chemistry.** 1957

Technological Trends and National Policy, Including the Social Implications of New Inventions. Report of the Subcommittee on Technology to the National Resources Committee. 1937

Thompson, John S. **History of Composing Machines.** 1904

Thompson, Robert Luther. **Wiring a Continent:** The History of the Telegraph Industry in the United States, 1832-1866. 1947

Tilley, Nannie May. **The Bright-Tobacco Industry, 1860-1929.** 1948

Tooker, Elva. **Nathan Trotter:** Philadelphia Merchant, 1787-1853. 1955

Turck, J. A. V. **Origin of Modern Calculating Machines.** 1921

Tyler, David Budlong. **Steam Conquers the Atlantic.** 1939

Wheeler, Gervase. **Homes for the People,** In Suburb and Country. 1855